# Digital Electronics

## Principles & Applications

### Seventh Edition

# Roger L. Tokheim

 **Higher Education**

Boston   Burr Ridge, IL   Dubuque, IA   New York   San Francisco   St. Louis
Bangkok   Bogotá   Caracas   Kuala Lumpur   Lisbon   London   Madrid   Mexico City
Milan   Montreal   New Delhi   Santiago   Seoul   Singapore   Sydney   Taipei   Toronto

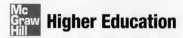

8 9 0 CTP/CTP 1 0 9 8 7 6 5 4 3 2

ISBN 978–0–07–312634–0
MHID 0–07–312634–9

Publisher: *David T. Culverwell*
Sponsoring Editor: *Thomas Casson*
Developmental Editor: *Jonathan Plant*
Senior Marketing Manager: *Nancy Bradshaw*
Project Manager: *Joyce Watters*
Lead Production Supervisor: *Sandy Ludovissy*
Senior Media Producer: *Damian Moshak*
Designer: *John Joran*
(USE) Cover Image: *©Digital Vision / Getty Images, Silicon wafer held by wafer tweezers*
Senior Photo Research Coordinator: *John C. Leland*
Photo Research: *Pam Carley*
Compositor: *Techbooks*
Typeface: *11/13 Times Roman*
Printer: *CTPS*

The credits section for this book begins on page 525 and is considered an extension of the copyright page.

**Library of Congress Cataloging-in-Publication Data**

Tokheim, Roger L.
  Digital electronics : principles & applications / Roger Tokheim. – 7th ed.
     p. cm. – (Basic skills in electricity & electronics series)
  Includes index.
   ISBN  978-0-07-312634-0 — ISBN  0-07-312634-9 (hard copy : alk. paper)
    1. Digital electronics–Textbooks. I. Title. II. Title: Digital electronics, principles and applications.
TK7868.D5T65 2008
621.381–dc22

                                                        2006035517

www.mhhe.com

# Contents

# Editors' Foreword

The McGraw-Hill Higher Education *Basic Skills in Electricity and Electronics* series has been designed to provide entry-level competencies in a wide range of occupations in the electrical and electronic fields. The series consists of coordinated instructional materials designed especially for the career-oriented student. A textbook, an experiments manual, and an instructor productivity center support each major subject area covered in the series. All of these focus on the theory, practices, applications, and experiences necessary for those preparing to enter technical careers.

There are two fundamental considerations in the preparation of a series like this: the needs of the learner and needs of the employer. This series meets these needs in an expert fashion. The authors and editors have drawn upon their broad teaching and technical experiences to accurately interpret and meet the needs of the student. The needs of business and industry have been identified through personal interviews, industry publications, government occupational trend reports, and reports by industry associations.

The processes used to produce and refine the series have been ongoing. Technological change is rapid and the content has been revised to focus on current trends.

Refinements in pedagogy have been defined and implemented based on classroom testing and feedback from students and instructors using the series. Every effort has been made to offer the best possible learning materials. These include animated PowerPoint presentations, circuit files for simulation, a test generator with correlated test banks, dedicated Web sites for both students and instructors, basic instrumentation labs, and other items as well. All of these are well coordinated and have been prepared by the authors.

The widespread acceptance of the *Basic Skills in Electricity and Electronics* series and the positive responses from users confirm the basic soundness in content and design of all of the components as well as their effectiveness as teaching and learning tools. Instructors will find the texts and manuals in each of the subject areas logically structured, well-paced, and developed around a framework of modern objectives. Students will find the materials to be readable, lucidly illustrated, and interesting. They will also find a generous amount of self-study, review items, and examples to help them determine their own progress.

*Charles A. Schuler, Project Editor*

## Basic Skills in Electricity and Electronics

Charles A. Schuler, Project Editor

### *New Editions in This Series*
*Electricity: Principles and Applications, Seventh Edition,* Richard J. Fowler
*Electronics: Principles and Applications, Seventh Edition,* Charles A. Schuler
*Digital Electronics: Principles and Applications, Seventh Edition,* Roger L. Tokheim

### *Other Series Titles Available:*
*Communication Electronics, Third Edition,* Louis E. Frenzel
*Microprocessors: Principles and Applications, Second Edition,* Charles M. Gilmore
*Industrial Electronics,* Frank D. Petruzella
*Mathematics for Electronics,* Harry Forster, Jr.

# Preface

*Digital Electronics: Principles and Applications*, seventh edition, is an easy-to-read introductory text for students new to the field of electronics. Providing entry-level knowledge and skills for a wide range of occupations is the goal of this textbook and its ancillary materials. Prerequisites are general math and introductory electricity/electronics. Binary math, Boolean concepts, simple programming, and various codes are introduced and explained as needed. Concepts are connected to practical applications, and a systems approach is followed that reflects current practice in industry. Earlier editions of the text have been used successfully in a wide range of programs: Electronic Technology, Electrical Trades & Apprenticeship Training, Computer Repair, Communications Electronics, and Computer Science, to name a few. This concise and practical text can be used in any program needing a quick and readable overview of digital principles.

## New to this Edition

While the level and approach of the book remains the same, a number of important updates and improvements have been made for this new edition. These include the following:

- Early and simplified introduction to lab and test equipment
- Updated coverage of memory and storage technology
- Expanded coverage of digital/computer systems
- DSP in a digital camera
- Expanded coverage of data transmission
- Expanded programming of microcontrollers (BASIC Stamp 2 Modules)
- More complete MultiSIM circuit simulation files

## Learning Features

As in previous versions, the new seventh edition includes a tried-and-true learning system shared by other texts in McGraw-Hill's Basic Skills in Electricity & Electronics ("BSEE") Series. The intent is to present basic information in the most understandable way possible for examples, illustrations, and self-tests to make the learning process easier, and to make retention of major concepts possible. The features include:

- Simple objectives
- Topics broken into short sections
- Simple, full-color illustrations and photographs
- Self-Tests for chapter sections
- Chapter review questions
- Critical Thinking questions
- Answers to Self-Test questions

## Students Resources

A full complement of student resources accompanies the main text, providing an entire learning system that includes hands-on work, computer simulation routines, study resources, visual aids, and additional test questions. The specific items available are:

- **Experiments Manual** with correlated hands-on-activities for each chapter (including hardware experiments, troubleshooting, and design activities, and computer simulations). Equipment and software is available from several sources. Each unit of the manual includes a Self-Test.
- **Student CD-ROM**, bound in with the text and Experiments Manual, contains comprehensive MultiSim files keyed to seventh edition circuits; the MultiSim Primer (by Patrick Hoppe of Gateway Technical College) which provides a tutorial on the software for beginners, with step-by-step explanations, screen captures and numerous examples from digital electronics for MultiSim; student PowerPoint presentations for review, and self-study; special presentations on Breadboarding, Soldering, and Circuit Interrupters; the Circuits Solver program; and additional items for self-study and review.
- **Online Learning Center website ("OLC")** www.mhhe.com/tokheim7e includes chapter study resources, links to industry and association sites, the MultiSim Primer, and other Student CD-ROM resources in online form.

## Instructor Resources

- **Instructor's Productivity Center CD-ROM ("IPC")** contains the complete Instructor's Manual in .pdf form, including part/equipment lists, performance objectives listing, textbook answer book, experiments manual answer book, instrumentation guide, and overhead masters; PowerPoint slide presentations for all text chapters, and special Power-Point presentations on Breadboarding, Soldering, Circuit interrupters (GFCI and AFCI), and Instrumentation. Also included are the Hewlett-Packard instrumentation simulator, and a test generator with full test banks for each chapter. The MultiSim Primer by Patrick Hoppe is also included—it provides a good practical introduction for those starting to use MultiSim for digital applications.

## About the Author

Over decades, Roger L. Tokheim has published many textbooks, lab manuals, and Schaum's Outline series books in the areas of digital electronics and microprocessors. His books have been translated into nine languages. He taught technical subjects including electronics for more than 35 years in public schools.

# Acknowledgments

Appreciation is given to many instructors, students, and industry representatives who contributed to this book. Special thanks is given to Darrell Klotzbach, software engineer at Adobe Systems, Inc, for his help on several sections including DSP and digital camera application, JTAG, and data transmission. Thanks also to family members Marshall, Rachael, Dan, and Carrie for their help on this project.

The author and publisher would also like to thank the reviewers who helped evaluate this new edition; their time and expertise is much appreciated.

Jon Brutlag
Chippewa Valley Technical College (WI)

Ronald G. Dreucci
California University of Pennsylvania (PA)

Larry E. Dukes
Wichita Technical Institute (KS)

Robbie Edens
ECPI College of Technology (SC)

Harmit Kaur
Sinclair Community College (OH)

Randy Owens
Henderson Community College (KY)

Andrew F. Volper
San Diego JATC (CA)

# Safety

Electric and electronic circuits can be dangerous. Safe practices are necessary to prevent electrical shock, fires, explosions, mechanical damage, and injuries resulting from the improper use of tools.

Perhaps the greatest hazard is electrical shock. A current through the human body in excess of 10 milliamperes can paralyze the victim and make it impossible to let go of a "live" conductor or component. Ten milliamperes is a rather small amount of current flow: It is only *ten one-thousandths* of an ampere. An ordinary flashlight can provide more than 100 times that amount of current!

Flashlight cells and batteries are safe to handle because the resistance of human skin is normally high enough to keep the current flow very small. For example, touching an ordinary 1.5-V cell produces a current flow in the microampere range (a microampere is one-millionth of an ampere). The amount of current is too small to be noticed.

High voltage, one the other hand, can force enough current through the skin to produce a shock. If the current approaches 100 milliamperes or more, the shock can be fatal. Thus, the danger of shock increases with voltage. Those who work with high voltage must be properly trained and equipped.

When human skin is moist or cut, its resistance to the flow of electricity can drop drastically. When this happens, even moderate voltages may cause a serious shock. Experienced technicians know this, and they also know that so-called low-voltage equipment may have a high-voltage section or two. In other words, they do not practice two methods of working with circuits: one for high voltage and one for low voltage. They follow safe procedures at all times. They do not assume protective devices are working. They do not assume a circuit is off even though the switch is in the OFF position. They know the switch could be defective.

Even a low-voltage, high-current-capacity system like an automotive electrical system can be quite dangerous. Short-circuiting such a system with a ring or metal watchband can cause very severe burns—especially when the ring or band welds to the points being shorted.

As your knowledge and experience grow, you will learn many specific safe procedures for dealing with electricity and electronics. In the meantime:

1. Always follow procedures.
2. Use service manuals as often as possible. They often contain specific safety information. Read, and comply with, all appropriate material safety data sheets.
3. Investigate before you act.
4. When in doubt, *do not act*. Ask your instructor or supervisor.

## General Safety Rules for Electricity and Electronics

Safe practices will protect you and your fellow workers. Study the following rules. Discuss them with others, and ask your instructor about any you do not understand.

1. Do not work when you are tired or taking medicine that makes you drowsy.
2. Do not work in poor light.
3. Do not work in damp areas or with wet shoes or clothing.
4. Use approved tools, equipment, and protective devices.
5. Avoid wearing rings, bracelets, and similar metal items when working around exposed electric circuits.
6. Never assume that a circuit is off. Double-check it with an instrument that you are sure is operational.
7. Some situations require a "buddy system" to guarantee that power will not be turned on while a technician is still working on a circuit.
8. Never tamper with or try to override safety devices such as an interlock (a type of switch that automatically removes power when a door is opened or a panel removed).
9. Keep tools and test equipment clean and in good working condition. Replace insulated probes and leads at the first sign of deterioration.
10. Some devices, such as capacitors, can store a *lethal* charge. They may store this charge for long periods of time. You must be certain these devices are discharged before working around them.

11. Do not remove grounds and do note use adaptors that defeat the equipment ground.

12. Use only an approved fire extinguisher for electrical and electronic equipment. Water can conduct electricity and may severely damage equipment. Carbon dioxide ($CO_2$) or halogenated-type extinguishers are usually preferred. Form-type extinguishers may also be desired in *some* cases. Commercial fire extinguishers are rated for the type of fires for which they are effective. Use only those rated for the proper working conditions.

13. Follow directions when using solvents and other chemicals. They may be toxic, flammable, or may damage certain materials such as plastics. Always read and follow the appropriate material safety data sheets.

14. A few materials used in electronic equipment are toxic. Examples include tantalum capacitors and beryllium oxide transistor cases. These devices should not be crushed or abraded, and you should wash your hands thoroughly after handling them. Other materials (such as heat shrink tubing) may produce irritating fumes if overheated. Always read and follow the appropriate material safety data sheets.

15. Certain circuit components affect the safe performance of equipment and systems. Use only exact or approved replacement parts.

16. Use protective clothing and safety glasses when handling high-vacuum devices such as picture tubes and cathode-ray tubes.

17. Don't work on equipment before your know proper procedures and area aware of any potential safety hazards.

18. Many accidents have been caused by people rushing and cutting corners. Take the time required to protect yourself and others. Running, horseplay, and practical jokes are strictly forbidden in shops and laboratories.

19. Never look directly into light-emitting diodes or fiber-optic cables; some light sources, although invisible, can cause serious eye damage.

Circuits and equipment must be treated with respect. Learn how they work and the proper way of working on them. Always practice safety: your health and life depend on it.

Electronics workers use specialized safety knowledge.

# Digital Electronics

## Chapter Objectives

*This chapter will help you to:*

1. *Identify* several characteristics of digital circuits as opposed to linear (analog) circuits.
2. *Classify* devices as using digital, analog, or a combination of technologies.
3. *Differentiate* between digital and analog signals and *identify* the HIGH and LOW portions of the digital waveform.
4. *List* three types of multivibrators and *describe* the general purpose of each type of circuit.
5. *Analyze* simple logic-level indicator circuits.
6. *Cite* several reasons for using digital circuits.
7. *Write* several limitations of digital circuits.
8. *Demonstrate* the use of several lab instruments.

Engineers generally classify electronic circuits as being either analog or digital in nature. Historically, most electronics products contained analog circuitry. Most newly designed electronic devices contain digital circuitry. This chapters introduces you to the world of digital electronics.

What are the clues that an electronic product *contains digital circuitry*? Signs that a device contains digital circuitry include:

1. Does it have an alphanumeric (shows letters and numbers) display?
2. Does it have a memory or can it store information?
3. Can the device be programmed?

If the answer to any one of the three questions is yes, then the product probably contains digital circuitry.

Digital circuitry is quickly becoming pervasive because of its *advantages* over analog including:

1. Generally, digital circuits are easier to design using modern integrated circuits (ICs).
2. Information storage is easy to implement with digital.
3. Devices can be made programmable with digital.
4. More accuracy and precision is possible.
5. Digital circuitry is less affected by unwanted electrical interference called noise.

All persons working in electronics must have knowledge of digital electronic circuits. You will use simple integrated circuits and displays to demonstrate the principles of digital electronics.

**Identifying digital products**

**Advantages of digital**

## 1-1 What Is a Digital Signal?

**Analog signal**

In your experience with electricity and electronics you have probably used analog circuits. The circuit in Fig. 1-1(a) puts out an *analog signal* or voltage. As the wiper on the potentiometer is moved upward, the voltage from points A to B *gradually* increases. When the wiper is moved downward, the voltage gradually decreases from 5 to 0 volts (V). The waveform diagram in Fig. 1-1(b) is a graph of the analog output. On the left side the voltage from A to B is gradually increasing to 5 V; on the right side the voltage is gradually decreasing to 0 V. By stopping the potentiometer wiper at any midpoint, we can get an output voltage anywhere between 0 and 5 V. An analog device, then, is one that has a signal which *varies continuously* in step with the input.

**Signal**

A digital device operates with a digital signal. Figure 1-2(a) pictures a square-wave generator. The generator produces a square waveform that is displayed on the oscilloscope. The digital signal is only at +5 V *or* at 0 V, as diagrammed in Fig. 1-2(b). The voltage at point A moves from 0 to 5 V. The voltage then stays at +5 V for a time. At point B the voltage drops immediately from +5 to 0 V. The voltage then stays at 0 V for a time. Only two voltages are present in a digital electronic circuit. In the waveform diagram in Fig. 1-2(b) these voltages are labeled *HIGH and LOW*. The HIGH voltage is +5 V; the LOW voltage is 0 V. Later we shall call the HIGH voltage (+5 V) a logical 1 and the LOW voltage (0 V) a logical 0. Circuits that handle only HIGH and LOW signals are called *digital circuits*.

**HIGH and LOW signals**

**Digital circuits**

The digital signal in Fig. 1-2(b) could also be generated by a simple on-off switch. A digital signal could also be generated by a transistor turning on and off. Digital electronic signals are usually generated and processed by integrated circuits (ICs).

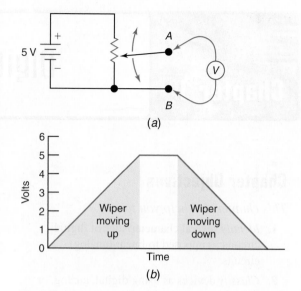

**Fig. 1-1** (a) Analog output from a potentiometer. (b) Analog signal waveform.

Both analog and digital signals are represented in graph form in Figs. 1-1 and 1-2. A *signal* can be defined as useful information transmitted within, to, or from electronics circuits. Signals are commonly represented as a voltage varying with time, as they are in Figs. 1-1 and 1-2. However, a signal could be an electrical current that either varies continuously (analog) or has an on-off (HIGH-LOW) characteristic (digital). Within most digital circuits, it is customary to represent signals in the voltage versus time format. When digital circuits are interfaced with nondigital devices such as lamps and motors, then the signal can be thought of as current versus time.

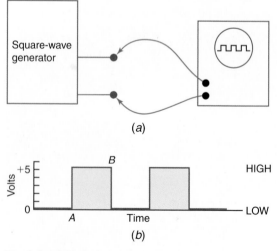

**Fig. 1-2** (a) Digital signal displayed on scope. (b) Digital signal waveform.

(a)

(b)

**Fig. 1-3** (*a*) Analog meter. (*b*) Digital multimeter (DMM). Courtesy Fluke Corporation. Reproduced with permission

The standard *volt-ohm-millimeter* (VOM) shown in Fig. 1-3(*a*) is an example of an *analog* measuring device. As the voltage, resistance, or current being measured by the VOM increases, the needle *gradually and continuously moves* up the scale. A *digital multimeter* (DMM) is shown in Fig. 1-3(*b*). This is an example of a *digital* measuring device. As the current, resist-

ance, or voltage being measured by the DMM increases, the display *jumps upward in small steps*. The DMM is an example of digital circuitry taking over tasks previously performed only by analog devices. This *trend toward digital circuitry* is growing. Currently, the modern technician's bench probably has both a VOM and a DMM.

**Volt-ohm-millimeter**

**Digital multimeter**

**Trend toward digital circuitry**

---

✓ *Self-Test*

*Supply the missing word in each statement.*

1. Refer to Fig. 1-2. The +5-V level of the _____ (analog, digital) signal could also be called a logical 1 or a _____ (HIGH, LOW).
2. A(n) _____ (analog, digital) device is one that has a signal which varies continuously in step with the input.
3. Refer to Fig. 1-4. The input to the electronic block is classified as a(n) _____ (analog, digital) signal.
4. Refer to Fig. 1-4. The *output* from the electronic block is classified as a(n) _____ (analog, digital) signal.
5. An analog circuit is one that processes analog signals while a digital circuit processes _____ signals.

INPUT → Electronic function → OUTPUT

**Fig. 1-4** Block diagram of electronic circuit shaping a sine wave into a square wave.

A photographic history of the computer. One of the first computers was the Eniac (*upper left*), developed in the 1940s. The 1970s marked the expanded use of the computer by businesses. The mainframe computer (*upper right*) was the tool of the time. In the 1980s personal computers such as the Apple IIe (*lower left*) brought computers into our homes and schools. Today, personal computers can go anywhere, as laptop computers (*lower right*) increase in popularity.

## 1-2 Why Use Digital Circuits?

Electronic designers and technicians must have a working knowledge of both analog and digital systems. The designer must decide if the system will use analog or digital techniques or a combination of both. The technicians must build a prototype or troubleshoot and repair digital, analog, and combined systems.

*Analog electronic systems* have been more popular in the past. "Real-world" information dealing with time, speed, weight, pressure, light intensity, and position measurements are all *analog* in nature.

A simple analog electronic system for measuring the amount of liquid in a tank is illustrated in Fig. 1-5. The input to the system is a varying resistance. The processing proceeds according to the Ohm's law formula, $I = V/R$. The output indicator is an ammeter which is calibrated as a water tank gage. In the analog system in Fig. 1-5 as the water rises, the input resistance drops. Decreasing the resistance $R$ causes an increase in current ($I$). Increased current causes the ammeter (water tank gage) to read higher.

The analog system in Fig. 1-5 is simple and efficient. The gage in Fig. 1-5 gives an indication

**Analog electronic system**

**4**    **Chapter 1**   Digital Electronics

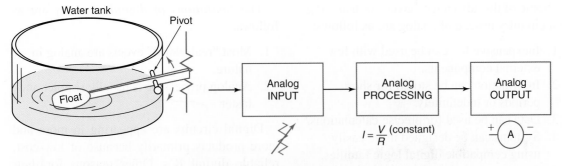

**Fig. 1-5** Analog system used to interpret float level in water tank.

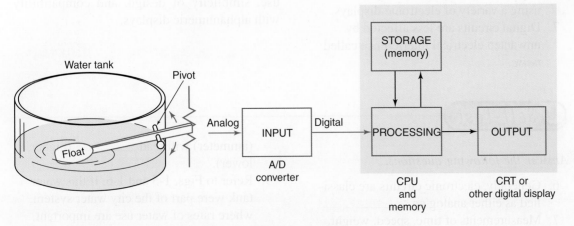

**Fig. 1-6** Digital system used to interpret float level in water tank.

of the water level in the tank. If more information is required about the water level, then a digital system such as the one shown in Fig. 1-6 might be used.

Digital systems are required when data must be stored, used for calculations, or displayed as numbers and/or letters. A somewhat more complex arrangement for measuring the amount of liquid in a water tank is the digital system shown in Fig. 1-6. The input is still a variable resistance as it was in the analog system. The resistance is converted into numbers by the *analog-to-digital (A/D) converter*. The *central processing unit (CPU)* of a computer can manipulate the input data, output the information, store the information, calculate things such as flow rates in and out, calculate the time until the tank is full (or empty) based on flow rates, and so forth. Digital systems are valuable when calculations, data manipulations, data storage, and alphanumeric outputs are required.

**History of Electronics**

**William (Bill) H. Gates III.** The Chairman and Chief Software Architect of Microsoft Corporation, Gates began programming computers at the age of 13. In 1974, while he was a college student, Gates developed a version of BASIC, the programming language for the first microcomputer. Believing that the personal computer would eventually be found on every office desktop and in every home, Gates and Paul Allen formed Microsoft in 1975. Since then, Microsoft has been a leading developer of computer software.

Some of the advantages given for using digital circuitry instead of analog are as follows:

1. Inexpensive ICs can be used with few external components.
2. Information can be stored for short periods or indefinitely.
3. Data can be used for precise calculations.
4. Systems can be designed more easily using compatible digital logic families.
5. Systems can be programmed and show some manner of "intelligence."
6. Alphanumeric information can be viewed using a variety of electronic displays.
7. Digital circuits are less affected by unwanted electrical interference called *noise*.

The *limitations of digital circuitry* are as follows:

1. Most "real-world" events are analog in nature.
2. Analog processing is usually simpler and faster.

Digital circuits are appearing in more and more products primarily because of low-cost, reliable digital ICs. Other reasons for their growing popularity are accuracy, added stability, computer compatibility, memory, ease of use, simplicity of design, and compatibility with alphanumeric displays.

## ✓ Self-Test

*Answer the following questions.*

6. Generally, electronic circuits are classified as either analog or _____.
7. Measurements of time, speed, weight, pressure, light intensity, and position are _____ (analog, digital) in nature.
8. Refer to Fig. 1-5. As the water level drops, the input resistance increases. This causes the current $I$ to _____ (decrease, increase) and the water level gage (ammeter) will read _____ (higher, lower).
9. Refer to Figs. 1-5 and 1-6. If this water tank were part of the city water system, where rates of water use are important, the system in Fig. _____ (1-5, 1-6) would be most appropriate.
10. True or false. The most important reason why digital circuitry is becoming more popular is because digital circuits are usually simpler and faster than analog circuits.

## 1-3 How Do You Generate a Digital Signal?

Digital signals are composed of two well-defined voltage levels. Most of the voltage levels used in this class will be about +3 V to +5 V for HIGH and near 0 V (GND) for LOW. These are commonly called *TTL voltage levels* because they are used with the *transistor-transistor logic* family of ICs.

### Generating a Digital Signal

A TTL digital signal could be made manually by using a mechanical switch. Consider the simple circuit shown in Fig. 1-7(*a*). As the blade of the single-pole, double-throw (SPDT) switch is moved up and down, it produces the *digital waveform* shown at the right. At time period $t_1$, the voltage is 0 V, or LOW. At $t_2$ the voltage is +5 V, or HIGH. At $t_3$, the voltage is again 0 V, or LOW, and at $t_4$, it is again +5 V, or HIGH.

The action of the switch causing the LOW, HIGH, LOW, HIGH waveform in Fig. 1-7(*a*) is called *toggling*. By definition, to *toggle* (the verb) means to switch over to an opposite state. As an example in Fig. 1-7(*a*), if the switch moves from LOW to HIGH we say the output has toggled. Again if the switch moves from HIGH to LOW we say the output has again toggled.

One problem with a mechanical switch is *contact bounce*. If we could look very carefully at a switch toggling from LOW to HIGH, it might look like the waveform in Fig. 1-7(*b*).

**TTL voltage levels**

**Transistor-transistor logic**

**Contact bounce**

**Digital waveform**

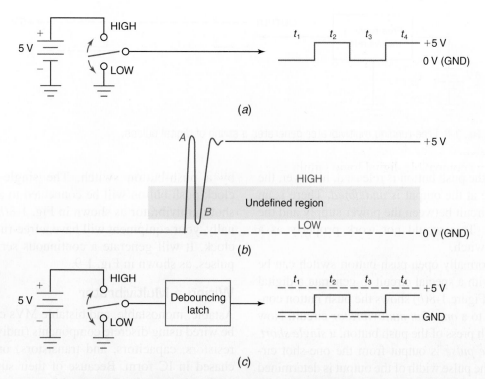

**Fig. 1-7** (a) Generating a digital signal with a switch. (b) Waveform of contact bounce caused by a mechanical switch. (c) Adding a debouncing latch to a simple switch to condition the digital signal.

The waveform first goes directly from LOW to HIGH (see *A*) but then, because of contact bounce, drops to LOW (see *B*) and then back to HIGH again. Although this happens in a very short time, digital circuits are fast enough to see this as a LOW, HIGH, LOW, HIGH waveform. Note that Fig. 1-7(*b*) shows that there is actually a range of voltages that are defined HIGH and LOW. The *undefined region* between HIGH and LOW may cause trouble in digital circuits and should be avoided.

To cure the problem illustrated in Fig. 1-7(*b*), mechanical switches are sometimes *debounced*. A block diagram of a *debounced logic switch* is shown in Fig. 1-7(*c*). Note the use of the debouncing circuit, or latch. Many of the

mechanical logic switches you will use on laboratory equipment will have been debounced with latch circuits. Latches are sometimes called *flip-flops*. Notice in Fig. 1-7(*c*) that the output of the latch during time period $t_1$ is LOW but not quite 0 V. During $t_2$ the output of the latch is HIGH even though it is something less than a full +5 V. Likewise $t_3$ is LOW and $t_4$ is HIGH in Fig. 1-7(*c*).

It might be suggested that a push-button switch be used to make a digital signal. If the button is pressed, a HIGH should be generated. If the push button is released, a LOW should be generated. Consider the simple circuit in Fig. 1-8(*a*). When the push button is pressed, a HIGH of about +5 V is generated at the output.

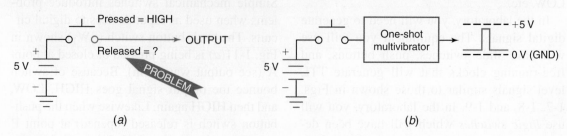

**Fig. 1-8** (a) Push button will not generate a digital signal. (b) Push button used to trigger a one-shot multivibrator for a single-pulse digital signal.

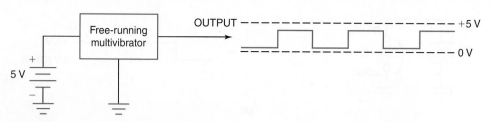

**Fig. 1-9** Free-running multivibrator generates a string of digital pulses.

When the push button is released, however, the voltage at the output is *undefined*. There is an open circuit between the power supply and the output. This would not work properly as a logic switch.

A normally open push-button switch can be used with a special circuit to generate a digital pulse. Figure 1-8(*b*) shows the push button connected to a *one-shot multivibrator* circuit. Now for each press of the push button, a *single short, positive pulse* is output from the one-shot circuit. The pulse width of the output is determined by the design of the multivibrator and *not* by how long you hold down the push button.

### Multivibrator Circuits

Both the latch circuit and the one-shot circuit were used earlier. Both are classified as *multivibrator* (MV) circuits. The latch is also called a flip-flop or a *bistable multivibrator*. The one-shot is also called the *monostable multivibrator*. A third type of MV circuit is the *astable multivibrator*. This is also called a *free-running multivibrator*. In many digital circuits it may be referred to simply as the *clock*.

The free-running MV oscillates by itself without the need for external switching or an external signal. A block diagram of a free-running MV is shown in Fig. 1-9. The free-running MV generates a continuous series of TTL level pulses. The output in Fig. 1-9 alternately toggles from LOW to HIGH, HIGH to LOW, etc.

In the laboratory, you will need to generate digital signals. The equipment you will use will have slide switches, push buttons, and free-running clocks that will generate TTL level signals similar to those shown in Figs. 1-7, 1-8, and 1-9. In the laboratory, you will use *logic switches* which will have been debounced using a latch circuit as in Fig. 1-7(*c*). You will also use a *single-pulse clock* triggered

by a push-button switch. The single-pulse clock push button will be connected to a one-shot multivibrator as shown in Fig. 1-8(*b*). Finally, your equipment will have a free-running clock. It will generate a continuous series of pulses, as shown in Fig. 1-9.

### Wiring a Multivibrator

Astable, monostable, and bistable MVs can all be wired using discrete components (individual resistors, capacitors, and transistors) or purchased in IC form. Because of their superior performance, ease of use, and low cost, the IC forms of these circuits will be used in this course. A schematic diagram for a practical free-running clock circuit is shown in Fig. 1-10(*a*). This clock circuit produces a low-frequency (1- to 2-Hz) TTL level output. The heart of the free-running clock circuit is a common 555 timer IC. Note that several resistors, a capacitor, and a power supply must also be used in the circuit.

A typical breadboard wiring of this free-running clock is sketched in Fig. 1-10(*b*). Notice the use of a solderless breadboard. Also note that pin 1 on the IC is immediately counterclockwise from the notch or dot near the end of the eight-pin IC. The wiring diagram in Fig. 1-10(*b*) is shown for your convenience. You will normally have to wire circuits on solderless breadboards directly from the schematic diagram.

### Wiring a Debounced Switch

Simple mechanical switches introduce problems when used as input devices to digital circuits. The push-button switch (SW$_1$) shown in Fig. 1-11(*a*) is being pressed or closed at point A (see output waveform). Because of switch bounce the output signal goes HIGH, LOW, and then HIGH again. Likewise when the push-button switch is released (opened) at point B more bouncing occurs. Switch bounce from input switches must be eliminated.

One-shot
multivibrator

Multivibrator
types: astable,
bistable, and
monostable

Free-running MV
(clock)

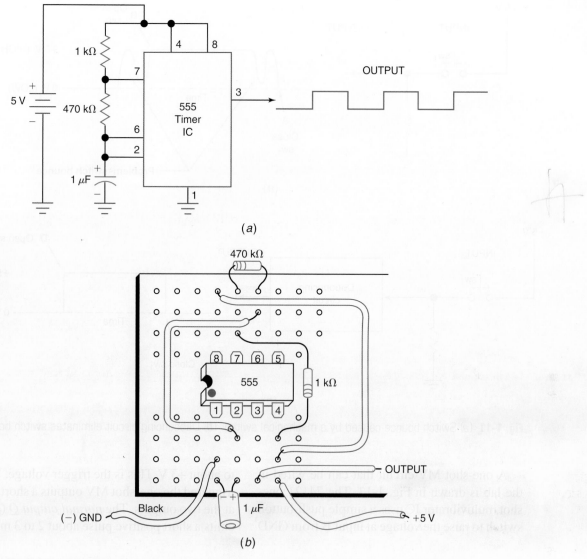

$(a)$

$(b)$

**Fig. 1-10** $(a)$ Schematic diagram of a free-running clock using a 555 timer IC. $(b)$ Wiring the free-running clock circuit on a solderless breadboard.

To solve the problem of switch bounce a *debouncing circuit* has been added in Fig. 1-11($b$). Now when the push-button switch is closed at point C (see output waveform), no bouncing occurs and the output toggles from LOW to HIGH. Likewise when $SW_1$ is opened at point D, no bouncing is observed on the waveform and the output toggles from HIGH to LOW.

An input switch with a debouncing circuit attached is drawn in Fig. 1-12. Observe that the 555 timer IC is at the heart of the debouncing circuit. When pushbutton switch $SW_1$ is closed (see point E on waveform) the output toggles from LOW to HIGH. Later when $SW_1$ is opened (see point F on wave-

form) the output of the 555 timer IC remains HIGH for a delay period. After the delay period (about 1 second for this circuit) the output toggles from HIGH to LOW. The delay period can be adjusted by changing the capacitance value of capacitor $C_2$. Decreasing the capacitance value of $C_2$ will decrease the delay time at the output while increasing $C_2$ will increase the delay. time

## Wiring a One-Shot Multivibrator

A *one-shot multivibrator (MV)* is also called a monostable multivibrator. The one-shot circuit responds to an input trigger pulse with an output pulse of a given width or time duration.

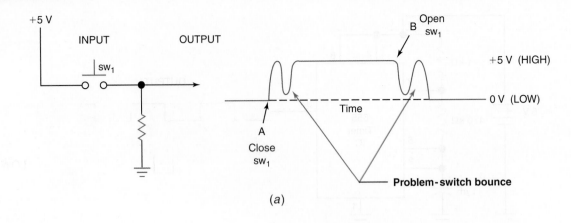

(a)

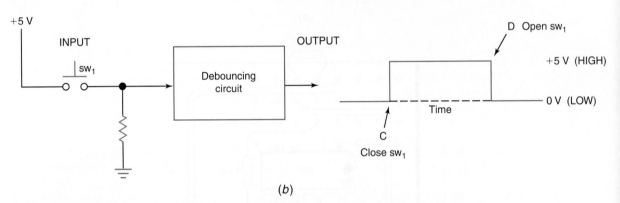

(b)

**Fig. 1-11** (*a*) Switch bounce caused by a mechanical switch. (*b*) Debouncing circuit eliminates switch bounce.

A one-shot MV circuit that can be wired in the lab is drawn in Fig. 1-13. The 74121 one-shot multivibrator IC uses a simple push-button switch to raise the voltage at input B from GND to about +3 V. This is the trigger voltage. When triggered the one-shot MV outputs a short pulse at the two outputs. The *normal output Q* (pin 6) emits a short positive pulse about 2 to 3 msec in

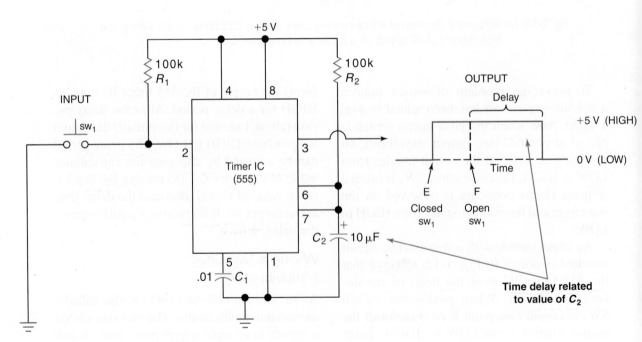

**Fig. 1-12** Switch debouncing circuit.

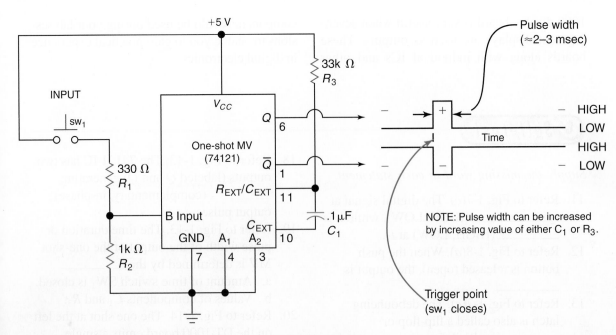

**Fig. 1-13** One-shot multivibrator circuit using the 74121 TTL IC.

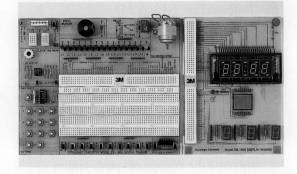

NOTE: Pulse width can be increased by increasing value of either $C_1$ or $R_3$.

duration. The *complementary output* $\overline{Q}$ emits the opposite output or a short negative pulse. On digital devices called flip-flops the outputs are commonly labeled $Q$ and $\overline{Q}$ (say not $Q$) and their outputs are always opposite or complementary. On complementary outputs, if $Q$ is HIGH then $\overline{Q}$ is LOW but if $Q$ is LOW then $\overline{Q}$ is HIGH. The outputs of the 74121 one-shot multivibrator IC come directly from an internal flip-flop and are therefore labeled $Q$ and $\overline{Q}$.

The *pulse width* generated by a one-shot multivibrator is dependent on the design of the MV and not how long the input switch is pressed. The pulse width of the one-shot MV sketched in Fig. 1-13 can be increased by increasing the value of capacitor $C_1$ and/or resistor $R_3$. Decreasing the values of capacitor $C_1$ and resistor $R_3$ will decrease the pulse width.

As a practical matter, the input switch in Fig. 1-13 may have to be debounced or the multivibrator IC could emit more than a single pulse. Using a good quality "snap-action" push-button switch may also help avoid the problem of false triggering by the one-shot MV circuit.

## Digital Trainer

A typical digital trainer used during lab sessions is featured in Fig. 1-14. The photograph actually shows a pair of pc boards specifically

designed to be used with this textbook's companion experiments manual. Dynalogic's DT-1000 digital trainer board on the left includes a solderless breadboard for hooking up circuits. It also includes input devices such as 12 logic switches (two are debounced), a keypad, a one-shot MV, and a variable frequency clock (astable MV). Output devices mounted on the DT-1000 trainer board include 16 LED output indicators, a piezo buzzer, a relay, and a small dc motor. Power connections are available on the upper left of the DT-1000 digital trainer board. On the right in Fig. 1-14 is a second pc board which contains sophisticated LED, LCD, and VF displays. Dynalogic's DB-

**Fig. 1-14** Digital trainer and display boards used to set up lab experiments.

Visit the website for Lucent Technologies for more information on digital circuits and components.

1000 display board is very useful when seven-segment displays are used as outputs. These boards along with individual ICs and other components could be used during your lab sessions to enable you to gain practical experience in digital electronics.

**Output indicators**

## 1-4 How Do You Test for a Digital Signal?

In the last section you generated digital signals using various MV circuits. These are the methods you will use in the laboratory to generate *input signals* for the digital circuits constructed. In this section, several simple methods of testing the *outputs* of digital circuits will be discussed.

Consider the circuit in Fig. 1-15(a). The *input* is provided by a simple SPDT switch and power supply. The *output indicator* is an LED (light-emitting diode). The 150-Ω resistor limits the current through the LED to a safe level. When the switch in Fig. 1-15(a) is in the HIGH (up) position, +5 V is applied to the anode end of the LED. The LED is forward-biased, current flows upward, and the LED lights. With the switch in the LOW (down) position, both the anode and cathode ends of the LED are grounded and it does not light. Using this indicator, a light means HIGH and no light generally means LOW.

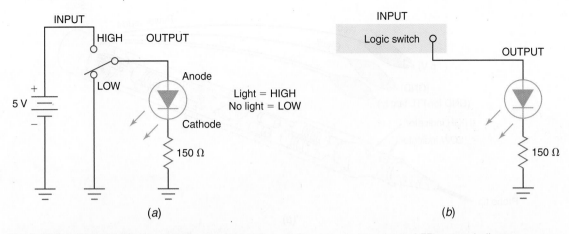

(a)                                                                    (b)

**Fig. 1-15** (a) Simple LED output indicator. (b) Logic switch connected to simple LED output indicator.

The simple LED output indicator is shown again in Fig. 1-15(b). This time a simplified diagram of a logic switch forms the input. The logic switch acts like the switch in Fig. 1-15(a) except it is debounced. The output indicator is again the LED with a series-limiting resistor. When the input logic switch in Fig. 1-15(b) generates a LOW, the LED will not light. However, when the logic switch produces a HIGH, the LED will light.

Another LED output indicator is illustrated in Fig. 1-16. The LED acts exactly the same as the one shown previously. It lights to indicate a logical HIGH and does not light to indicate a LOW. The LED in Fig. 1-16 is driven by a transistor instead of directly by the input. The transistorized circuit in Fig. 1-16 holds an advantage over the direct-drive circuit in that it draws less current from the switch or an output of the digital circuit under test. Light-emitting diode output indicators wired like the one shown in Fig. 1-16 may be found in your laboratory equipment.

Consider the output indicator circuit using two LEDs shown in Fig. 1-17. When the input is HIGH (+5 V), the bottom LED lights while the top LED does not light. When the input is LOW (GND), only the top LED lights. If point Y in the circuit in Fig. 1-17 enters the undefined region between HIGH and LOW or is not connected to a point in the circuit, both LEDs light.

Output voltages from a digital circuit can be measured with a standard voltmeter. With the TTL family of ICs, a voltage from 0 to 0.8 V is considered a LOW. A voltage from 2 to 5 V is considered a HIGH. Voltages between about

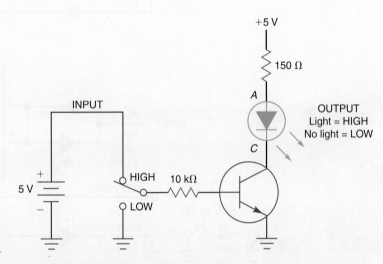

**Fig. 1-16** Transistor-driven LED output indicator.

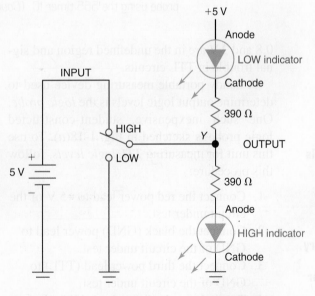

**Fig. 1-17** LED output indicators that will show LOW, HIGH, and undefined logic levels.

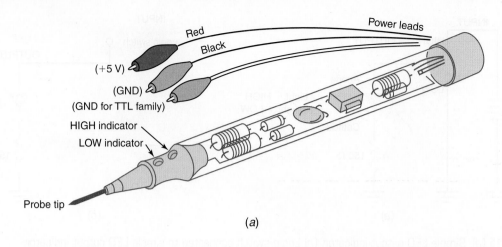

(a)

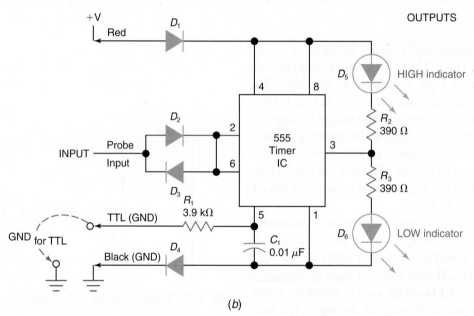

(b)

**Fig. 1-18** (a) Sketch of a student-constructed logic probe. (b) Schematic diagram of the logic probe using the 555 timer IC. (*Courtesy of Electronic Kourseware Interactive, Inc.*)

0.8 and 2 V are in the undefined region and signal trouble in TTL circuits.

A handy portable measuring device used to determine output logic levels is the *logic probe*. One such inexpensive, student-constructed logic probe is sketched in Fig. 1-18(a). To use this unit for measuring *TTL logic levels*, follow this procedure:

1. Connect the red power lead to +5 V of the circuit under test.
2. Connect the black (GND) power lead to GND of the circuit under test.
3. Connect the third power lead (TTL) to GND of the circuit under test.
4. Touch the probe tip to the point in the digital circuit to be tested.

5. One or the other LED indicator shown in Fig. 1-18(a) should light. If both light, the tip is disconnected from the circuit or the point is in the undefined region between HIGH and LOW.

The logic probe in Fig. 1-18(a) can also be used with the *CMOS logic family* of ICs; CMOS is short for *complementary metal oxide semiconductor*. If the logic probe is to be used to measure *CMOS logic levels*, the TTL family power lead is left *disconnected*. Then the red power lead is connected to positive (+) of the power supply while the black lead (GND) goes to GND. Touch the probe tip to the test point in the CMOS digital circuit and the LED indicators tell its CMOS logic level, either HIGH or LOW.

**Logic probe**

**TTL logic levels**

**CMOS logic family**

**Complementary metal oxide semiconductor**

**CMOS logic levels**

A schematic diagram of the logic probe is shown in Fig. 1-18(*b*). The 555 timer IC is being used in this circuit. The 555 IC uses power supply voltages ranging from about 5 V up to 18 V. TTL circuits operate on 5 V, whereas some CMOS circuits operate on voltages as high as 15 V. The three power supply connections are shown at the left in Fig. 1-18(*b*). The red lead goes to the positive of the supply while the black lead goes to GND. If the circuit under test is TTL, the TTL (GND) lead is also connected to GND. If a CMOS circuit is being tested, the TTL (GND) power lead is left disconnected. The input to the logic probe is shown on the left, entering pins 2 and 6 of the 555 IC. If this voltage is LOW, the *bottom* LED ($D_6$) lights. If the input voltage is HIGH, the *top* LED ($D_5$) lights. With the input probe disconnected, both LEDs light. Note that pin 3 of the 555 IC (output of 555) always goes to the *opposite* logic level of the input. Therefore, if the input (pins 2 and 6) is HIGH, the output of the 555 IC (pin 3) goes LOW. This in turn activates and lights the top LED (the HIGH indicator).

The four silicon diodes ($D_1$ to $D_4$) in Fig. 1-18(*b*) protect the IC from reverse polarity. Capacitor $C_1$ prevents transient voltages from affecting the *logic probe* when the TTL (GND) lead is not connected. Pin 5 of the 555 IC is grounded through resistor $R_1$. This resistor "adjusts" for TTL instead of CMOS logic levels.

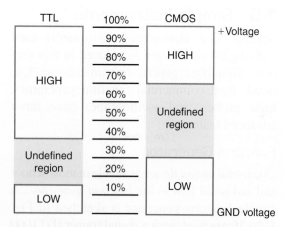

**Fig. 1-19** Defining logic levels for the TTL and CMOS families of digital ICs.

The logic probe in Fig. 1-18 responds to voltage levels differently in the CMOS and TTL modes. Figure 1-19 shows the TTL and CMOS logic levels in terms of percentage of total supply voltage. For TTL, which uses a +5-V supply, this logic probe will indicate a HIGH for 2 V or greater. For TTL, it will indicate a LOW for 0.8 V or less.

In the laboratory, you may be asked to construct a logic probe like the one in Fig. 1-18 or your instructor might furnish you with a commercial logic probe for checking digital circuits. The operating instructions are different for each logic probe. Read the instruction manual on the unit you will be using.

**Defining TTL and CMOS logic levels**

**Logic probe**

---

**Self-Test**

*Supply the missing word or words in each statement.*

24. Refer to Fig. 1-15. If the input is HIGH, the LED will _____ (light, not light) because the diode is _____ (forward-, reverse-) biased.

25. Refer to Fig. 1-16. If the input is LOW, the transistor is turned _____ (off, on) and the LED _____ (does, does not) light.

26. Refer to Fig. 1-17. If the input is HIGH, the _____ (bottom, top) LED lights because its _____ (cathode, anode)

has +5 V applied, forward biasing the diode.

27. The student-constructed _____ in Fig. 1-18 can be used to test either TTL or _____ digital circuits.

28. Refer to Fig. 1-19 and assume a 5-V power supply. In a TTL circuit, a voltage of 2.5 V would be considered a(n) _____ (HIGH, LOW, undefined) logic level.

29. Refer to Fig. 1-19 and assume a 10-V power supply. In a CMOS circuit, a voltage of 2 V would be considered a(n) _____ (HIGH, LOW, undefined) logic level.

## 1-5 Simple Instruments

Several basic commercial instruments used with digital circuits are introduced in this section. Simplified generic instruments are featured. Real commercial function generators, logic probes, and oscilloscopes have more advanced features.

### Function Generator

One useful *output device* available in most school and industrial labs is the *function generator*. A simple function generator is sketched in Fig. 1-20. If you work with a digital trainer (DT1000 trainer in Fig. 1-14) in your school lab, it may contain outputs like those of a function generator.

To use the function generator, you first select the *shape* of the waveform. A square-wave would be selected when working with most digital circuits. Second, the *frequency* (in hertz or Hz) may be selected using the range switch along with the variable multiplier dial. Third, the output voltage is selected. This function generator features two separate voltage outputs (5V TTL and Variable). The 5V TTL output is handy for driving many TTL logic circuits. If you use the variable output on the function generator, the amplitude knob will adjust the output voltage.

What is the *shape* and *frequency* of the waveform being generated by the function generator pictured in Fig. 1-20? The shape knob is in the square-wave position. The range frequency se-

lector knob points at 10 Hz. The multiplier frequency dial points at 1. The output frequency is 10 Hz (range × multiplier = frequency or 10 × 1 = 10 Hz). In Fig. 1-20, the output is taken from the 5V TTL output of the instrument. This output will directly drive a TTL logic circuit.

### Logic Probe

The most basic instrument for testing the digital logic levels is the *logic probe*. A simple logic probe is sketched in Fig. 1-21. The slide switch is used to select the type of logic family under test, either TTL or CMOS. The logic probe pictured in Fig. 1-21 is set to test a TTL-type digital circuit. Typically, two leads provide power to the logic probe. The red lead is connected to the positive (+) of the power supply and the black lead is connected to negative (−) or GND of the power supply. After powering the logic probe, the needlelike probe is touched to the test point in the circuit. Either the HIGH (red LED) or LOW (green LED) indicator will light. If neither or both indicators light, it usually means the voltage is somewhere between HIGH and LOW (undefined region). As a reminder, definitions for *HIGH, LOW,* and *undefined* logic levels are detailed in Fig. 1-19.

The logic probe will be a useful tool when you wire and test digital circuits in the school lab. Read the operating instructions for your specific logic probe.

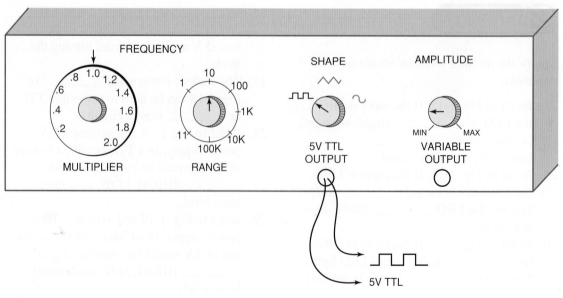

**Fig. 1-20** Function generator.

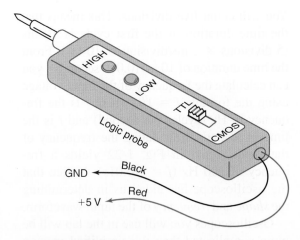

**Fig. 1-21** Logic probe.

## Oscilloscope

The oscilloscope is a very versatile piece of test equipment. A simplified generic oscilloscope, or "scope", is sketched in Fig. 1-22. The basic function of the oscilloscope is to *graph time versus input voltage*. Time is the horizontal distance on the screen and voltage is the vertical deflection. Oscilloscopes work best on signals that repeat over and over.

Consider the digital signal of 4V p-p, 100 Hz shown entering the scope's input in Fig. 1-22. The *horizontal sweep time* knob on the scope has been set to 2 msec (2 milliseconds = 0.002 seconds). This will cause a dot of light to move across the screen from left to right at 2 ms per division (20 ms to cross the entire screen). The dot of light will then jump back to the left end of the screen to start the process again. The *vertical deflection* knob on the scope is set at 1 V per division. A signal voltage of 0 V to +4 V is entering the input. Starting at the left edge of the scope face, first, the dot is deflected four divisions (1 V per division) upward for the first 5 ms. Second, the input voltage drops to 0 V and

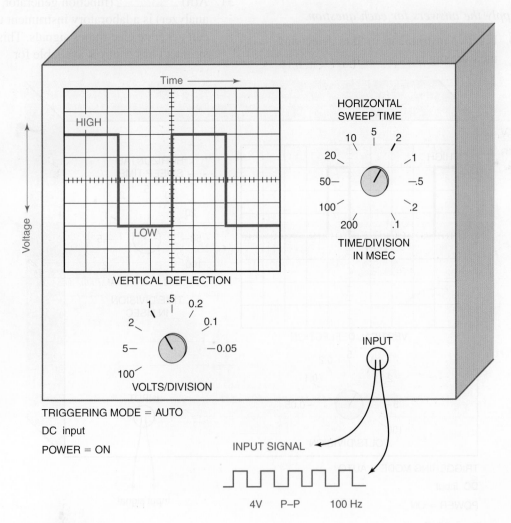

**Fig. 1-22** Oscilloscope.

the lighted dot traces the bottom line for 5 ms. Third, the input voltage jumps to +4 V and with the second upper trace. Fourth, the voltage drops to 0 V with the second lower trace. Finally, the lighted dot jumps back to the left of the screen then repeats. TTL logic levels are labeled on Fig. 1-22 as HIGH (+4 V) and LOW (0 V).

Consider the wave traced on the oscilloscope's screen in Fig. 1-22. The shape of the signal is a square wave. Square waves are useful in digital electronics. A careful look will show that *two* waveforms are displayed on the screen. We say that two *cycles* are displayed.

Look at the waveform on the scope in Fig. 1-22. What is the *time duration* for one cycle? You will count five divisions. This means that the time duration of the first cycle is 10 ms (5 divisions × 2 ms/division = 10 ms). From the time duration of 10 ms (0.010 seconds), you can calculate the frequency of the input voltage using the formula $f = 1/t$, where $f$ is the frequency in Hz (cycles per second) and $t$ is the time in seconds. Calculating the frequency of the input signal in Fig. 1-22 yields a frequency of 100 Hz ($f = 1/0.01s$). Notice that the oscilloscope has aided us in determining the *shape* and *frequency* of the input waveform.

Oscilloscopes you will use in the lab will be more complicated than the simplified version shown in Fig. 1-22. However, the basic function of the scope has been illustrated.

## ✓ Self-Test

*Supply the answers for each question.*

30. List two instruments used to detect and measure digital signals. DMM, fg.

31. A(n) ____f g____ (function generator, logic analyzer) is a laboratory instrument that can generate electronics signals. This instrument has controls available for

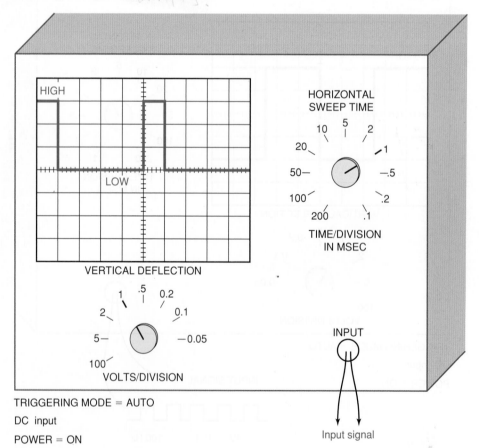

**Fig. 1-23** Oscilloscope problem.

varying the voltage, shape, and frequency of the output signal.

32. Refer to Fig. 1-23. How many cycles are displayed on the screen of the oscilloscope? *2*

33. Refer to Fig. 1-23. What is the time duration for one cycle? *5ms*

34. Refer to Fig. 1-23. What is the frequency of the input signal? *200*

35. Refer to Fig. 1-23. What is the peak-to-peak (p-p) voltage of the input signal? *1,5*

36. Refer to Fig. 1-23. The digital input signal stays HIGH for _____ (1 ms, 10 ms) and goes LOW for _____ (4 ms, 8 ms).

# Chapter 1 Summary and Review

## Summary

1. Analog signals vary gradually and continuously, while digital signals produce discrete voltage levels commonly referred to as HIGH and LOW.
2. Most modern electronics equipment contains both analog and digital circuitry.
3. Logic levels are different for various digital logic families, such as TTL and CMOS. These logic levels are commonly referred to as HIGH, LOW, and undefined. Figure 1-19 details these TTL and CMOS logic levels.
4. Digital circuits have become very popular because of the availability of low-cost digital ICs. Other advantages of digital circuitry are computer compatibility, memory, ease of use, simplicity of design, accuracy, and stability.
5. Bistable, monostable, and astable multivibrators are used to generate digital signals. These are sometimes called latches, one-shot, and free-running multivibrators, respectively.
6. Logic level indicators may take the form of simple LED and resistor circuits, voltmeters, or logic probes. Light-emitting diode logic level indicators will probably be found on your laboratory equipment.
7. A function generator is a lab instrument used to generate electronic signals. The operator can vary an output signal's voltage, frequency, and shape.
8. An oscilloscope is a widely used test and troubleshooting instrument used to graph signals. Oscilloscopes are useful in showing waveform shape, time duration, and frequency of repetitive signals.

## Chapter Review Questions

*Answer the following questions.*

1-1. Define the following:
   a. Analog signal
   b. Digital signal
1-2. Draw a square-wave digital signal. Label the bottom "0 V" and the top "+5 V." Label the HIGH and LOW on the waveform. Label the logical 1 and logical 0 on the waveform.
1-3. List two devices that contain digital circuits which do mathematical calculations.
1-4. Refer to Fig. 1-6. The processing, storage of data, and output in this system consist mostly of _____ (analog, digital) circuits.
1-5. Traditionally, most consumer electronics devices (TVs, radios and phones) have used _____ (analog, digital) circuitry.
1-6. Unwanted electrical interference in electronic circuitry is commonly called _____ (gobo, noise).

1-7. An electronic product that has an alphanumeric display and is programmable, and can store information for sure contains _____ (analog, digital) circuitry.
1-8. Digital circuitry is becoming more pervasive because information storage is easy, the device can be programmed, and _____ (greater, less) accuracy and precision is possible.
1-9. Refer to Fig. 1-7. When using a SPDT switch to produce a digital signal, a _____ latch is used to condition the output.
1-10. Refer to Fig. 1-8. A _____-_____, multivibrator is commonly used to condition the output of a push-button switch when generating a single digital pulse.
1-11. An astable, or _____-_____, multivibrator produces a string of digital pulses.
1-12. The circuit in Fig. 1-10 is classed as a(n) _____ (astable, bistable) multivibrator.

1-13. Refer to Fig. 1-14. The one-shot MV on the DT-1000 trainer produces a _____ (series of pulses, single pulse) when the push button is pressed once.

1-14. Refer to Fig. 1-14. The clock on the DT-1000 trainer produces a _____ (continuous series of pulses, single pulse) and the circuit can be referred to as a(n) _____ (astable, monostable) multivibrator.

1-15. Refer to Fig. 1-14. The two solid-state logic switches on the DT-1000 trainer are _____ (analog, debounced).

1-16. Refer to Fig. 1-14. The DB-1000 board features what *three* types of seven-segment displays?

1-17. The LED in Fig. 1-15(*b*) lights when the input logic switch is _____ (HIGH, LOW).

1-18. Refer to Fig. 1-17. The _____ (bottom, top) LED lights when the input switch is LOW.

1-19. The logic probe in Fig. 1-18 can be used with which two digital logic families?

1-20. Refer to Fig. 1-18(*b*). If the input is LOW, pin 3 of the 555 IC will be _____ (HIGH, LOW). This causes the _____ (bottom, top) LED to light.

1-21. Refer to Fig. 1-19 and assume a 5-V power supply. In a TTL circuit, a voltage of 1.2 V would be considered a(n) _____ (HIGH, LOW, undefined) logic level.

1-22. Refer to Fig. 1-19 and assume a 10-V power supply. In a CMOS circuit, a voltage of 9 V would be considered a(n) _____ (HIGH, LOW, undefined) logic level.

1-23. Refer to Fig. 1-19 and assume a 10-V power supply. In a CMOS circuit, a voltage of 0.5 V would be considered a(n) _____ (HIGH, LOW, undefined) logic level.

1-24. When referring to digital ICs, TTL stands for _____.

1-25. When referring to digital ICs, CMOS stands for _____.

1-26. Refer to Fig. 1-12. The 555 timer IC is wired to function as a(n) _____ (astable MV, switch debouncing) circuit.

1-27. Refer to Fig. 1-12. Increasing the capacitance value of C2 will _____ (decrease, increase) the time delay of the output waveform.

1-28. Refer to Fig. 1-13. The 74121 IC can be described as a(n) _____ (astable, bistable, monostable) multivibrator.

1-29. Refer to Fig. 1-13. Activating input switch $SW_1$ causes a short _____ (negative pulse, positive pulse) to be emitted from the normal Q output of the 74121 one-shot multivibrator IC.

1-30. A flip-flop is the name for a device classified as a(n) _____ (astable, bistable, monostable) multivibrator.

1-31. The repeated action of a switch or other device causing a LOW, HIGH, LOW, HIGH output is called _____ (complementing, toggling).

1-32. Refer to Fig. 1-20. The output from the function generator is a 5 V TTL square-wave signal that has a frequency of _____ Hz.

1-33. Refer to Fig. 1-21. This logic probe can test either TTL or _____ (CMOS, PPC) logic circuits.

1-34. The advantage of the oscilloscope in testing and troubleshooting is that voltage, time duration, frequency, and the _____ (quantum level, shape) of a waveform can be easily determined.

## Critical Thinking Questions

1-1. List several advantages of digital over analog circuits.

1-2. When you are looking at electronic equipment, what are some clues that might indicate it contains at least some digital circuitry?

1-3. Refer to Fig. 1-7(*a*). What is the main drawback of this circuit for generating a digital signal?

1-4. Refer to Fig. 1-8(*a*). What is the difficulty with this circuit for generating a digital signal?

1-5. Refer to Fig. 1-24. From the oscilloscope settings and display, determine the following signal characteristics:
   a. Voltage (peak-to-peak)
   b. Waveform shape
   c. Time duration (one cycle)
   d. Frequency ($f = 1/t$)

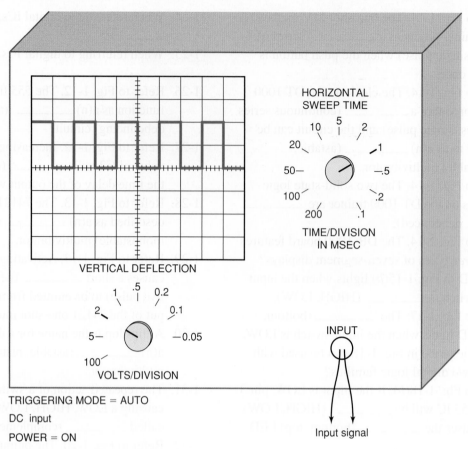

HORIZONTAL
SWEEP TIME

10   5   2
20       1
50      —.5
100
200   .1

TIME/DIVISION
IN MSEC

VERTICAL DEFLECTION

1   .5   0.2
2        0.1
5—       —0.05
100
VOLTS/DIVISION

INPUT

TRIGGERING MODE = AUTO
DC input
POWER = ON

Input signal

**Fig. 1-24** Oscilloscope problem.

1-6. At the option of your instructor, use circuit simulation software to (1) draw a free-running clock circuit using a 555 timer IC such as that pictured in Fig. 1-25, (2) test the operation of the clock circuit, and (3) determine the approximate frequency of the clock using the time period measurements from the scope and the formula $f = 1/t$.

1-7. At the option of your instructor, use circuit simulation software to (1) draw the clock circuit in Fig. 1-25 as in question 1-6, (2) change the resistance of $R_2$ to 100 k$\Omega$, (3) test the operation of the clock circuit, and (4) determine the approximate frequency of the clock using the time period measurements from the scope and the formula $f = 1/t$.

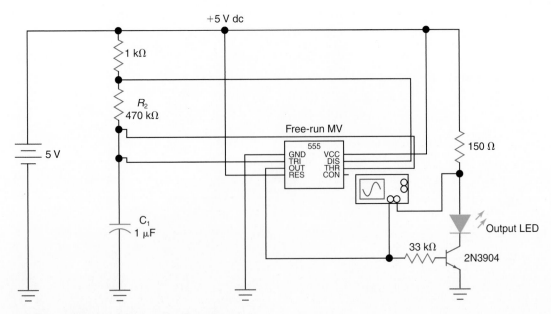

**Fig. 1-25** Circuit simulation problem-clock circuit.

## Answers to Self-Tests

1. digital, HIGH
2. analog
3. analog
4. digital
5. digital
6. digital
7. analog
8. decrease, lower
9. 1–6
10. F

11. HIGH, LOW
12. undefined
13. bistable
14. monostable
15. timer, astable
16. debounce
17. one-shot
18. complementary
19. b
20. monostable

21. astable
22. LED (logic indicators), piezo buzzer, dc motor, relay
23. LED, LCD, VF
24. light, forward-
25. off, does not
26. bottom, anode
27. logic probe, CMOS
28. HIGH

29. LOW
30. logic probe, oscilloscope
31. function generator
32. 2
33. 5 ms or 0.005 s
34. 200 Hz
35. 3 V
36. 1 ms, 4 ms

# Numbers We Use in Digital Electronics

## Chapter Objectives

*This chapter will help you to:*

1. *Demonstrate* understanding of the idea of place value in the decimal, binary, octal, and hexadecimal number systems.
2. *Convert* binary numbers to decimal and decimal numbers to binary.
3. *Convert* hexadecimal numbers to binary, binary to hexadecimal, hexadecimal to decimal, and decimal numbers to hexadecimal.
4. *Convert* octal numbers to binary, binary to octal, octal to decimal, and decimal numbers to octal.
5. *Use* terms such as bit, nibble, byte, and word when describing data groupings.

**M**ost people understand us when we say we have nine pennies. The number 9 is part of the *decimal* number system we use every day. But digital electronic devices use a "strange" number system called *binary*. Digital computers and many other digital systems use other number systems called *hexadecimal* and *octal*. Men and women who work in electronics must know how to convert numbers from the everyday decimal system to the binary, hexadecimal, and octal systems.

Besides decimal, binary, hexadecimal, and octal, many other codes are used in digital electronics. Some of these codes include *binary coded decimal (BCD),* the *Gray code,* and the *ASCII code.* Arithmetic circuits represent positive and negative binary numbers using *2s complement numbers.* Many of these specialized codes will be studied in later chapters.

## 2-1 Counting in Decimal and Binary

A number system is a code that uses symbols to refer to a number of items. The *decimal number system* uses the symbols 0, 1, 2, 3, 4, 5, 6, 7, 8, and 9. The decimal number system contains 10 symbols and is sometimes called the *base 10 system.* The *binary number system* uses only the two symbols 0 and 1 and is sometimes called the *base 2 system.*

Figure 2-1 compares a number of coins with the symbols we use for counting. The decimal symbols that we commonly use for counting from 0 to 9 are shown in the left column; the right column has the symbols we use to count nine coins in the binary system. Notice that the 0 and 1 count in binary is the same as in decimal counting. To represent two coins, the binary number 10 (say "one zero") is used. To

**Decimal number system**

**Base 10 system**

**Binary number system**

**Base 2 system**

represent three coins, the binary number 11 (say "one one") is used. To represent nine coins, the binary number 1001 (say "one zero zero one") is used.

For your work in digital electronics you should memorize the binary symbols used to count to at least 15.

**✔Self-Test**

*Supply the missing word or number in each statement.*

1. The binary number system is sometimes called the _____ system.
2. Decimal 8 equals _____ in binary.
3. The binary number 0110 equals _____ in decimal.
4. The binary number 1001 equals _____ in decimal.

| COINS | DECIMAL SYMBOL | BINARY SYMBOL |
|---|---|---|
| No coins | 0 | 0 |
| ● | 1 | 1 |
| ●● | 2 | 10 |
| ●●● | 3 | 11 |
| ●●●● | 4 | 100 |
| ●●●●● | 5 | 101 |
| ●●●●●● | 6 | 110 |
| ●●●●●●● | 7 | 111 |
| ●●●●●●●● | 8 | 1000 |
| ●●●●●●●●● | 9 | 1001 |

**Fig. 2-1** Symbols for counting.

**Place value**

## 2-2   Place Value

The clerk at the local store totals your bill and asks you for $2.43. We all know that this amount equals 243 cents. Instead of paying with 243 pennies, however, you could probably give the clerk the money shown in Fig. 2-2: two

| | HUNDREDS | TENS | UNITS |
|---|---|---|---|
| 648 = | 600 + | 40 + | 8 |

**Fig. 2-3** Place value in the decimal system.

$1 dollar bills, four dimes, and three pennies. This money example illustrates the very important idea of *place value*.

Consider the decimal number 648 in Fig. 2-3. The digit 6 represents 600 because of its placement three positions left of the decimal point. The digit 4 represents 40 because of its placement two positions left of the decimal point. The digit 8 represents eight units because of its placement one position left of the decimal point. The total number 648, then, represents six hundred and forty-eight units. This is an example of place value in the decimal number system.

The binary number system also uses the idea of place value. What does the binary number 1101 (say "one one zero one") mean? Figure 2-4 shows that the digit 1 nearest the binary

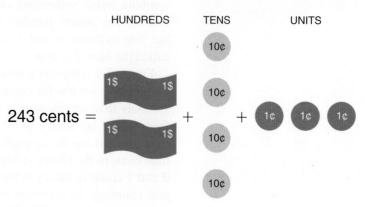

**Fig. 2-2** An example of place value.

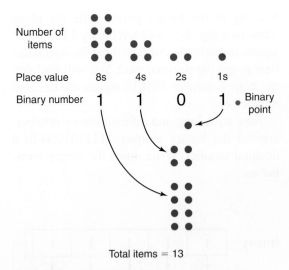

Number of items

Place value   8s   4s   2s   1s

Binary number   **1**   **1**   **0**   **1** .   Binary point

Total items = 13

**Fig. 2-4** Place value in the binary number system.

point is the units, or 1s, position, and so we add one item. The digit 0 in the 2s position tells us we have no 2s. The digit 1 in the 4s position tells us to add four items. The digit 1 in the 8s position tells us to add eight more items. When we count all the items, we find that the binary number 1101 represents 13 items.

How about the binary number 1100 (say "one one zero zero")? Using the system from Fig. 2-4, we find that we have the following:

| 8s | 4s | 2s | 1s | place value |
|----|----|----|----|-------------|
| yes | yes | no | no | binary |
| (1) | (1) | (0) | (0) | number |
| •• | •• | | | number of items |
| •• | •• | | | |
| •• | •• | | | |
| •• | | | | |

The binary number 1100, then, represents 12 items.

Figure 2-5 shows each place in the value of the binary system. Notice that each place value is determined by multiplying the value to the right by 2. The term "base 2" for binary comes from this idea.

Many times the weight or value of each place in the binary number system is referred to as a *power of 2*. In Fig. 2-5, place values for a binary number are shown in decimal and also in powers of 2. For instance, the 8s place is the same as the $2^3$ position, the 32s place the same as the $2^5$ position and so on.

Recall that $2^4$ means $2 \times 2 \times 2 \times 2$ which equals 16. From Fig. 2-5 it is noted that the fifth place left of the binary point is $2^4$ or the 16s place.

**Powers of 2**

---

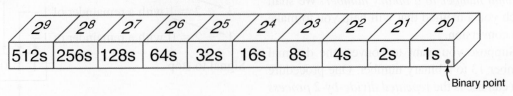

| $2^9$ | $2^8$ | $2^7$ | $2^6$ | $2^5$ | $2^4$ | $2^3$ | $2^2$ | $2^1$ | $2^0$ |
|-------|-------|-------|-------|-------|-------|-------|-------|-------|-------|
| 512s | 256s | 128s | 64s | 32s | 16s | 8s | 4s | 2s | 1s . |

Binary point

**Fig. 2-5** Value of the places left of the binary point.

---

**✓ Self-Test**

*Supply the missing number in each statement.*

5. The 1 in the binary number 1000 has a place value of _____ in decimal.
6. The binary number 1010 equals _____ in decimal.
7. The binary number 100000 equals _____ in decimal.

8. The number $2^7$ equals _____ in decimal.
9. The binary number 1111 1111 equals _____ in decimal.
10. The first place left of the binary point has a value of 1 or _____ ($2^0$, $2^1$).
11. The expression $2^6$ means $2 \times 2 \times 2 \times 2 \times 2 \times 2$ which equals _____ in decimal.

✓

## 2-3 Binary to Decimal Conversion

While working with digital equipment you will have to convert from the *binary code to decimal numbers*. If you are given the binary number 110011, what would you say it equals in decimal? First write down the binary number as:

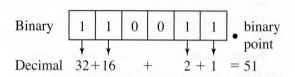

Start at the *binary point* and work to the left. For each binary 1, place the decimal value of that position (see Fig. 2-5) below the binary digit. Add the four decimal numbers to find the decimal equivalent. You will find that binary 110011 equals the decimal number 51.

Another practical problem is to convert the binary number 101010 to a decimal number. Again write down the binary number as:

| Binary | 1 | 0 | 1 | 0 | 1 | 0 | . |
|--------|---|---|---|---|---|---|---|

Decimal    32   +   8   +   2      =   42

Starting at the binary point, write the place value (see Fig. 2-5) for each binary 1 below the square in decimals. Add the three decimal numbers to get the decimal total. You will find that the binary number 101010 equals the decimal number 42.

Now try a long and difficult binary number: convert the binary number 1111101000 to a decimal number. Write down the binary number as:

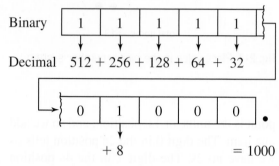

From Fig. 2-5, convert each binary 1 to its correct decimal value. Add the decimal values to get the decimal total. The binary number 1111101000 equals the decimal number 1000.

**Self-Test**

*Supply the missing number in each statement.*

12. The binary number 1111 equals _____ in decimal.

13. The binary number 100010 equals _____ in decimal.
14. The binary number 1000001010 equals _____ in decimal.

## 2-4 Decimal to Binary Conversion

Many times while working with digital electronic equipment you must be able to convert a *decimal number to a binary number*. We shall teach you a method that will help you to make this conversion.

Suppose you want to convert the decimal number 13 to a binary number. One procedure you can use is the *repeated divide-by-2 process* shown next:

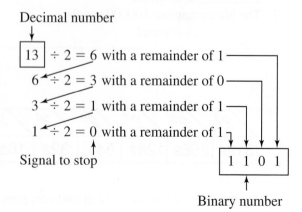

Notice that 13 is first divided by 2, giving a quotient of 6 with a remainder of 1. This remainder becomes the 1s place in the binary number. The 6 is then divided by 2, giving a quotient of 3 with a remainder of 0. This remainder becomes the 2s place in the binary number. The 3 is then divided by 2, giving a quotient of 1 with a remainder of 1. This remainder becomes the 4s place in the binary number. The 1 is then divided by 2, giving a quotient of 0 with a remainder of 1. This remainder becomes the 8s place in the binary number. When the quotient becomes 0 you stop the divide-by-2 process. The decimal number 13 has been converted to the binary number 1101.

Practice this procedure by converting the decimal number 37 to a binary number. Follow the procedure used before:

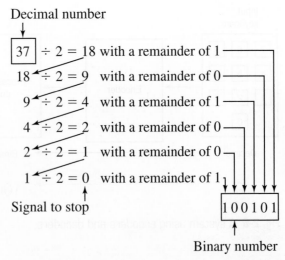

Notice that you stop dividing by 2 when the quotient becomes 0. According to this procedure, the decimal number 37 is equal to the binary number 100101.

## ✓ Self-Test

*Supply the missing number in each statement.*

15. The decimal number 39 equals _____ in binary.

16. The decimal number 100 equals _____ in binary.

17. The decimal number 133 equals _____ in binary.

## 2-5 Electronic Translators

If you were to try to communicate with a French-speaking person who did not know the English language, you would need someone to *translate* the English into French and then the French into English. A similar problem exists in digital electronics. Almost all digital circuits (calculators, computers) understand only binary numbers. But most people understand only decimal numbers. Thus we must have electronic devices that can translate from decimal to binary numbers and from binary to decimal numbers.

Figure 2-6 diagrams a typical system that might be used to translate from decimal to binary numbers and back to decimals. The device that translates from the keyboard decimal numbers to binary is called an *encoder;* the device labeled *decoder* translates from binary to decimal numbers.

The bottom of Fig. 2-6 shows a typical conversion. If you press the decimal number 9 on the keyboard, the encoder will convert the 9 into the binary number 1001. The decoder will convert the binary 1001 into the decimal number 9 on the output display.

Encoders and decoders are very common electronic circuits in all digital devices. A pocket calculator, for instance, must have encoders and decoders to translate electronically from decimal to binary numbers and back to decimal. When you press the number 9 on the keyboard the number appears on the output display.

In modern electronic systems, the encoding and decoding may be performed by *hardware* as suggested in Fig. 2-6, or by computer programs or *software*. In computer jargon to *encrypt* means to encode. Likewise, in computer software to decode means to convert

**Electronic translators**

**Encoders**

**Decoders**

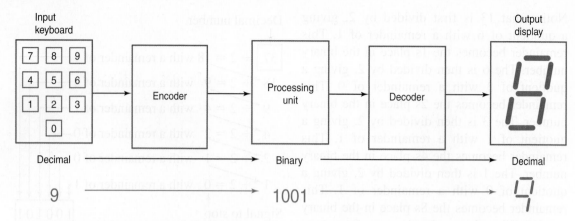

**Fig. 2-6** A system using encoders and decoders.

unreadable or encrypted codes into readable numbers or text. In electronics hardware, to decode means to translate from one code to another. Usually an electronic decoder converts encrypted codes into a more readable form.

You can buy encoders and decoders that translate from any of the commonly used codes in digital electronics. Most of the encoders and decoders you will use will be packaged as single ICs.

### General Definitions:

Here are some general definitions:

*Decode (verb).* To translate from an encrypted code into a more readable form,

for example converting binary code to decimal.

*Decoder (noun).* A logic device that translates from binary code to decimal. Generally, it makes the conversion from processed data in a digital system to a more readable form such as alphanumeric.

*Encode (verb).* To translate or encrypt, for example converting a decimal input into a binary code.

*Encoder (noun).* A logic device that translates from decimal to another code such as binary. Generally, it converts input information to a code useful to digital circuits.

---

✔ **Self-Test**

*Supply the missing word in each statement.*

18. A(n) _____ is an electronic device that translates a decimal input number to binary.
19. The processing unit of a calculator outputs binary. This binary is converted to a decimal output display by an electronic device called a(n) _____.

20. To translate or encrypt from a readable form of data to a binary-coded form is called _____ (decoding, encoding).
21. To translate from an encrypted code (such as binary) to a more readable form (such as decimal) is called _____ (decoding, encoding).

---

## 2-6 Hexadecimal Numbers

The *hexadecimal number system* uses the 16 symbols 0, 1, 2, 3, 4, 5, 6, 7, 8, 9, A, B, C, D, E, and F and is referred to as the *base 16 system.* Figure 2-7 shows the equivalent binary and hexadecimal representations for the decimal

numbers 0 through 17. The letter "A" stands for decimal 10, "B" for decimal 11, and so on. The advantage of the hexadecimal system is its usefulness in converting directly from a 4-bit binary number. For instance, hexadecimal F stands for the 4-bit binary number 1111. *Hexadecimal*

| Decimal | Binary | Hexadecimal |
|---------|--------|-------------|
| 0 | 0000 | 0 |
| 1 | 0001 | 1 |
| 2 | 0010 | 2 |
| 3 | 0011 | 3 |
| 4 | 0100 | 4 |
| 5 | 0101 | 5 |
| 6 | 0110 | 6 |
| 7 | 0111 | 7 |
| 8 | 1000 | 8 |
| 9 | 1001 | 9 |
| 10 | 1010 | A |
| 11 | 1011 | B |
| 12 | 1100 | C |
| 13 | 1101 | D |
| 14 | 1110 | E |
| 15 | 1111 | F |
| 16 | 10000 | 10 |
| 17 | 10001 | 11 |

**Fig. 2-7** Binary and hexadecimal equivalents to decimal numbers.

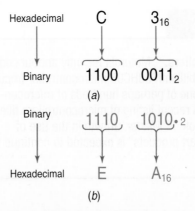

**Fig. 2-8** (a) Converting a hexadecimal number to binary. (b) Converting a binary number to hexadecimal.

*notation* is typically used to represent a binary number. For instance, the hexadecimal number A6 would represent the 8-bit binary number 10100110. Hexadecimal notation is widely used in *microprocessor-based systems* to represent 4, 8-, 16-, 32-, or 64-bit binary numbers.

The number 10 represents how many objects? It can be observed from the table shown in Fig. 2-7 that the number 10 could mean ten objects, two objects, or sixteen objects depending on the base of the number. *Subscripts* are sometimes added to a number to indicate the base of the number. Using subscripts, the number $10_{10}$ represents ten objects. The subscript (10 in this example) indicates it is a *base 10,* or decimal number. Using subscripts, the number $10_2$ represents two objects since this is in binary (*base 2*). Again using subscripts, the number $10_{16}$ represents sixteen objects since this is in hexadecimal (*base 16*).

Converting *hexadecimal numbers to binary* and binary numbers to hexadecimal is a common task when working with microprocessors and microcontrollers. Consider converting $C3_{16}$ to its binary equivalent. In Fig. 2-8(a), each hexadecimal digit is converted to its 4-bit binary equivalent (see Fig. 2-7). Hexadecimal C equals the 4-bit binary number 1100, while $3_{16}$ equals 0011. Combining the binary groups yields $C3_{16} = 11000011_2$.

Now reverse the process and convert the binary number 11101010 to its hexadecimal equivalent. The simple process is detailed in Fig. 2-8(b). The binary number is divided into 4-bit groups starting at the binary point. Next, each 4-bit group is translated into its equivalent hexadecimal number with the help of the table shown in Fig. 2-7. The example in Fig. 2-8(b) shows $11101010_2 = EA_{16}$.

Consider converting hexadecimal $2DB_{16}$ to its decimal equivalent. The place values for the first three places in the hexadecimal number are shown across the top in Fig. 2-9 as 256s, 16s, and 1s. In Fig. 2-9 there are eleven 1s. There are thirteen 16s, which equals 208. There are two 256s, which equals 512. Adding 11 + 208 + 512 will equal $731_{10}$. The example given in Fig. 2-9 shows that $2DB_{16} = 731_{10}$.

Now reverse the process and convert the decimal number 47 to its hexadecimal equivalent. Figure 2-10 details the *repeated divide-by-16 process*. The decimal number 47 is first divided by 16, resulting in a quotient of 2 with a remainder of 15. The remainder of 15 (F in hexadecimal) becomes the least significant digit

| Place value | 256s | 16s | 1s |
|-------------|------|-----|-----|
| Hexadecimal | 2 | D | $B_{16}$ |
| | 256 | 16 | 1 |
| | × 2 | × 13 | × 11 |
| | 512 | 208 | 11 |
| Decimal | 512 | + 208 | + 11 = $731_{10}$ |

**Fig. 2-9** Converting a hexadecimal number to decimal.

## ABOUT ELECTRONICS

**Growth of Microcontrollers.** Motorola recently announced the shipment of its five billionth 68HC05 microcontroller (mcu). This microcontroller is one of perhaps hundreds of microcontrollers on the market. A recent listing of microcontrollers filled over 60 pages in the book, *IC Master*. Growth in the use of microcontrollers in "smart products" is expected to continue to rise rapidly.

$$47_{10} \div 16 = 2 \quad \text{remainder of } 15$$

$$2 \div 16 = 0 \quad \text{remainder of } 2$$

$$47_{10} = 2\,F_{16}$$

**Fig. 2-10** Converting a decimal number to hexadecimal using the repeated divide-by-16 process.

(LSD) of the hexadecimal number. The quotient (2 in this example) is transferred to the dividend position and divided by 16. This results in a quotient of 0 with a remainder of 2. The 2 becomes the next digit in the hexadecimal number. The process is complete because the integer part of the quotient is 0. The divide-by-16 process shown in Fig. 2-10 converts $47_{10}$ to its hexadecimal equivalent of $2F_{16}$.

### Self-Test

*Supply the missing number in each statement.*

22. Decimal 15 equals ________ in hexadecimal.
23. Hexadecimal A6 equals ________ in binary.
24. Binary 11110 equals ________ in hexadecimal.
25. Hexadecimal 1F6 equals ________ in decimal.
26. Decimal 63 equals ________ in hexadecimal.

## 2-7 Octal Numbers

**Octal number system**

Some older computer systems use octal numbers to represent binary information. The *octal number system* uses the eight symbols 0, 1, 2, 3, 4, 5, 6, and 7. Octal numbers are also referred to as *base 8* numbers. The table shown in Fig. 2-11 gives the equivalent binary and octal representations for decimal numbers 0 through 17. The advantage of the octal system is its usefulness in converting directly from a 3-bit binary number. Octal notation is used to represent binary numbers.

**Octal-to-binary conversion**

**Octal-to-decimal conversion**

Converting *octal numbers to binary* is a common operation when using some computer systems. Consider converting the octal number $67_8$ (say "six seven base eight") to its binary equivalent. In Fig. 2-12(*a*), each octal digit is converted to its 3-bit binary equivalent. Octal 6 equals 110, while 7 equals 111. Combining the binary groups yields $67_8 = 110111_2$.

| Decimal | Binary | Octal |
|---|---|---|
| 0 | 000 | 0 |
| 1 | 001 | 1 |
| 2 | 010 | 2 |
| 3 | 011 | 3 |
| 4 | 100 | 4 |
| 5 | 101 | 5 |
| 6 | 110 | 6 |
| 7 | 111 | 7 |
| 8 | 001 000 | 10 |
| 9 | 001 001 | 11 |
| 10 | 001 010 | 12 |
| 11 | 001 011 | 13 |
| 12 | 001 100 | 14 |
| 13 | 001 101 | 15 |
| 14 | 001 110 | 16 |
| 15 | 001 111 | 17 |
| 16 | 010 000 | 20 |
| 17 | 010 001 | 21 |

**Fig. 2-11** Binary and octal equivalents to decimal numbers.

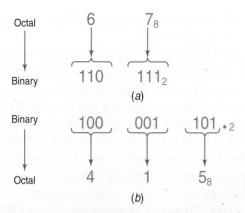

Octal    6     7$_8$

Binary   110   111$_2$

(a)

Binary   100   001   101$\cdot_2$

Octal    4     1     5$_8$

(b)

**Fig. 2-12** (a) Converting an octal number to binary. (b) Converting a binary number to octal.

Now reverse the process and convert binary 100001101 to its octal equivalent. The simple process is detailed in Fig. 2-12(b). The binary number is divided into 3-bit groups (100 001 101) starting at the binary point. Next, each 3-bit group is translated into its equivalent octal number. The example in Fig. 2-12(b) shows $100\ 001\ 101_2 = 415_8$.

Consider converting octal 415 (say "four one five base eight") to its decimal equivalent. The place values for the first three places in the octal number are shown across the top in Fig. 2-13 as 64s, 8s, and 1s. There are five 1s and one 8. There are four 64s, which equals 256. Adding $5 + 8 + 256 = 269_{10}$. The example in Fig. 2-13 shows that $415_8 = 269_{10}$.

Now reverse the process and convert the decimal number 498 to its octal equivalent. Figure 2-14

details the *repeated divide-by-8 process*. The decimal number 498 is first divided by 8, resulting in a quotient of 62 with a remainder of 2. The remainder (2) becomes the LSD of the octal number. The quotient (62 in this example) is transferred to the dividend position and divided by 8. This results in a quotient of 7 with a remainder of 6. The 6 becomes the next digit in the octal number. The last quotient (7 in this example) is transferred to the dividend position and divided by 8. The quotient is 0 with a remainder of 7. The 7 is the most significant digit (MSD) in the octal number. The divide-by-8 process shown in Fig. 2-14 converts $498_{10}$ to its octal equivalent of $762_8$. Note that the signal to end the repeated divide-by-8 process is when the quotient becomes 0.

Technicians, engineers, and programmers must be able to convert between number systems. Many commercial calculators can aid in making binary, octal, hexadecimal, and decimal conversions. These calculators also perform arithmetic operations on binary, octal, and hexadecimal as well as decimal numbers.

Most home and school computers feature various calculators. When working with varied number systems, select the *scientific calculator*. This will allow you to make number system conversions (between binary, octal, hex, and decimal). The scientific calculator will also allow arithmetic calculations (add, subtract, etc.) in various number systems.

**Repeated divide-by-8 process**

| Place value: | 64s | 8s | 1s |
|---|---|---|---|
| Octal | 4 | 1 | 5$_8$ |
| | 64 | 8 | 1 |
| | × 4 | × 1 | × 5 |
| | 256 | 8 | 5 |
| Decimal | 256 | + 8 | + 5 = 269$_{10}$ |

**Fig. 2-13** Converting an octal number to decimal.

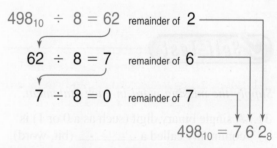

$498_{10} \div 8 = 62$   remainder of 2

$62 \div 8 = 7$   remainder of 6

$7 \div 8 = 0$   remainder of 7

$498_{10} = 7\ 6\ 2_8$

**Fig. 2-14** Converting a decimal number to octal using the repeated divide-by-8 process.

**Decimal-to-octal conversion**

---

**Self-Test**

*Supply the missing number in each statement.*

27. Octal 73 equals _111 011_ in binary.

28. Binary 100000 equals _40_ in octal.

29. Octal 753 equals _491_ in decimal.

30. Decimal 63 equals _77_ in octal.

## 2-8 Bits, Bytes, Nibbles and Word Size

**bit**

A single binary number (either a 0 or a 1) is called a *bit*. Bit is an abbreviation for a *binary digit*. The bit is the smallest unit of data in a digital system. Physically in a digital circuit a single bit is commonly represented by a HIGH or LOW voltage. On a magnetic storage medium (such as a floppy disk) a bit is a small section that can be either a 1 or 0. On an optical disk (such as a CD-ROM) a bit is a small area that either is a pit or no pit for a 1 or 0.

**byte**

All but the simplest digital devices handle larger data groups which in computer jargon may be called *words*. For most computer systems, the width of the main data bus is the same as the *word size*. For instance, a microprocessor or microcontroller might operate on and store 8-bit groups as a single unit of data. Many common microprocessors have word lengths of 8-, 16-, 32-, or 64-bits. A 16-bit piece of data is commonly referred to as a

**Word size**

**nibble**

*word*. A *double-word* contains 32 bits. A *quad-word* features 64 bits.

An 8-bit group of data that represents a number, letter, punctuation mark, control character, or some operation code (op code) in a digital device is called a *byte*. For instance, the hexadecimal number 4F is shorthand for the byte 0100 1111. A byte is an abbreviation for *binary term*. A byte represents a small amount of information and in memory devices we talk of kilobytes ($2^{10}$ or 1024 bytes), megabytes ($2^{20}$ or 1,048,576 bytes), or gigabytes ($2^{30}$ or 1,073, 741,824 bytes) of storage.

A simple digital device might be designed to handle a 4-bit group of data. A half-byte or a 4-bit group of data is called a *nibble*. For instance, the hexadecimal number C is shorthand for the nibble 1100.

In summary, the common names for binary digit groupings are

| Bit | 1 bit (such as 0 or 1) |
|---|---|
| Nibble | 4 bits (such as 1010) |
| Byte | 8 bits (such as 1110 11111) |
| Word | 16 bits (such as 1100 0011 1111 1010) |
| Double-word | 32 bits (such as 1001 1100 1111 0001 0000 1111 1010 0001) |
| Quad-word | 64 bits (such as 1110 1100 1000 0000 0111 0011 1001 1000 0011 0000 1111 1110 1001 0111 0101 0001) |

### Self-Test

*Supply the missing word in each statement.*

31. A single binary digit (such as a 0 or 1) is commonly called a _____ (bit, word).
32. An 8-bit data group that represents a number, letter, punctuation mark, or control character is commonly referred to as a _____ (byte, nibble).
33. A 4-bit data group that represents some number or code is called a _____ (nibble, octet).
34. The length of data groups in a computer system is commonly referred to as its _____ (memory, word) size.
35. A 32-bit data group in a computer system is commonly referred to as a(n)_____ (double-word, nibble).
36. A word in a computer system most often suggests a _____ (16-bit, 64-bit) data group.

# Chapter 2 Summary and Review

## Summary

1. The decimal number system contains 10 symbols: 0, 1, 2, 3, 4, 5, 6, 7, 8, and 9.
2. The binary number system contains two symbols: 0 and 1.
3. The place values left on the binary point in binary are 64, 32, 16, 8, 4, 2, and 1.
4. All men and women in the field of digital electronics must be able to convert binary to decimal numbers and decimal to binary numbers.
5. Encoders are electronic circuits that convert decimal numbers to binary numbers.
6. Decoders are electronic circuits that convert binary numbers to decimal numbers.
7. By general definition, to encode means to convert a readable code (such as decimal) to an encrypted code (such as binary).
8. By general definition, to decode means to convert a machine code (such as binary) to a more readable form (such as alphanumeric).
9. The hexadecimal number system contains 16 symbols: 0, 1, 2, 3, 4, 5, 6, 7, 8, 9, A, B, C, D, E, and F.
10. Hexadecimal digits are widely used to represent binary numbers in the computer field.
11. The octal number system uses eight symbols: 0, 1, 2, 3, 4, 5, 6, and 7. Octal numbers are used to represent binary numbers in a few computer systems.
12. Data groupings have common names including bit, nibble (4 bits), byte (8 bits), word (16 bits), double-word (32 bits), and quad-word (64 bits).

## Chapter Review Questions

*Answer the following questions.*

2-1. How would you say the decimal number 1001?

2-2. How would you say the binary number 1001?

2-3. Convert the binary numbers in **a** to **j** to decimal numbers:
   a. 1        f. 10000
   b. 100      g. 10101
   c. 101      h. 11111
   d. 1011     i. 11001100
   e. 1000     j. 11111111

2-4. Convert the decimal numbers in **a** to **j** to binary numbers:
   a. 0        f. 64
   b. 1        g. 69
   c. 18       h. 128
   d. 25       i. 145
   e. 32       j. 1001

2-5. Encode the decimal numbers in **a** to **f** to binary numbers:
   a. 9        d. 13
   b. 3        e. 10
   c. 15       f. 2

2-6. Decode the binary numbers in **a** to **f** to decimal numbers:
   a. 0010     d.   0111
   b. 1011     e.   0110
   c. 1110     f.   0000

2-7. What is the job (function) of an encoder?

2-8. What is the job (function) of a decoder?

2-9. Write the decimal numbers from 0 to 15 in binary.

2-10. Convert the hexadecimal numbers in **a** to **d** to binary numbers:
   a. 8A       c. 6C
   b. B7       d. FF

2-11. Convert the binary numbers in **a** to **d** to hexadecimal numbers:
   a. 01011110     c. 11011011
   b. 00011111     d. 00110000

2-12. Hexadecimal 3E6 = _____$_{10}$.

2-13. Decimal 4095 = _____$_{16}$.

2-14. Octal 156 = _____$_{10}$.

2-15. Decimal 391 = _____$_{8}$.

2-16. A single 0 or 1 is commonly called a _____ (bit, word).

2-17. An 8-bit group of 1s and 0s, which represents a number, letter, or op code, is commonly called a _____ (byte, nibble).

2-18. A nibble is a term that describes a _____ (4-bit, 12-bit) data group.

2-19. Microprocessor-based systems (such as a computer) commonly identify the size of a data group as _____ (file, word) length.

2-20. To encrypt data from a readable form (such as alphanumeric) to machine code useable by digital circuits is called _____ (encoding, interfacing).

## Critical Thinking Questions

2-1. If the digital circuits in a computer only respond to binary numbers, why are octal and hexadecimal numbers used extensively by computer specialists?

2-2. In a digital system such as a microcomputer, it is common to consider an 8-bit group (called a *byte*) as having a meaning. Predict some of the possible meanings of a byte (such as 11011011$_{2}$) in a microcomputer.

2-3. At the option of your instructor, use circuit simulation software to (a) draw the logic diagram of the decimal-to-binary encoder circuit sketched in Fig. 2-15, (b) operate the circuit, and (c) demonstrate the decimal-to-binary encoder simulation to your instructor.

2-4. At the option of your instructor, use circuit simulation software to (a) draw the logic diagram of the binary-to-decimal decoder circuit shown in Fig. 2-16 on page 38, (b) operate the circuit, and (c) demonstrate the binary-to-decimal decoder simulation to your instructor.

2-5. At the direction of your instructor, use a scientific calculator to convert from one number system to another. Show the instructor your procedure and results.

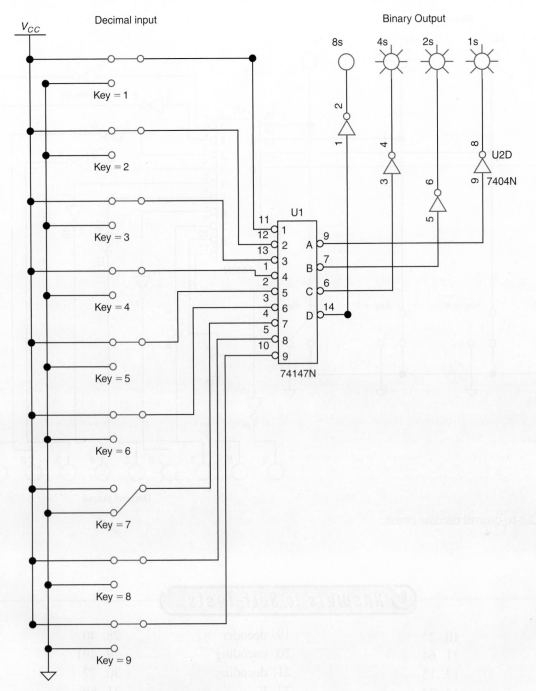

**Fig. 2-15** Decimal-to-binary encoder circuit.

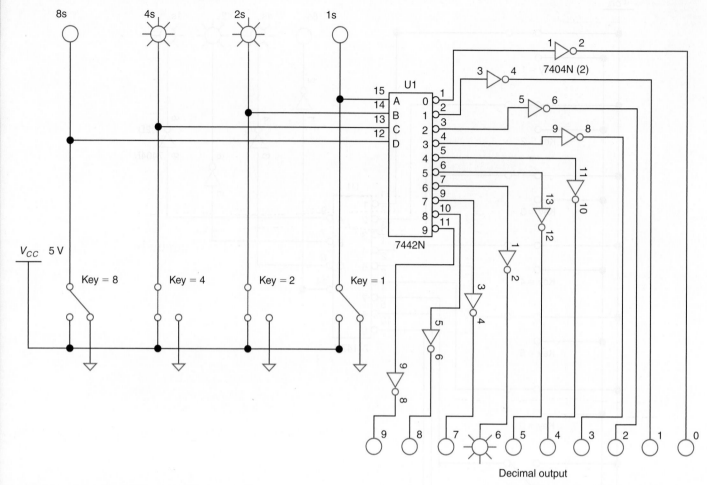

**Fig. 2-16** Binary (BCD)-to-decimal decoder circuit.

## Answers to Self-Tests

1. base 2
2. 1000
3. 6
4. 7
5. 8 or ($2^3$)
6. 10
7. 32
8. 128
9. 255

10. $2^0$
11. 64
12. 15
13. 34
14. 522
15. 100111
16. 1100100
17. 10000101
18. encoder

19. decoder
20. encoding
21. decoding
22. F
23. 10100110
24. 1E
25. 502
26. 3F
27. 111011

28. 40
29. 491
30. 77
31. bit
32. byte
33. nibble
34. word
35. double-word
36. 16-bit

# Chapter 3

# Logic Gates

## Chapter Objectives

*This chapter will help you to:*

1. *Memorize* the name, symbol, truth table, function, and Boolean expression for the eight basic logic gates.
2. *Draw* a logic diagram of any of the eight basic logic functions using only NAND gates.
3. *Convert* one type of basic gate to any other logic function using inverters.
4. *Sketch* logic diagrams illustrating how two-input gates could be used to create gates with more inputs.
5. *Memorize* the inverted input forms of the NAND and NOR gates.
6. *Identify* pin numbers and manufacturer's markings on both TTL and CMOS dual-in-line package ICs.
7. *Troubleshoot* simple logic gate circuits.
8. *Recognize* new logic gate symbols used in dependency notation (IEEE standard 91-1984).
9. *Analyze* the operation of several simple logic gate applications.
10. *Recognize* the use of bubbles attached to a traditional logic symbol showing active-low inputs or outputs.
11. *Program* several logic functions using a BASIC Stamp microcontroller module.

Computers, calculators, and other digital devices are sometimes looked upon by the general public as being magical. Actually, digital electronic devices are extremely *logical* in their operation. The basic building block of any digital circuit is a *logic gate*. Persons working in digital electronics understand and use logic gates every day. Remember that logic gates are the building blocks for even the most complex computers. Logic gates can be constructed by using simple switches, relays, vacuum tubes, transistors and diodes, or ICs. Because of their availability, wide use, and low cost, ICs will be used to construct digital circuits. A variety of *logic gates* are available in all logic families including TTL and CMOS.

**Logic gates**

The task performed by a logic gate is called its *logic function*. Logic functions can be implemented by hardware (logic gates) or by programming devices such as microcontrollers or computers.

**Logic function**

## 3-1 The AND Gate

The AND gate is sometimes called the "*all or nothing gate*." Fig. 3-1 shows the basic idea of the AND gate using simple switches.

**"All or nothing" gate**

What must be done in Fig. 3-1 to get the output lamp ($L_1$) to light? You must close *both* switches $A$ and $B$ to get the lamp to light. You could say that switch $A$ *and* switch $B$ must be closed to get the output to light. The AND gates you will operate most often are constructed of diodes and transistors and packaged inside an IC. To show the AND gate we use the *logic symbol* in Fig. 3-2. This standard AND gate symbol is used whether we are using relays, switches, pneumatic circuits, discrete diodes and transistors, or ICs. This is the symbol

**AND gate logic symbol**

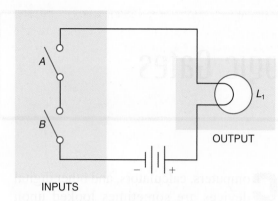

Fig. 3-1 AND circuit using switches.

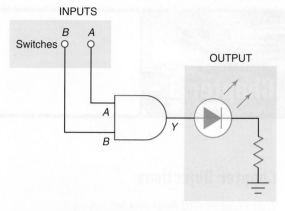

Fig. 3-3 Practical AND gate circuit.

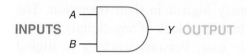

INPUTS ────── Y OUTPUT

Fig. 3-2 AND gate logic symbol.

**Positive logic**

**Truth table**

**AND function**

you will memorize and use from now on for AND gates.

The term "logic" is usually used to refer to a decision-making process. A logic gate, then, is a circuit that can decide to say yes or no at the output based upon the inputs. We already determined that the AND gate circuit in Fig. 3-1 says yes (light on) at the output only when we have a yes (switches closed) at *both* inputs.

Now let us consider an actual circuit similar to one you will set up in the laboratory. The AND gate in Fig. 3-3 is connected to input switches *A* and *B*. The output indicator is an LED. If a LOW voltage (GND) appears at inputs *A* and *B*, then the output LED is *not lit*. This situation is illustrated in line 1 of the truth table in Fig. 3-4. Notice also in line 1 that the inputs and outputs are given as *binary digits*. Line 1 indicates that if the inputs are binary 0 and 0, then the output will be a binary 0. Carefully look over the four combinations of

switches *A* and *B* in Fig. 3-4. Notice that only binary 1s at both *A* and *B* will produce a binary 1 at the output (see the last line of the table).

It is a +5 V compared to GND appearing at *A*, *B*, or *Y* that is called a binary 1 or a HIGH voltage. A binary 0, or LOW voltage, is defined as a GND voltage (near 0 V compared to GND) appearing at *A*, *B*, or *Y*. We are using *positive logic* because it takes a *positive* 5 V to produce what we call a binary 1. You will use positive logic in most of your work in digital electronics.

The table in Fig. 3-4 is called a *truth table*. The truth table for the AND gate gives all the possible input combinations of *A* and *B* and the resulting outputs. Thus the truth table defines the exact operation of the AND gate. The truth table in Fig. 3-4 is said to describe the AND *function*. You should memorize the truth table for the AND function. The unique output from the AND gate is a HIGH only when all inputs are HIGH. The output column in Fig. 3-4 shows that *only* the last line in the AND truth table generates a 1 while the rest of the outputs are 0.

So far you have memorized the logic symbol and the truth table for the AND gate. Now you will learn a shorthand method of writing the

| INPUTS | | | | OUTPUT | |
|---|---|---|---|---|---|
| **B** | | **A** | | **Y** | |
| Switch voltage | Binary | Switch voltage | Binary | Light | Binary |
| LOW | 0 | LOW | 0 | No | 0 |
| LOW | 0 | HIGH | 1 | No | 0 |
| HIGH | 1 | LOW | 0 | No | 0 |
| HIGH | 1 | HIGH | 1 | Yes | 1 |

Fig. 3-4 AND truth table.

| In the English language | Input *A* is ANDed with input *B* to get output *Y*. |
|---|---|
| As a Boolean expression | $A \cdot B = Y$<br>↖ AND symbol |
| As a logic symbol |  |
| As a truth table | <table><tr><th>A</th><th>B</th><th>Y</th></tr><tr><td>0</td><td>0</td><td>0</td></tr><tr><td>0</td><td>1</td><td>0</td></tr><tr><td>1</td><td>0</td><td>0</td></tr><tr><td>1</td><td>1</td><td>1</td></tr></table> |

**Fig. 3-5** Four ways to express the logical ANDing of *A* and *B*.

statement "input *A* is ANDed with input *B* to get output *Y*." The short method used to represent this statement is called its *Boolean expression* ("Boolean" from *Boolean algebra*—the algebra of logic). The Boolean expression is a universal language used by engineers and technicians in digital electronics. Figure 3-5 shows the ways to express that input *A* is ANDed with input *B* to produce output *Y*. The top expression in Fig. 3-5 is how you would tell someone in the English

language that you are ANDing inputs *A* and *B* to get output *Y*. Next in Fig. 3-5 you see the Boolean expression for ANDing inputs *A* and *B*. Note that a multiplication dot ($\cdot$) is used to symbolize the AND function in Boolean expressions. In common practice the Boolean expression $A \cdot B = Y$ can be simplified to $AB = Y$. Both $A \cdot B = Y$ or $AB = Y$ describe the two-input AND function.

Examine the output column of the AND truth table in Fig. 3-5. You will notice that the *unique output* is the last line when output *Y* is HIGH. The AND function's unique output is a HIGH output *only* when all inputs are HIGH.

## Summary

The four commonly used methods of expressing the ANDing of inputs *A* and *B* are detailed in Fig. 3-5. All these methods are widely used and must be learned by persons working in the electronics industry.

The term *logic function* implies the logical relationship between inputs and output while *logic gate* suggests the physical implementation. We might say that an AND logic gate (IC) performs the AND function.

The unique output of the AND function is a HIGH output only when all inputs are HIGH.

**Boolean expression**
**Boolean algebra**

## ✔ Self-Test

*Answer the following questions.*

1. Write the Boolean expression for a two-input AND gate.
2. Refer to Fig. 3-3. When both inputs are HIGH, output *Y* will be _____ (HIGH, LOW) and the LED will _____ (light, not light).
3. Refer to Fig. 3-6. The output of the AND gate at time period $t_1$ is a logical _____ (0, 1).
4. Refer to Fig. 3-6. The output of the AND gate at time period $t_2$ is a logical _____ (0, 1).
5. Refer to Fig. 3-6. The output of the AND gate at time period $t_3$ is a logical _____ (0, 1).
6. Refer to Fig. 3-6. The output of the AND gate at time period $t_4$ is a logical _____ (0, 1).
7. The unique output of an AND gate is a _____ (HIGH, LOW) output only when all inputs are HIGH.

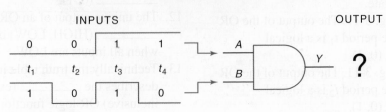

**Fig. 3-6** Pulse train problem.

## 3-2 The OR Gate

**OR gate**

**"Any or all" gate**

The OR gate is sometimes called the *"any or all gate."* Figure 3-7 illustrates the basic idea of the OR gate using simple switches. Looking at the circuit in Fig. 3-7, you can see that the output lamp will light when *either* or *both* of the input switches are closed but not when both are open. A truth table for the OR circuit is shown in Fig. 3-8. The truth table lists the switch and light conditions for the OR gate circuit in Fig. 3-7. The truth table in Fig. 3-8 is said to describe the *inclusive* OR *function.* The unique output from the OR gate is a LOW only when all inputs are LOW. The output column in

**Inclusive OR function**

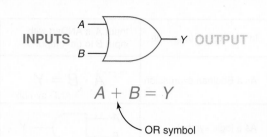

$$A + B = Y$$

OR symbol

**Fig. 3-9** OR logic gate symbol and Boolean expression.

Fig. 3-8 shows that *only* the first line in the OR truth table generates a 0 while all others are 1.

The logic symbol for the OR gate is diagrammed in Fig. 3-9. Notice in the logic diagram that inputs $A$ and $B$ are being ORed to produce an output $Y$. The engineer's Boolean expression for the OR function is also illustrated in Fig. 3-9. Note that the plus ($+$) sign is the Boolean symbol for OR.

You should memorize the logic symbol, Boolean expression, and truth table for the OR gate.

A brief summary of the OR function is shown in Fig. 3-10. It lists four methods of describing the logical ORing of two variables (A and B).

The unique output of the OR function is a LOW output only when all inputs are LOW. Examining the $Y$ output column of the OR truth table in Fig. 3-10 shows that the first line describes the unique output condition for this logic gate.

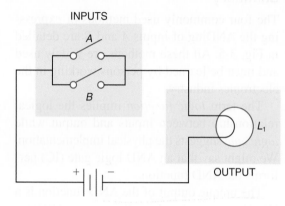

**Fig. 3-7** OR circuit using switches.

|  | INPUTS |  |  | OUTPUT |  |
|---|---|---|---|---|---|
|  | **A** |  | **B** |  | **Y** |
| **Switch** | **Binary** | **Switch** | **Binary** | **Light** | **Binary** |
| Open | 0 | Open | 0 | No | 0 |
| Open | 0 | Closed | 1 | Yes | 1 |
| Closed | 1 | Open | 0 | Yes | 1 |
| Closed | 1 | Closed | 1 | Yes | 1 |

**OR gate truth table**

**Fig. 3-8** OR gate truth table.

### Self-Test

*Answer the following questions.*

8. Write the Boolean expression for a two-input OR gate.
9. Refer to Fig. 3-11. The output of the OR gate at time period $t_1$ is a logical _____ (0, 1).
10. Refer to Fig. 3-11. The output of the OR gate at time period $t_2$ is a logical _____ (0, 1).
11. Refer to Fig. 3-11. The output of the OR gate at time period $t_3$ is a logical _____ (0, 1).
12. The unique output of an OR gate is a _____ (HIGH, LOW) output only when all inputs are LOW.
13. Technically the truth table in Fig. 3-10 describes the _____ (exclusive, inclusive) OR logic function.

| In the English language | Input *A* is ORed with input *B* to yield output *Y*. |
|---|---|
| As a Boolean expression | $A + B = Y$ <br> └─ OR symbol |
| As a logic symbol | 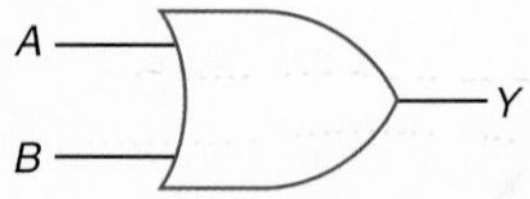 *A* ───┐<br>       ╲─── *Y*<br>*B* ───┘ |
| As a truth table | *A*  *B*  *Y* <br> 0  0  0 <br> 0  1  1 <br> 1  0  1 <br> 1  1  1 |

**Fig. 3-10** Four ways to express the logical ORing of inputs *A* and *B*.

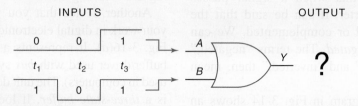

INPUTS

0    0    1
$t_1$    $t_2$    $t_3$
1    0    1

OUTPUT

?

**Fig. 3-11** Pulse train problem.

## 3-3  The Inverter and Buffer

All the gates so far have had at least two inputs and one output. The *NOT circuit*, however, has only one input and one output. The NOT circuit is often called an *inverter*. The job of the NOT circuit (inverter) is to give an output that is not the same as the input. The logic symbol for the inverter (NOT circuit) is shown in Fig. 3-12.

If we were to put in a logic 1 at input *A* in Fig. 3-12, we would get out the opposite, or a logical 0, at output *Y*. We say that the inverter *complements* or *inverts* the input. Figure 3-12 also shows how we would write a Boolean expression for the NOT, or INVERT function.

Notice the use of the overbar ( ¯ ) symbol above the output to show that *A* has been inverted or complemented. We say that the Boolean term "$\overline{A}$" would be "not *A*."

An alternative NOT symbol used in Boolean expressions is also shown in Fig. 3-12. Notice the use of the apostrophe symbol to show that *A* has been inverted or complemented. We would say that the Boolean term *A'* would be "not *A*" or "*A* not." The use of the overbar is the preferred NOT symbol but the apostrophe is common when Boolean expressions are shown on a computer screen as in circuit simulation programs.

**NOT circuit**

**Inverter**

**Complements**

**Inverts**

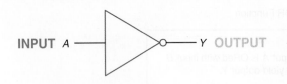

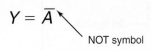

$$Y = \overline{A}$$

NOT symbol

$$Y = A'$$

Alternate NOT symbol

**Fig. 3-12** A logic symbol and Boolean expression for an inverter.

The truth table for the inverter is shown in Fig. 3-13. If the voltage at the input of the inverter is LOW, then the output voltage is HIGH. However, if the input voltage is HIGH, then the output is LOW. As you learned, the output is always opposite the input. The truth table also gives the characteristics of the inverter in terms of binary 0s and 1s.

You learned that when a signal passes through an inverter, it can be said that the input is inverted or complemented. We can also say it is *negated*. The terms "negated," "complemented," and "inverted," then, mean the same thing.

The logic diagram in Fig. 3-14 shows an arrangement where input $A$ is run through two inverters. Input $A$ is first inverted to produce a "not $A$" ($\overline{A}$) and then inverted a second time for a "double not $A$" ($\overline{\overline{A}}$). In terms of binary digits, we find that when the input 1 is inverted twice, we end up with the original digit. Therefore we find that $\overline{\overline{A}}$ equals $A$. Thus, a Boolean term with two bars over it is equal to the term under the two bars, as shown at the bottom of Fig. 3-14.

| INPUT | | OUTPUT | |
|---|---|---|---|
| **A** | | **Y** | |
| **Voltages** | **Binary** | **Voltages** | **Binary** |
| LOW | 0 | HIGH | 1 |
| HIGH | 1 | LOW | 0 |

**Fig. 3-13** Truth table for an inverter.

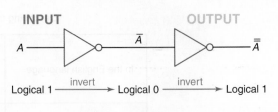

Therefore $\overline{\overline{A}} = A$

**Fig. 3-14** Effect of double inverting.

Two symbols found in logic diagrams that look something like an inverter symbol are pictured in Fig. 3-15. The logic symbol in Fig. 3-15(*a*) is an alternative symbol for an inverter and performs the NOT function. The placement of the "invert bubble" on the left side of the inverter symbol in Fig. 3-15(*a*) might suggest that this is an active LOW input.

The symbol in Fig. 3-15(*b*) is that of a *noninverting buffer/driver*. The noninverting buffer serves no logical purpose (it does not invert), but is used to supply greater *drive current* at its output than is normal for a regular gate. Since regular digital ICs have limited drive current capabilities, the noninverting buffer/driver is important when interfacing ICs with other devices such as LEDs, lamps, and others. Buffer/drivers are available in both noninverting and inverting form.

Another device that you will encounter in your work in digital electronics is the symbol in Fig. 3-16(*a*). It represents a common type of buffer/driver used with *bus* systems (busses are used in computers). The unit detailed in Fig. 3-16 is a *three-state buffer*. It looks like a regular

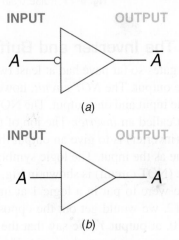

**Fig. 3-15** (*a*) Alternative inverter logic symbol (note bubble at input). (*b*) Non-inverting buffer/driver logic symbol.

buffer/driver device except that it has an extra *control input.* According to the truth table for the three-state buffer in Fig. 3-16(*b*), when the control (C) input goes HIGH, its output goes to a *high-impedance state* (high-Z state). In the high-impedance state, the output will act like an open switch between the output of the buffer and the bus. In its high-impedance state, the output of the buffer has no effect on the logic level on the bus to which it is connected. This allows several logic devices with three-state outputs to be connected to a bus at the same time, however, only one three-state device can be active at a time.

When the control input to the three-state buffer goes LOW (see Fig. 3-16), the buffer will transfer true (noninverted) data from input to output.

In summary, you now know the logic symbols, Boolean expression, and truth table for the inverter or NOT gate. Second, you can recognize the symbol for a noninverting buffer/driver and know its purpose is to drive LEDs, lamps, and so forth. Third, you can recognize a three-state buffer/driver and know it is used when connecting to bus systems.

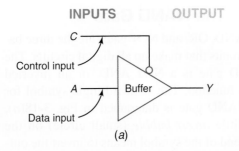

INPUTS          OUTPUT

*(a)*

| INPUTS | | OUTPUT |
|---|---|---|
| *C* | *A* | *Y* |
| L | L | L |
| L | H | H |
| H | X | (Z) |
| Control input | Data input | |

L = Low voltage level
H = High voltage level
X = Don't care (Does not affect output)
(Z) = High impedance

*(b)*

**Fig. 3-16** Noninverting buffer/driver (three-state output). (*a*) Logic symbol for three-state buffer; (*b*) truth table for three-state buffer.

## Self-Test

*Answer the following questions.*

14. Refer to Fig. 3-12. If input *A* is HIGH, output *Y* from the inverter will be _____.

15. Refer to Fig. 3-14. If input *A* to the left inverter is LOW, the output from the right inverter will be _____.

16. Write the Boolean expression used to describe the action of an inverter.

17. List two words that are used to mean "inverted."

18. Refer to Fig. 3-15(*b*). If input *A* is LOW, the output from this buffer will be

_____.

19. Refer to Fig. 3-17. The output of the three-state buffer during time period $t_1$ is _____ (HIGH, LOW, high-impedance).

20. Refer to Fig. 3-17. The output of the three-state buffer during time period $t_2$ is _____ (HIGH, LOW, high-impedance).

21. Refer to Fig. 3-17. The output of the three-state buffer during time period $t_3$ is _____ (HIGH, LOW, high-impedance).

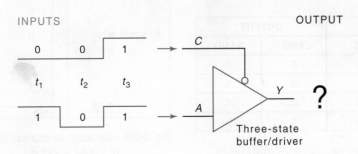

**Fig. 3-17** Pulse train problem.

## 3-4 The NAND Gate

**NAND truth table**

**NAND gate**

**NAND gate logic symbol**

**Invert bubble**

**NAND logic function**

**NAND Boolean expression**

The AND, OR, and NOT gates are the three basic circuits that make up all digital circuits. The NAND gate is a NOT AND, or an inverted AND function. The standard logic symbol for the NAND gate is diagramed in Fig. 3-18(a). The little *invert bubble* (small circle) on the right end of the symbol means to invert the output of AND.

Figure 3-18(b) shows a separate AND gate and inverter being used to produce the *NAND logic function*. Also notice that the Boolean expressions for the AND gate $(A \cdot B)$ and the NAND $(\overline{A \cdot B})$ are shown on the logic diagram in Fig. 3-18(b).

The truth table for the NAND gate is shown at the right in Fig. 3-19. Notice that the truth table for the NAND gate is developed by *inverting the outputs* of the AND gate. The AND gate outputs are also given in the table for reference.

Do you know the logic symbol, Boolean expression, and truth table for the NAND gate?

You must commit these to memory. The unique output from the NAND gate is a LOW only when all inputs are HIGH. The NAND output column in Fig. 3-19 shows that *only* the last line in the truth table generates a 0 while all other outputs are 1.

A brief summary of the NAND function is shown in Fig. 3-20. It lists four methods of describing the logical NANDing of two variables (*A* and *B*). Several alternative methods of writing the *NAND Boolean expression* are given in Fig. 3-20. The first two are traditional Boolean expressions using long overbars while the last $[(AB)' = Y]$ is a computer version used to represent the NAND function.

The unique output of the NAND function is a LOW output only when all inputs are HIGH. Examining the *Y* output column of the NAND truth table in Fig. 3-20 shows that the last line describes the unique output condition for this logic gate.

Describing the NAND Function

| In the English language | Input *A* is NANDed with input *B* yielding output *Y*. |
|---|---|
| As a Boolean expression | NOT symbol $\overline{A \cdot B} = Y$ or AND symbol $\overline{AB} = Y$ or $(AB)' = Y$ |
| As a logic symbol | A ──┐ Y; B ──┘ |
| As a truth table | see table below |

| A | B | Y |
|---|---|---|
| 0 | 0 | 1 |
| 0 | 1 | 1 |
| 1 | 0 | 1 |
| 1 | 1 | 0 |

**Fig. 3-20** Four ways to express the logical NANDing of inputs *A* and *B*.

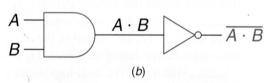

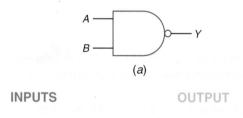

**Fig. 3-18** (a) NAND gate logic symbol. (b) A Boolean expression for the output of a NAND gate.

| INPUTS | | OUTPUT | |
|---|---|---|---|
| **B** | **A** | **AND** | **NAND** |
| 0 | 0 | 0 | 1 |
| 0 | 1 | 0 | 1 |
| 1 | 0 | 0 | 1 |
| 1 | 1 | 1 | 0 |

**Fig. 3-19** Truth tables for AND and NAND gates.

*Answer the following questions.*

22. Write a Boolean expression for a two-input NAND gate.
23. Refer to Fig. 3-21. The output of the NAND gate at time period $t_1$ is a logical _____ (0, 1).
24. Refer to Fig. 3-21. The output of the NAND gate at time period $t_2$ is a logical _____ (0, 1).
25. Refer to Fig. 3-21. The output of the NAND gate at time period $t_3$ is a logical _____ (0, 1).
26. The unique output of an NAND gate is a _____ (HIGH, LOW) output only when all inputs are HIGH.

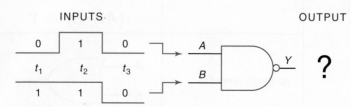

**Fig. 3-21** Pulse train problem.

## 3-5  The NOR Gate

The NOR gate is actually a NOT OR gate. In other words, the output of an OR gate is inverted to form a NOR gate. The *logic symbol* for the NOR gate is diagramed in Fig. 3-22(*a*). Note that the NOR symbol is an OR symbol with a small *invert bubble* (small circle) on the right side. The NOR function is being performed by an OR gate and an inverter in Fig. 3-22(*b*). The Boolean expression for the OR function $(A + B)$ is shown, The Boolean expression for the final NOR function is $\overline{A + B}$.

The truth table for the NOR gate is shown in Fig. 3-23. Notice that the NOR gate truth table is just the complement of the output of the OR gate. The output of the OR gate is also included in the truth table in Fig. 3-23 for reference.

You now should memorize the symbol, Boolean expression, and truth table for the NOR gate. You will encounter these items often in your work in digital electronics. The unique output from the NOR gate is a HIGH only when all inputs are LOW. The output column in Fig. 3-23 shows that *only* the first line in the NOR *truth table* generates a 1 while all other outputs are 0.

A brief summary of the NOR function is shown in Fig. 3-24. It lists four methods of describing the logical NORing of two variables

**NOR gate**

**NOR gate logic symbol**

**Invert bubble**

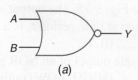

(a)

**INPUTS**          **OUTPUT**

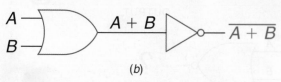

(b)

**Fig. 3-22** (*a*) NOR gate logic symbol. (*b*) Boolean expression for the output of a NOR gate.

| INPUTS | | OUTPUT | |
|---|---|---|---|
| **B** | **A** | **OR** | **NOR** |
| 0 | 0 | 0 | 1 |
| 0 | 1 | 1 | 0 |
| 1 | 0 | 1 | 0 |
| 1 | 1 | 1 | 0 |

**Fig. 3-23** Truth table for OR and NOR gates.

**NOR Boolean expression**

(*A* and *B*). Several alternative methods of writing the NOR *Boolean expression* are given in Fig. 3-24. The first is the traditional Boolean expression using a long overbar while the last $(A + B)' = Y$ is a computer version representing the NOR function.

The unique output of the NOR function is a HIGH output only when all inputs are LOW.

Describing the NOR Function

| In the English language | Input *A* is NORed with input *B* yielding output *Y*. |
|---|---|
| As a Boolean expression | $\overline{A + B} = Y$ or $(A + B)' = Y$ <br> NOT symbol / OR symbol |
| As a logic symbol | (logic gate symbol with inputs *A*, *B* and output *Y*) |
| As a truth table | *A*  *B*  *Y* <br> 0  0  1 <br> 0  1  0 <br> 1  0  0 <br> 1  1  0 |

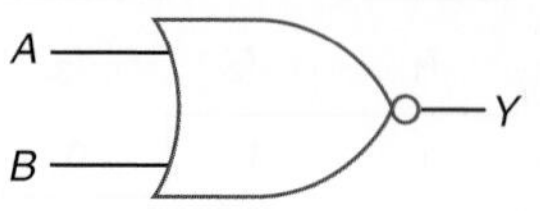

**NOR truth table**

**Fig. 3-24** Four ways to express the logical NORing of inputs *A* and *B*.

---

✔ *Self-Test*

*Answer the following questions.*

27. Write a Boolean expression for a two-input NOR gate.
28. Refer to Fig. 3-25. The output of the NOR gate at time period $t_1$ is a logical _____ (0, 1).
29. Refer to Fig. 3-25. The output of the NOR gate at time period $t_2$, is a logical _____ (0, 1).
30. Refer to Fig. 3-25. The output of the NOR gate at time period $t_3$ is a logical _____ (0, 1).
31. The unique output of an NOR gate is a _____ (HIGH, LOW) output only when all inputs are LOW.

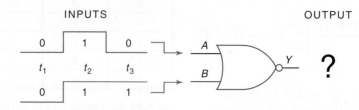

**Fig. 3-25** Pulse train problem.

## 3-6  The Exclusive OR Gate

The *exclusive* OR *gate* is sometimes referred to as the "*odd but not even gate*." The term "exclusive OR gate" is often shortened to "XOR gate." The logic symbol for the *XOR gate* is diagramed in Fig. 3-26(*a*); the Boolean expression for the *XOR function* is illustrated in Fig. 3-26(*b*). The symbol ⊕ means the terms are XORed together.

The output for the XOR gate is shown at the right in Fig. 3-27. Notice that if any but not all of the inputs are 1, then the output will be a binary, or logical, 1. The OR gate truth table is also given in Fig. 3-27, so that you may compare the OR gate truth table with the XOR gate *truth table*.

The unique characteristic of the XOR gate is that it produces a HIGH output only when an *odd number of HIGH inputs are present*. To demonstrate this idea, Fig. 3-28 depicts a three-input XOR gate logic symbol, a Boolean expression, and a truth table. In Fig. 3-28(*b*) the three-input XOR function is described in the output column (*Y*). The HIGH outputs are generated only when an odd number of HIGH inputs are present (lines 2, 3, 5, and 8 in the truth table). If an even number of HIGH inputs to the XOR gate are present the output will be

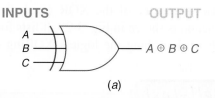

INPUTS    OUTPUT

$A \oplus B \oplus C$

(*a*)

Exclusive OR gate

XOR gate

XOR function

*odd but not even*

XOR truth table

3-input **XOR**

| INPUTS | | | OUTPUT |
|---|---|---|---|
| *C* | *B* | *A* | *Y* |
| 0 | 0 | 0 | 0 |
| 0 | 0 | 1 | 1 |
| 0 | 1 | 0 | 1 |
| 0 | 1 | 1 | 0 |
| 1 | 0 | 0 | 1 |
| 1 | 0 | 1 | 0 |
| 1 | 1 | 0 | 0 |
| 1 | 1 | 1 | 1 |

(*b*)

**Fig. 3-28** (*a*) Three-input XOR gate symbol and Boolean expression. (*b*) Truth table for three-input XOR gate.

LOW (lines 1, 4, 6, and 7 in the truth table). The XOR gates are used in a variety of arithmetic circuits.

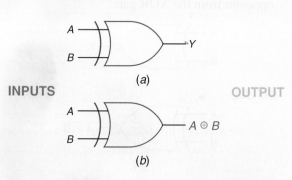

(*a*)

INPUTS

OUTPUT

$A \oplus B$

(*b*)

**Fig. 3-26** (*a*) XOR gate logic symbol. (*b*) Boolean expression for the output of an XOR gate.

| INPUTS | | OUTPUT | |
|---|---|---|---|
| *B* | *A* | OR | XOR |
| 0 | 0 | 0 | 0 |
| 0 | 1 | 1 | 1 |
| 1 | 0 | 1 | 1 |
| 1 | 1 | 1 | 0 |

**Fig. 3-27** Truth table for OR and XOR gates.

Describing the XOR Function

| In the English language | Inputs *A*, *B*, and *C* are XORed yielding output *Y*. |
|---|---|
| As a Boolean expression | $A \oplus B \oplus C = Y$ <br> XOR symbol |
| As a logic symbol | *A* *B* *C* — *Y* |
| As a truth table | <table><tr><td>*A*</td><td>*B*</td><td>*C*</td><td>*Y*</td></tr><tr><td>0</td><td>0</td><td>0</td><td>0</td></tr><tr><td>0</td><td>0</td><td>1</td><td>1</td></tr><tr><td>0</td><td>1</td><td>0</td><td>1</td></tr><tr><td>0</td><td>1</td><td>1</td><td>0</td></tr><tr><td>1</td><td>0</td><td>0</td><td>1</td></tr><tr><td>1</td><td>0</td><td>1</td><td>0</td></tr><tr><td>1</td><td>1</td><td>0</td><td>0</td></tr><tr><td>1</td><td>1</td><td>1</td><td>1</td></tr></table> |

XOR gate logic symbol

XOR Boolean expression

**Fig. 3-29** Four ways to express the XORing of inputs *A*, *B*, and *C*.

A brief summary of the XOR (exclusive-OR) function is shown in Fig. 3-29. It lists four methods of describing the logical XORing of three variables ($A$, $B$ and $C$). The unique output of the XOR gate is a HIGH only when an *odd number* of inputs are HIGH.

*Answer the following questions.*

32. Write a Boolean expression for a three-input XOR gate.
33. Refer to Fig. 3-30. The output of the XOR gate at time period $t_1$ is a logical _____ (0, 1).
34. Refer to Fig. 3-30. The output of the XOR gate at time period $t_2$ is a logical _____ (0, 1).
35. Refer to Fig. 3-30. The output of the

XOR gate at time period $t_3$ is a logical _____ (0, 1).
36. Refer to Fig. 3-30. The output of the XOR gate at time period $t_4$ is a logical _____ (0, 1).
37. Refer to Fig. 3-30. The output of the XOR gate at time period $t_5$ is a logical _____ (0, 1).
38. The unique output of an XOR gate is a HIGH when an _____ (even, odd) number of inputs is HIGH.

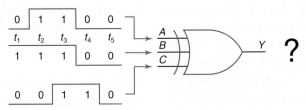

**Fig. 3-30** Pulse train problem.

## 3-7 The Exclusive NOR Gate

**Exclusive NOR gate**

**XNOR gate logic symbol**

**XNOR function**

**XNOR truth table**

**XNOR Boolean expression**

The term "exclusive NOR gate" is often shortened to "XNOR gate." The *logic symbol* for the XNOR gate is shown in Fig. 3-31(*a*). Notice that it is the XOR symbol with the added invert bubble on the output side. Figure 3-31(*b*) illustrates one of the Boolean expressions used for the XNOR *function*. Observe that the Boolean expression for the XNOR gate is $\overline{A \oplus B}$. The bar over the $A \oplus B$ expression tells us we have inverted the output of the XOR gate. Examine the truth table in Fig. 3-31(*c*). Notice that the output of the XNOR gate is the complement of the XOR *truth table*. The XOR gate output is also shown in the table in Fig. 3-31(*c*).

You now will have mastered the logic symbol, truth table, and *Boolean expression* for the XNOR gate.

A brief summary of the XNOR (exclusive-NOR) function is shown in Fig. 3-32. It lists four methods of describing the logical XNOR-ing of three variables ($A$, $B$ and $C$). The unique

output of the XNOR gate is a LOW only when an *odd number* of inputs are HIGH which is the opposite from the XOR gate.

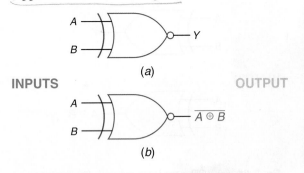

| INPUTS | | OUTPUT | |
|---|---|---|---|
| *A* | *B* | XOR | XNOR |
| 0 | 0 | 0 | 1 |
| 0 | 1 | 1 | 0 |
| 1 | 0 | 1 | 0 |
| 1 | 1 | 0 | 1 |

(*c*)

**Fig. 3-31** (*a*) XNOR gate logic symbol. (*b*) Boolean expression for the output of an XNOR gate. (*c*) Truth table for XOR and XNOR gates.

### Describing the XNOR Function

| In the English language | Inputs $A$, $B$, and $C$ are XNORed yielding output $Y$. |
|---|---|
| As a Boolean expression | NOT symbol $\overline{A \oplus B \oplus C} = Y$ XOR symbol |
| As a logic symbol | $A$ $B$ $C$ → $Y$ |

| A | B | C | Y |
|---|---|---|---|
| 0 | 0 | 0 | 1 |
| 0 | 0 | 1 | 0 |
| 0 | 1 | 0 | 0 |
| 0 | 1 | 1 | 1 |
| 1 | 0 | 0 | 0 |
| 1 | 0 | 1 | 1 |
| 1 | 1 | 0 | 1 |
| 1 | 1 | 1 | 0 |

As a truth table

**Fig. 3-32** Four ways to express the logical XNORing of inputs $A$, $B$, and $C$.

*(handwritten note: XNOR is a High out even but not odd)*

## Self-Test

*Answer the following questions.*

39. Write a Boolean expression for a three-input XNOR gate.
40. Refer to Fig. 3-33. The output of the XNOR gate at time period $t_1$ is a logical _____ (0, 1).
41. Refer to Fig. 3-33. The output of the XNOR gate at time period $t_2$ is a logical _____ (0, 1).
42. Refer to Fig. 3-33. The output of the XNOR gate at time period $t_3$ is a logical _____ (0, 1).
43. Refer to Fig. 3-33. The output of the XNOR gate at time period $t_4$ is a logical _____ (0, 1).
44. Refer to Fig. 3-33. The output of the XNOR gate at time period $t_5$ is a logical _____ (0, 1).
45. The unique output of an XNOR gate is a _____ (HIGH, LOW) when an odd number of inputs are HIGH.

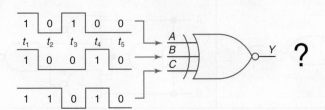

| 1 | 0 | 1 | 0 | 0 |
|---|---|---|---|---|
| $t_1$ | $t_2$ | $t_3$ | $t_4$ | $t_5$ |
| 1 | 0 | 0 | 1 | 0 |

| 1 | 1 | 0 | 1 | 0 |

$A$ $B$ $C$ → $Y$ **?**

**Fig. 3-33** Pulse train problem.

## 3-8 The NAND Gate as a Universal Gate

So far in this chapter you have learned the basic building blocks used in all digital circuits. You also have learned about the seven types of gating circuits and now know the characteristics of the AND, OR, NAND, NOR, XOR, and XNOR gates and the inverter. You can buy ICs that perform any of these seven basic functions.

In looking through manufacturers' literature you will find that NAND gates seem to be more widely available than many other types of gates. Because of the NAND gate's wide use,

we shall show how it can be used to make other types of gates. We shall be using the NAND gate as a "universal gate."

The chart in Fig. 3-34 shows how you would wire NAND gates to create any of the other basic logic functions. The logic function to be performed is listed in the left column of the table; the customary symbol for that function is listed in the center column. In the right column of Fig. 3-34 is a symbol diagram of how NAND gates would be wired to perform the logic function. The chart in Fig. 3-34 need *not* be memorized, but it may be referred to as needed in your future work in digital electronics.

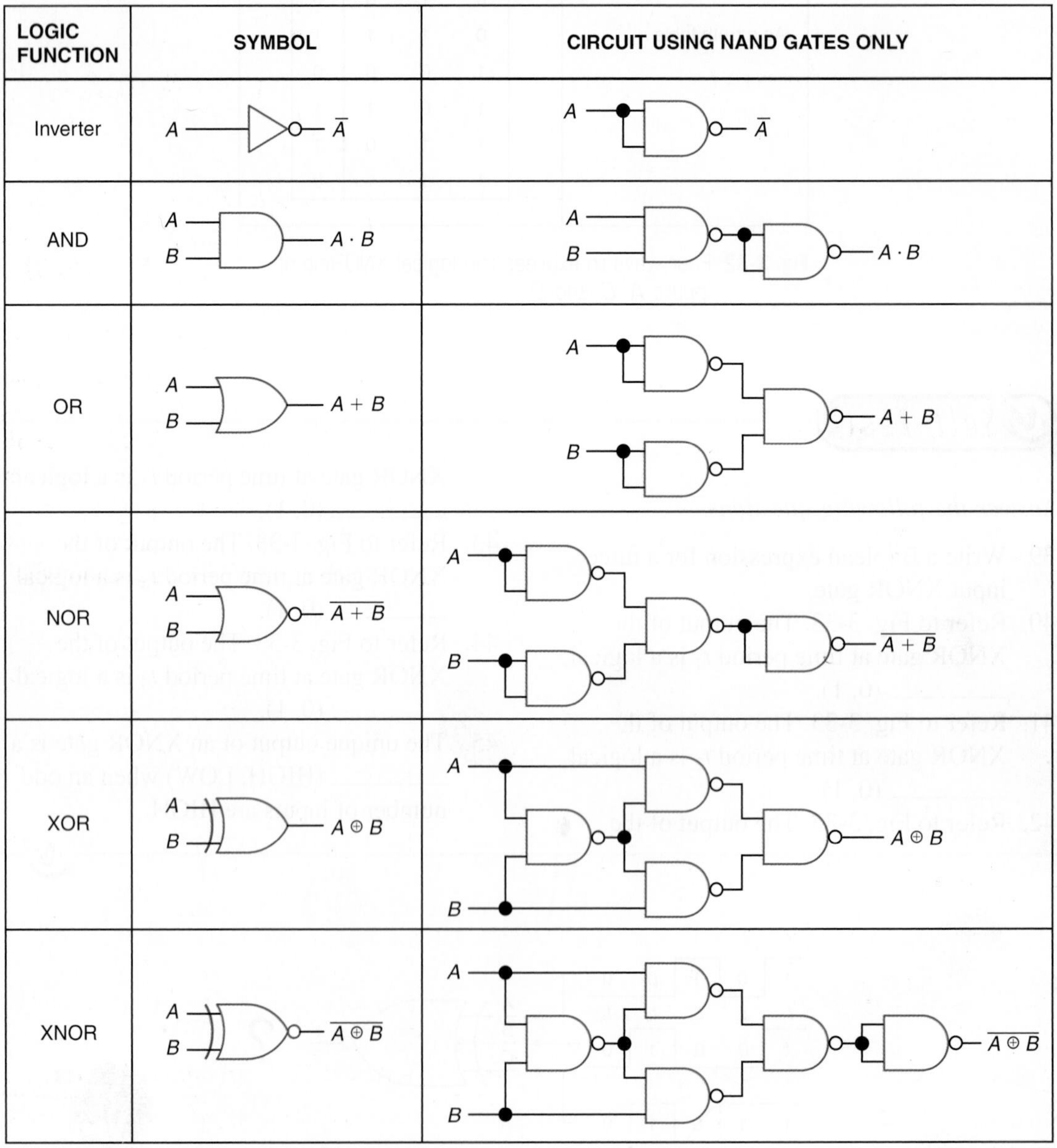

| LOGIC FUNCTION | SYMBOL | CIRCUIT USING NAND GATES ONLY |
|---|---|---|
| Inverter | $A \longrightarrow \overline{A}$ | $A \longrightarrow \overline{A}$ |
| AND | $A$, $B$ — $A \cdot B$ | $A$, $B$ — $A \cdot B$ |
| OR | $A$, $B$ — $A + B$ | $A$, $B$ — $A + B$ |
| NOR | $A$, $B$ — $\overline{A + B}$ | $A$, $B$ — $\overline{A + B}$ |
| XOR | $A$, $B$ — $A \oplus B$ | $A$, $B$ — $A \oplus B$ |
| XNOR | $A$, $B$ — $\overline{A \oplus B}$ | $A$, $B$ — $\overline{A \oplus B}$ |

**Fig. 3-34** Substituting NAND gates.

OR + NAND = NOR
XOR + NAND = XNOR

*Answer the following questions.*

46. The NAND gate can perform the invert function if the inputs are _____ (connected together, left open).

47. How many two-input NAND gates must be used to produce the two-input OR function?

## 3-9 Gates with More Than Two Inputs

Figure 3-35(*a*) shows a *three-input AND gate*. The Boolean expression for the three-input AND gate is $A \cdot B \cdot C = Y$, as illustrated in Fig. 3-35(*b*). All the possible combinations of inputs *A, B,* and *C* are given in the truth table in Fig. 3-35(*c*); the outputs for the three-input AND gate are tabulated in the right column of the truth table. Notice that with three inputs the possible combinations in the truth table have increased to eight ($2^3$).

How could you produce a three-input AND gate as illustrated in Fig. 3-35 if you have only two-input AND gates available? The solution is given in Fig. 3-36(*a*). Note the wiring of the two-input AND gates on the right side of the diagram to form a three-input AND gate. Figure 3-36(*b*) illustrates how a *four-input AND gate* could be wired by using available two-input AND gates.

The logic symbol for a four-input OR gate is illustrated in Fig. 3-37(*a*). The Boolean expression for the *four-input* OR *gate* is $A + B + C + D = Y$. This Boolean expression is written in Fig. 3-37(*b*). Read the Boolean expression $A + B + C + D = Y$ as "input *A* or input *B* or input *C* or input *D* will equal output *Y*." Remember that the $+$ symbol means the logic function OR in Boolean expressions. The truth table for the four-input OR gate is shown in Fig. 3-37(*c*). Notice that because of the four inputs there are 16 possible combinations ($2^4$) of *A, B, C,* and *D*. To wire the four-input OR gate, you could buy the correct gate from a manufacturer of digital logic circuits or you could use two-input OR gates to wire the four-input OR gate. Figure 3-38(*a*) diagrams how you could wire a

**Gates with more than two inputs**

**Three-input AND gate**

**Four-input OR gate**

**Four-input AND gate**

**Four-input OR gate**

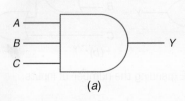

(*a*)

$$A \cdot B \cdot C = Y$$

(*b*)

| INPUTS | | | OUTPUT |
|---|---|---|---|
| *A* | *B* | *C* | *Y* |
| 0 | 0 | 0 | 0 |
| 0 | 0 | 1 | 0 |
| 0 | 1 | 0 | 0 |
| 0 | 1 | 1 | 0 |
| 1 | 0 | 0 | 0 |
| 1 | 0 | 1 | 0 |
| 1 | 1 | 0 | 0 |
| 1 | 1 | 1 | 1 |

(*c*)

**Fig. 3-35** Three-input AND gate. (*a*) Logic symbol. (*b*) Boolean expression. (*c*) Truth table.

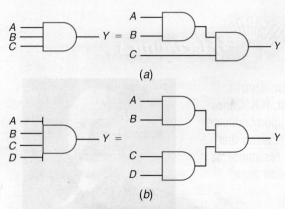

**Fig. 3-36** Expanding the number of inputs. (*a*) Using two AND gates to wire three-input AND. (*b*) Using three AND gates to wire four-input AND.

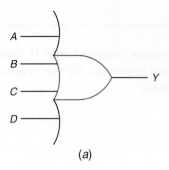

(a)

**Four-input NAND gate**

$$A + B + C + D = Y$$

(b)

| INPUTS | | | | OUTPUT |
|---|---|---|---|---|
| **A** | **B** | **C** | **D** | **Y** |
| 0 | 0 | 0 | 0 | 0 |
| 0 | 0 | 0 | 1 | 1 |
| 0 | 0 | 1 | 0 | 1 |
| 0 | 0 | 1 | 1 | 1 |
| 0 | 1 | 0 | 0 | 1 |
| 0 | 1 | 0 | 1 | 1 |
| 0 | 1 | 1 | 0 | 1 |
| 0 | 1 | 1 | 1 | 1 |
| 1 | 0 | 0 | 0 | 1 |
| 1 | 0 | 0 | 1 | 1 |
| 1 | 0 | 1 | 0 | 1 |
| 1 | 0 | 1 | 1 | 1 |
| 1 | 1 | 0 | 0 | 1 |
| 1 | 1 | 0 | 1 | 1 |
| 1 | 1 | 1 | 0 | 1 |
| 1 | 1 | 1 | 1 | 1 |

(c)

**Fig. 3-37** Four-input OR gate. (*a*) Logic symbol illustrating the method used to show extra inputs beyond the width of the symbol. (*b*) Boolean expression. (*c*) Truth table.

four-input OR gate using two-input OR gates. Figure 3-38(*b*) shows how to convert two-input OR gates into a three-input OR gate. Notice that the *pattern* of connecting both OR and AND gates to expand the number of inputs is the same (compare Figs. 3-36 and 3-38).

Expanding the number of inputs of a NAND gate is somewhat more difficult than expanding AND and OR gates. Figure 3-39 shows how a *four-input NAND gate* can be wired using 2 two-input NAND gates and 1 two-input OR gate.

You frequently will run into gates that have from two to as many as eight and more inputs. The basics covered in this section are a handy reference when you need to expand the number of inputs to a gate.

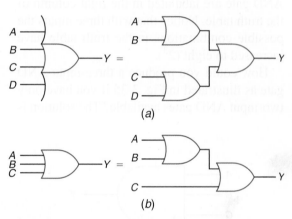

**Fig. 3-38** Expanding the number of inputs.

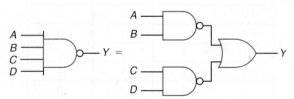

**Fig. 3-39** Expanding the number of inputs.

*Answer the following questions.*

48. Write the Boolean expression for a three-input NAND gate.
49. The truth table for a three-input NAND gate would have _____ lines to include all the possible input combinations.
50. Write the Boolean expression for a four-input NOR gate.
51. The truth table for a five-input NOR gate would contain _____ lines to include all the possible input combinations.

## 3-10 Using Inverters to Convert Gates

Frequently it is convenient to convert a basic gate such as an AND, OR, NAND, or NOR to another logic function. This can be done easily with the use of inverters. The chart in Fig. 3-40 is a handy guide for converting any given gate to any other logic function. Look over the chart: notice that in the top section only the outputs are inverted. Inverting the outputs leads to rather predictable results, shown on the right side of the chart.

The center section of the chart shows only the gate inputs being inverted. For instance, if you invert both inputs of an OR gate, the gate generates the NAND function. This fact is emphasized in Fig. 3-41(*a*). Notice that the *invert bubbles* have been added to the inputs of the OR gate in Fig. 3-41(*a*), which converts the OR gate to a NAND function. Also, in the center section of the chart the inputs of the AND gate are being inverted. This is redrawn in Fig. 3-41(*b*). Notice that the invert bubbles at the input of the AND gate convert it into a NOR function. The new symbols at the left (with the input invert bubbles) in Fig. 3-41 are used in some logic diagrams in place of the more standard NAND and NOR logic symbols at the right. Be aware of these new symbols because you will run into them in your future work in digital electronics.

Figure 3-42 illustrates how adding inverters (invert bubbles) to a logic symbol is described in Boolean expression form. Consider the NAND symbol at the left in Fig. 3-42(*a*) as an AND with an inverter attached to the output.

The Boolean expression for the AND gate alone is $A \cdot B = Y$. Adding the inverter to the *output* of the AND gate in Fig. 3-42(*a*) is symbolized in the Boolean expression as a *long overbar* as $\overline{A \cdot B} = Y$. At the right in Fig. 3-42(*a*) is a simple truth table describing the NAND logic function.

Next consider the alternative NAND symbol in Fig. 3-42(*b*). Notice that the inverters (invert bubbles) are attached to the inputs of the OR symbol. Inverters at an input are symbolized with a short overbar as shown in Fig. 3-42(*b*). The $\overline{A} + \overline{B} = Y$ expression describes the alternative NAND logic symbol with its logic function shown in the NAND truth table at the right. The two Boolean expressions $\overline{A \cdot B} = Y$ and $\overline{A} + \overline{B} = Y$ both describe the NAND logic function. The two logic symbols at the left in Fig. 3-42 both produce the NAND truth table. Applying *DeMorgan's theorem* (part of Boolean algebra) is a systematic way of converting simple logic functions to fundamental AND or OR circuits. DeMorgan's theorem will be covered in some detail in Chapter 4.

The bottom section of the chart in Fig. 3-40 shows both the inputs and the outputs being inverted. Notice that by using inverters at the inputs and outputs you can convert back and forth from AND to OR and from NAND to NOR.

With the 12 conversions shown in the chart in Fig. 3-40 you can convert any basic gate (AND, OR, NAND, and NOR) to any other gate with just the use of inverters. You will *not* need to memorize the chart in Fig. 3-40, but remember it for reference.

**Using inverters to convert gates**

**Alternate NAND symbol**

**Alternate NOR symbol**

**Invert bubbles**

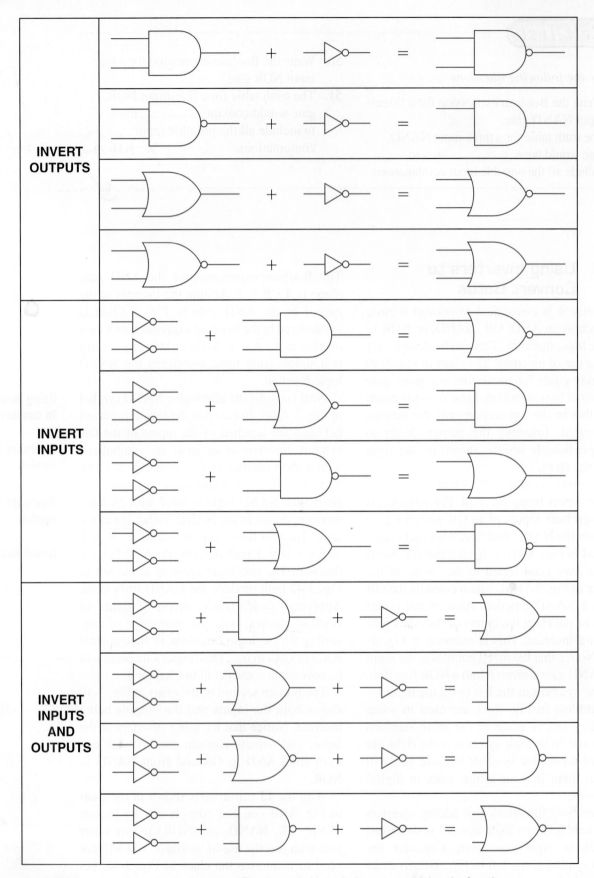

**Fig. 3-40** Gate conversions using inverters. The + symbol here indicates combining the functions.

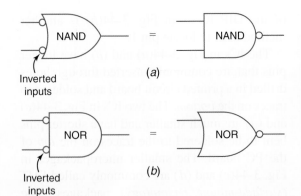

Inverted inputs

(a)

Inverted inputs

(b)

**Fig. 3-41** Common alternative logic symbols. (a) NAND symbols. (b) NOR symbols. *Note:* Invert bubbles at inputs commonly mean an active-LOW input.

**ABOUT ELECTRONICS**

**Electronic Highways.** *Intelligent Vehicle Highway Systems* (IVHS) are currently being developed. These electronic highways are able to react quickly to changing traffic conditions. In Japan, drivers can access roadside beacons to get traffic information, individual communications, and directions. Monitoring systems that allow controllers to quickly reroute traffic from backed-up areas are also being developed.

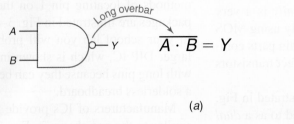

Long overbar

$$\overline{A \cdot B} = Y$$

| A | B | Y |
|---|---|---|
| 0 | 0 | 1 |
| 0 | 1 | 1 |
| 1 | 0 | 1 |
| 1 | 1 | 0 |

(a)

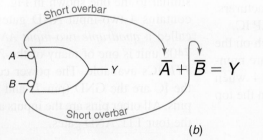

Short overbar

Short overbar

$$\overline{A} + \overline{B} = Y$$

| A | B | Y |
|---|---|---|
| 0 | 0 | 1 |
| 0 | 1 | 1 |
| 1 | 0 | 1 |
| 1 | 1 | 0 |

(b)

**Fig. 3-42** (a) NAND logic symbols. (b) Boolean expressions, and truth tables.

**✔ Self-Test**

*Supply the missing word in each sentence.*

52. The OR gate can be converted to the NAND function by adding _____ to the inputs of the OR gate.

53. Adding inverters to the inputs of the AND gate produces the _____ logic function.

54. Adding an inverter to the output of an AND gate produces the _____ logic function.

55. Adding inverters to all inputs and outputs of an AND gate produces the _____ logic function.

56. Write the Boolean expression for the standard NOR logic symbol shown in Fig. 3-43(a) (use long overbar).

57. Write the Boolean expression that best describes the alternative NOR logic symbol shown in Fig. 3-43(b) (use two short overbars).

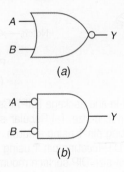

(a)

(b)

**Fig. 3-43** NOR logic symbols.

## 3-11 Practical TTL Logic Gates

The popularity of digital circuits is due partly to the availability of inexpensive ICs. Manufacturers have developed many *families of digital ICs*. These families are groups of devices that can be used together. The ICs in a family are said to be compatible and can be easily connected to one another.

One group of families is manufactured using *bipolar technology*. These ICs contain parts comparable to discrete bipolar transistors, diodes, and resistors. Another group of digital IC families uses *metal oxide semiconductor (MOS) technology*. In the laboratory, you will probably have an opportunity to use both TTL and CMOS ICs. The *CMOS family* is a very low power and widely used family using MOS technology. The CMOS ICs contain parts comparable to insulated-gate field-effect transistors (IGFETs).

A traditional type of IC is illustrated in Fig. 3-44(*a*). This case style is referred to as a *dual in-line package* (DIP) by IC manufacturers. This particular IC is called a 14-pin DIP IC.

Just *counterclockwise* from the notch on the IC in Fig. 3-44(*a*) is pin 1. The pins are numbered counterclockwise from 1 to 14 when viewed from the top of the IC. A dot on the top of the DIP IC as in Fig. 3-44(*b*) is another method used to locate pin 1.

The ICs in Fig. 3-44(*a*) and (*b*) have longer pins that are commonly inserted through holes drilled in a printed circuit board and soldered to traces on the *bottom*. The two ICs in Fig. 3-44(*c*) and (*d*) are much smaller and have shorter pins bent to be soldered to the traces on the *top* of the PC board. The smaller micropackages in Fig. 3-44(*c*) and (*d*) are commonly called *SMT (surface-mount technology)* packages. The SMT packages are typically much smaller in order to save PC board space and are easier to align when being positioned and soldered using automated manufacturing equipment. Two methods of locating pin 1 on the small SMT packages are illustrated in Fig. 3-44(*c*) and (*d*). In your school lab you will probably use the larger DIP IC, which is shown in Fig. 3-44(*a*) with long pins because they can be inserted into a solderless breadboard.

Manufacturers of ICs provide pin diagrams similar to the one shown in Fig. 3-45. This IC contains 4 two-input AND gates. Thus, it is called a *quadruple two-input AND gate*. This 7408 unit is one of many of the *7400 series of TTL* ICs available. The power connections to the IC are the GND (pin 7) and $V_{CC}$ (pin 14) pins. All other pins are the inputs and outputs to the four TTL AND gates.

**Families of digital ICs**

**SMT**

**Bipolar technology**

**Metal-oxide semiconductor (MOS) technology**

**CMOS family**

**Dual-in-line package (DIP)**

**7400 series of TTL ICs**
**Pin diagram**

For related information, visit the website for Fairchild Semiconductor. www.fair-childsemi.com

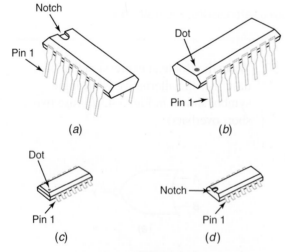

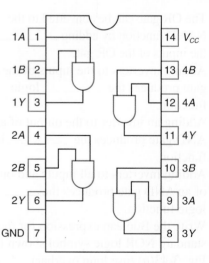

**Fig. 3-44** Dual-in-line package (DIP) ICs. Regular and micro size. (*a*) Regular-size DIP-locating pin 1 using a notch. (*b*) Regular-size DIP-locating pin 1 using a dot. (*c*) Micro-size DIP surface mount IC locating pin 1 using a dot. (*d*) Micro-size DIP surface mount IC locating pin 1 using a notch.

**Fig. 3-45** Pin diagram for the 7408 digital IC.

Given the logic diagram in Fig. 3-46(a), wire this circuit using the 7408 TTL IC. A wiring diagram for the circuit is shown in Fig. 3-46(b). A 5-V dc regulated power supply is typically used with TTL devices. The positive ($V_{CC}$) and negative (GND) power connections are made to pins 14 and 7 of the IC. Input switches (A and B) are wired to pins 1 and 2 of the 7408 IC. Notice that if a switch is in the *up position*, a logical 1 (+5 V) is applied to the input of the AND gate. If a switch is in the down position, however, a logical 0 is applied to the input. At the right in Fig. 3-46(b), an LED and 150-Ω limiting resistor are connected to GND. If the output at pin 3 is HIGH (near +5 V), current will flow through the LED.

When the LED is lit, it indicates a HIGH output from the AND gate.

The top of a typical *TTL digital IC* is shown in Fig. 3-47(a). The block form of the letters "NS" on the top of the IC shows the manufacturer as National Semiconductor. The DM7408N part number can be divided into sections as shown in Fig. 3-47(b). The prefix "DM" is the manufacturer's code (National Semiconductor uses the letters "DM" as a prefix). The core part number is 7408, which is a quadruple two-input AND gate TTL IC. This core part number is the same from manufacturer to manufacturer. The trailing letter "N" (the suffix) is a code used by several manufacturers to designate the DIP.

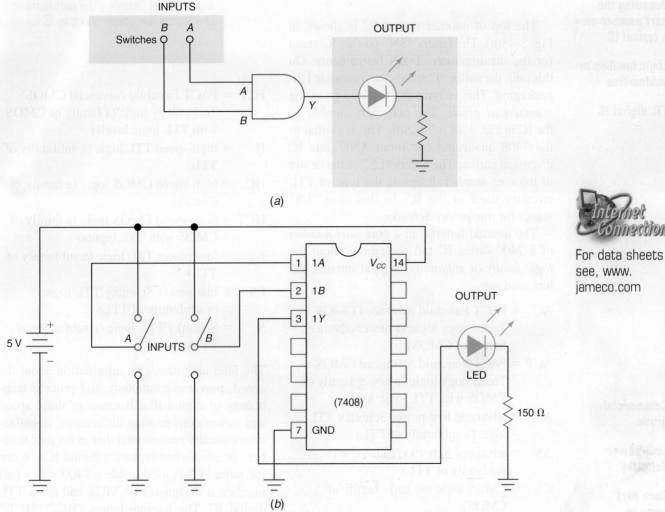

**Fig. 3-46** (a) Logic diagram for a two-input AND gate circuit. (b) Wiring diagram to implement the two-input AND function.

Internet Connection

For data sheets see, www. jameco.com

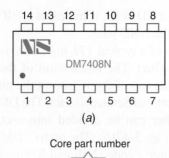

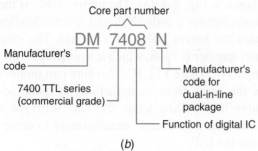

Core part number

DM **7408** N

Manufacturer's code ──┘

7400 TTL series (commercial grade) ──┘

Manufacturer's code for dual-in-line package ──┘

Function of digital IC ──┘

*(b)*

**Fig. 3-47** (*a*) Marking on a typical digital IC. (*b*) Decoding the part number on a typical IC.

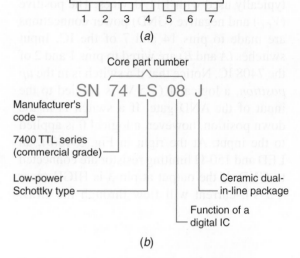

Core part number

SN 74 **LS** 08 J

Manufacturer's code ──┘

7400 TTL series (commercial grade) ──┘

Low-power Schottky type ──┘

Ceramic dual-in-line package ──┘

Function of a digital IC ──┘

*(b)*

**Fig. 3-48** (*a*) Markings on a Texas Instruments digital IC. (*b*) Decoding the part number of a typical low-power Schottky IC.

---

**Marking on a typical digital IC**

**Decoding the part number on a typical IC**

**Logic families or subfamilies**

**TTL digital IC**

**Commercial grade**

**Low-power Schottky**

**Core part number**

**CMOS ICs**

---

The top of another digital IC is shown in Fig. 3-48(*a*). The letters "SN" on this IC stand for the manufacturer, Texas Instruments. On this unit, the suffix "J" stands for a ceramic DIP packaging. This is typically referred to as the *commercial grade*. The core part number of the IC in Fig. 3-48 is 74LS08. This is similar to the 7408 quadruple two-input AND gate IC discussed earlier. The letters "LS" in the center of the core number designate the type of TTL circuitry used in the IC. In this case "LS" stands for *low-power Schottky.*

The internal letter(s) in a *core part number* of a 7400 series IC tell something about the *logic family* or *subfamily.* Typical internal letters used are:

AC  = FACT Fairchild Advanced CMOS Technology logic (a newer advanced family of CMOS)

ACT = FACT Fairchild Advanced CMOS Technology logic (a newer family of CMOS with TTL logic levels)

ALS = advanced low-power Schottky TTL logic (a subfamily of TTL)

AS  = advanced Schottky TTL logic (a subfamily of TTL)

C   = CMOS logic (an early family of CMOS)

F   = FAST Fairchild Advanced Schottky TTL logic (a new subfamily of TTL)

FCT = FACT Fairchild Advanced CMOS Technology logic (a family of CMOS with TTL logic levels)

H   = high-speed TTL logic (a subfamily of TTL)

HC  = high-speed CMOS logic (a family of CMOS)

HCT = high-speed CMOS logic (a family of CMOS with TTL inputs)

L   = low-power TTL logic (a subfamily of TTL)

LS  = low-power Schottky TTL logic (a subfamily of TTL)

S   = Schottky TTL logic (a subfamily of TTL)

The internal letters give information about the speed, power consumption, and process technology of digital ICs. Because of these speed and power consumption differences, manufacturers usually recommend that exact part numbers be used when replacing digital ICs. When the letter "C" is used inside a 7400 series part number, it designates a CMOS and not a TTL digital IC. The internal letters "HC," "HCT," "AC," "ACT," and "FCT" also designate *CMOS ICs.*

Data manuals from manufacturers contain much valuable information on digital ICs. They contain pin diagrams and packaging information. Data manuals also contain details on part numbering and other valuable data for the technician, student, or engineer. Manufacturer's websites usually have data sheets available for download at no cost.

www.ti.com

## ✔ Self-Test

*Answer the following questions.*

58. List two popular digital IC families.
59. Refer to Fig. 3-44(*a*). This IC uses the popular case style called the _____ package.
60. A _____ -V dc power supply is used with TTL ICs. The $V_{CC}$ pin is connected to the _____ (−, +) of the supply.
61. Refer to Fig. 3-46(*b*). How is the 7408 IC described by the manufacturer?

62. What can you tell about a digital IC that has the marking "74LS08N" printed on the top?
63. A digital IC with markings of 74F08 on top would be a quad 2-input AND gate from what modern TTL subfamily?
64. A digital IC with markings of 74ACT08 on top would be a quad 2-input AND gate using _____ (CMOS, TTL) technology and supporting TTL logic levels.

## 3-12 Practical CMOS Logic Gates

The older 7400 series of TTL logic devices has been extremely popular for many decades. One of its disadvantages is its higher power consumption. In the late 1960s, manufacturers developed CMOS digital ICs which consume little power and were perfect for battery-operated electronic devices. CMOS stands for *complementary metal oxide semiconductor.*

Several families of compatible CMOS ICs have been developed. The first was the *4000 series*. Next came the *74C00 series* and more recently the *74HC00 series* of CMOS digital ICs. In 1985, the *FACT (Fairchild Advanced CMOS Technology) 74AC00 series*, 74ACT00 series, and 74FCT00 series of extremely fast, low-power CMOS digital ICs were introduced by Fairchild. Many large-scale integrated (LSI) circuits such as digital wristwatch and calculator chips are also manufactured using the CMOS technology.

A typical 4000 series CMOS IC is pictured in Fig. 3-49(*a*). Note that pin 1 is marked as such on the top of the IC immediately counterclockwise from the notch. The CD4081BE part number can be divided into sections as shown in Fig. 3-49(*b*). The prefix CD is the manufacturer's code for CMOS digital ICs. The core part number is 4081B, which stands for a CMOS quadruple two-input AND gate IC. This core part number is almost always the same from manufacturer to manufacturer. The trailing letter "E" is the manufacturer's packaging code for a plastic DIP IC. The letter "B" is a "buffered version" of the original 4000A series. The buffering provides the 4000B series devices with greater output drive and some protection from static electricity.

Figure 3-49(*c*) is a pin diagram for the CD4081BE CMOS quad two-input AND gate IC. The power connections are $V_{DD}$ (positive voltage) and $V_{SS}$ (GND or negative voltage). The labeling of the power connections on TTL and 4000 series CMOS ICs is different. This difference can be observed by comparing Figs. 3-45 and 3-49(*c*).

Given the schematic diagram in Fig. 3-50(*a*), wire this circuit using the 4081B CMOS IC. A

**Practical CMOS logic gates**

**Complementary metal oxide semiconductor**

**4000 series**

**74C00 series**

**74HC00 series**

**FACT series**

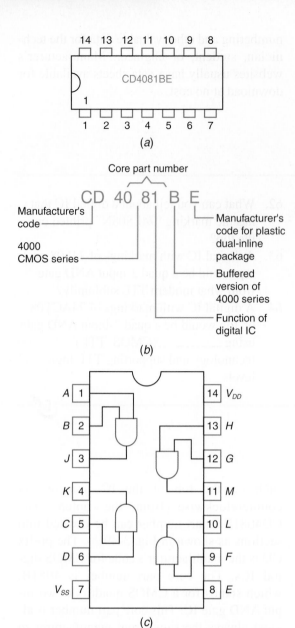

Fig. 3-49 (a) Markings on CMOS digital IC. (b) Decoding the part number on a typical 4000B series CMOS IC. (c) Pin diagram for the 4081B CMOS IC.

wiring diagram for the circuit is shown in Fig. 3-50(b). A 5-V dc power supply is shown but the 4000 series CMOS ICs can use voltages from 3 to 18 V dc. Care is taken in removing the 4081 from its conductive foam storage because CMOS ICs can be damaged by static charges. Do not touch the pins when inserting the 4081 CMOS IC in a socket or mounting board. $V_{DD}$ and $V_{SS}$ power connections are made with the power off. *When using CMOS,*

**Cautions when using CMOS ICs**

**Unused inputs**

*all unused inputs are tied to GND or $V_{DD}$.* In this example, unused inputs (C, D, E, F, H, G) are grounded. The output of the AND gate (pin 3) is connected to the driver transistor. The transistor turns the LED on when pin 3 is HIGH or off when the output is LOW. Finally, inputs A and B are connected to input switches.

When the input switches in Fig. 3-50(b) are in the up position, they generate a HIGH input. A LOW input is generated when the switches are in the down position. Two LOW inputs to the AND gate produce a LOW output at pin 3 of the IC. The LOW output turns off the transistor and the LED does not light. Two HIGH inputs to the AND gate produce a HIGH output at pin 3. The HIGH (about +5 V) output at the base of Q1 turns on the transistor and the LED lights. The 4081 CMOS IC will generate a two-input AND truth table.

Several families of CMOS digital ICs are available. A 4000 series IC was used as an example in this section. The newer 74HC00 series CMOS digital ICs have gained favor because they are somewhat more compatible with the popular TTL logic. The 74HC00 series of CMOS ICs also has more drive capabilities than the older 4000 and 74C00 series units and operates at high frequencies. The "HC" in the 74HC00 series part number stands for high-speed CMOS.

The FACT Fairchild Advanced CMOS Technology logic series is a later CMOS family of ICs. It includes the 74AC00-, 74ACT00-, 74ACTQ00-, 74FCT00-, and 74FCTA00-series subfamilies of CMOS digital ICs. The FACT logic family has outstanding operating characteristics exceeding all CMOS and most TTL subfamilies. For direct replacement of 74LS00 and 74ALS00 TTL series ICs, the 74ACT00-, 74ACTQ00-, 74FCT00-, and 74FCTA00-series circuits with TTL-type input voltage characteristics are included in the FACT CMOS family. FACT logic devices are ideal for portable systems because of their extremely low power consumption and excellent high-speed characteristics.

Caution must be used so that static charges do not damage CMOS ICs. All unused inputs to CMOS gates must be grounded or connected to $V_{DD}$. Most important, *input voltages must not exceed the GND ($V_{SS}$) to $V_{DD}$ voltage.*

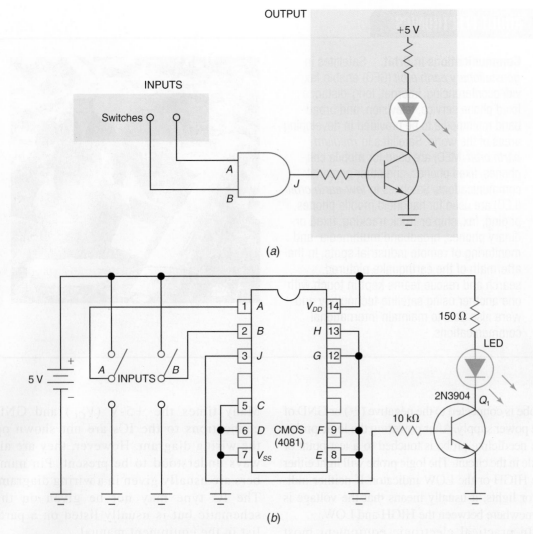

Fig. 3-50 (a) Logic diagram for a two-input AND gate circuit. (b) Wiring diagram using the 4081 CMOS IC to implement the two-input AND function.

**✓ Self-Test**

*Answer the following questions.*

65. The primary advantage of CMOS digital ICs is their _____ (high, low) power consumption.
66. While TTL must use a 5-V power supply, 4000 series CMOS ICs can operate on dc voltages from _____ to _____ V.
67. Refer to Fig. 3-49. How is this 4081B IC described by the manufacturer?
68. What rule (dealing with unused inputs) must be followed when wiring CMOS ICs?

## 3-13 Troubleshooting Simple Gate Circuits

The most basic piece of test equipment used in digital troubleshooting is the *logic probe*. A simple logic probe is pictured in Fig. 3-51. The slide switch on the unit is used to select the type of logic family under test, either TTL or CMOS. The logic probe in Fig. 3-51 is set to test a TTL type of digital circuit in this example. Typically, two leads provide power to the logic probe. The red lead is connected to the positive (+) of the power supply while the black lead of the logic

**Logic probe**

probe is connected to the negative (−) or GND of the power supply. After powering the logic probe, the needlelike probe is touched to a test point or node in the circuit. The logic probe will light either the HIGH or the LOW indicator. If neither indicator lights, it usually means that the voltage is somewhere between the HIGH and LOW.

In practical electronic equipment most digital ICs are mounted on a *printed circuit (PC) board*. An example is shown in Fig. 3-52($a$). Also available to the student or technician is a circuit wiring diagram or schematic similar to that in Fig. 3-52($b$).

**Printed circuit (PC) board**

**Steps in troubleshooting**

**Floating inputs**

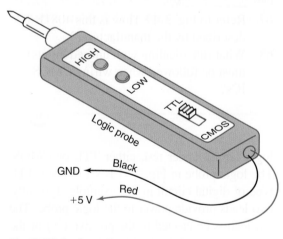

**Fig. 3-51** Logic probe.

Many times the $+5$-V ($V_{CC}$) and GND connections to the ICs are not shown on the wiring diagram. However, they are always understood to be present. Pin numbers are usually given in a wiring diagram. The IC type may not be given on the schematic but is usually listed on a parts list in the equipment manual.

The first step in troubleshooting is to use your senses. *Feel* the flat top of the ICs to determine if they are hot. Some ICs operate cool, others run slightly warm. CMOS ICs should always be cool. *Look* for broken connections, solder bridges, broken PC board traces, and bent IC pins. *Smell* for possible overheating. *Look* for signs of excessive heat, such as discoloration or charring.

The next step in troubleshooting might involve checking whether each IC has power. With the logic probe connected to power, check at the points labeled $A$, $B$ (the $V_{CC}$ pin), $C$, and $D$ in Fig. 3-52($a$). Nodes $A$ and $B$ should give a bright HIGH light on the logic probe. Nodes $C$ and $D$ should give a bright LOW light on the logic probe.

The next step might be to trace the path of logic through the circuit. The circuit is equal to a three-input AND gate in this example (Fig. 3-52). Its unique state is a HIGH when all inputs are HIGH. Check pins 1, 2, and 5 of the IC in Fig. 3-52(*a*) with the logic probe. Manipulate the equipment to get all inputs HIGH. When all inputs are HIGH, the output (pin 6 of the IC) should be HIGH and the circuit LED should light. If the unique state works, try several other input combinations and verify their proper operation.

Refer to Fig. 3-52(*a*). Assume a HIGH reading at node *A* and a LOW reading on the logic probe at node *B* (pin 14 of the IC). This probably means an open circuit in the PC board trace or a faulty solder joint between points *A* and *B*. If DIP IC sockets are used, the thin part of the IC pin can be bent. This common difficulty causes an open between the IC pin and the socket and PC board trace.

Refer to Fig. 3-52(*a*). Assume LOW readings at pins 1, 2, and 3 with no reading (neither LED on the logic probe lit) at pin 4. No reading on most logic probes means a voltage between LOW and HIGH (perhaps 1 to 2 V in TTL). This input (pin 4) is floating (not connected) and is considered to be a HIGH by the TTL circuitry inside the 7408 IC. The output of the first AND gate (pin 3) is supposed to pull the input to the second AND gate (pin 4) LOW. If it does not, the fault could be in the PC board trace, solder connections, or a bent IC pin. Internal opens and shorts also occur in digital ICs.

Troubleshooting a comparable CMOS circuit proceeds in the same way with a few exceptions. The logic probe must be set to CMOS instead of TTL. Floating inputs on CMOS ICs can harm the IC. A LOW in CMOS is defined as approximately 0 to 20 percent of the power supply voltage. A HIGH in CMOS is defined as approximately 80 to 100 percent of supply voltage.

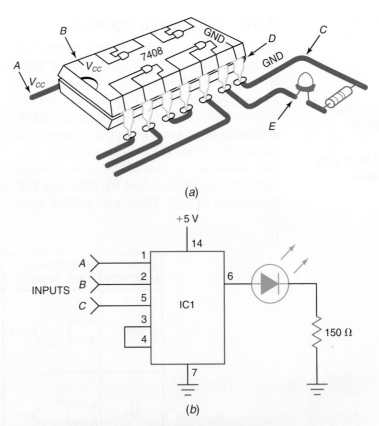

(a)

(b)

**Fig. 3-52** (*a*) Digital IC mounted on a printed circuit (PC) board. (*b*) Wiring or schematic diagram of a digital gating circuit.

## Summary

Troubleshooting *first* involves using your senses. *Second,* check with a logic probe to see if each IC has power. *Third,* determine the exact job of the gating circuit and test for the unique output conditions. *Finally,* check other input and output conditions. Short-circuit conditions can occur inside ICs as well as in the wiring. Digital ICs should be replaced with exact subfamily replacements when possible.

---

**Self-Test**

*Answer the following questions.*

69. Refer to Fig. 3-51. With what two logic families can this logic probe be used?
70. What is the first step in troubleshooting gating circuits using TTL ICs?
71. What is the second step in troubleshooting?
72. Floating inputs to CMOS ICs are _____ (allowed, not allowed).

## 3-14 IEEE Logic Symbols

IEEE standard symbols

IEEE logic gate symbols

The logic gate symbols you have memorized are the traditional ones recognized by all workers in the electronics industry. These symbols are very useful in that they have distinctive shapes. Manufacturers' data manuals include traditional logic symbols and are recently including the newer *IEEE functional logic symbols*. These newer symbols are in accordance with ANSI/IEEE Standard 91-1984 and IEC Publication 617-12. These newer IEEE symbols are commonly referred to as "*dependency notation.*" For simple gating circuits, the traditional logic symbols are probably preferred, but the IEEE standard symbols have advantages as ICs become more complicated. Most military contracts call for the use of *IEEE standard symbols*.

Figure 3-53 shows the traditional logic symbols and their IEEE counterparts. All IEEE logic symbols are rectangular. There is an identifying character or symbol inside the rectangle.

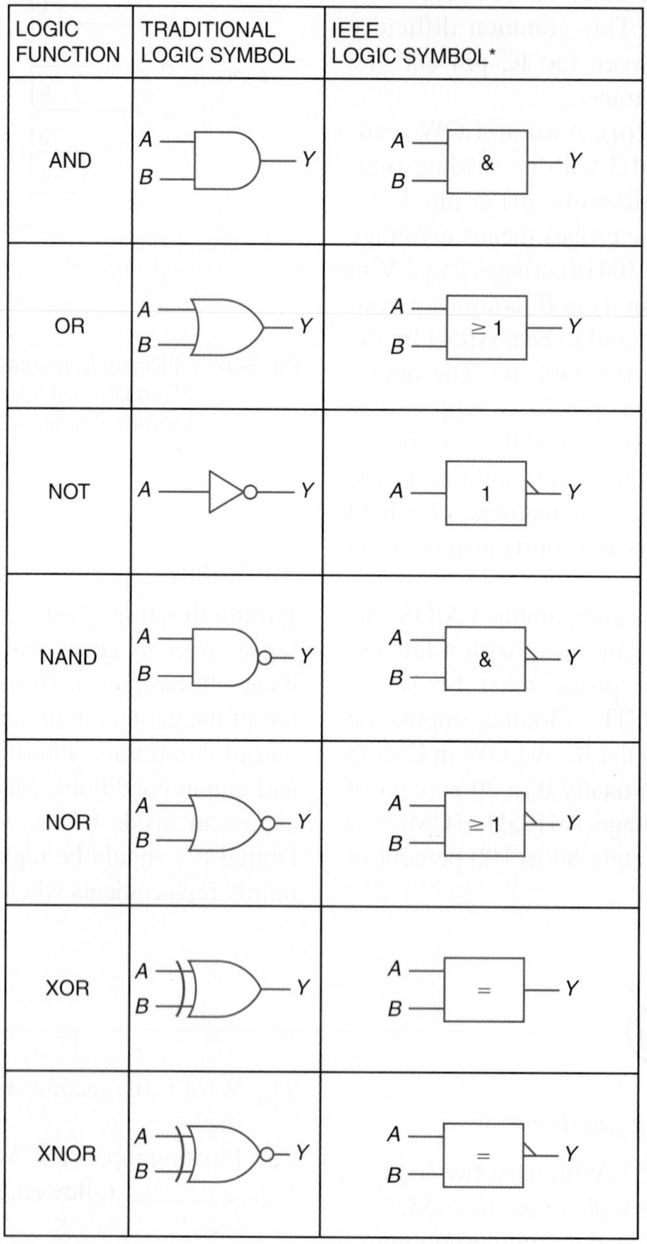

| LOGIC FUNCTION | TRADITIONAL LOGIC SYMBOL | IEEE LOGIC SYMBOL* |
|---|---|---|
| AND | | |
| OR | | |
| NOT | | |
| NAND | | |
| NOR | | |
| XOR | | |
| XNOR | | |

*ANSI/IEEE Standard 91-1984 and IEC Publication 617-12.

**Fig. 3-53** Comparing traditional and IEEE logic gate symbols.

For instance, notice in Fig. 3-53 that the ampersand (&) character is printed inside the IEEE standard AND gate symbol. Characters *outside* the rectangle are not part of the standard symbol and may vary from manufacturer to manufacturer. The invert bubble on traditional logic symbols (NOT, NAND, NOR, and XOR) is replaced with a right triangle on corresponding IEEE standard symbols. The IEEE right triangle can also be used on inputs to signify an active-low input. You have memorized the traditional logic gate symbols. You will not have to memorize the IEEE logic gate symbols but should be aware they exist.

Recent manufacturers' data manuals will probably give both the traditional and IEEE functional logic symbols for a particular IC. For instance, logic symbols for the 7408 quadruple two-input AND gate are illustrated in Fig. 3-54. The traditional logic diagram for the 7408 IC is shown in Fig. 3-54(*a*). The IEEE logic diagram for the 7408 IC is reproduced in Fig. 3-54(*b*). Note in the IEEE symbol for the 7408 IC that only the top AND gate contains the & symbol, but it is understood that the lower three rectangles also represent two-input AND gates.

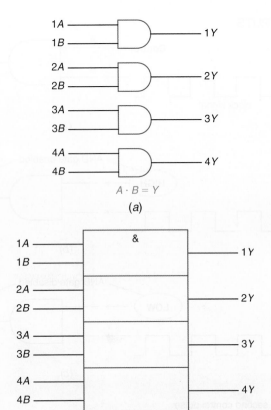

This symbol is in accordance with ANSI/IEEE Standard 91-1984 and IEC Publication 617-12.

(*b*)

**Fig. 3-54** Logic symbol for 7408 quadruple two-input AND gate. (*a*) Traditional symbol (most common). (*b*) IEEE functional logic symbol (newer method).

---

## Self-Test

*Answer the following questions.*

73. Draw the IEEE standard logic symbol for a three-input AND gate.
74. Draw the IEEE standard logic symbol for a three-input OR gate.
75. Draw the IEEE standard logic symbol for a three-input NAND gate.

76. The right triangle on IEEE symbols replaces the invert _____ on traditional logic symbols.
77. For simple gating circuits, the _____ (IEEE standard, traditional) logic symbols are probably preferred because of their distinctive shapes.

---

## 3-15 Simple Logic Gate Applications

Consider the use of the AND gate in Fig. 3-55(*a*). Input *A* is the input which *controls* whether the clock signal is blocked or passed through the AND gate to output *Y*. The clock waveform is considered to be continuous. If the

control input to the AND gate goes HIGH, the gate is said to be *enabled*. This means that the clock signal passes through the gate to the output with no change. The AND gate is shown in its enabled mode in Fig. 3-55(*b*). If the control input to the AND gate goes LOW, the gate is said to be *disabled*. Being disabled means the output of the AND gates stays LOW and the

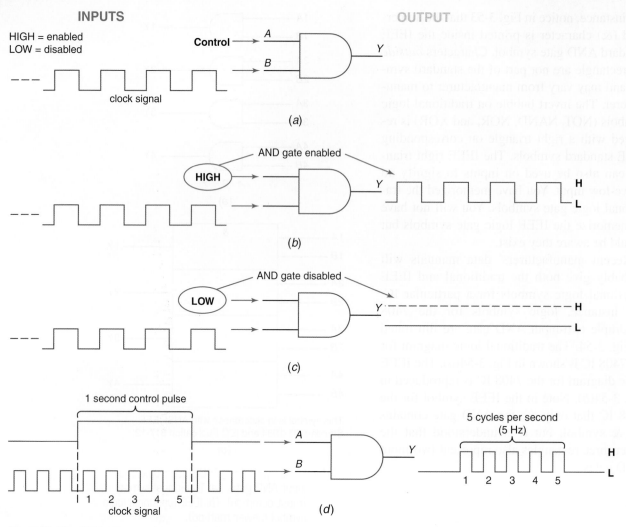

HIGH = enabled
LOW = disabled

Control → A
B
Y

clock signal

(a)

AND gate enabled

HIGH

Y

H
L

(b)

AND gate disabled

LOW

Y

H
L

(c)

1 second control pulse

A
B
Y

5 cycles per second
(5 Hz)

H
L

1  2  3  4  5

1  2  3  4  5
clock signal

(d)

**Fig. 3-55** The AND gate used as a control gate.

clock signal is blocked from passing through to the output. The AND gate is shown in its disabled mode in Fig. 3-55(c).

The control input to the AND gate in Fig. 3-53 is said to an *active-HIGH* input. By definition, an active-HIGH input is a digital input which executes its function when a HIGH is present. The job or function of the gate in Fig. 3-55 was to pass (not block) the clock signal.

The AND gate in Fig. 3-55(d) serves as a special control gate. This circuit is a very fundamental *frequency counter* circuit. The control pulse at input A to the AND is exactly 1 second allowing the clock signal to pass through the gate for only 1s. In this example 5 pulses pass through the AND gate from input B to output Y during the one second. Counting the pulses at the output of the gate in Fig. 3-55(d) means that the clock signal must be 5 cycles per second (5 Hz)

Consider the use of a pushbutton switch to activate the clear (CLR) input of an 8-bit binary counter IC in Fig. 3-56. With $SW_1$ open, the *pull-up resistor $R_1$* pulls the input of the inverter HIGH. The output of the inverter is LOW at this time and the *CLR* input to the counter IC is not active (disabled). Pressing input switch $SW_1$ applies a LOW to the input of the inverter whose output goes HIGH enabling the *CLR* input of the counter. This clears the counters output to 00000000. The bubble at the input of the inverter in Fig. 3-56 ($IC_1$) indicates that the active state is a LOW while the lack of a bubble on the binary counter $IC_2$ symbol indicates an active-HIGH input.

Consider the simplified automobile alarm system shown in Fig. 3-57(a). The alarm will sound when any one or all of the door-mounted normally-closed (NC) pushbutton switches are

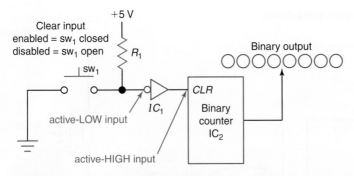

Fig. 3-56 Active-LOW and active-HIGH inputs.

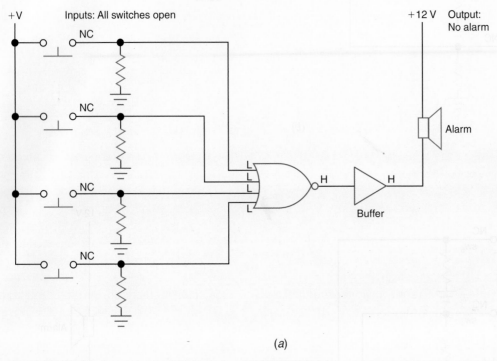

(a)

Fig. 3-57 Simple alarm circuit. (a) No alarm with all inputs open.

released (closed) by a door opening. Each input to the NOR gate has a *pull-down resistor* attached to pull the inputs to the gate LOW when the switches are open. The bubble at the output of the NOR gate suggests that it has an *active-LOW* output. The NOR gate in Fig. 3-57(a) has active-HIGH inputs. With all the auto doors closed and all input switches open as shown in Fig. 3-57(a), the inputs to the NOR gate are LLLL causing a HIGH output. The alarm is disabled.

If any car door opens, the door mounted switch springs closed as shown in Fig. 3-57(b). The inputs to the NOR gate are HLLL causing a LOW output. The noninverting buffer also

outputs a LOW which turns on the alarm. The alarm sounds. The buffer provides extra current to drive the alarm device.

To disable the alarm system a switch SW$_1$ has been added along with an OR gate in Fig. 3-57(c). The OR gate is redrawn to look like a AND symbol with inverted inputs and output. This arrangement produces the OR function (see the conversion chart in Fig. 3-40). The reason the alternative symbol was used is to suggest that it takes two LOW inputs to generate a LOW output and sound the alarm. The two bubbles at the inputs to the alternative OR symbol means it takes a LOW from the ON/OFF switch as well as a LOW from the

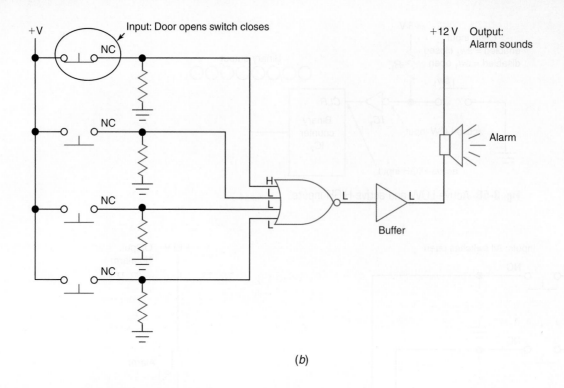

(b)

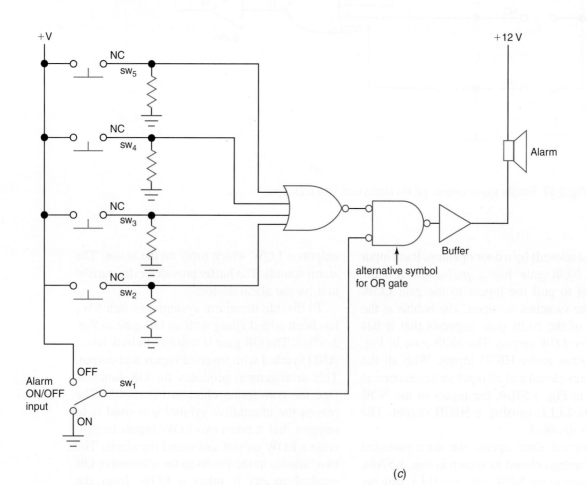

(c)

**Fig. 3-57 (continued)** (b) Alarm sound with top input switch closed. (c) Adding ON/OFF switch to alarm.

NOR gate to produce an active-LOW output turning on the alarm. The alarm is turned off or disabled by placing SW$_1$ in the OFF position which passes a HIGH to the OR gate. A HIGH at any input to an OR gate will always generate a HIGH output and disable the alarm.

This example was given to alert students that traditional logic symbols as well as their alternative symbols appear in manufacturer's literature.

![Self-Test]

*Answer the following questions.*

78. Refer to Fig. 3-55(c). When the control input to the AND is LOW, the gate is _____ (disabled, enabled) and the clock signal is blocked from passing through to the output.
79. Refer to Fig. 3-55(b). When the control input to the AND is HIGH, the gate is enabled and the clock signal is _____ (blocked from passing through, passed through) to the output.
80. Refer to Fig. 3-55(d). The AND gate along with a 1 second positive control pulse demonstrates the concept used in an electronic lab instrument called a _____ (digital multimeter, frequency counter).
81. Refer to Fig. 3-56. To clear the binary counter to binary 00000000, the push-button is _____ (pressed, released) causing the input to the inverter IC to go _____ (HIGH, LOW) driving the CLR input to IC2 _____ (HIGH, LOW).
82. Refer to Fig. 3-56. The clear or CLR pin to the binary counter IC is an _____ (active-HIGH, active-LOW) input.
83. Many times logic symbols attach a small _____ to show either an active-LOW input or an active-LOW output.
84. Refer to Fig. 3-57(c). If switch SW1 is LOW and switch SW$_2$ is closed by the opening of a car door, the alarm _____ (does not sound, sounds).
85. Refer to Fig. 3-57(c). If switch SW1 is HIGH and both SW1 and SW2 are closed by the opening of car doors, the alarm _____ (does not sound, sounds).

## 3-16 Logic Functions Using Software (BASIC Stamp Module)

It is common for logic functions (AND, OR, XOR, etc.) to be programmed using software. In this section, we will program logic functions using a high-level language called PBASIC (a version of BASIC used by Parallax, Inc.). The programmable hardware device used in these examples will be the BS2 BASIC Stamp Microcontroller Module by Parallax, Inc. The hardware needed to program the BASIC Stamp Microcontroller Module is sketched in Fig. 3-58 (a). The hardware includes the BASIC Stamp 2 module, a PC system, a serial cable for downloading, and assorted electronic components (switches, resistors, and an LED). The actual BS2 BASIC Stamp IC is sketched in Fig. 3-58(b). Notice that the BS2 module takes the form of a 24-pin DIP IC. The BS2 module is manufactured using several components including a PIC16C57 microcontroller with PBASIC interpreter in firmware, EEPROM program memory, and other parts.

The procedure for programming the BASIC Stamp Module to operate as a 2-input AND gate is represented in Fig. 3-58. The steps in wiring and programming the BASIC Stamp 2 Module are:

1. Refer to Fig. 3-58. Wire the two active-HIGH pushbutton switches and connect them to ports P11 and P12. Wire the red LED output indicator and connect it to port P1. The ports will be defined as either an output or inputs in the PBASIC program.

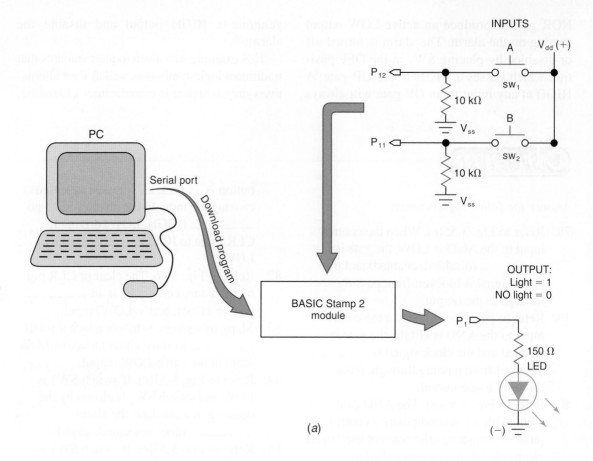

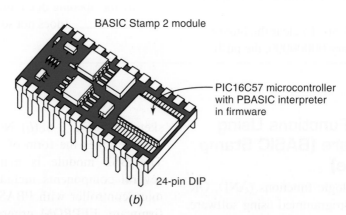

**Fig. 3-58** (*a*) BASIC Stamp 2 module wired as a 2-input AND gate. (*b*) Physical appearance of the BS2 BASIC Stamp by Parallax, Inc.

2. Load the PBASIC text editor program (version for the BS2 IC) into the PC. Type your PBASIC program describing the 2-input AND logic function. A PBASIC program titled **'2-input AND function** is listed in the shaded box below.

3. Attach a serial cable between the PC and the BASIC Stamp 2 development board

(such as the Board of Education by Parallax, Inc.).

4. With the BASIC Stamp 2 Module turned on, download your PBASIC program from the PC to BS2 module using the RUN command.

5. Disconnect the serial cable from the BS2 module.

6. Test the 2-input AND program by pressing the input switches. The output indicator (red LED) will light only when both input switches are activated (pressed). The PBASIC program stored in EEPROM program memory in the BASIC Stamp 2 Module will start each time the BS2 IC is turned on.

## PBASIC Program- 2-input AND Function

Consider the PBASIC program titled '2-input AND function. Line 1 starts with an apostrophe ('), which means this is a *remark statement*. Remark statements are used to clarify the program and are not executed by the microcontroller. Lines 2–4 are lines of code to *declare variables* that will be used later in the program. As an example, line 2 reads—**A VAR Bit**. This tells the microcontroller that A is a variable name that will hold only 1 bit (a 0 or 1). Lines 5–7 are lines of code that *declare* which ports are used as inputs and which is an output. As an example, line 5 reads—**INPUT 11**. This

informs the microcontroller that port 11 (P11) will be used as an input in this program. Another example in line 7 reads **OUTPUT 1**, which declares that port 1 will be used as an output. Notice that line 7 code is followed by *a remark statement*—'**Declare port 1 as an output**. The remark statements at the right in this PBASIC program are not required but they aid in understanding the purpose for lines of code.

Next consider the main routine with the starts with the **Ckswitch:** line of code. In PBASIC, any word followed by a colon (:) is called a *label*. A label is a reference point in the program that usually locates the beginning of a main or subroutine.

In the '**2-input AND function** sample program, the label **Ckswitch:** is the starting point in the main routine used to check the condition of input switches A and B and logically AND the inputs. The **Ckswitch:** routine repeats continuously because either lines 14 (**GOTO Ckswitch**) or 18 (**GOTO Ckswitch**) will always return the program to the beginning of the **Ckswitch:** main routine.

| '2-input AND function | 'Title of program (Fig. 3-58) | L1 |
|---|---|---|
| A       VAR       Bit | 'Declare A as variable, 1 bit | L2 |
| B       VAR       Bit | 'Declare B as variable, 1 bit | L3 |
| Y       VAR       Bit | 'Declare Y as variable, 1 bit | L4 |
| INPUT 11 | 'Declare port 11 as an input | L5 |
| INPUT 12 | 'Declare port 12 as an input | L6 |
| OUTPUT 1 | 'Declare port 1 as an output | L7 |
| Ckswitch: | 'Label for check switch routine | L8 |
|     OUT1 = 0 | 'Initialize: port 1 at 0, red LED off | L9 |
|     A = IN12 | 'Assign value: port 12 input to variable A | L10 |
|     B = IN11 | 'Assign value: port 11 input to variable B | L11 |
|     Y = A & B | 'Assign value: A ANDed with B to variable Y ← **Line 12** | L12 |
|     If Y = 1 THEN Red | 'If Y = 1 then goto red subroutine, otherwise next line | L13 |
| GOTO Ckswitch | 'Go to Ckswitch—begin check switch routine again | L14 |
| Red: | 'Label for lighting red LED, means HIGH output | L15 |
|     OUT1 = 1 | 'Output P1 goes HIGH, lights red LED | L16 |
|     PAUSE 100 | 'Pause for 100ms (milliseconds) | L17 |
| GOTO Ckswitch | 'Go to Ckswitch: begin check switch routine again | L18 |

Line 9 of the PBASIC program initializes or turns off the output LED. The **OUT1 = 0** statement causes port 1 (P1) of the BS2 IC to go LOW. Lines 10 and 11 assign the current binary value at input ports 11 (P11) and 12 (P12) to variables B and A. For instance, if both input switches are pressed, then both variable B and A would hold logical 1.

Line 12 of the PBASIC program is code that logically ANDs the values in variables A and B. As an example, if both inputs are HIGH, then variable Y = 1. Line 13 is an IF-THEN statement used for making decisions. If Y = 1, then the PBASIC statement **IF Y = 1 THEN Red** will cause the program to jump to the **Red:** label or the subroutine that lights the red LED. If Y = 0, then the first section of PBASIC statement **IF Y = 1 THEN Red** is false. This will cause the program to proceed to the next line of code (line 14—**GOTO Ckswitch**). Line 14 (**GOTO Ckswitch**) sends the program back to the beginning of the main routine with the label of **Ckswitch:**.

The **Red:** subroutine in the PBASIC program **'2-input AND function** causes the port 1 (pin P1) of the BS2 IC to go HIGH using the **OUT1 = 1** statement. This turns on and lights the red LED. Line 17 (**PAUSE 100**) causes the LED to stay on for an extra 100 ms (milliseconds). Line 18 (**GOTO Ckswitch**) sends the program back to the main routine labeled **Ckswitch:**

The PBASIC program **'2-input AND function** runs continuously while the BS2 BASIC Stamp 2 module is powered. The PBASIC program is held in EEPROM program memory for future use. Turning the BS2 OFF and then ON again will restart the program. The current PBASIC listing can be changed by downloading a different program.

### Programming Other Logic Functions

Other logic functions can also be programmed using PBASIC and the BS2 BASIC Stamp module. These include OR, NOT, NAND, NOR, XOR, and XNOR. The next PBASIC program titled **'2-input OR function** is used with the hardware from Fig. 3-58 and operates like a 2-input OR gate. This program listing looks almost the same as the earlier PBASIC program except for the title line (**'2-input OR function**) and line 12 (**Y = A | B**).

Line 12 of the **'2-input OR function** program shows inputs A and B being ORed with

| | | | | |
|---|---|---|---|---|
| '2-input OR function | | | 'Title of program (Fig. 3-58) | L1 |
| | | | | |
| A | VAR | Bit | 'Declare A as variable, 1 bit | L2 |
| B | VAR | Bit | 'Declare B as variable, 1 bit | L3 |
| Y | VAR | Bit | 'Declare Y as variable, 1 bit | L4 |
| | | | | |
| INPUT 11 | | | 'Declare port 11 as an input | L5 |
| INPUT 12 | | | 'Declare port 12 as an input | L6 |
| OUTPUT 1 | | | 'Declare port 1 as an output | L7 |
| | | | | |
| Ckswitch: | | | 'Label for check switch routine | L8 |
| | OUT1 = 0 | | 'Initialize: port 1 at 0, red LED off | L9 |
| | A = IN12 | | 'Assign value: port 12 input to variable A | L10 |
| | B = IN11 | | 'Assign value: port 11 input to variable B | L11 |
| | Y = A \| B | | 'Assign value: A ORed with B to variable Y ← **Line 12** | L12 |
| | If Y = 1 THEN Red | | 'If Y = 1 then goto red subroutine, otherwise next line | L13 |
| GOTO Ckswitch | | | 'Go to Ckswitch—begin check switch routine again | L14 |
| | | | | |
| Red: | | | 'Label for lighting red LED, means HIGH output | L15 |
| | OUT1 = 1 | | 'Output P1 goes HIGH, lights red LED | L16 |
| | PAUSE 100 | | 'Pause for 100ms (milliseconds) | L17 |
| GOTO Ckswitch | | | 'Go to Ckswitch: begin check switch routine again | L18 |

| LOGIC FUNCTION | BOOLEAN EXPRESSION | PBASIC CODE (BS2 IC) |
|:---:|:---:|:---:|
| AND | $A \cdot B = Y$ | $Y = A \ \& \ B$ |
| OR | $A + B = Y$ | $Y = A \mid B$ |
| NOT | $A = \overline{A}$ | $Y = \sim (A)$ |
| NAND | $\overline{A \cdot B} = Y$ | $Y = \sim (A \ \& \ B)$ |
| NOR | $\overline{A + B} = Y$ | $Y = \sim (A \mid B)$ |
| XOR | $A \oplus B = Y$ | $Y = A \ ^\wedge \ B$ |
| XNOR | $\overline{A \oplus B} = Y$ | $Y = \sim (A \ ^\wedge \ B)$ |

Fig. 3-59 Logic functions implemented using PBASIC code with the BASIC Stamp 2 module by Parallax, Inc.

the resulting output being assigned to variable Y. The symbol for the OR function in PBASIC is the vertical line (|) and not the plus sign (+) that is used in traditional Boolean expressions.

The chart in Fig. 3-59 details the PBASIC code used to generate logic *functions using the BS2 BASIC Stamp module*. Notice the use of unique symbols in PBASIC to define AND, OR, NOT, and XOR logic functions. The ampersand (&) symbol is used for AND, and the vertical line (|) denotes the OR logic function. The tilde (~) is used for NOT. The circumflex accent (^) symbol is used to show the XOR logic function.

From Fig. 3-59, notice the use of both the tilde (~) and ampersand (&) symbols in the NAND function. An example of the 2-input NAND function would be Y = ~(A & B). Likewise in PBASIC code, both the tilde (~) and vertical line (|) symbols are used in the NOR function. An example of the 2-input NOR function would be Y = ~(A | B).

From Fig. 3-59, notice the use of the circumflex accent (^) symbol to define the exclusive-OR (XOR) logic function. The PBASIC code for the 2-input XOR logic function would be Y = A ^ B. In PBASIC, both the tilde (~) and circumflex accent (^) symbols are used to describe the XNOR logic function. An example would be Y = ~(A ^ B), which describes A XNORed with B and the output being assigned to Y.

## Self-Test

*Answer the following questions.*

86. The BS2 IC by Parallax is described as a BASIC Stamp _____ (microcontroller, multiplexer) module.

87. The BASIC Stamp 2 module can be programmed in a high-level language called FORTRAN by its manufacturer (T or F).

88. The PBASIC assignment statment Y = A | B | C is for the 3-input _____ (OR, XOR) logic function.

89. Write the PBASIC assignment statement that would describe the 2-input NAND logic function.

90. Write the PBASIC assignment statement that would describe the Boolean expression $A \cdot B = Y$.

91. Write the PBASIC assignment statement that would describe the Boolean expression $\overline{A \oplus B} = Y$.

92. Write the PBASIC assignment statement that would describe the Boolean expression $\overline{A + B} = Y$.

93. In Fig. 3-58 *(a)*, if input P12 is HIGH and P11 is LOW and the PBASIC program titled '**2-input OR function** is loaded into the BS2 IC, then output P1 will be _____ (HIGH, LOW), and the LED will _____ (light, not light).

94. In Fig. 3-58 *(a)*, if input P12 is HIGH and P11 is LOW and the PBASIC program titled '**2-input AND function** is loaded into the BS2 IC, then output P1 will be _____ (HIGH, LOW), and the red LED will _____ (light, not light).

# Chapter 3 Summary and Review

## Summary

1. Binary logic gates are the basic building blocks for all digital circuits.
2. Figure 3-60 shows a summary of the seven basic logic gates. This information should be memorized.
3. NAND gates are widely employed and can be used to make other logic gates.
4. Logic gates are often needed with 2 to 10 inputs. Several gates may be connected in the proper manner to get more inputs.
5. AND, OR, NAND, and NOR gates can be converted back and forth by using inverters.

| LOGIC FUNCTION | LOGIC SYMBOL | BOOLEAN EXPRESSION | TRUTH TABLE | | |
|---|---|---|---|---|---|
| | | | INPUTS | | OUTPUT |
| | | | B | A | Y |
| AND | A, B → Y | $A \cdot B = Y$ | 0 | 0 | 0 |
| | | | 0 | 1 | 0 |
| | | | 1 | 0 | 0 |
| | | | 1 | 1 | 1 |
| OR | A, B → Y | $A + B = Y$ | 0 | 0 | 0 |
| | | | 0 | 1 | 1 |
| | | | 1 | 0 | 1 |
| | | | 1 | 1 | 1 |
| Inverter | A → $\overline{A}$ | $A = \overline{A}$ | | 0 | 1 |
| | | | | 1 | 0 |
| NAND | A, B → Y | $\overline{A \cdot B} = Y$ | 0 | 0 | 1 |
| | | | 0 | 1 | 1 |
| | | | 1 | 0 | 1 |
| | | | 1 | 1 | 0 |
| NOR | A, B → Y | $\overline{A + B} = Y$ | 0 | 0 | 1 |
| | | | 0 | 1 | 0 |
| | | | 1 | 0 | 0 |
| | | | 1 | 1 | 0 |
| XOR | A, B → Y | $A \oplus B = Y$ | 0 | 0 | 0 |
| | | | 0 | 1 | 1 |
| | | | 1 | 0 | 1 |
| | | | 1 | 1 | 0 |
| XNOR | A, B → Y | $\overline{A \oplus B} = Y$ | 0 | 0 | 1 |
| | | | 0 | 1 | 0 |
| | | | 1 | 0 | 0 |
| | | | 1 | 1 | 1 |

**Fig. 3-60** Summary of basic logic gates.

6. Logic gates are commonly packaged in DIP ICs. The larger traditional DIP ICs are used on through-the-hole printed circuit boards. Modern small-sized DIP ICs are used for surface mounting.

7. Both TTL and CMOS digital ICs are used in very small systems. Modern high-speed low-power CMOS ICs (such as FACT series) are used in many new designs.

8. Very low power consumption is an advantage of CMOS digital ICs.

9. A technical person's knowledge of the normal operation of a circuit, powers of observation and skill in the use and interpretation of test data are all important in troubleshooting.

10. Logic symbols sometimes have small bubbles attached. These bubbles usually indicate that these pins are active-LOW inputs or outputs.

11. When using CMOS ICs, all unused inputs must go to $V_{DD}$ or GND. Care must be exercised in storing and handling CMOS ICs to avoid static electricity. Input voltages to a CMOS IC must never exceed the power supply voltages.

12. The logic probe, knowledge of the circuit, and your senses of sight, smell, and touch are basic tools used in troubleshooting gating circuits.

13. Figure 3-53 compares the traditional logic gate symbols with the newer IEEE standard logic symbols.

14. Logic functions can be implemented by hard-wiring logic gates or by programming various programmable devices.

15. The chart in Fig. 3-59 shows the PBASIC (Parallax, Inc.'s version of BASIC for the BS2 IC) code used in programming the logic functions AND, OR, NOT, NAND, NOR, XOR, and XNOR. This code is executed using a device called a microcontroller (BASIC Stamp 2 module).

## Chapter Review Questions

*Answer the following questions.*

3-1. Draw the traditional logic symbols for **a** to **j** (label inputs *A, B, C, D* and outputs *Y*):
   a. Two-input AND gate
   b. Three-input OR gate
   c. Inverter (two symbols)
   d. Two-input XOR gate
   e. Four-input NAND gate
   f. Two-input NOR gate
   g. Two-input XNOR gate
   h. Two-input NAND gate (special symbol)
   i. Two-input NOR gate (special symbol)
   j. Buffer (noninverting)
   k. Three-state buffer (noninverting)

3-2. Write the Boolean expression for the following (label inputs A, B, C, D, and outputs Y):
   a. Three-input AND function
   b. Two-input NOR function
   c. Three-input XOR function
   d. Four-input XNOR function
   e. Two-input NAND function

3-3. Draw the truth table for the following (label inputs A, B, C, and outputs Y):
   a. Three-input OR
   b. Three-input NAND

   c. Three-input XOR
   d. Two-input NOR
   e. Two-input XNOR

3-4. Look at the chart in Fig. 3-60. Which logic gate always responds with an output of logical 0 only when all inputs are HIGH?

3-5. Which logic gate might be called the "all or nothing gate"?

3-6. Which logic gate might be called the "any or all gate"?

3-7. Which logic circuit complements the input?

3-8. Which logic gate might be called the "any but not all gate"?

3-9. The unique output of a(n) _____ (AND, NAND) gate is a HIGH only when all inputs are HIGH.

3-10. The unique output of a(n) _____ (NAND, OR) gate is a LOW only when all inputs are LOW.

3-11. The unique output of a(n) _____ (NOR, XOR) gate is a HIGH only when an odd number of inputs are HIGH.

3-12. The unique output of a(n) _____ (NAND, OR) gate is a LOW only when all inputs are HIGH.

3-13. The unique output of a(n) _____ (NAND, NOR) gate is a HIGH only when all inputs are LOW.

3-14. Given an AND gate and inverters, draw how you would produce a NOR function.

3-15. Given a NAND gate and inverters, draw how you would produce an OR function.

3-16. Given a NAND gate and inverters, draw how you would produce an AND function.

3-17. Given 4 two-input AND gates, draw how you would produce a five-input AND gate.

3-18. Given several two-input NAND and OR gates, draw how you would produce a four-input NAND gate.

3-19. Switches arranged in series (see Fig. 3-1) act like what type of logic gate?

3-20. Switches arranged in parallel (see Fig. 3-7) act like what type of logic gate?

3-21. Figure 3-44(*b*) illustrates a(n) _____ (8, 16) pin _____ (three letters) IC.

3-22. Draw a wiring diagram similar to Fig. 3-46(*b*) for a circuit that will perform the three-input AND function. Use a 7408 IC, a 5-V dc power supply, three input switches, and an output indicator.

3-23. The PC board pad labeled _____ (*A, C*) is pin 1 of the IC in Fig. 3-61.

3-24. The PC board pad labeled _____ (letter) is the GND pin on the 7408 IC in Fig. 3-61.

Fig. 3-62 Top view of a typical digital IC.

3-25. The PC board pad labeled _____ (letter) is the $V_{CC}$ pin on the 7408 IC in Fig. 3-61.

3-26. The unit shown in Fig. 3-62 is a _____ (low-power, standard) TTL 14-pin DIP IC.

3-27. Pin 1 of the IC shown in Fig. 3-62 is labeled with the letter _____ .

3-28. The pin labeled with a "C" on the IC in Fig. 3-62 is pin number _____ .

3-29. Figure 3-52(*b*) is an example of a _____ (logic, wiring) diagram that might be used by service personnel.

3-30. Refer to Fig. 3-52(*a*). If all input pins (1, 2, and 5) are HIGH and output pin 6 is HIGH but point *E* is LOW, the LED _____ (will, will not) light and the circuit _____ (is, is not) working properly.

3-31. Refer to Fig. 3-52(*a*). List several possible problems if pin 6 is HIGH but point *E* is LOW.

3-32. Refer to Fig. 3-52(*a*). An *internal open* between the output of the first AND gate and pin 3 might give neither a HIGH nor LOW indication on the logic probe. This means that both pins 3 and 4 are floating _____ (HIGH, LOW).

3-33. Refer to Fig. 3-63. The core number of this IC is _____, which means it is a _____ (CMOS, TTL) logic device.

3-34. Pin 1 of the IC shown in Fig. 3-63 is labeled with the letter _____.

3-35. What precaution should be taken when storing the DIP IC like the one in Fig. 3-63?

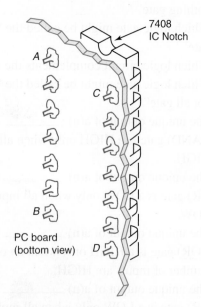

Fig. 3-61 An IC soldered to a PC board.

Fig. 3-63 Top view of a typical digital IC.

3-36. Draw the IEEE standard logic symbol for a three-input NOR gate.

3-37. Draw the IEEE standard logic symbol for a three-input XNOR gate.

3-38. The right _____ (circle, triangle) at the output of an IEEE standard NAND logic symbol signifies to invert the output of the AND function.

3-39. The IEEE standard AND logic symbol uses the _____ sign to signify the AND function.

3-40. A microcontroller (such as the BASIC Stamp 2 module) can be programmed to perform logic functions (AND, OR, etc.). (T or F)

3-41. The _____ (BS2, BX10) BASIC Stamp module by Parallax is programmed in a high-level computer language called PBASIC (a version of BASIC).

3-42. When programming the BASIC Stamp 2 module, the PBASIC code used to represent the Boolean expression A + B = Y (2-input OR) is _____ (Y = A OR B, Y = A|B).

3-43. When programming the BS2 microcontroller module, the PBASIC code used to represent the 2-input NAND function would be _____ [Y = A + B, Y = ~(A & B)].

3-44. The PBASIC code Y = A ^ B ^ C represents the assignment statement for the 3-input _____ (AND, XOR) logic function.

## Critical Thinking Questions

3-1. What three-input logic gate would you use in your design if you require a HIGH output *only* when all three input switches go HIGH?

3-2. What four-input logic gate would you use in your design if you require a HIGH output only when an *odd* number of input switches are HIGH?

3-3. Refer to Fig. 3-41(*a*). Explain why the OR gate with inverted inputs produces the NAND function.

3-4. Inverting both inputs of a two-input NAND gate produces a circuit that generates the _____ logic function.

3-5. Inverting both inputs and the output of a two-input OR gate produces a circuit that generates the _____ logic function.

3-6. Refer to Fig. 3-50. If input A is HIGH and input B is LOW, output J (pin 3) will be _____ (HIGH, LOW). Transistor $Q_1$ is turned _____ (off, on) and the LED _____ (does, does not) light.

3-7. Refer to Fig. 3-50. Why are pins 5, 6, 8, 9, 12, and 13 grounded in this circuit?

3-8. Refer to Fig. 3-52. If the 7408 TTL IC developed an internal "short circuit," the top of the IC would probably feel _____ (hot, cool) to the touch.

3-9. Draw a logic diagram (use AND and inverter symbols) for the Boolean expression $\overline{A} \cdot \overline{B} = Y$.

3-10. The Boolean expression $\overline{A} \cdot \overline{B} = Y$ is one representation of the _____ (NAND, NOR) logic function.

3-11. Draw a waveform to represent the logic levels (*H* and *L*) at output Y of the AND gate in Fig. 3-64.

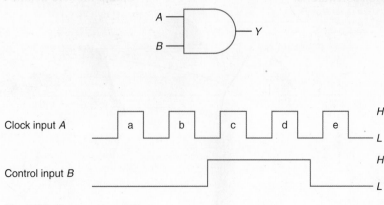

Fig. 3-64 Pulse train problem.

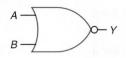

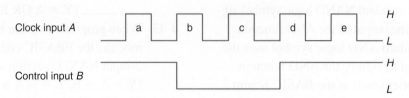

Clock input *A*

Control input *B*

**Fig. 3-65** Pulse train problem.

3-12. Draw a waveform to represent the logic levels (*H* and *L*) at output *Y* of the NOR gating circuit in Fig. 3-65.

3-13. Prove to your instructor that both logic diagrams drawn in Fig. 3-41(*a*) will generate a 2-input NAND truth table. (HINT: think of the bubbles as inverters). You may use one of the following methods (ask instructor):
  a. Wire and test logic circuits in hardware, or
  b. Wire and test logic circuits using a computer circuit-simulation software, or
  c. Use a series of truth tables to make your proof.

3-14. Prove to your instructor that both logic diagrams drawn in Fig. 3-41(*b*) will generate a 2-input NOR truth table. (*Hint:* think of the bubbles as inverters) You may use one of the following methods (ask instructor):
  a. Wire and test logic circuits in hardware, or
  b. Wire and test logic circuits using a computer circuit-simulation software, or

  c. Use a series of truth tables to make your proof.

3-15. Refer to Fig. 3-66(*a*). The normally open switches are wired _____ (active-HIGH, active-LOW) inputs.

3-16. Refer to Fig. 3-66(*a*). The LED at P1 will light when port 1 becomes _____ (HIGH, LOW).

3-17. Refer to Fig. 3-66. What three lines of PBASIC computer code declare which BS2 IC ports are inputs?

3-18. Refer to Fig. 3-66. What is the purpose of Line 11 of the PBASIC code?

3-19. Refer to Fig. 3-66. If all inputs to the BASIC Stamp 2 module are HIGH, the output will be _____ (HIGH, LOW), and the red LED will _____ (light, not light).

3-20. Answer selected questions asked by your instructor about the PBASIC program and the programming and operation of the BS2 BASIC Stamp module detailed in Fig. 3-66.

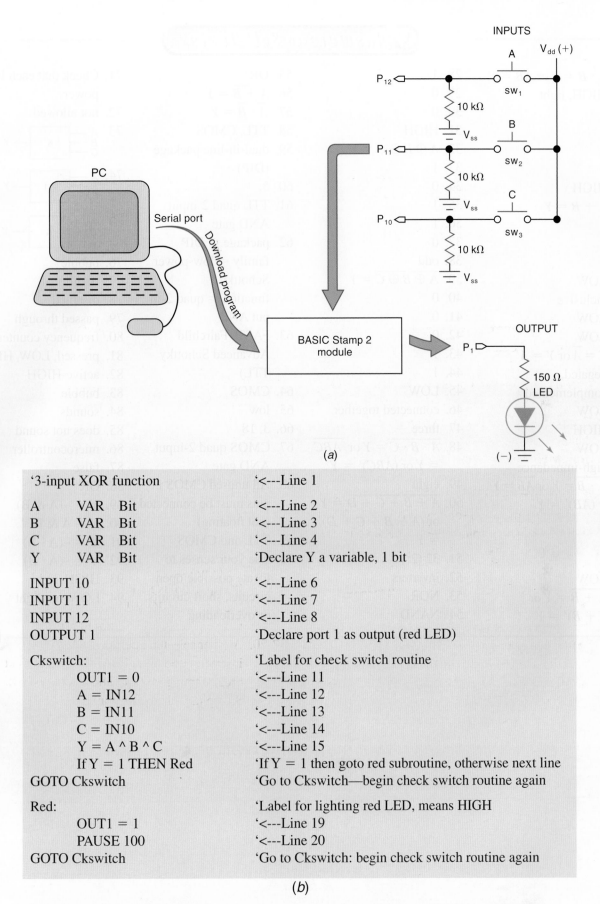

'3-input XOR function        '<---Line 1

A      VAR    Bit             '<---Line 2
B      VAR    Bit             '<---Line 3
C      VAR    Bit             '<---Line 4
Y      VAR    Bit             'Declare Y a variable, 1 bit

INPUT 10                      '<---Line 6
INPUT 11                      '<---Line 7
INPUT 12                      '<---Line 8
OUTPUT 1                      'Declare port 1 as output (red LED)

Ckswitch:                     'Label for check switch routine
        OUT1 = 0              '<---Line 11
        A = IN12              '<---Line 12
        B = IN11              '<---Line 13
        C = IN10              '<---Line 14
        Y = A ^ B ^ C         '<---Line 15
        If Y = 1 THEN Red     'If Y = 1 then goto red subroutine, otherwise next line
GOTO Ckswitch                 'Go to Ckswitch—begin check switch routine again

Red:                          'Label for lighting red LED, means HIGH
        OUT1 = 1              '<---Line 19
        PAUSE 100             '<---Line 20
GOTO Ckswitch                 'Go to Ckswitch: begin check switch routine again

(b)

**Fig. 3-66** (*a*) Wiring of BS2 BASIC Stamp module used with '3-input XOR function program. (*b*) PBASIC program loaded into BS2 module.

1. $A \cdot B = Y$ or $AB = Y$
2. HIGH, light
3. 0
4. 0
5. 1
6. 0
7. HIGH
8. $A + B = Y$
9. 1
10. 0
11. 1
12. LOW
13. inclusive
14. LOW
15. LOW
16. $Y = \overline{A}$ or $Y = A'$
17. negated, complemented
18. LOW
19. HIGH
20. LOW
21. high-impedance
22. $\overline{A \cdot B} = Y$ or $\overline{AB} = Y$ or $(AB)' = Y$
23. 1
24. 0
25. 1
26. LOW
27. $\overline{A + B} = Y$ or $(A + B)' = Y$

28. 1
29. 0
30. 0
31. HIGH
32. $A \oplus B \oplus C = Y$
33. 1
34. 0
35. 1
36. 1
37. 0
38. odd
39. $\overline{A \oplus B \oplus C} = Y$
40. 0
41. 0
42. 0
43. 1
44. 1
45. LOW
46. connected together
47. three
48. $\overline{A \cdot B \cdot C} = Y$ or $\overline{ABC} = Y$ or $(ABC)' = Y$
49. eight
50. $\overline{A + B + C + D} = Y$ or $(A + B + C + D)' = Y$
51. 32 ($2^5$)
52. inverters
53. NOR
54. NAND

55. OR
56. $\overline{A + B} = Y$
57. $\overline{A \cdot B} = Y$
58. TTL, CMOS
59. dual-in-line package (DIP)
60. 5, +
61. TTL quad 2-input AND gate
62. package = DIP family = low-power Schottky function = quad 2-input AND
63. FAST (Fairchild Advanced Schottky TTL)
64. CMOS
65. low
66. 3, 18
67. CMOS quad 2-input AND gate
68. All unused CMOS inputs must be connected (not floating)
69. TTL and CMOS
70. Use your senses to locate possible open circuits, short circuits, or overloading

71. Check that each IC has power
72. not allowed
73.
74.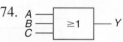
75.
76. bubble
77. traditional
78. disabled
79. passed through
80. frequency counter
81. pressed, LOW, HIGH
82. active-HIGH
83. bubble
84. sounds
85. does not sound
86. microcontroller
87. false
88. OR
89. Y = ~(A & B)
90. Y = A & B
91. Y = ~(A ^ B)
92. Y = ~(A | B)
93. HIGH, light
94. LOW, not light

# Combining Logic Gates

## Chapter Objectives

*This chapter will help you to:*

1. *Draw* logic diagrams from minterm and maxterm Boolean expressions.

2. *Design* a logic diagram from a truth table by first developing a minterm Boolean expression and then drawing the AND-OR logic diagram.

3. *Reduce* a minterm Boolean expression to its simplest form using two-, three-, four-, and five-variable Karnaugh maps.

4. *Simplify* AND-OR logic circuits using NAND gates.

5. *Convert* back and forth from Boolean expression to truth table to logic symbol diagram using computer simulation (such as the Logic Converter instrument from Electronic Workbench® or MultiSIM®).

6. *Solve* logic problems using data selectors.

7. *Understand* the fundamentals of selected programmable logic devices (PLDs).

8. *Convert* minterm-to-maxterm and maxterm-to-minterm Boolean expressions using De Morgan's theorems.

9. *Use* a "keyboard version" of Boolean expressions.

10. *Program* several logic functions using a BASIC Stamp microcontroller module.

Earlier you memorized the symbol, truth table, and Boolean expression for each logic gate. These gates are the building blocks for more complicated digital devices. In this chapter you will use your knowledge of gate symbols, truth tables, and Boolean expressions to solve real-world problems in electronics.

You will be connecting gates to form what engineers refer to as *combinational logic circuits*. By definition, combinational logic is an interconnection of logic gates to generate a specified logic function where the inputs result in an immediate output; having no memory or storage capabilities. This is also sometimes called *combinatorial logic*. Digital circuits that have a memory or storage capability are called *sequential logic circuits* and will be studied later.

You will be combining gates (ANDs, ORs) and inverters to solve logic problems that do not require memory. The "tools of the trade" for solving combinational logic problems are: truth tables, Boolean expressions, and logic symbols. Do you know your truth tables, Boolean expressions, and logic symbols? An understanding of combination logic is knowledge required of all who work as a technician, troubleshooter, designer, or engineer in electronics.

To gain maximum experience you should try to implement your combinational logic circuits in hardware in the laboratory. Logic gates are packaged in inexpensive, easy-to-use integrated circuits (ICs). Also, your combinational logic circuits can be tested using circuit simulation software on your computer.

Solve combinational logic problems with programming. Try programming a simple programmable logic device such as an

**Combinational logic circuits**

**Sequential logic circuits**

inexpensive PAL or GAL if your lab has PLD programming equipment. Finally, solve real-world combinational logic functions by programming a microcontroller using a PC and the BASIC Stamp 2 module.

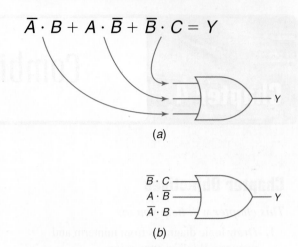

$$\overline{A} \cdot B + A \cdot \overline{B} + \overline{B} \cdot C = Y$$

(a)

(b)

**Fig. 4-2** Step 1 in constructing a logic circuit.

## 4-1 Constructing Circuits from Boolean Expressions

We use Boolean expressions to guide us in building logic circuits. Suppose you are given the Boolean expression $A + B + C = Y$ (read as "$A$ or $B$ or $C$ equals output $Y$") and told to build a circuit that will perform this logic function. Looking at the expression, notice that each input must be ORed to get output $Y$. Figure 4-1 illustrates the *gate* needed to do the job.

Now suppose you are given the Boolean expression $\overline{A} \cdot B + A \cdot \overline{B} + \overline{B} \cdot C = Y$ (read as "not $A$ and $B$, or $A$ and not $B$, or not $B$ and $C$ equals output $Y$"). How would you construct a circuit that will do the job of this expression? The first step is to look at the Boolean expression and note that you must OR $\overline{A} \cdot B$ with $A \cdot \overline{B}$ with $\overline{B} \cdot C$. Figure 4-2(*a*) shows that a three-input OR gate will form the output $Y$. This may be redrawn as in Fig. 4-2(*b*).

The second step used in constructing a logic circuit from the given Boolean expression $\overline{A} \cdot B + A \cdot \overline{B} + \overline{B} \cdot C = Y$ is shown in Fig. 4-3. Notice in Fig. 4-3(*a*) that an AND gate has been added to feed the $\overline{B} \cdot C$ to the OR gate and an inverter has been added to form the $\overline{B}$ for the input to AND gate 2. Figure 4-3(*b*) adds AND gate 3 to form the $A \cdot \overline{B}$ input to the OR gate. Finally, Fig. 4-3(*c*) adds AND gate 4 and inverter 6 to form the $\overline{A} \cdot B$ input to the OR gate. Figure 4-3(*c*) is the circuit that would be constructed to perform the required logic given in the Boolean expression $\overline{A} \cdot B + A \cdot \overline{B} + \overline{B} \cdot C = Y$.

Notice that we started at the output of the logic circuit and worked toward the inputs. You have now experienced how combinational logic circuits are constructed from Boolean expressions.

Boolean expressions come in two forms. The *sum-of-products (SOP) form* is the type we saw in Fig. 4-2. Another example of this form is $A \cdot B + B \cdot C = Y$. The other Boolean expression form is the *product-of-sums (POS)*; an example is $(D + E) \cdot (E + F) = Y$. The sum-of-products form

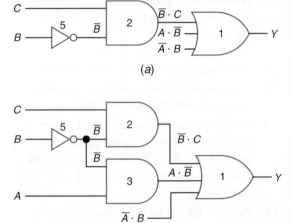

(a)

(b)

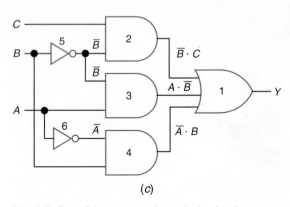

(c)

**Fig. 4-3** Step 2 in constructing a logic circuit.

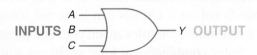

INPUTS $\begin{matrix} A \\ B \\ C \end{matrix}$ ————— $Y$ OUTPUT

**Fig. 4-1** Logic diagram for Boolean expression $A + B + C = Y$.

is called the *minterm form* in engineering texts. The product-of-sums form is called the *maxterm form* by engineers, technicians, and scientists.

Computer *circuit simulation software,* such as Electronics Workbench or MultiSIM, will draw a logic diagram from a Boolean expression.

This software can draw logic diagrams from either minterm or maxterm Boolean expressions. Professionals in digital design will commonly use computer circuit simulations. Your instructor may have you use circuit simulation software in the lab.

**Minterm form**

**Maxterm form**

**Circuit simulation software**

---

## ✓ Self-Test

*Answer the following questions.*

1. Construct logic circuits using AND, OR, and NOT gates for the following minterm Boolean expressions:
   a. $\overline{A} \cdot \overline{B} + A \cdot B = Y$
   b. $\overline{A} \cdot \overline{C} + A \cdot B \cdot C = Y$
   c. $A \cdot D + \overline{B} \cdot \overline{D} + C \cdot \overline{D} = Y$
2. A minterm Boolean expression is also called the _____ form.

3. A maxterm Boolean expression is also called the _____ form.
4. The minterm Boolean expression $A \cdot D + \overline{B} \cdot \overline{D} + C \cdot \overline{D} = Y$ has a pattern that is called the _____ (product-of-sums, sum-of-products) form.
5. The maxterm Boolean expression $(A + D) \cdot (B + \overline{C}) \cdot (A + C) = Y$ has a pattern that is called the _____ (product-of-sums, sum-of-products) form.

---

## 4-2 Drawing a Circuit from a Maxterm Boolean Expression

Suppose you are given the maxterm Boolean expression $(A + B + C) \cdot (\overline{A} + \overline{B}) = Y$. The first step in constructing a logic circuit for this Boolean expression is shown in Fig. 4-4(*a*). Notice that the terms $(A + B + C)$ and $(\overline{A} + \overline{B})$ are ANDed together to form output $Y$. Figure 4-4(*b*) shows the logic circuit redrawn. The second step in drawing the logic circuit is shown in Fig. 4-5. The $(\overline{A} + \overline{B})$ part of the expression is

produced by adding OR gate 2 and inverters 3 and 4, as illustrated in Fig. 4-5(*a*). Then, the expression $(A + B + C)$ is delivered to the AND gate by OR gate 5 in Fig. 4-5(*b*). The logic circuit shown in Fig. 4-5(*b*) is the complete logic circuit for the maxterm Boolean expression $(A + B + C) \cdot (\overline{A} + \overline{B}) = Y$.

In summary, we work from right to left (from output to input) when converting a

$$(A + B + C) \cdot (\overline{A} + \overline{B}) = Y$$

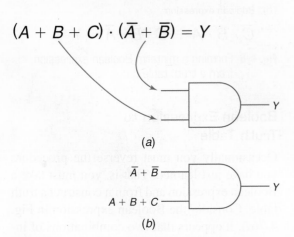

(a)

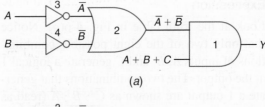

(b)

**Fig. 4-4** Step 1 in constructing a product-of-sums logic circuit.

(a)

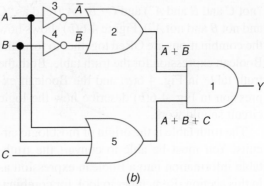

(b)

**Fig. 4-5** Step 2 in constructing a product-of-sums logic circuit.

Boolean expression to a logic circuit. Notice that we use only AND, OR, and NOT gates when constructing combinational logic circuits. Maxterm and minterm Boolean expressions both can be converted to logic circuits. Minterm expressions create AND-OR logic circuits similar to that in Fig. 4-3(c), whereas maxterm expressions create OR-AND logic circuits similar to that in Fig. 4-5(b).

You now should be able to identify minterm and maxterm Boolean expressions, and you should be able to convert Boolean expressions to combinational logic circuits by using AND, OR, and NOT gates.

*Answer the following questions.*

6. Construct a logic circuit using AND, OR, and NOT gates from the following Boolean expressions:
   a. $(A + B) \cdot (\overline{A} + \overline{B}) = Y$
   b. $(\overline{A} + B) \cdot \overline{C} = Y$.
   c. $(A + B) \cdot (\overline{C} + \overline{D}) \cdot (\overline{A} + C) = Y$
7. Refer to question 6. These Boolean expressions are _____ (maxterm, minterm).
8. Refer to question 6. These Boolean expressions are in _____ (product-of-sums, sum-of-products) form.
9. Maxterm Boolean expressions are used to create _____ (AND-OR, OR-AND) logic circuits.

## 4-3 Truth Tables and Boolean Expressions

Boolean expressions are a convenient method of describing how a logic circuit operates. The *truth table* is another precise method of describing how a logic circuit works. As you work in digital electronics, you may have to convert information from truth-table form to a Boolean expression.

### Truth Table to Boolean Expression

Look at the truth table in Fig. 4-6(a). Notice that only two of the eight possible combinations of inputs A, B, and C generate a logical 1 at the output. The two combinations that generate a 1 output are shown as $\overline{C} \cdot B \cdot A$ (read as "not C and B and A") and $C \cdot \overline{B} \cdot \overline{A}$ (read as "C and not B and not A"). Figure 4-6(b) shows how the combinations are ORed together to form the Boolean expression for the truth table. Both the truth table in Fig. 4-6(a) and the Boolean expression in Fig. 4-6(b) describe how the logic circuit should work.

The truth table is the origin of most logic circuits. You must be able to convert the truth-table information into a Boolean expression as in this section. Remember to look for combinations of variables that generate a logical 1 output in the truth table.

Truth table

| INPUTS | | | OUTPUT |
|---|---|---|---|
| C | B | A | Y |
| 0 | 0 | 0 | 0 |
| 0 | 0 | 1 | 0 |
| 0 | 1 | 0 | 0 |
| 0 | 1 | 1 | 1 |
| 1 | 0 | 0 | 1 |
| 1 | 0 | 1 | 0 |
| 1 | 1 | 0 | 0 |
| 1 | 1 | 1 | 0 |

$\overline{C} \cdot B \cdot A = 1$

$C \cdot \overline{B} \cdot \overline{A} = 1$

(a)

(b) Boolean expression

$$\overline{C} \cdot B \cdot A + C \cdot \overline{B} \cdot \overline{A} = Y$$

**Fig. 4-6** Forming a minterm Boolean expression from a truth table.

### Boolean Expression to Truth Table

Occasionally you must reverse the procedure you have just learned. That is, you must take a Boolean expression and from it construct a truth table. Consider the Boolean expression in Fig. 4-7(a). It appears that two combinations of inputs A, B, and C generate a logical 1 at the output. In Fig. 4-7(b) we find the correct combina-

Forming a Boolean expression from a truth table

(a) Boolean expression

$$\overline{C} \cdot B \cdot \overline{A} + C \cdot \overline{B} \cdot A = Y$$

Truth table

| INPUTS | | | OUTPUT |
|---|---|---|---|
| C | B | A | Y |
| 0 | 0 | 0 | 0 |
| 0 | 0 | 1 | 0 |
| 0 | 1 | 0 | 1 |
| 0 | 1 | 1 | 0 |
| 1 | 0 | 0 | 0 |
| 1 | 0 | 1 | 1 |
| 1 | 1 | 0 | 0 |
| 1 | 1 | 1 | 0 |

(b)

**Fig. 4-7** Constructing a truth table from a minterm Boolean expression.

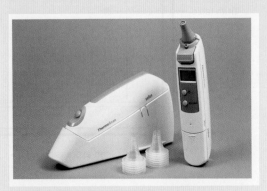

## ABOUT ELECTRONICS

**Electronic Thermometers.** Today, taking a temperature is not the challenge it was for previous generations. The Braun ThermaScan ear thermometer takes a reading in just one second. This is possible because the thermometer is able to read the infrared heat emitted from the eardrum and surrounding tissue. Advanced electronics then "translate" this signal to a temperature that appears on the digital readout.

tions of A, B, and C that are given in the Boolean expression and mark a 1 in the output column. All other outputs in the truth table are 0. The Boolean expression in Fig. 4-7(a) and the truth table in Fig. 4-7(b) both accurately describe the operation of the same logic circuit.

Suppose you are given the Boolean expression in Fig. 4-8(a). At first glance it seems that this would produce two outputs with a logical 1. However, if you look closely at Fig. 4-8(b) you will see that the Boolean expression $\overline{C} \cdot \overline{A}$

(a) Boolean expression

$$\overline{C} \cdot \overline{A} + C \cdot B \cdot A = Y$$

Truth table

| INPUTS | | | OUTPUT |
|---|---|---|---|
| C | B | A | Y |
| 0 | 0 | 0 | 1 |
| 0 | 0 | 1 | 0 |
| 0 | 1 | 0 | 1 |
| 0 | 1 | 1 | 0 |
| 1 | 0 | 0 | 0 |
| 1 | 0 | 1 | 0 |
| 1 | 1 | 0 | 0 |
| 1 | 1 | 1 | 1 |

(b)

**Fig. 4-8** Constructing a truth table from a minterm Boolean expression.

$+ C \cdot B \cdot A = Y$ actually generates three logical 1s in the output column. The "trick" illustrated in Fig. 4-8 should make you very cautious. Make sure you have all the combinations that generate a logical 1 in the truth table. The Boolean expression in Fig. 4-8(a) and the truth table in Fig. 4-8(b) both describe the same logic circuit or logic function.

You have now converted truth tables to Boolean expressions and Boolean expressions to truth tables. You were reminded that the Boolean expressions you worked with were minterm Boolean expressions. The procedure for producing maxterm Boolean expressions from a truth table is quite different.

### Circuit Simulation Conversions

Circuit simulation software running on modern computers can accurately convert Boolean expressions to truth tables or truth tables to Boolean expressions. We will demonstrate the use of one popular electronic circuit simulator.

One easy-to-use circuit simulator is Electronics Workbench® (EWB) or MultiSIM®. The EWB software contains an instrument called a *logic converter,* shown in Fig. 4-9(a). To use this EWB instrument to convert a Boolean expression to a truth table, you would take the following steps:

**Constructing a truth table from a Boolean expression**

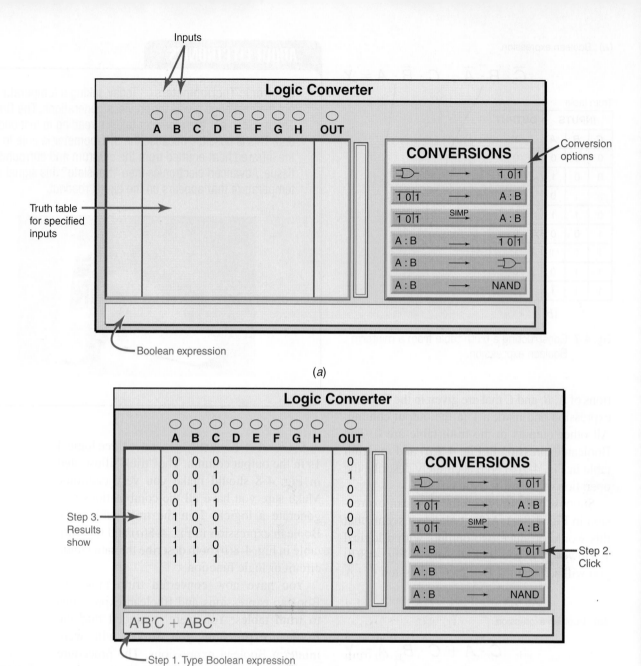

**Inputs**

**Logic Converter**

A B C D E F G H OUT

Truth table for specified inputs

**CONVERSIONS**

SIMP

1 0 1
A : B

Conversion options

Boolean expression

(a)

**Logic Converter**

A B C D E F G H OUT

| A | B | C | | | | | | OUT |
|---|---|---|---|---|---|---|---|-----|
| 0 | 0 | 0 | | | | | | 0 |
| 0 | 0 | 1 | | | | | | 1 |
| 0 | 1 | 0 | | | | | | 0 |
| 0 | 1 | 1 | | | | | | 0 |
| 1 | 0 | 0 | | | | | | 0 |
| 1 | 0 | 1 | | | | | | 0 |
| 1 | 1 | 0 | | | | | | 1 |
| 1 | 1 | 1 | | | | | | 0 |

Step 3. Results show

**CONVERSIONS**

SIMP

Step 2. Click

A'B'C + ABC'

Step 1. Type Boolean expression

(b)

**Fig. 4-9** Logic converter instrument from an electronic circuit simulator. (a) Logic converter instrument layout. (b) The three steps in converting a Boolean expression to a truth table.

Step 1. Type the expression in the bottom section (see Fig. 4-9(b)).

Step 2. Activate the Boolean expression to a truth-table option (see Fig. 4-9(b)).

Step 3. View the resulting truth table on the computer monitor (see Fig. 4-9(b)).

The Boolean expression $A'B'C + ABC'$ entered in step 1, Fig. 4-9(b), is a shortened "keyboard version" of the $C \cdot \overline{B} \cdot \overline{A} + \overline{C} \cdot B \cdot A = Y$ in Fig. 4-6(b). It is important to recognize

that $A'B'C + ABC'$ equals $C \cdot \overline{B} \cdot \overline{A} + \overline{C} \cdot B \cdot A = Y$. The apostrophe in the "keyboard version" of a Boolean expression means the same as an overbar over that letter. Therefore $A'$ (say A not) means the same as $\overline{A}$ (say A not). Notice that the *order* that the variables appear in the Boolean expression are reversed. This difference in order has no effect on the logic function. Therefore, $ABC$ means the same as $CBA$. Also notice that the AND dot between variables has been eliminated so that $A \cdot B \cdot C$ can be shortened to $ABC$.

Compare the output columns in Figs. 4-6(*a*) and 4-9(*b*). Both these truth tables describe the same logic function although the output columns seem different. This is because the order of the input variables are listed as *CBA* in Fig. 4-6(*a*) whereas they appear as *ABC* in Fig. 4-9(*b*). Truth table line 5 (100) in Fig. 4-6(*a*) is the same as line 2 (001) in Fig. 4-9(*b*). This demonstrates that the headings on truth tables and Boolean expression representations vary.

Workers in electronics will become familiar with several methods of labeling truth tables and variations in Boolean expressions.

Electronic circuit simulators such as EWB can commonly handle either minterm or maxterm Boolean expressions. Observe from Fig. 4-9(*a*) that five other logic conversions are available using this version of EWB. Your instructor may have you use the many features available on your electronic circuit simulation software.

## Self-Test

*Answer the following questions.*

10. Refer to Fig. 4-6(*a*). Assume that only the bottom two lines of the truth table produce an output of 1 (all other outputs = 0). Write the sum-of-products Boolean expression for this situation.
11. Refer to Fig. 4-6(*a*). The Boolean expression $\overline{C} \cdot \overline{B} \cdot \overline{A} + \overline{C} \cdot \overline{B} \cdot A = Y$ produces a truth table that has HIGH outputs in what two lines?
12. Construct a truth table for the Boolean expression $C \cdot B \cdot \overline{A} + C \cdot \overline{B} \cdot A = Y$.
13. The procedure illustrated in Fig. 4-6 converts a truth table to a _____ (maxterm, minterm) Boolean expression.

14. The procedure illustrated in Figs. 4-7 and 4-8 converts a _____ (maxterm, minterm) Boolean expression to a truth table.
15. Write the "keyboard version" of the Boolean expression $\overline{C} \cdot \overline{B} \cdot A + B \cdot \overline{A} = Y$.
16. The Boolean expression $A \cdot B \cdot C = Y$ means the same as $ABC = Y$. (*T* or *F*)
17. The Boolean expression $A \cdot B \cdot C = Y$ will generate the same truth table as $C \cdot B \cdot A = Y$. (*T* or *F*)
18. $A'C' + AB = Y$ is the "keyboard version" of the traditional Boolean expression $(\overline{A} + \overline{C}) \cdot (A + B) = Y$. (*T* or *F*)

## 4-4 Sample Problem

The procedures in Secs. 4-1 to 4-3 are useful skills as you work in digital electronics. To assist you in developing your skills, we shall take an everyday logic problem and work from truth table to Boolean expression to logic circuit as shown in Fig. 4-10.

Let us assume that we are designing a simple *electronic lock*. The lock will open only when certain switches are activated. Figure 4-10(*a*) is the truth table for the electronic lock. Notice that the two combinations of input switches, *A*, *B*, and *C* generate a 1 at the output. A HIGH (or 1) output will open the lock. Figure 4-10(*b*) shows how we form the minterm Boolean expression for the electronic lock circuit. The logic circuit in Fig. 4-10(*c*) is then drawn from the Boolean expression. Look over the sample problem in Fig. 4-10 and be sure you can follow how we converted from the truth table to the Boolean expression and then to the logic circuit.

Many electronic circuit simulation programs can handle these conversions. For instance, the logic converter instrument in Electronics Workbench® or MultiSIM® could make these conversions. The logic converter instrument from Electronics Workbench® or MultiSIM® will be used to solve the lock problem presented earlier. The steps in solving this lock problem using this software are represented in Fig. 4-11. These steps are:

Step 1: Fill out the lock problem truth table [see Fig. 4-11(*a*)].

Step 2: Activate the truth table to Boolean expression [see Fig. 4-11(*a*)].

The resulting Boolean expression will be $A'B'C + ABC$.

Step 3: Activate the Boolean expression to logic circuit button [see Fig. 4-11(*b*)].

The resulting AND-OR logic will be displayed on the EWB screen.

**Electronic lock**

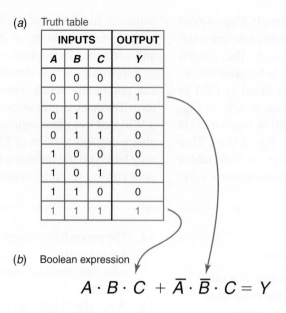

(a) Truth table

| INPUTS | | | OUTPUT |
|---|---|---|---|
| **A** | **B** | **C** | **Y** |
| 0 | 0 | 0 | 0 |
| 0 | 0 | 1 | 1 |
| 0 | 1 | 0 | 0 |
| 0 | 1 | 1 | 0 |
| 1 | 0 | 0 | 0 |
| 1 | 0 | 1 | 0 |
| 1 | 1 | 0 | 0 |
| 1 | 1 | 1 | 1 |

(b)  Boolean expression

$$A \cdot B \cdot C + \overline{A} \cdot \overline{B} \cdot C = Y$$

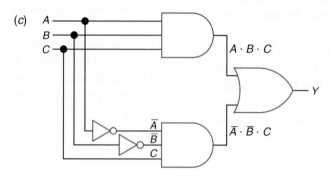

(c)

**Fig. 4-10** Electronic lock problem. (a) Truth table. (b) Boolean expression. (c) Logic circuit.

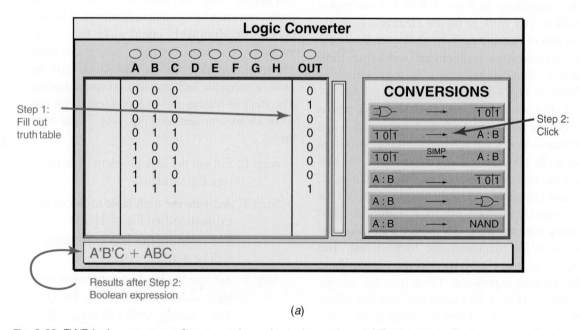

(a)

**Fig. 4-11** EWB logic converter software used to solve logic problem. (a) Truth table to Boolean expression.

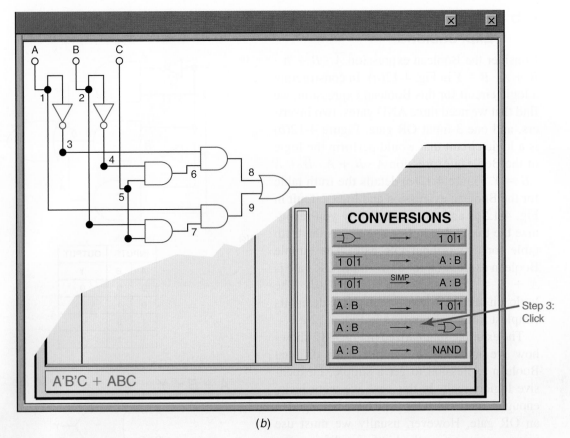

A'B'C + ABC

*(b)*

**Fig. 4-11** (*b*) Boolean expression to logic circuit conversion.

You should now be able to solve a logic problem like the one described in this section. You can solve these problems either by hand (Fig. 4-10) or by using simulation software (Fig. 4-11). The following test will give you some practice in solving problems dealing with truth tables, Boolean expressions, and combinational logic circuits.

**Self-Test**

*Answer the following questions.*

19. Using the truth table below for an electronic lock, write the minterm Boolean expression for this truth table.
20. From the Boolean expression developed in question 19, draw a logic symbol diagram for the electronic lock problem.

| INPUT SWITCHES | | | OUTPUT |
|---|---|---|---|
| C | B | A | Y |
| 0 | 0 | 0 | 0 |
| 0 | 0 | 1 | 0 |
| 0 | 1 | 0 | 1 |
| 0 | 1 | 1 | 0 |
| 1 | 0 | 0 | 0 |
| 1 | 0 | 1 | 1 |
| 1 | 1 | 0 | 0 |
| 1 | 1 | 1 | 0 |

Truth Table for Question 19—
Lock Problem

## 4-5 Simplifying Boolean Expressions

Consider the Boolean expression $\overline{A} \cdot B + A \cdot \overline{B} + A \cdot B = Y$ in Fig. 4-12(a). In constructing a logic circuit for this Boolean expression, we find that we need three AND gates, two inverters, and one 3-input OR gate. Figure 4-12(b) is a logic circuit that would perform the logic of the Boolean expression $\overline{A} \cdot B + A \cdot \overline{B} + A \cdot B = Y$. Figure 4-12(c) details the truth table for the Boolean expression and logic circuit in Fig. 4-12(a) and (b). Immediately you recognize the truth table in Fig. 4-12(c) as the truth table for a two-input OR gate. The simple Boolean expression for a two-input OR gate is $A + B = Y$, as shown in Fig. 4-12(d). The logic circuit for a two-input OR gate in its simplest form is diagramed in Fig. 4-12(e).

The example summarized in Fig. 4-12 shows how we must try to simplify our original Boolean expression to get a simple, inexpensive logic circuit. In this case we were lucky enough to notice that the truth table belonged to an OR gate. However, usually we must use more systematic methods of simplifying our Boolean expression. Such methods include applying Boolean algebra, *Karnaugh mapping*, and computer simulations.

Boolean algebra was originated by George Boole (1815–1864). Boole's algebra was adapted in the 1930s for use in digital logic circuits; it is the basis for the tricks we shall use to simplify Boolean expressions. Only selected topics in Boolean algebra are covered in this text. Many of you who continue on in digital electronics and engineering will study Boolean algebra in detail.

Karnaugh mapping, an easy-to-use graphic method of simplifying Boolean expressions, is

**Boolean algebra**

**Karnaugh mapping**

**Tabular method of simplification**

**Quine-McCluskey method**

(a) Original Boolean expression

$$\overline{A} \cdot B + A \cdot \overline{B} + A \cdot B = Y$$

(b)

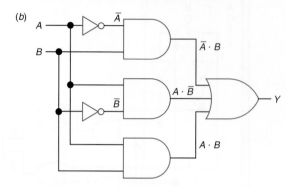

(c)      Truth table

| INPUTS | | OUTPUT |
|---|---|---|
| **A** | **B** | **Y** |
| 0 | 0 | 0 |
| 0 | 1 | 1 |
| 1 | 0 | 1 |
| 1 | 1 | 1 |

(d)   Simplified Boolean expression

$$A + B = Y$$

(e)

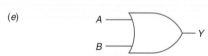

**Fig. 4-12** Simplifying Boolean expressions.
(a) Unsimplified Boolean expression.
(b) Complex logic diagram. (c) Truth table.
(d) Simplified Boolean expression: two-input OR by inspection. (e) Simple logic diagram.

covered in detail in Secs. 4-6 to 4-10. Several other simplification methods are available, including Veitch diagrams, Venn diagrams, and the *tabular method of simplification*. The tabular method used by computer software such as the Electronics Workbench® is called the *Quine-McCluskey method*.

## Self-Test

*Supply the missing word or words in each statement.*

21. The logic circuits in Fig. 4-12(b) and (e) produce _____ (different, identical) truth tables.
22. Boolean expressions can many times be simplified by inspection or by using

methods that include _____ algebra or _____ mapping.
23. Karnaugh mapping is a systematic graphic method of logic circuit simplification but the _____ _____ method is better suited for computer simplification.

# 4-6 Karnaugh Maps

In 1953 Maurice Karnaugh published an article about his system of mapping and thus simplifying Boolean expressions. Figure 4-13 illustrates a Karnaugh map. The four squares (1, 2, 3, 4) represent the four possible combinations of $A$ and $B$ in a two-variable truth table. Square 1 in the Karnaugh map, then, stands for $\overline{A} \cdot \overline{B}$, square 2 for $\overline{A} \cdot B$, and so forth.

Let us map the familiar problem from Fig. 4-12. The original Boolean expression $\overline{A} \cdot B + A \cdot \overline{B} + A \cdot B = Y$ is rewritten in Fig. 4-14(a) for your convenience. Next, a 1 is placed in each square of the Karnaugh map, as shown in Fig. 4-14(b). The filled-in *Karnaugh map* is now ready for *looping*. The looping technique is shown in Fig. 4-15. *Adjacent 1s* are *looped together* in groups of two, four, or eight. Looping continues until all 1s are included inside a loop. Each loop represents a new term in the simplified Boolean expression. Notice that we have two loops in Fig. 4-15. These two loops mean that we shall have two terms ORed together in our new simplified Boolean expression.

Now let us *simplify the Boolean expression* based upon the two loops that are redrawn in Fig. 4-16. First the bottom loop: notice that an $A$ is included along with a $B$ and a $\overline{B}$. The $B$ and $\overline{B}$ terms can be *eliminated* according to the rules of Boolean algebra. This leaves the $A$ term in the bottom loop. Likewise, the vertical loop contains an $A$ and a $\overline{A}$, which are eliminated, leaving only a $B$ term. The leftover $A$ and $B$ terms are then ORed together, giving the simplified Boolean expression $A + B = Y$.

The procedure for simplifying a Boolean expression sounds complicated. Actually, this

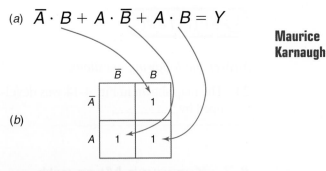

Fig. 4-14 Marking 1s on a Karnaugh map.

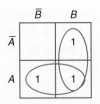

Fig. 4-15 Looping 1s together on a Karnaugh map.

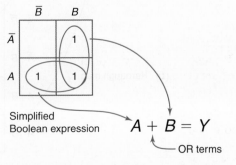

Fig. 4-16 Simplifying a Boolean expression from a Karnaugh map.

procedure is quite easy after some practice. Here is a summary of the six steps:

1. Start with a minterm Boolean expression.
2. Record 1s on a Karnaugh map.
3. Loop adjacent 1s (loops of two, four, or eight squares).
4. Simplify by dropping terms that contain a term and its complement within a loop.
5. OR the remaining terms (one term per loop).
6. Write the simplified minterm Boolean expression.

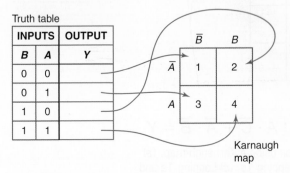

Truth table

| INPUTS | | OUTPUT |
|---|---|---|
| **B** | **A** | **Y** |
| 0 | 0 | |
| 0 | 1 | |
| 1 | 0 | |
| 1 | 1 | |

Fig. 4-13 The meaning of squares in a Karnaugh map.

*Answer the following questions.*

24. The map shown in Fig. 4-14 was developed by _____.

25. List the six steps used in simplifying a Boolean expression using a Karnaugh map.

## 4-7 Karnaugh Maps with Three Variables

**Three-variable Karnaugh map**

Consider the unsimplified Boolean expression $A \cdot \overline{B} \cdot \overline{C} + \overline{A} \cdot \overline{B} \cdot \overline{C} + \overline{A} \cdot \overline{B} \cdot C + A \cdot B \cdot \overline{C} = Y$, as given in Fig. 4-17(a). A *three-variable Karnaugh map* is illustrated in Fig. 4-17(b). Notice the eight possible combinations of $A$, $B$, and $C$, which are represented by the eight squares in the map. Tabulated on the map are four 1s, which represent each of the four terms in the original

Boolean expression. The Karnaugh map with loops is redrawn in Fig. 4-17(c). Adjacent groups of two 1s are looped. The bottom loop contains both a $B$ and a $\overline{B}$. The $B$ and $\overline{B}$ terms are eliminated. The bottom loop still contains the $A$ and $\overline{C}$, giving the $A \cdot \overline{C}$ term. The upper loop contains both a $C$ and a $\overline{C}$. The $C$ and $\overline{C}$ terms are eliminated, leaving the $\overline{A} \cdot \overline{B}$ term. A minterm Boolean expression is formed by adding the OR symbol. The simplified Boolean expression is written in Fig. 4-17(d) as $A \cdot \overline{C} + \overline{A} \cdot \overline{B} = Y$.

**Simplifying a Boolean expression**

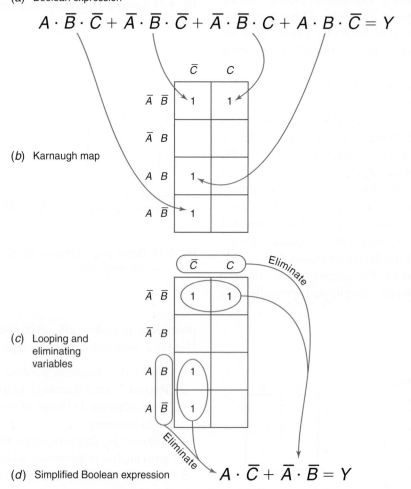

**Fig. 4-17** Simplifying a Boolean expression using a Karnaugh map. (*a*) Unsimplified expression. (*b*) Mapping 1s. (*c*) Looping 1s and eliminating variables. (*d*) Forming simplified minterm expression.

You can see that the simplified Boolean expression in Fig. 4-17 would take fewer electronic parts than the original expression. Remember that the much different looking simplified Boolean expression produces the same truth table as the original Boolean expression.

It is critical that the Karnaugh map be prepared just as the one shown in Fig. 4-17 was. Note that as you progress downward on the left side of the map, only one variable changes for each step. At the top left $\overline{A}\,\overline{B}$ is listed, while directly below is $\overline{A}\,B$ ($\overline{B}$ changed to $B$). Then, progressing downward from $\overline{A}\,B$ to $AB$ the $\overline{A}$ term is changed to $A$. Finally, moving downward from $AB$ to $A\overline{B}$ the $B$ term is changed to $\overline{B}$. The Karnaugh map will not work properly if it is not laid out correctly.

*Answer the following questions.*

26. Simplify the Boolean expression $\overline{A} \cdot \overline{B} \cdot C + \overline{A} \cdot B \cdot C + A \cdot \overline{B} \cdot \overline{C} + A \cdot \overline{B} \cdot C = Y$ by:
    a. Plotting 1s on a three-variable Karnaugh map
    b. Looping groups of two or four 1s
    c. Eliminating variables whose complement appears within the loop(s)
    d. Writing the simplified minterm Boolean expression

27. Simplify the Boolean expression $\overline{A} \cdot B \cdot \overline{C} + \overline{A} \cdot B \cdot C + A \cdot B \cdot \overline{C} + A \cdot B \cdot C = Y$ by:
    a. Plotting 1s on a three-variable Karnaugh map
    b. Looping groups of two or four 1s
    c. Eliminating variables whose complement appears within the loop(s)
    d. Writing the simplified minterm Boolean expression

## 4-8 Karnaugh Maps with Four Variables

The truth table for four variables has 16 ($2^4$) possible combinations. Simplifying a Boolean expression that has four variables sounds complicated, but a Karnaugh map makes the job of simplifying easy.

Consider the Boolean expression $A \cdot \overline{B} \cdot \overline{C} \cdot \overline{D} + \overline{A} \cdot B \cdot \overline{C} \cdot D + \overline{A} \cdot \overline{B} \cdot \overline{C} \cdot D + \overline{A} \cdot \overline{B} \cdot C \cdot D + \overline{A} \cdot B \cdot C \cdot D + A \cdot \overline{B} \cdot \overline{C} \cdot D = Y$, as in Fig. 4-18(a). The *four-variable Karnaugh map* in Fig. 4-18(b) gives the 16 possible combinations of $A$, $B$, $C$, and $D$. These are represented in the 16 squares of the map. Tabulated on the map are six 1s, which represent the six terms in the original Boolean expression. The Karnaugh map is redrawn in Fig. 4-18(c). Adjacent groups of two 1s and four 1s are looped. The bottom loop of two 1s eliminates the $D$ and $\overline{D}$ terms. The bottom loop then produces the $A \cdot \overline{B} \cdot \overline{C}$ term. The upper loop of four 1s eliminates the $C$ and $\overline{C}$ and $B$ and $\overline{B}$ terms. The upper loop then produces the $\overline{A} \cdot D$ term. The $A \cdot \overline{B} \cdot \overline{C}$ and $\overline{A} \cdot D$ terms are then ORed together. The simplified minterm Boolean expression is written in Fig. 4-18(d) as $A \cdot \overline{B} \cdot \overline{C} + \overline{A} \cdot D = Y$.

Observe that the same procedure and rules are used for simplifying Boolean expressions with two, three, or four variables and that larger loops in a Karnaugh map eliminate more variables. You must take care to make sure that the maps look just like the ones in Figs. 4-17 and 4-18.

**Karnaugh Maps with Four Variables**

*Answer the following questions.*

28. Simplify the Boolean expression $\overline{A} \cdot B \cdot \overline{C} \cdot \overline{D} + A \cdot B \cdot \overline{C} \cdot \overline{D} + \overline{A} \cdot B \cdot \overline{C} \cdot D + A \cdot B \cdot \overline{C} \cdot D + A \cdot \overline{B} \cdot C \cdot D + A \cdot \overline{B} \cdot C \cdot \overline{D} = Y$ by:
    a. Plotting 1s on a four-variable Karnaugh map
    b. Looping groups of two or four 1s
    c. Eliminating variables whose complements appear within loops
    d. Writing the simplified minterm Boolean expression

29. Simplify the Boolean expression $\overline{A} \cdot \overline{B} \cdot \overline{C} \cdot \overline{D} + \overline{A} \cdot \overline{B} \cdot \overline{C} \cdot D + \overline{A} \cdot B \cdot \overline{C} \cdot \overline{D} + \overline{A} \cdot B \cdot \overline{C} \cdot D + A \cdot B \cdot C \cdot D + A \cdot B \cdot \overline{C} \cdot \overline{D} = Y$ by:
   a. Plotting 1s on a four-variable Karnaugh map

   b. Looping groups of two or four 1s
   c. Eliminating variables whose complement appears within the loop(s)
   d. Writing the simplified minterm Boolean expression

(a) Boolean expression

(b) Karnaugh map

(c) Eliminating variables by looping

(d) Simplified Boolean expression

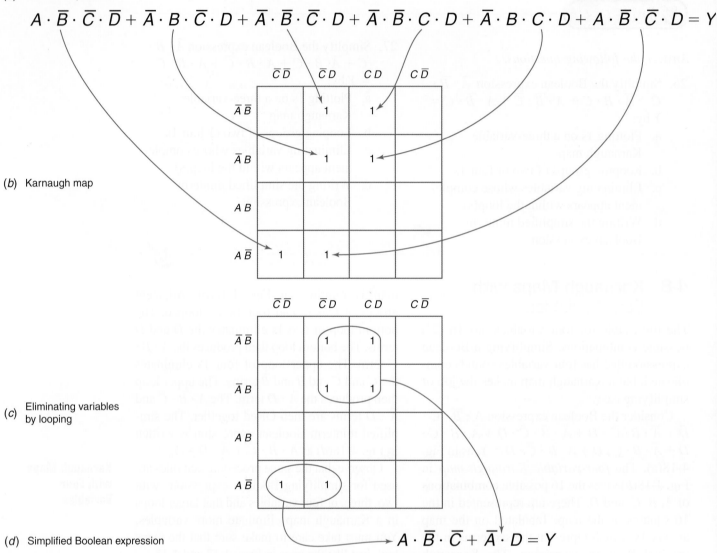

**Fig. 4-18** Simplifying a four-variable Boolean expression using a Karnaugh map.

## 4-9  More Karnaugh Maps

**K map looping variations**

This section presents some sample Karnaugh maps. Notice the unusual looping procedures used on most maps in this section.

Consider the Boolean expression in Fig. 4-19(a). The four terms are shown as four 1s on the Karnaugh map in Fig. 4-19(b). The

correct looping procedure is shown. Notice that the Karnaugh map is considered to be wrapped in a cylinder, with the left side adjacent to the right side. Also notice the elimination of the $A$ and $\overline{A}$ and $C$ and $\overline{C}$ terms. The simplified Boolean expression of $B \cdot \overline{D} = Y$ is shown in Fig. 4-19(c).

(a) Boolean expression

$$A \cdot B \cdot \overline{C} \cdot \overline{D} + \overline{A} \cdot B \cdot \overline{C} \cdot \overline{D} +$$
$$\overline{A} \cdot B \cdot C \cdot \overline{D} + A \cdot B \cdot C \cdot \overline{D} = Y$$

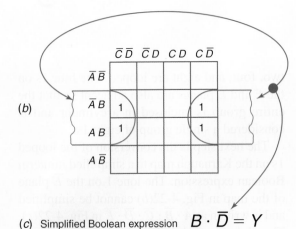

(b)

(c) Simplified Boolean expression $\quad B \cdot \overline{D} = Y$

**Fig. 4-19** Simplifying a Boolean expression by considering the map as a vertical cylinder. In this way, the four 1s can be looped.

Another unusual looping variation is illustrated in Fig. 4-20(a). Notice that, while looping, the top and bottom of the map are adjacent to one another, as if rolled into a cylinder. The simplified Boolean expression for this map is given as $\overline{B} \cdot \overline{C}$ = $Y$ in Fig. 4-20(b). The $A$ and $\overline{A}$ as well as the $D$ and $\overline{D}$ terms have been eliminated in Fig. 4-20.

Figure 4-21(a) shows still another unusual looping pattern. The four corners of the Karnaugh map are considered connected, as if the map were formed into a ball. The four corners are then adjacent and may be formed into one loop as shown. The simplified Boolean expression is $\overline{B} \cdot \overline{D} = Y$, given in Fig. 4-21(b). In this example, the $A$ and $\overline{A}$ as well as the $C$ and $\overline{C}$ terms have been eliminated.

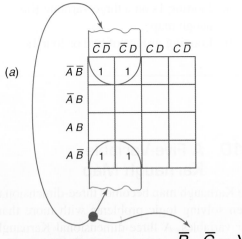

(a)

(b) Simplified Boolean expression $\quad \overline{B} \cdot \overline{C} = Y$

**Fig. 4-20** Simplifying a Boolean expression by considering the map as a horizontal cylinder. In this way, the four 1s can be looped.

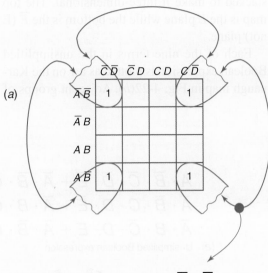

(a)

(b) Simplified Boolean expression $\quad \overline{B} \cdot \overline{D} = Y$

**Fig. 4-21** Simplifying a Boolean expression by thinking of the Karnaugh map as a ball. In this way, the 1s at the four corners can be enclosed in a single loop.

---

**✓ Self-Test**

*Answer the following questions.*

30. Simplify the following Boolean expression $\overline{A} \cdot B \cdot \overline{C} \cdot \overline{D} + \overline{A} \cdot B \cdot \overline{C} \cdot D + \overline{A} \cdot \overline{B} \cdot C \cdot D + \overline{A} \cdot B \cdot C \cdot \overline{D} + A \cdot \overline{B} \cdot \overline{C} \cdot D + A \cdot \overline{B} \cdot C \cdot D = Y$ by:
    a. Plotting 1s on a four-variable Karnaugh map

b. Looping groups of two or four 1s
c. Eliminating variables whose complements appear within loops
d. Writing the simplified minterm Boolean expression

31. Simplify the following Boolean expression $\overline{A} \cdot \overline{B} \cdot \overline{C} + \overline{A} \cdot \overline{B} \cdot C + A \cdot \overline{B} \cdot \overline{C} + A \cdot \overline{B} \cdot C + A \cdot B \cdot C = Y$ by:

a. Plotting 1s on a three-variable Karnaugh map

b. Looping groups of two or four 1s

c. Eliminating variables whose complements appear within loops

d. Writing the simplified minterm Boolean expression

## 4-10 A Five-Variable Karnaugh Map

**Five-variable Karnaugh map**

**Three-dimensional Karnaugh map**

**Looping cylinder**

The Karnaugh map becomes three-dimensional when solving logic problems with more than four variables. A three-dimensional Karnaugh map will be used in this section.

A five-variable unsimplified Boolean expression is given in Fig. 4-22(a). A five-variable Karnaugh map is drawn in Fig. 4-22(b). Notice that it has 2 four-variable Karnaugh maps stacked to make it three-dimensional. The top map is the $E$ plane while the bottom is the $\overline{E}$ (E not) plane.

Each of the nine terms in the unsimplified Boolean expression is plotted as a 1 on the Karnaugh map in Fig. 4-22(b). Adjacent groups of

two, four, and eight are looped. The four 1s on the $E$ and $\overline{E}$ planes are also adjacent so that the entire group is enclosed in a cylinder and is considered a single group of eight 1s.

The next step is the conversion of the looped 1s on the Karnaugh map to a simplified minterm Boolean expression. The lone 1 on the $\overline{E}$ plane of the map in Fig. 4-22(b) cannot be simplified and is written as $A \cdot \overline{B} \cdot \overline{C} \cdot \overline{D} \cdot \overline{E}$ in Fig. 4-22(c). The eight 1s enclosed in the *looping cylinder* can be simplified. The $E$ and $\overline{E}$, the $C$ and $\overline{C}$, and the $B$ and $\overline{B}$ variables are eliminated leaving the term $\overline{A} \cdot D$. The terms $A \cdot \overline{B} \cdot \overline{C} \cdot \overline{D} \cdot \overline{E}$ and $\overline{A} \cdot D$ are ORed yielding the simplified minterm Boolean expression shown in Fig. 4-22(c) as $A \cdot \overline{B} \cdot \overline{C} \cdot \overline{D} \cdot \overline{E} + \overline{A} \cdot D = Y$.

$$A \cdot \overline{B} \cdot \overline{C} \cdot \overline{D} \cdot \overline{E} + \overline{A} \cdot \overline{B} \cdot \overline{C} \cdot D \cdot \overline{E} + \overline{A} \cdot B \cdot \overline{C} \cdot D \cdot \overline{E} +$$
$$\overline{A} \cdot \overline{B} \cdot C \cdot D \cdot \overline{E} + \overline{A} \cdot B \cdot C \cdot D \cdot \overline{E} + \overline{A} \cdot \overline{B} \cdot \overline{C} \cdot D \cdot E +$$
$$\overline{A} \cdot B \cdot \overline{C} \cdot D \cdot E + \overline{A} \cdot \overline{B} \cdot C \cdot D \cdot E + \overline{A} \cdot B \cdot C \cdot D \cdot E = Y$$

(a) Unsimplified Boolean expression

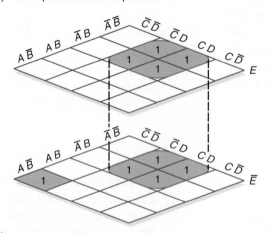

(b) Karnaugh map. Plotting 1s and looping

$$A \cdot \overline{B} \cdot \overline{C} \cdot \overline{D} \cdot \overline{E} + \overline{A} \cdot D = Y$$

(c) Simplified Boolean expression

**Fig. 4-22** Using a five-variable Karnaugh map to simplify a Boolean expression.

*Answer the following questions.*

32. Simplify the Boolean expression $A \cdot \overline{B} \cdot \overline{C} \cdot \overline{D} \cdot \overline{E} + A \cdot \overline{B} \cdot \overline{C} \cdot D \cdot \overline{E} + A \cdot \overline{B} \cdot \overline{C} \cdot \overline{D} \cdot E + A \cdot \overline{B} \cdot \overline{C} \cdot D \cdot E + \overline{A} \cdot B \cdot C \cdot D \cdot E + \overline{A} \cdot \overline{B} \cdot C \cdot D \cdot E = Y$ by:
   a. Plotting 1s on a five-variable Karnaugh map

b. Looping groups of two, four, or eight adjacent 1s

c. Eliminating variables whose complements appear within loops or cylinders

d. Writing the simplified minterm Boolean expression

## 4-11   Using NAND Logic

Earlier you learned that the NAND gate can be used as a universal gate. In this section, you will see how NAND gates are used in wiring combinational logic circuits. NAND gates might be used because they are easy to use and readily available.

Suppose your supervisor gives you the Boolean expression $A \cdot B + A \cdot \overline{C} = Y$, as shown in Fig. 4-23(*a*). You are told to solve this

logic problem at the least cost. You first draw the logic circuit for the Boolean expression shown in Fig. 4-23(*b*), using AND gates, an OR gate, and an inverter. Checking manufacturer's data manuals, you determine that you must use three different ICs to do the job.

Your supervisor suggests that you try *using NAND logic*. You redraw your logic circuit to look like the NAND-NAND logic circuit in Fig. 4-23(*c*). Upon checking a catalog, you find

**Using NAND logic**

(*a*) $A \cdot B + A \cdot \overline{C} = Y$

(*b*)

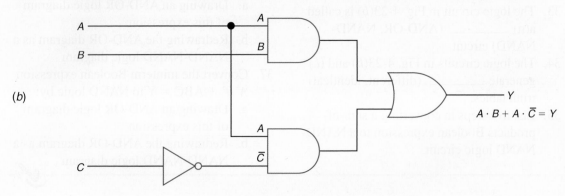

(*c*)

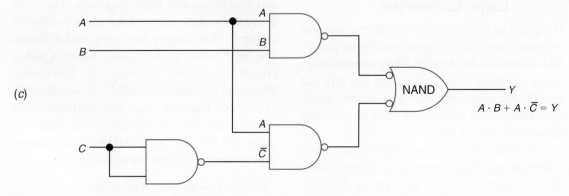

**Fig. 4-23** Using NAND gates in logic circuits. (*a*) Boolean expression. (*b*) AND-OR logic circuit. (*c*) Equivalent NAND-NAND logic circuit.

**AND-OR logic circuit**

**NAND-NAND logic circuit**

you need only one IC that contains the four NAND gates to do the job. Recall that the OR symbol with invert bubbles at the inputs is another symbol for a NAND gate. You finally test the circuit in Fig. 4-23(c) and find that it performs the logic $A \cdot B + A \cdot \overline{C} = Y$. Your supervisor is pleased you have found a circuit that requires only one IC, as compared to the circuit in Fig. 4-23(b), which uses three ICs.

Remembering this trick will help you appreciate *why* NAND gates are used in many logic circuits. If your future job is in digital circuit design, this can be a useful tool for making your final circuit the best for the least cost.

You may have questioned why the NAND gates in Fig. 4-23(c) could be substituted for the AND and OR gates in Fig. 4-23(b). If you look carefully at 4-23(c), you will see two AND symbols feeding into an OR symbol. From previous experience we know that if we invert twice we

have the original logic state. Hence the two invert bubbles in Fig. 4-23(c) between the AND and OR symbols cancel one another. Because the two invert bubbles cancel one another, we end up with two AND gates feeding an OR gate.

In summary, using NAND gates involves these steps:

1. Start with a minterm (sum-of-products) Boolean expression.
2. Draw the AND-OR logic diagram using AND, OR, and NOT symbols.
3. Substitute NAND symbols for each AND and OR symbol, keeping all connections the same.
4. Substitute NAND symbols with all inputs tied together for each inverter.
5. Test the logic circuit containing all NAND gates to determine if it generates the proper truth table.

*Answer the following questions.*

33. The logic circuit in Fig. 4-23(b) is called a(n) _____ (AND-OR, NAND-NAND) circuit.
34. The logic circuits in Fig. 4-23(b) and (c) generate _____(different, identical) truth tables.
35. List five steps in converting a sum-of-products Boolean expression to a NAND-NAND logic circuit.

36. Convert the minterm Boolean expression $\overline{A} \cdot \overline{B} + A \cdot B = Y$ to NAND logic by:
   a. Drawing an AND-OR logic diagram of this expression
   b. Redrawing the AND-OR diagram as a NAND-NAND logic diagram
37. Convert the minterm Boolean expression $A'B' + ABC = Y$ to NAND logic by:
   a. Drawing an AND-OR logic diagram of this expression
   b. Redrawing the AND-OR diagram as a NAND-NAND logic diagram ✔

## 4-12 Computer Simulations— Logic Converter

Designers and engineers have used professional computer simulation software running on powerful workstations for decades. More recently easy-to-use electronic circuit simulators that will run on a PC (personal computer) have become available. Inexpensive educational versions of circuit simulation software are very user-friendly.

Recall that three methods used to describe a combinational logic circuit is by its truth table, Boolean expression, or logic symbol diagram. A useful computer simulation instrument called a logic converter will convert back and

forth between truth table, Boolean expressions, and combinational logic diagrams. The logic converter makes many of the tasks performed earlier in this chapter fast, easy, and accurate. The logic converter instrument, which is part of circuit simulation software by Electronics Workbench® and MultiSIM® is sketched in Fig. 4-24. The tasks that this instrument can perform are listed as buttons on the right side under the title CONVERSIONS. The conversion options are (from top to bottom):

1. Logic diagram to truth table.
2. Truth table to unsimplified Boolean expression.

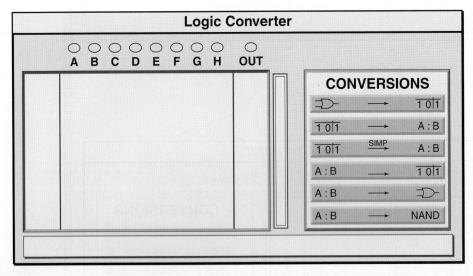

**Logic Converter**

A B C D E F G H OUT

**CONVERSIONS**

**Fig. 4-24** Logic Converter screen (from Electronics Workbench® or MultiSIM®)

3. Truth table to simplified Boolean expression.
4. Boolean expression to truth table.
5. Boolean expression to logic diagram using AND, OR, and NOT gates.
6. Boolean expression to logic diagram using NAND gates only.

You will notice that these are the same subjects covered earlier in the chapter.

An experiment using most of the conversion functions of the Logic Converter is illustrated in Fig. 4-25.

*Step 1* in the experiment is to draw the logic symbol diagram and connect it to the Logic Converter as shown in Fig. 4-25(*a*). You will notice that this is an AND-OR pattern of logic gates that is equivalent to a minterm or sum-of-products Boolean expression.

*Step 2* shows the Logic Converter enlarged on the screen and the top button (logic diagram to truth table) being activated. The results of this conversion are shown in Fig. 4-25(*b*) as an equivalent 4-input truth table.

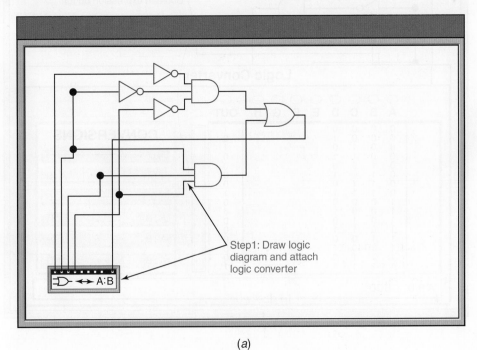

Step1: Draw logic diagram and attach logic converter

(*a*)

**Fig. 4-25** (*a*) -Step 1. Draw logic diagram.

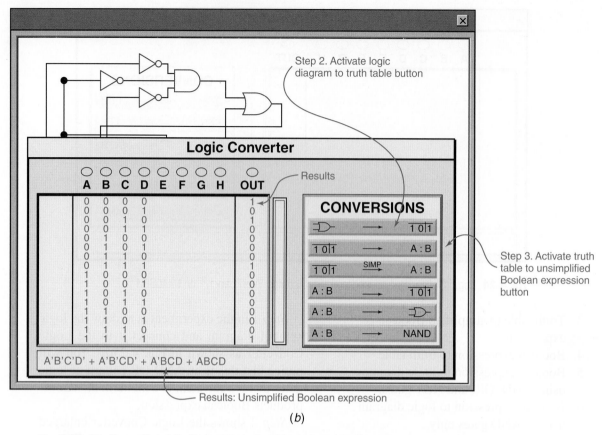

(b)

**Fig. 4-25** (b) -Steps 2 and 3. Generate truth table and unsimplified Boolean expression.

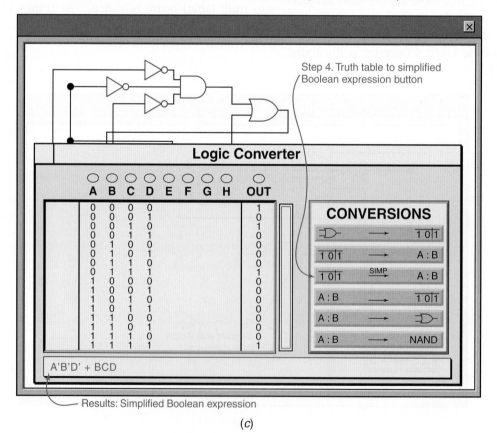

(c)

**Fig. 4-25** (c) -Step 4. Generate simplified Boolean expression.

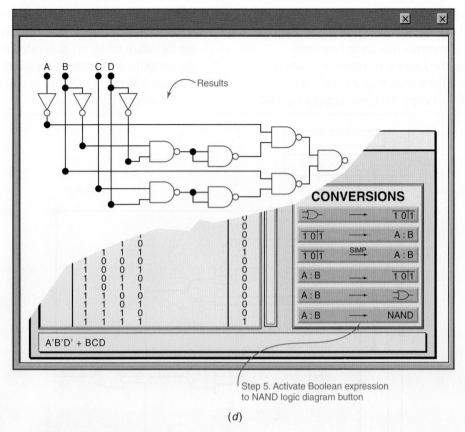

Step 5. Activate Boolean expression
to NAND logic diagram button

(d)

**Fig. 4-25** (d) -Step 5. Generate the NAND logic diagram.

*Step 3* illustrates the second button from the top (truth table to unsimplified Boolean expression) on the Logic Converter being activated. The result of this conversion is shown near the bottom of the screen in Fig. 4-25(b). The unsimplified Boolean expression is shown in its "keyboard" version as $A'B'C'D' + A'B'CD' + A'BCD + ABCD$.

*Step 4* illustrates the third button from the top (truth table to simplified Boolean expression) on the Logic Converter being activated. The result of this conversion is shown near the bottom of the screen in Fig. 4-25(c). The simplified Boolean expression is shown in its keyboard version as $A'B'D' + BCD$.

*Step 5* illustrates the bottom button (Boolean expression to NAND logic gate diagram) on the Logic Converter being activated. The result of this conversion is shown as a NAND-NAND logic circuit near the upper left of the screen in Fig. 4-25(d).

In summary, modern computer simulations such as the Logic Converter instrument we observed makes the task of converting back and forth between representations of logic functions easier, more accurate, and less time-consuming. Computer software and simulations are commonly used in the development stage of digital circuitry.

**Self-Test**

*Answer the following questions with the aid of the Logic Converter from Electronics Workbench® or MultiSIM®.*

38. Using the Logic Converter instrument from Electronics Workbench® or Multi-SIM® (a) draw the AND-OR logic

diagram in Fig. 4-26, (b) copy its 4-variable truth table, (c) generate its simplified minterm Boolean expression, and

39. Using the Logic Converter instrument from Electronics Workbench® or Multi-SIM® (a) enter the minterm Boolean expression $AC'D + BD'$, (b) redraw the

4-variable truth table for this expression, and (c) copy the AND-OR logic diagram that represents this logic function.

40. Using the Logic Converter instrument from Electronics Workbench® or Multi-SIM® (a) copy the truth table in Fig. 4-27 into the Logic Converter, (b) generate and write the unsimplified Boolean expression for this truth table, (c) generate and write the simplified Boolean expression, and (d) sketch the AND-OR logic diagram for the simplified expression.

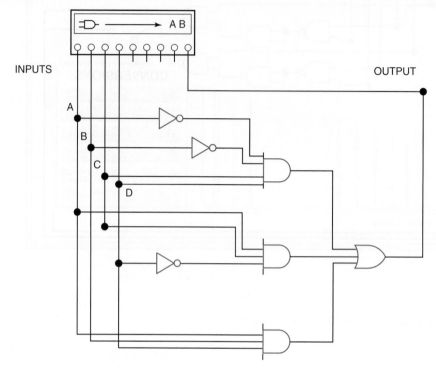

INPUTS

OUTPUT

A

B

C

D

**Fig. 4-26** Logic converter problem.

| INPUTS | | | | OUTPUT |
|---|---|---|---|---|
| A | B | C | D | Y |
| 0 | 0 | 0 | 0 | 1 |
| 0 | 0 | 0 | 1 | 0 |
| 0 | 0 | 1 | 0 | 1 |
| 0 | 0 | 1 | 1 | 0 |
| 0 | 1 | 0 | 0 | 0 |
| 0 | 1 | 0 | 1 | 0 |
| 0 | 1 | 1 | 0 | 0 |
| 0 | 1 | 1 | 1 | 0 |
| 1 | 0 | 0 | 0 | 0 |
| 1 | 0 | 0 | 1 | 0 |
| 1 | 0 | 1 | 0 | 0 |
| 1 | 0 | 1 | 1 | 0 |
| 1 | 1 | 0 | 0 | 0 |
| 1 | 1 | 0 | 1 | 0 |
| 1 | 1 | 1 | 0 | 1 |
| 1 | 1 | 1 | 1 | 1 |

**1-of-8 data selector**
**Rotary switch**

**Data Selector**

**Fig. 4-27** Truth table.

## 4-13 Solving Logic Problems—Data Selectors

Manufacturers of ICs have simplified the job of solving simple combinational logic problems by producing *data selectors*. A data selector is often a *one-package solution* to a complicated logic problem. The data selector actually contains a rather large number of gates packaged inside a single IC.

A *1-of-8 data selector* is illustrated in Fig. 4-28. Notice the eight *data inputs* numbered from 0 to 7 on the left. Also notice the three *data selector inputs* labeled *A*, *B*, and *C* at the bottom of the data selector. The output of the data selector is labeled *W*.

The basic job the data selector performs is transferring data from a *given* data input (0 to 7) to the output (*W*). Which data input is selected is determined by which binary number you place on the data selector inputs at the bottom (see Fig. 4-28). The data selector in Fig. 4-28 functions in the same manner as a

rotary switch. Figure 4-29 shows the data at input 3 being transferred to the output by the *rotary switch* contacts. In like manner the data from data input 3 in Fig. 4-28 is being transferred to output *W* of the data selector. In the rotary switch you must mechanically change the switch position to transfer data from another input. In the 1-of-8 data selector in Fig. 4-28, you need only change the binary input at the data selector inputs to transfer data from another data input to the output. Remember that the data selector operates somewhat as a rotary switch in transferring

logical 0s or 1s from a given input to the single output.

Now you will learn how data selectors can be used to solve logic problems. Consider the *simplified* Boolean expression shown in Fig. 4-30(*a*) on page 106. For your convenience a logic circuit for this complicated Boolean expression is drawn in Fig. 4-30(*b*). Using standard ICs, we probably would have to use from six to nine IC packages to solve this problem. This would be quite expensive because of the cost of the ICs and PC board space.

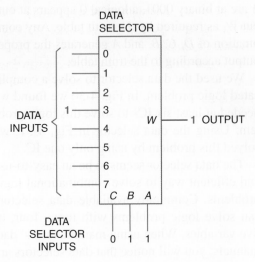

**Fig. 4-28** Logic symbol for a 1-of-8 data selector.

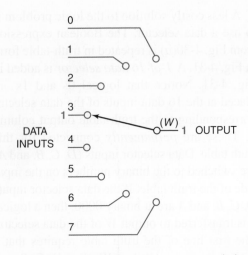

**Fig. 4-29** Single-pole eight-position rotary switch works as a data selector.

(a) Simplified Boolean expression

$$A \cdot B \cdot C \cdot D + \overline{A} \cdot \overline{B} \cdot \overline{C} \cdot \overline{D} + A \cdot \overline{B} \cdot \overline{C} \cdot D + A \cdot B \cdot \overline{C} \cdot \overline{D} +$$
$$\overline{A} \cdot B \cdot C \cdot \overline{D} + \overline{A} \cdot B \cdot \overline{C} \cdot D + \overline{A} \cdot \overline{B} \cdot C \cdot D = Y$$

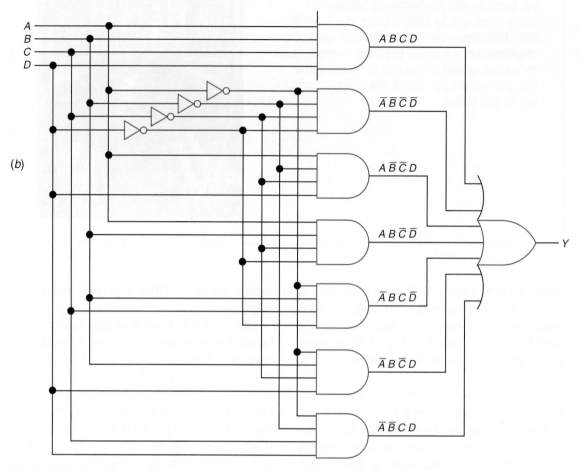

**Fig. 4-30** (a) Simplified Boolean expression. (b) Logic circuit for Boolean expression.

**1-of-16 data selector**

**Multiplexers**

A less costly solution to the logic problem is to use a data selector. The Boolean expression from Fig. 4-30(a) is repeated in truth-table form in Fig. 4-31. A *1-of-16 data selector* is added in Fig. 4-31. Notice that logical 0s and 1s are placed at the 16 data inputs of the data selector corresponding to the truth-table output column *Y*. These are *permanently* connected for this truth table. Data selector inputs (*D, C, B,* and *A*) are switched to the binary numbers on the input side of the truth table. If the data selector inputs *D, C, B,* and *A* are at binary 0000, then a logical 1 is transferred to output *W* of the data selector. The first line of the truth table requires that a logical 1 appear at output *W* when *D, C, B,* and *A* are all 0s. If data selector inputs *D, C, B,* and

*A* are at binary 0001, a logical 0 appears at output *W,* as required by the truth table. Any combination of *D, C, B,* and *A* generates the proper output according to the truth table.

We used the data selector to solve a complicated logic problem. In Fig. 4-30 we found we needed at least six ICs to solve this logic problem. Using the data selector in Fig. 4-31, we solved this problem by using only one IC.

The data selector seems to be an easy-to-use and efficient way to solve combinational logic problems. Commonly available data selectors can solve logic problems with three, four, or five variables. When using manufacturers' data manuals, you will notice that data selectors are also called *multiplexers.*

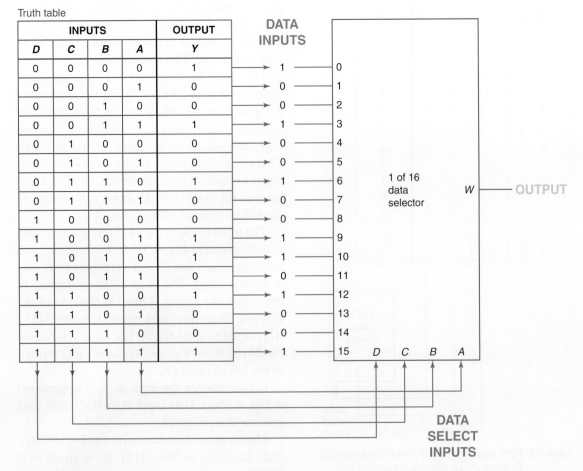

(a)                                                    (b)

Truth table

| INPUTS | | | | OUTPUT |
|---|---|---|---|---|
| D | C | B | A | Y |
| 0 | 0 | 0 | 0 | 1 |
| 0 | 0 | 0 | 1 | 0 |
| 0 | 0 | 1 | 0 | 0 |
| 0 | 0 | 1 | 1 | 1 |
| 0 | 1 | 0 | 0 | 0 |
| 0 | 1 | 0 | 1 | 0 |
| 0 | 1 | 1 | 0 | 1 |
| 0 | 1 | 1 | 1 | 0 |
| 1 | 0 | 0 | 0 | 0 |
| 1 | 0 | 0 | 1 | 1 |
| 1 | 0 | 1 | 0 | 1 |
| 1 | 0 | 1 | 1 | 0 |
| 1 | 1 | 0 | 0 | 1 |
| 1 | 1 | 0 | 1 | 0 |
| 1 | 1 | 1 | 0 | 0 |
| 1 | 1 | 1 | 1 | 1 |

**Fig. 4-31** Solving logic problem with a data selector IC.

---

**Self-Test**

*Supply the missing word, letter, or number in each statement.*

41. Figure 4-28 illustrates the logic symbol for a 1-of-8 _____.

42. Refer to Fig. 4-28. If all data select inputs are HIGH, data at input _____ (number) is selected and transferred to output _____ (letter) of the data selector.

43. The action of a data selector is many times compared to that of a mechanical _____ switch.

44. Refer to Fig. 4-31. If all data select inputs are HIGH, data from input _____ (number) will be transferred to output W. Under these conditions, output W will be _____ (HIGH, LOW).

45. Data selector ICs might also be listed in catalogs as _____ (counters, multiplexers) suggesting another use of these devices.

---

## 4-14 More Data Selector Problems

The previous section used a 1-of-16 data selector to solve a four-variable logic problem. A similar logic problem can be solved using a less expensive, 1-of-8 data selector. This is done by using what is sometimes called the *folding technique*.

Consider the four-variable truth table shown in Fig. 4-32. Note that the pattern of inputs

**Solving logic problem with a data selector**

**Folding technique**

| Line Number | INPUTS | | | | OUTPUT |
|---|---|---|---|---|---|
| | D | C | B | A | Y |
| 0 | 0 | 0 | 0 | 0 | 0 |
| 1 | 0 | 0 | 0 | 1 | 1 |
| 2 | 0 | 0 | 1 | 0 | 0 |
| 3 | 0 | 0 | 1 | 1 | 1 |
| 4 | 0 | 1 | 0 | 0 | 0 |
| 5 | 0 | 1 | 0 | 1 | 0 |
| 6 | 0 | 1 | 1 | 0 | 1 |
| 7 | 0 | 1 | 1 | 1 | 0 |
| 8 | 1 | 0 | 0 | 0 | 0 |
| 9 | 1 | 0 | 0 | 1 | 1 |
| 10 | 1 | 0 | 1 | 0 | 1 |
| 11 | 1 | 0 | 1 | 1 | 0 |
| 12 | 1 | 1 | 0 | 0 | 1 |
| 13 | 1 | 1 | 0 | 1 | 0 |
| 14 | 1 | 1 | 1 | 0 | 0 |
| 15 | 1 | 1 | 1 | 1 | 0 |

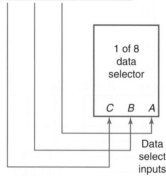

**Fig. 4-32** First step in using a 1-of-8 data selector to solve a four-variable logic problem.

$C$, $B$, and $A$ is the same in lines 0 through 7 as in lines 8 through 15. These areas are circled on the truth table in Fig. 4-32. To solve this logic problem using a 1-of-8 data selector, inputs $C$, $B$, and $A$ are connected to the data select inputs of the unit. This is shown in the lower part of Fig. 4-32.

The eight data inputs ($D_0$ to $D_7$) shown in Fig. 4-33($i$), must now be determined one by one. The input to $D_0$ on the 74151 1-of-8 data selector IC is determined in Fig. 4-33($a$). The truth table from Fig. 4-32 is folded over to compare lines 0 and 8. Inputs $C$, $B$, and $A$ (which connect to the 74151 IC's data select or inputs) are each 000. If input $D$ is 0 or 1, output $Y$ is always 0 according to Fig. 4-33($a$). Therefore, a logical 0 (GND) is applied to the $D_0$ input to the 74151 1-of-8 data selector IC. This is shown in Fig. 4-33($i$).

The input to $D_1$ of the 74151 IC is determined in Fig. 4-33($b$). The folding technique is used to compare lines 1 and 9 of the truth table. Inputs $C$, $B$, and $A$ must be the same. Whether input $D$ is 0 or 1, output $Y$ is always 1. Therefore, a logical 1 (+5 V) is applied to the $D_1$ input to the 74151 data selector IC. This is shown in Fig. 4-33($i$).

The input to $D_2$ of the 74151 IC is determined in Fig. 4-33($c$). Folding the truth table compares lines 2 and 10. Inputs $C$, $B$, and $A$ are the same. The outputs are different. In each case, the output is the same as the $D$ input. Therefore, data input $D_2$ to the 74151 IC is equal to input $D$ from the truth table. The logic symbol for the 74151 IC shows a $D$ written to the left of input $D_2$ [Fig. 4-33($i$).]

The input to $D_3$ of the 74151 data selector IC is determined in Fig. 4-33($d$). Folding the truth table compares lines 3 and 11. Inputs $C$, $B$, and $A$ are the same. The outputs are different. In each case, the output is the complement of the $D$ input. Therefore, data input $D_3$ to the 74151 IC is equal to a not $D$ ($\overline{D}$). The logic symbol for the 74151 IC in Fig. 4-33($i$) shows a $\overline{D}$ written to the left of input $D_3$.

In like manner, the input to $D_4$ is determined in Fig. 4-33($e$). Data input $D_4$ to the 1-of-8 data selector is equal to $D$.

The input to $D_5$ is determined in Fig. 4-33($f$). Data input $D_5$ to the 74151 IC is equal to 0 (GND).

The input to $D_6$ is determined in Fig. 4-33($g$). Data input $D_6$ to the 1-of-8 data selector is equal to $\overline{D}$ (not $D$).

Finally, the input to $D_7$ is determined in Fig. 4-33($h$). Data input $D_7$ of the 74151 data selector IC is equal to 0 (GND).

Note in Fig. 4-33($i$) that data inputs $D_0$, $D_5$, and $D_7$ of the 74151 IC are permanently grounded. Data input $D_1$ is permanently connected to +5 V. Data inputs $D_2$ and $D_4$ are connected directly to input $D$ from the truth table. Data inputs $D_3$ and $D_6$ are connected through an inverter to the complement of input $D$. The enable, or strobe, input to the 74151 1-of-8 data selector must be held LOW (at logical 0) for the unit to operate. The small bubble on the logic symbol in Fig. 4-33($i$) means that the enable input is an active LOW input.

The data selector (multiplexer) has been used as a *universal logic element* in the last two sections. It is a simple, low-cost solution to many logic problems with from three to five input variables.

**Universal logic element**

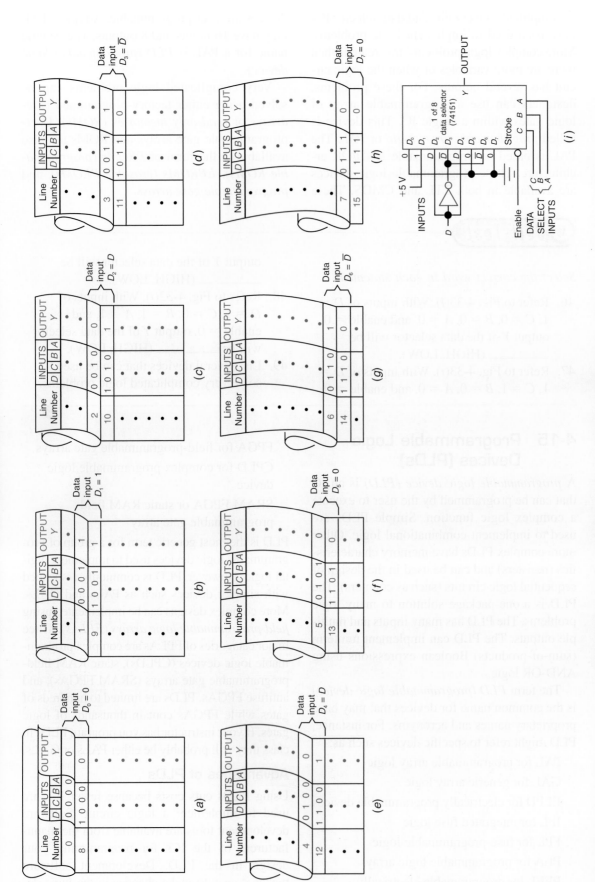

**Fig. 4-33** Second step in using a 1-of-8 data selector to solve a four-variable logic problem using the "folding" technique. (a) Determining data to be placed at input $D_0$. (b) Determining data to be placed at input $D_1$. (c) Determining data to be placed at input $D_2$. (d) Determining data to be placed at input $D_3$. (e) Determining data to be placed at input $D_4$. (f) Determining data to be placed at input $D_5$. (g) Determining data to be placed at input $D_6$. (h) Determining data to be placed at input $D_7$. (i) Solution to four-variable logic problem posed in truth table.

Simplified gate circuits and data selector ICs have been used to implement logic problems. More complex logic problems are created when there are more variables or when the logic circuit has several outputs. For these problems, designers can use a programmable array of logic gates within a single IC. This device is called *programmable array logic* or *PAL*. The PAL is based on programmable AND/OR architecture. These programmable logic devices are available in both TTL and CMOS. These devices are user-programmable. A typical PAL may have 16 inputs and 8 outputs. The generic name for a PAL is *PLD* (*programmable logic device*).

Very complicated logic problems can be solved using either factory-programmed *gate-arrays* or *read-only memories (ROMs)*. User-programmable gate arrays and ROMs are also available in the form of *PROMs (programmable ROMs), EPROMs (erasable PROMs)*, and *programmable gate arrays*.

**Programmable logic device (PLD)**

![Self-Test]

*Select the correct word in each statement.*

46. Refer to Fig. 4-33(*i*). With inputs of $D = 1$, $C = 0$, $B = 0$, $A = 0$, and enable $= 0$, output $Y$ of the data selector will be _____ (HIGH, LOW).

47. Refer to Fig. 4-33(*i*). With inputs of $D = 1$, $C = 1$, $B = 0$, $A = 0$, and enable $= 0$, output $Y$ of the data selector will be _____ (HIGH, LOW).

48. Refer to Fig. 4-32(*i*). With inputs of $D = 1$, $C = 0$, $B = 1$, $A = 1$, and enable $= 0$, output $Y$ of the data selector will be _____ (HIGH, LOW).

49. List several devices that can be used to solve very complicated logic problems.

## 4-15 Programmable Logic Devices (PLDs)

A *programmable logic device (PLD)* is an IC that can be programmed by the user to execute a complex logic function. Simple PLDs are used to implement combinational logic. Other more complex PLDs have memory characteristics (registers) and can be used in the design of sequential logic circuits (such as counters). The PLD is a one-package solution to many logic problems. The PLD has many inputs and multiple outputs. The PLD can implement minterm (sum-of-products) Boolean expressions using AND-OR logic.

The term *PLD (programmable logic device)* is the common name for devices that may have proprietary names and acronyms. For instance, PLD might refer to specific devices such as:

PAL for programmable array logic

GAL for generic array logic

ELPD for electrically programmable devices

IFL for integrated fuse logic

FPL for fuse-programmable logic

PLA for programmable logic arrays

PEEL for programmable electrically erasable logic

FPGA for field-programmable gate arrays

CPLD for complex programmable logic device

SRAM FPGA or static RAM field-programmable gate array

PLD is the most generic term for a group of programmable logic devices used to implement digital logic. However, PLD is commonly associated with simpler devices such as PALs and GALs. More complex designs can be implemented using *field-programmable logic arrays (FPLAs)*. Three major catagories of FPLAs are complex programmable logic devices (CPLDs), static RAM field-programmable gate arrays (SRAM FPGAs), and antifuse FPGAs. PLDs are limited to hundreds of gates while FPGAs contain thousands of logic gates. If your instructor has you program PLDs in class they will probably be either PALs or GALs.

### Advantages of PLDs

Using PLDs cuts costs because fewer ICs are used to implement a logic circuit. Software development tools are available from the manufacturers of the ICs for programming your design in the PLD. Development software makes it easy to make changes in the logic design. Other advantages of PLDs are the lower cost of inventory because they are somewhat generic logic devices. Upgrades and modifications are more easily made in prototypes and products using programmable logic devices. The PLD is a very reliable component. Proprietary logic designs can be more easily hidden from competitors. PLDs are inexpensive because they are available from many sources and are manufactured in large quantities. For instance, a recent catalog lists the cost of a simple PAL at less than one dollar even when ordered in small quantities.

## Programming PLDs

The PLD is commonly programmed in the local development lab, school lab, or shop and not at the manufacturer. Development software is available from several manufacturers for PLDs. Some common development software used by schools might include:

- ABEL software from Lattice Semiconductor Corp.
- CUPL software from Logical Devices, Inc.

Many manufacturers allow for downloading a version of their development software for temporary use by students, engineers and designers.

A common system used in schools and small labs for programming PLDs is sketched in Fig. 4-34. The system includes a PC (personal computer), development software, a IC programmer (IC burner), and a cable to connect the IC programmer to the PC (serial cable shown).

The general steps in programming are shown in Fig. 4-34. Step 1 includes loading the development software. Step 2 would include entering the logic design as required by the development software and informing the software which device (for instance a PAL10H8 IC) you will use to implement the design. Development software will allow you to describe your logic circuit in at least three ways. They are by (1) Boolean expression (sum-of-products form), (2) truth table, or (3) logic diagram. Describing your logic circuit can also take other forms. Step 3 would have you compiling and simulating your design to check for proper operation. Step 4 includes

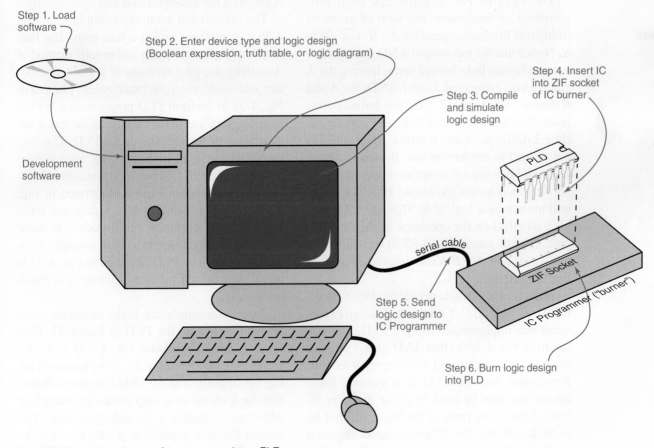

**Fig. 4-34** Typical equipment for programming a PLD.

Step 1. Load software

Development software

Step 2. Enter device type and logic design (Boolean expression, truth table, or logic diagram)

Step 3. Compile and simulate logic design

Step 4. Insert IC into ZIF socket of IC burner

PLD

ZIF Socket

IC Programmer ("burner")

serial cable

Step 5. Send logic design to IC Programmer

Step 6. Burn logic design into PLD

placing your PLD integrated circuit in the ZIF (zero insertion force) IC socket. Step 5 shows sending your design via the serial output cable to the IC Programmer. Step 6 would be to "burn" or program the PLD IC. In summary, Fig. 4-34 shows both the hardware and the general procedure for programming a PLD.

## What's Inside the PLD

A simplified version of a programmable logic device is detailed in Fig. 4-35($a$). Notice that it has the look of the AND-OR circuits you worked with earlier to implement a sum-of-product (minterm) Boolean expression. This simple logic circuit has two inputs and one output while a typical commercial PLD may have 12 inputs and 10 outputs, as is the case for the PAL12H10 IC. The simplified PLD sketched in Fig. 4-35($a$) has intact (not blown) fuses used for programming the AND gates. The OR gate is not programmable in the device. The PLD in Fig. 4-35($a$) is shown as it comes from the manufacturer- with all fuses intact (not blown). The PLD in Fig. 4-35($a$) needs to be programmed by burning open selected fuses.

**Fuse map**

The PLD in Fig. 4-35($b$) has been programmed to implement the sum-of-products (minterm) Boolean expression $A \cdot \overline{B} + \overline{A} \cdot B = Y$. Notice that the top 4-input AND gate (gate 1) has two fusible links burned open, leaving the $A$ and $\overline{B}$ terms connected. Gate 1 ANDs the $A$ and $\overline{B}$ terms. AND gate 2 has two burned-open fuses, leaving the $\overline{A}$ and $B$ inputs connected. Gate 2 ANDs the $\overline{A}$ and $B$ terms. AND gate 3 is not needed to implement this Boolean expression. All fuses are left intact as shown in Fig. 4-35($b$), which means the output of AND gate 3 will always be a logical 0. This logical 0 will have no effect on the operation of the final OR gate. The OR gate in Fig. 4-35($b$) logically ORs the $A \cdot \overline{B}$ and $\overline{A} \cdot B$ terms implementing the Boolean expression.

In the simple example detailed in Fig. 4-34($b$), the $A \cdot \overline{B} + \overline{A} \cdot B = Y$ minterm Boolean expression was implemented using a PLD. You can see from Fig. 4-35($b$) that AND gate 3 was not used in this circuit and this seems wasteful. Remember that the PLD is a generic logic device that can be used to solve many problems. Sometimes parts of the logic will not be used. Recall that the IC programmer depicted in Fig. 4-34 "burns open" selected fuses. The

IC programmer instrument is therefore commonly called the "PLD burner."

The sample problem in Fig. 4-35($b$) would not be solved using a PLD. Designers and engineers look to the most cost-effective method to implement electronic designs. The Boolean expression $A \cdot \overline{B} + \overline{A} \cdot B = Y$ describes the 2-input XOR function which might be implemented cheaper using a dedicated 2-input XOR gate IC.

An abbreviated notation system used with PLDs is illustrated in Fig. 4-36. Note that all AND and OR gates appear to have only one input, while in reality each AND gate has four inputs, and the OR gate has three inputs. The PLD represented in Fig. 4-36($a$) has all fuses intact before programming. The X mark at an intersection of lines represents an intact fuse when using the abbreviated notation system.

The Boolean expression $A \cdot \overline{B} + \overline{A} \cdot B = Y$ was implemented earlier in Fig. 4-35($b$). The same Boolean expression is implemented in Fig. 4-36($b$) but using the abbreviated notation system to describe the programming of the PLD. Notice that an X at an intersection of lines means an intact (not blown) fuse while no X means a burned-open fuse (no connection).

The abbreviated notation system used in Fig. 4-36 is sometimes called a *fuse map*. The fuse map is a graphic or "paper and pencil" method of describing the programming of a PLD. In practice you would use a computer system like that in Fig. 4-34 to perform PLD programming but the fuse maps are useful for visualizing the inside organization or architecture of the PLD. The fuse map also assists in understanding what is happening inside a PLD when it is programmed.

A more complex PLD is illustrated in Fig. 4-37. This PLD features four inputs and three outputs. It is common for decoders to have many inputs and outputs as they translate from code to code. The PLD sketched in Fig. 4-37 is not a commercial product because it is much too simple.

Three combinational logic problems have been solved using the PLD in Fig. 4-37. First the Boolean expression $\overline{A} \cdot B \cdot \overline{C} \cdot D + A \cdot B \cdot C \cdot D + \overline{A} \cdot B \cdot C \cdot D = Y_1$ is implemented using the upper group of AND-OR gates. Recall that the $X$ on the fuse map means an intact fuse while no $X$ means a burned-open fuse. The second Boolean expression $A \cdot B \cdot C \cdot D + \overline{A} \cdot B \cdot C \cdot \overline{D} = Y_2$ is implemented using the middle

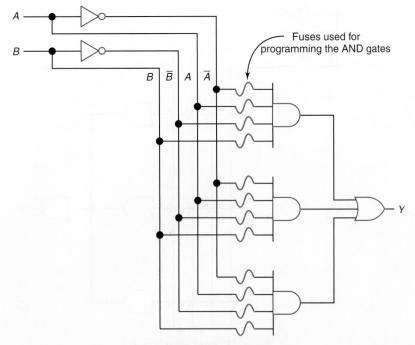

Fuses used for programming the AND gates

Fuses intact (as from manufacturer)

*(a)*

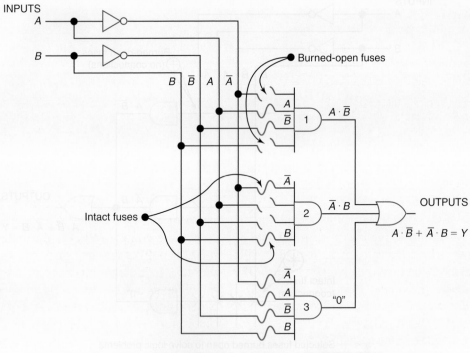

INPUTS

Burned-open fuses

$A \cdot \overline{B}$

1

$\overline{A} \cdot B$

2

OUTPUTS

$A \cdot \overline{B} + \overline{A} \cdot B = Y$

Intact fuses

"0"

3

Selected fuses burned open to solve logic problems

*(b)*

**Fig. 4-35** Simplified PLD.

group of AND-OR gates. Note that the bottom AND gate in the middle group is not needed. Therefore it has all eight fuses intact, which means it generates a logical 0 having no effect on the output of the OR gate. The third Boolean expression $\overline{A} \cdot \overline{B} \cdot C \cdot D + A \cdot B \cdot C \cdot \overline{D} + \overline{A} \cdot B \cdot \overline{C} \cdot \overline{D} = Y_3$ is implemented using the bottom group of AND-OR gates.

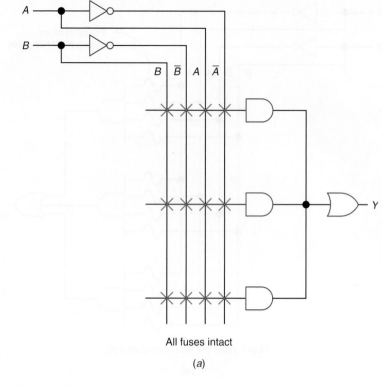

All fuses intact

(a)

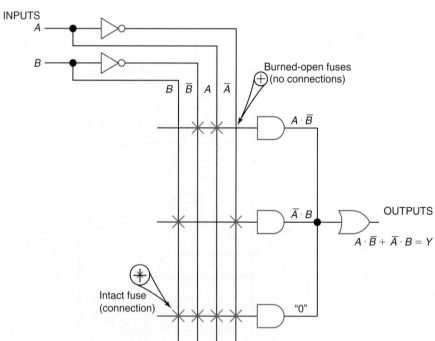

Selected fuses burned open to solve logic problems

(b)

**Fig. 4-36** Notation system used with PLDs.

**Field-programmable logic array (FPLA)**

A more complicated PLD architecture is suggested in Fig. 4-38. This PLD provides both programmable AND and OR arrays. The programmable logic devices studied earlier contained only programmable AND gates. This type of device is sometimes called a *field-programmable logic array (FPLA)*. Notice that all the links are intact (not burned) in this simplified example.

INPUTS

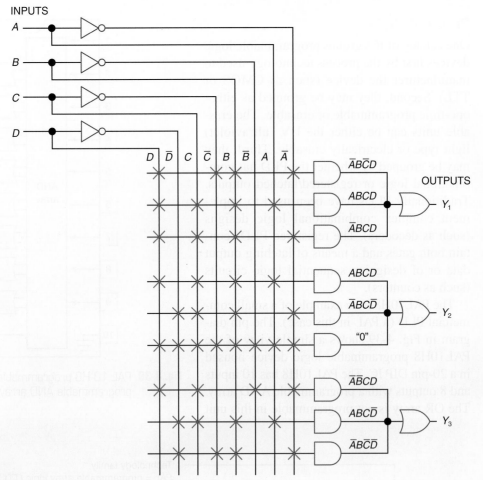

Fig. 4-37 Programming a PLD using a fuse map.

INPUTS

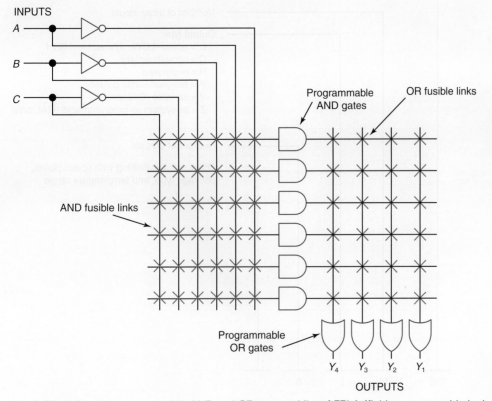

Fig. 4-38 PLD with programmable AND and OR arrays. Like of FPLA (field-programmable logic array).

## Practical PLDs

One catalog of ICs groups programmable logic devices first by the process technology used to manufacturer the device (such as CMOS or TTL). Second, they may be grouped as either one-time programmable or erasable. The erasable units can be either the UV (ultraviolet) light type or electrically erasable. Third, they may be grouped by whether the PLD has combinational logic or registered/latched outputs. Traditionally PLDs have been used to implement complex combinational logic designs (such as decoders). The registered PLDs contain both gates and a means of latching output data or of designing sequential logic circuits (such as counters).

The PAL10H8 is an example of a small commercial PLD (a PAL in this case). The pin diagram in Fig. 4-39 shows a simple view of the PAL10H8 programmable logic device housed in a 20-pin DIP IC. The PAL10H8 has 10 inputs and 8 outputs with a programmable AND array. The OR array is not programmable in this unit

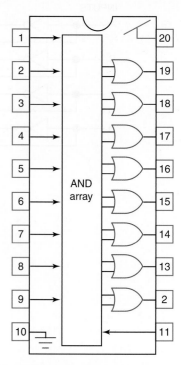

**Fig. 4-39** PAL 10 H8 programmable logic IC with a programmable AND array.

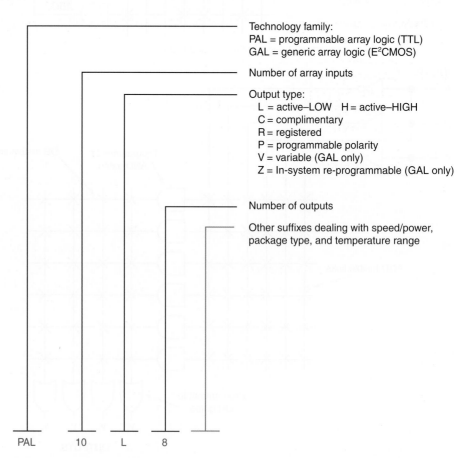

Technology family:
PAL = programmable array logic (TTL)
GAL = generic array logic (E$^2$CMOS)

Number of array inputs

Output type:
  L = active–LOW   H = active–HIGH
  C = complimentary
  R = registered
  P = programmable polarity
  V = variable (GAL only)
  Z = In-system re-programmable (GAL only)

Number of outputs

Other suffixes dealing with speed/power, package type, and temperature range

PAL      10      L      8

**Fig. 4-40** Decoding a PAL part number.

and has active-HIGH outputs. The PAL10H8 is also available in other IC packages.

If your school has programming equipment, you will probably use lowcost PALs with fusible links. PALs can be programmed only once. Your instructor may have you use slightly more expensive GALs which look like a PAL on the inside except the "fuses" are electronic cells (using $E^2CMOS$ technology) which can be turned on or off during programming. The GAL is useful because it can be erased and re-programmed.

PAL/GAL IC part identification guidelines are illustrated in Fig. 4-40. The first letters on the left identify the technology family used to manufacture the PLD. The older PAL uses the TTL technology. The newer GAL uses CMOS technology. Moving right, the next number (10 in this example) is the number of inputs to the AND array. Moving right, the next letter (L in this example) identifies the type of output (in this example the output is an active-LOW). Moving right, the next number (8 in this exam-

ple) is the number of outputs. Any trailing letters deal with speed/power, packaging, and the temperature range of the PLD. Some manufacturers may add to this list. Many PLDs allow output pins on the IC to be configured as either an input or output.

As an example, suppose a 20-pin DIP IC had PAL14H4 printed on its top. According to the guidelines from Fig. 4-40, this would be a PAL using TTL technology with 14 inputs and 4 outputs. It would have active-HIGH outputs. Recall that a PAL can be programmed only once. Data sheets must be looked at to find out more information on the IC.

A second example, suppose a 20-pin DIP IC has GAL16V8 printed on its top. According to the guidelines from Fig. 4-40, this would be a GAL using $E^2CMOS$ technology with up to 16 inputs and 8 outputs. The outputs can be configured as either inputs or outputs and recall that GAL technology allows the $E^2CMOS$ cells to be reprogrammed.

## ✔ Self-Test

*Answer the following questions.*

50. In electronics technology, the acronym PLD stands for _____.

51. In electronics technology, the acronym PAL stands for _____.

52. In electronics technology, the acronym GAL stands for _____.

53. In electronics technology, the acronym FPLA stands for _____.

54. Programmable logic devices from the PAL family are commonly used for implementing _____ (combinational, fuzzy) logic.

55. FPLAs, PALs, and GALs are commonly programmed _____ (by the manufacturer, by the local user).

56. Programming simple PALs consists of _____ (burning open selected fusible links, turning $E^2CMOS$ in the array either on or off).

57. The equipment needed to program PLDs includes a PC, development software, the correct PLD IC, a serial-cable, and an

instrument called a _____ (PLD burner or programmer, logic analyzer).

58. The development software used for programming a PLD allows your logic design to be entered in at least three forms. These include a truth table, a logic symbol diagram, or a _____ (Boolean expression, Winchester table).

59. PLDs such as the PAL you studied are organized to implement _____ (maxterm, sum-of-products) Boolean expressions using and AND-OR pattern of logic gates.

60. A programmable logic device IC with a number printed on top of PAL12H6 would be based on TTL technology, have _____ (6, 12) inputs, have _____ (6, 12) outputs with the outputs being _____ (active-HIGH, active-LOW).

61. A programmable logic device IC with a number printed on the top of GAL16V8 would be based on _____ (CMOS, TTL) technology.

62. A programmable logic device IC with a number printed on the top of GAL16V8 would have a maximum of _____ (8, 16) outputs.

63. Refer to Fig. 4-39. The PAL10H8 IC has a programmable _____ (AND, OR)

array and can implement sum-of-products Boolean expressions.

64. The PLD shown in fuse map form in Fig. 4-41 would implement what Boolean expression?

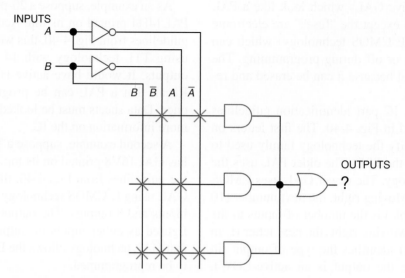

**Fig. 4-41** PLD fuse map.

## 4-16 Using De Morgan's Theorems

Boolean algebra, the algebra of logic circuits, has many laws or theorems. *De Morgan's theorems* are very useful. They allow us to convert back and forth from minterm to maxterm forms of Boolean expressions. They also allow us to eliminate long overbars that cover several variables.

De Morgan's theorems can be stated in the form shown in Fig. 4-42. The first theorem $(\overline{A + B} = \overline{A} \cdot \overline{B})$ shows that the long overbar

covering the $\overline{A + B}$ term can be eliminated. A simple example of the first theorem is shown in the example section [Fig. 4-42(*b*)] when the customary NOR logic symbol ($\overline{A + B} = Y$) is shown as equivalent to the alternative NOR symbol ($\overline{A} \cdot \overline{B} = Y$).

De Morgan's second theorem is stated in Fig. 4-42(*c*) as $\overline{A \cdot B} = \overline{A} + \overline{B}$. A simple example of the second theorem is shown in the example section where the customary NAND logic symbol $\overline{A \cdot B}$ is shown as equivalent to the alternative NAND symbol ($\overline{A} + \overline{B} = Y$).

(*a*)  First theorem

$$\overline{A + B} = \overline{A} \cdot \overline{B}$$

(*b*)  Example—first theorem

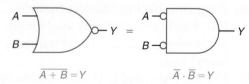

(*c*)  Second theorem

$$\overline{A \cdot B} = \overline{A} + \overline{B}$$

(*d*)  Example—second theorem

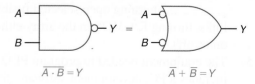

**Fig. 4-42** De Morgan's theorems and practical examples.

## Boolean Expressions— Keyboard Version

The long overbars in Boolean expressions (for example, $\overline{A \cdot B}$) are somewhat more difficult to show in the keyboard versions of an expression. For instance, the keyboard version of $\overline{A \cdot B}$ would be $(AB)'$. The apostrophe *outside* the parenthesis means a long overbar. Next consider the Boolean expression $A \cdot B \cdot \overline{C} + \overline{A} \cdot \overline{B} \cdot C = Y$. The keyboard version of this expression would be $((ABC') + (A'B'C))'$. The customary Boolean expression for the NOR function is $\overline{A + B}$ while the keyboard version could be typed as $(A + B)'$. Do not be surprised when working with some circuit simulation programs if the software converts a minterm to a maxterm or maxterm to a minterm type expression. For instance, it might convert the conventional NAND notation $(\overline{A \cdot B})$ to alternative NAND notation $(A' + B')$. Computer circuit simulation programs use De Morgan's theorems to make these conversions.

## Minterm-to-Maxterms or Maxterm-to-Minterms

Four steps are needed to convert a maxterm to minterm Boolean expression or from minterm to maxterm form. The four steps, which are based on De Morgan's theorems, are as follows:

Step 1.  Change all ORs to ANDs and all ANDs to ORs.

Step 2.  Complement each individual variable (add short overbars to each).

Step 3.  Complement the entire function (add long overbar to entire function).

Step 4.  Eliminate all groups of double overbars.

As an example, consider converting the customary NAND expression $(\overline{A \cdot B} = Y)$ to its alternative NAND form $(\overline{A} + \overline{B} = Y)$. Follow the four-step process in Fig. 4-43 to get familiar with the procedure. At the end of the procedure, the alternative NAND expression is shown as $\overline{A} + \overline{B} = Y$, but on the computer it would be represented as $A' + B' = Y$.

Now we will use the four-step procedure in converting a more complicated maxterm expression to its minterm form. Conversions from maxterm-to-minterm or minterm-to-maxterm form are commonly undertaken to get rid of long overbars in the Boolean expression. The

Begin.   Customary NAND expression.

$$\overline{A \cdot B} = Y$$

Step 1.  Change all ORs to ANDs and all ANDs to ORs.

$$\overline{A + B} = Y$$

Step 2.  Complement each individual variable (short overbar).

$$\overline{\overline{A} + \overline{B}} = Y$$

Step 3.  Complement the entire function (long overbar).

$$\overline{\overline{\overline{A} + \overline{B}}} = Y$$

Step 4.  Eliminate all groups of double overbars.

$$\overline{A} + \overline{B} = Y$$

End.    Alternative NAND expression.

$$\overline{A} + \overline{B} = Y$$

**Fig. 4-43** Four-step process using De Morgan's second theorem to convert conventional NAND to alternative NAND. Note that the long overbar is eliminated.

new example illustrated in Fig. 4-44 will change the maxterm expression $(\overline{A} + \overline{B} + \overline{C}) \cdot (A + B + \overline{C}) = Y$ to its minterm equivalent and eliminate the long overbar. Carefully follow the conversion process in Fig. 4-44. The

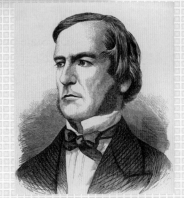

George Boole was born in Lincoln, England, on November 2, 1815. He was a self-taught mathematician who invented modern symbolic logic and pioneered the calculus of operators. Around 1850, George Boole created Boolean algebra, which underlies the theory of logic.

Augustus DeMorgan (1806–1871) was born in Madras Province, India. He taught mathematics at the University of London for 30 years, and published many texts on arithmetic, algebra, trigonometry, and calculus, and important treatises on the theory of probability and formal logic. DeMorgan contributed the method of changing from sum-of-products to product-of-sums.

Begin.   Maxterm expression.

$$(\overline{\overline{A} + \overline{B} + \overline{C}}) \cdot (A + B + \overline{C}) = Y$$

Step 1.   Change all ORs to ANDs and all ANDs to ORs.

$$\overline{\overline{A} \cdot \overline{B} \cdot \overline{C} + A \cdot B \cdot \overline{C}} = Y$$

Step 2.   Complement each individual variable (short overbars)

$$\overline{\overline{\overline{A}} \cdot \overline{\overline{B}} \cdot \overline{\overline{C}} + \overline{A} \cdot \overline{B} \cdot \overline{\overline{C}}} = Y$$

Step 3.   Complement the entire function (long overbar).

$$\overline{\overline{\overline{\overline{A}} \cdot \overline{\overline{B}} \cdot \overline{\overline{C}} + \overline{A} \cdot \overline{B} \cdot \overline{\overline{C}}}} = Y$$

Step 4.   Eliminate all groups of double overbars.

$$A \cdot B \cdot C + \overline{A} \cdot \overline{B} \cdot C = Y$$

End.   Minterm expression.

$$A \cdot B \cdot C + \overline{A} \cdot \overline{B} \cdot C = Y$$

**Fig. 4-44** Four-step process using De Morgan's theorems to convert from maxterm-to-minterm form. Note that the long overbar is eliminated.

result of this conversion yields the minterm form $A \cdot B \cdot C + \overline{A} \cdot \overline{B} \cdot C = Y$, which performs exactly the same logic function as the maxterm expression $(\overline{\overline{A} + \overline{B} + \overline{C}}) \cdot (A + B + \overline{C}) = Y$. The resulting minterm expression can be written in conventional form as $A \cdot B \cdot C + \overline{A} \cdot \overline{B} \cdot C = Y$ using overbars or in the shortened keyboard version $ABC + A'B'C = Y$ using apostrophes.

It must be understood that the logic diagrams that would be wired using the maxterm expression $(\overline{\overline{A} + \overline{B} + \overline{C}}) \cdot (A + B + \overline{C}) = Y$ or its equivalent minterm form $A \cdot B \cdot C + \overline{A} \cdot \overline{B} \cdot C = Y$ from Fig. 4-44 would *look different*, but they would generate the same truth table. It is said that they generate the same logic function.

In summary, De Morgan's theorems are useful for converting from maxterm-to-minterm or

minterm-to-maxterm form of Boolean expressions. We commonly make this conversion to eliminate long overbars in a Boolean expression. A second reason to use De Morgan's theorem might be to examine two different logic diagrams that perform the same logic function. One logic diagram might be simpler than the other.

## Self-Test

*Answer the following questions.*

65. State two of De Morgan's theorems.
66. Convert the maxterm Boolean expression $(A + \overline{B} + \overline{C}) \cdot (\overline{A} + B + \overline{C}) = Y$ to its minterm form. Show each step as is done in Fig. 4-44.
67. Convert the minterm Boolean expression $\overline{A} \cdot B \cdot C + \overline{A} \cdot \overline{B} \cdot \overline{C} = Y$ to its maxterm form. Show each step as is done in Fig. 4-44.
68. Write the Boolean expression

$\overline{A} \cdot B \cdot C + \overline{A} \cdot \overline{B} \cdot \overline{C} = Y$ in the keyboard version using apostrophes instead of overbars.
69. Draw a logic symbol diagram for the Boolean expression $(A'BC + A'B'C')' = Y$. HINT: Use a two-input NOR gate nearest output.
70. Draw a logic symbol diagram for the Boolean expression $((A + B + C + D)(A' + D)(A' + B' + C'))' = Y$. HINT: Use three-input NAND gate nearest output.

## 4-17 Solving a Logic Problem (BASIC Stamp Module)

It is common for logic functions to be programmed using software. In this section we will solve combinational logic problems using a high-level language called PBASIC (a version of BASIC used by Parallax, Inc.). The programmable hardware device used in these examples will be the BS2 BASIC Stamp Microcontroller Module by Parallax, Inc. The hardware includes the BASIC Stamp 2 module, a PC system, a serial cable for downloading, and assorted electronic components (switches, resistors, and LEDs).

The truth table in Fig. 4-45(*a*) details the logic problem to be solved. From the truth table it appears that there are three separate combinational logic problems with outputs labeled Y1, Y2, and Y3. The schematic diagram in Fig. 4-45(*b*) shows three active-HIGH input switches (A, B, and C) and three colored output indicators (LEDs). The programmable device used to solve this logic problem is the BASIC Stamp 2 microcontroller module.

The procedure for solving the logic problem with the use of the BASIC Stamp 2 Module is detailed below. The steps in wiring and programming the BASIC Stamp 2 Module are:

1. Refer to Fig. 4-45(*b*). Wire the three active-HIGH pushbutton switches and connect them to ports P10, P11, and P12. Wire the red, green, and yellow LED output indicators with limiting resistors and connect them to ports P1, P2, and P3 of the BASIC Stamp 2 Module. The ports will be defined as either outputs or inputs in the PBASIC program.
2. Load the PBASIC text editor program (version for the BS2 IC) into the PC. Type your PBASIC program describing the **'3in-3out logic problem.** A PBASIC program titled **'3in-3out logic problem** is listed in the shaded box.
3. Attach a serial cable between the PC and the BASIC Stamp 2 development board (such as the Board of Education by Parallax, Inc.).
4. With the BASIC Stamp 2 Module turned on, download your PBASIC program from the PC to BS2 module using the RUN command.
5. Disconnect the serial cable from the BS2 module.
6. Test the program by pressing the input switches (A, B, and C) while observing the outputs (red, green, and yellow LEDs). The PBASIC program stored in EEPROM program memory in the BASIC Stamp 2 Module will start each time the BS2 IC is turned on.

### PBASIC Program—3in-3out Logic Problem

Consider the PBASIC program titled **'3in-3out logic problem.** Line 1 starts with an apostrophe ('), which means this is a *remark statement.* Remark statements are used to clarify the program and are not executed by the microcontroller. Lines 2–7 are lines of code to *declare variables* that will be used later in the program. As an example, line 2 reads **A VAR Bit.** This tells the microcontroller that A is a variable name that will hold only 1 bit (a 0 or 1). Lines 8–13 are lines of code that *declare* which ports are used as inputs or outputs. As an example, line 9 reads **INPUT 11.** This informs the microcontroller that port 11 (P11) will be used as an input in this program. Another example in line 11 reads **OUTPUT 1,** which declares that port 1 will be used as an output. Notice that line 11 code is followed by a *remark statement* **'Declare port 1 as output Y1 (red LED).** The remark statements at the right in this PBASIC program are not required but they aid in understanding the purpose for lines of code.

Next consider the main routine with the starts with the **CkAllSwit:** line of code (L14). In PBASIC, any word with a colon (:) after it is called a *label.* A label is a reference point in the program that usually locates the starting point of a routine.

In the **'3in-3out logic problem** sample program, the label **CkAllSwit:** is the starting point in the main routine used to check the condition of input switches A, B, and C. The Boolean expression using variables A, B, and C is then evaluated. The **CkAllSwit:** routine repeats continuously because either lines 29 or 38 (**GOTO CkAllSwit**) will always return the program to the beginning of the **CkAllSwit:** routine.

Lines 15–17 of the PBASIC program initializes or turns off all three output LEDs. As an example, the OUT1 = 0 statement causes port

Truth Table

| INPUTS | | | OUTPUTS | | |
|---|---|---|---|---|---|
| **A** | **B** | **C** | Red **Y1** | Green **Y2** | Yellow **Y3** |
| 0 | 0 | 0 | 1 | 1 | 1 |
| 0 | 0 | 1 | 0 | 1 | 1 |
| 0 | 1 | 0 | 0 | 0 | 0 |
| 0 | 1 | 1 | 0 | 0 | 1 |
| 1 | 0 | 0 | 0 | 0 | 0 |
| 1 | 0 | 1 | 0 | 1 | 1 |
| 1 | 1 | 0 | 0 | 0 | 0 |
| 1 | 1 | 1 | 1 | 1 | 0 |

(a)

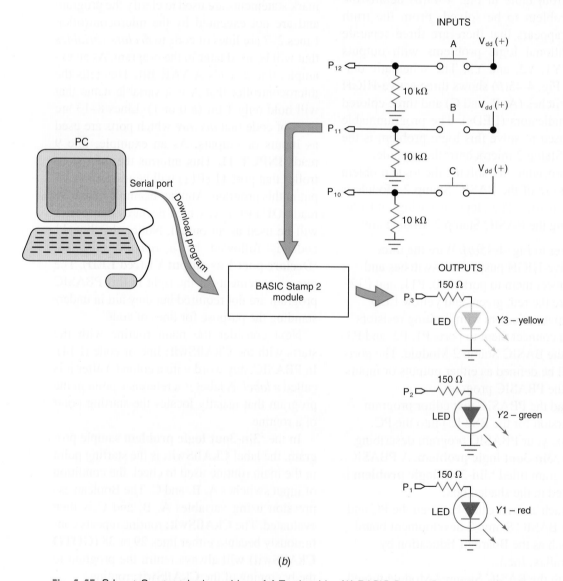

(b)

**Fig. 4-45** 3-Input 3-output logic problem. (a) Truth table. (b) BASIC Stamp 2 module wiring diagram.

|  |  |  |
|---|---|---|
| '3in-3out logic problem | 'Title of program (Fig. 4-45) | L1 |
| A VAR Bit | 'Declare A as variable, 1 bit | L2 |
| B VAR Bit | 'Declare B as variable, 1 bit | L3 |
| C VAR Bit | 'Declare C as variable, 1 bit | L4 |
| Y1 VAR Bit | 'Declare Y1 as variable, 1 bit | L5 |
| Y2 VAR Bit | 'Declare Y2 as variable, 1 bit | L6 |
| Y3 VAR Bit | | L7 |
| INPUT 10 | 'Declare port 10 as an input | L8 |
| INPUT 11 | 'Declare port 11 as an input | L9 |
| INPUT 12 | 'Declare port 12 as an input | L10 |
| OUTPUT 1 | 'Declare port 1 as an output Y1 (red LED) | L11 |
| OUTPUT 2 | 'Declare port 2 as output Y2 (green LED) | L12 |
| OUTPUT 3 | | L13 |
| CkAllSwit: | 'Label for main routine | L14 |
| OUT1 = 0 | 'Initialize port 1 at 0, red LED off | L15 |
| OUT2 = 0 | 'Initialize port 2 at 0, green LED off | L16 |
| OUT3 = 0 | 'Initialize port 3 at 0, yellow LED off | L17 |
| A = IN12 | 'Assign value: port 12 input to variable A | L18 |
| B = IN11 | 'Assign value: port 11 input to variable B | L19 |
| C = IN10 | 'Assign value: port 10 input to variable C | L20 |
| Y1 = (A&B&C) \| (~A&~B&~C) | 'Assign value of expression to variable Y1 | L21 |
| If Y1 = 1 THEN Red | 'If Y = 1 then go to Red:, otherwise next line | L22 |
| CkGreen: | | L23 |
| Y2 = (~A&~B) \| (A&C) | 'Assign value of expression to variable Y2 | L24 |
| If Y2 = 1 THEN Green | 'If Y2 = 1 then go to Green:, or next line | L25 |
| CkYellow: | | L26 |
| Y3 = (~A) \| (~B&C) | 'Assign value of expression to variable Y3 | L27 |
| If Y3 = 1 THEN Yellow | 'If Y3 = 1 then go to Yellow:, or next line | L28 |
| GOTO CkAllSwit | 'Go to CkAllSwit- start main routine | L29 |
| Red: | 'Label- light red LED subroutine | L30 |
| OUT1 = 1 | 'Output P1 goes HIGH, red LED lights | L31 |
| GOTO CkGreen | 'Go to CkGreen: | L32 |
| Green: | 'Label- light green LED subroutine | L33 |
| OUT2 = 1 | 'Output P2 goes HIGH, green LED lights | L34 |
| GOTO CkYellow | 'Go to CkYellow: | L35 |
| Yellow: | 'Label- light yellow LED subroutine | L36 |
| OUT3 = 1 | | L37 |
| GOTO CkAllSwit | 'Start main routine again at CkAllSwit: | L38 |

1 (P1) of the BS2 IC to go LOW. Lines 18–20 assign the current binary value at input ports 10 (P10), 11 (P11), and 12 (P12) to variables C, B, and A. For instance, if all input switches were pressed then all variables A, B, and C would all equal to logical 1.

Line 21 of the PBASIC program is code evaluates the Boolean expression **Y1 = (A&B&C) | (~A&~B&C)**. As an example, if all inputs are HIGH, then variable $Y1 = 1$ (see last line in truth table—Fig. 4-45). Line 22 is an IF-THEN statement used for making decisions.

If Y1 = 1 then the PBASIC statement **IF Y1 = 1 THEN Red** will cause the program to jump to the **Red:** label or the subroutine that lights the red LED. If Y1 = 0, then the first section of PBASIC statement **IF Y1 = 1 THEN Red** is false. The false will cause the program to proceed to the next line of code (line 23).

The **Red:** subroutine (Lines 30–32) in the PBASIC program **'3-in-3out logic problem** causes the port 1 (pin P1) of the BS2 IC to go HIGH using the **OUT1 = 1** statement. This turns on and lights the red LED. Line 32 (**GOTO CkGreen**) sends the program back to the routine labeled **CkGreen:** (Lines 23–25).

After the PBASIC program is downloaded to the BS2 BASIC Stamp 2 unit, the module wired as in Fig. 4-45(*b*) will perform the logic functions detailed in the truth table [Fig. 4-45(*a*).] You have programmed the logic functions called for in the truth table into the microcontroller module.

The PBASIC program **'3-in-3out logic problem** will run continuously while the BS2 BASIC Stamp 2 module is powered. The PBASIC program is held in EEPROM program memory for future use. Turning the BS2 OFF and then ON again will restart the program. Downloading a different PBASIC program to the BASIC Stamp module will erase the old program and start execution of the new listing.

![Self-Test]

*Answer the following questions.*

71. Refer to Fig. 4-45(*b*). Inputs A, B, and C are wired as _____ (active-HIGH, active-LOW) switches, which generate a HIGH when the pushbuttons are depressed.

72. Refer to Fig. 4-45(*b*). If the outputs from the BASIC Stamp 2 module are P3 = HIGH, P2 = LOW, and P1 = HIGH, which LED(s) will light?

73. Refer to Fig. 4-45(*a*). The logic function in output column Y1 can be described by the Boolean expression _____.

74. When using the BASIC Stamp 2 module, the program is typed on a PC using the PBASIC text editor and then _____ (downloaded, poured) through a serial cable to the microcontroller unit.

75. Refer to Fig. 4-45 and the **'3in-3out logic problem** listing. If only pushbuttons A and C are depressed, which LEDs will light?

76. The program line **Y2 = (~A&~B) | (A&C)** is the PBASIC version of what Boolean expression?

77. Refer to line 25 in the program **'3in-3out logic problem.** If variable Y2 = 0, then the next PBASIC program code executed would be_____ (line 26, line 33).

78. Refer to line 22 in the program **'3in-3out logic problem.** If variable Y1 = 1, then the next PBASIC program code executed would be _____ (line 23, line 30).

79. Refer to Fig. 4-45 and the **'3in-3out logic problem** listing. The BASIC Stamp 2 module knows that ports P10, P11, and P12 are _____ (inputs, outputs) because they are declared as such in the PBASIC program listing.

80. Refer to the **'3in-3out logic problem** listing. The main routine labeled **CkAllSwit:** in the PBASIC program begins with line 14 and ends with _____ (line 29, line 38) repeating over and over until power to the BASIC Stamp 2 module is turned off.

# Chapter 4 Summary and Review

## Summary

1. Combining gates in combinational logic circuits from Boolean expressions is a necessary skill for most competent technicians and engineers.
2. Workers in digital electronics must have an excellent knowledge of gate symbols, truth tables, and Boolean expressions and know how to convert from one form to another.
3. The minterm Boolean expression (sum-of-products form) might look like the expression in Fig. 4-46($a$). The Boolean expression $A \cdot B + \overline{A} \cdot \overline{C} = Y$ would be wired as shown in Fig. 4-46(b).
4. The pattern of gates shown in Fig. 4-46($b$) is called an AND-OR logic circuit.
5. The maxterm Boolean expression (product-of-sums form) might look like the expression in Fig. 4-46($c$). The Boolean expression $(A + \overline{C}) \cdot (\overline{A} + B) = Y$ would be wired as shown in Fig. 4-46($d$). This is an OR-AND logic circuit.
6. A Karnaugh map is a convenient method of simplifying Boolean expressions.
7. AND-OR logic circuits can be wired easily by using only NAND gates, as shown in Fig. 4-47.
8. Data selectors are a simple, one-package method of solving many gating problems. Less expensive data selectors can be used when the folding design technique is utilized.
9. Computer simulations can easily and accurately convert back and forth between Boolean expressions, truth tables, and logic diagrams. The simulations can also simplify Boolean expressions.
10. Programmable logic devices (PLDs) are inexpensive one-package solutions to many complex logic problems. In this chapter, simple PLDs are used to solve combinational logic problems but may also be applied to sequential logic designs.
11. De Morgan's theorems are useful in converting maxterm-to-minterm and minterm-to-maxterm Boolean expressions.
12. A keyboard version of a Boolean expression is used with computer systems. An example would be $\overline{A \cdot B} = Y$ is equivalent to $(A'B)' = Y$.

($a$) Minterm Boolean expression

$$A \cdot B + \overline{A} \cdot \overline{C} = Y$$

($b$)

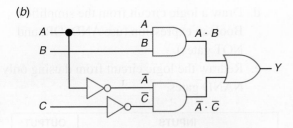

($c$) Maxterm Boolean expression

$$(A + \overline{C}) \cdot (\overline{A} + B) = Y$$

($d$)

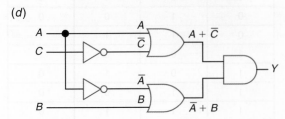

**Fig. 4-46** ($a$) Minterm expression. ($b$) AND-OR logic circuit. ($c$) Maxterm expression. ($d$) OR-AND logic circuit.

$$A \cdot \overline{B} + \overline{A} \cdot B = Y$$

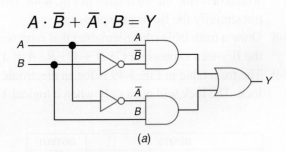

($a$)

$$A \cdot \overline{B} + \overline{A} \cdot B = Y$$

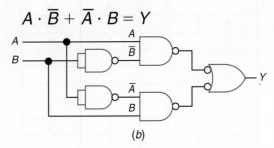

($b$)

**Fig. 4-47** ($a$) AND-OR logic circuit. ($b$) Equivalent NAND-NAND logic circuit.

13. BASIC Stamp modules are microcontroller-based devices that can generate logic functions. They are programmed using Boolean expressions. The programs are downloaded from the PC to the BASIC Stamp module. The PC is then disconnected and the BASIC Stamp module will execute the correct logic.

14. The traditional Boolean expression $\overline{A} \cdot \overline{B} + B \cdot C = Y$ would be coded in PBASIC as Y = (~A&~B) | (B&C) and its logic function implemented by a BASIC Stamp module.

## Chapter Review Questions

*Answer the following questions.*

4-1. Logic gate circuits whose outputs respond immediately (no memory characteristic) to a change at the inputs are called _____ (combinational, sequential)-logic circuits.

4-2. Draw a logic diagram for the Boolean expression $\overline{A} \cdot \overline{B} + B \cdot C = Y$. Use one OR gate, two AND gates, and two inverters.

4-3. The Boolean expression $\overline{A} \cdot \overline{B} + B \cdot C = Y$ is in _____ (product-of-sums, sum-of-products) form.

4-4. The Boolean expression $(A + B) \cdot (C + D) = Y$ is in _____ (product-of-sums, sum-of-products) form.

4-5. A Boolean expression in product-of-sums form is also called a _____ expression.

4-6. A Boolean expression in sum-of-products form is also called a _____ expression.

4-7. Write the minterm Boolean expression that would describe the truth table in Fig. 4-48. Do not simplify the Boolean expression.

4-8. Draw a truth table (three-variable) that represents the Boolean expression $\overline{C} \cdot \overline{B} + C \cdot \overline{B} \cdot A = Y$.

4-9. The truth table in Fig. 4-49 is for an electronic lock. The lock will open only when a logical 1 appears at the output. First, write the minterm Boolean expression for the lock. Second, draw the logic circuit for the lock (use AND, OR, and NOT gates).

4-10. List the six steps for simplifying a Boolean expression using a Karnaugh map as discussed in Sec. 4-6.

4-11. Use a Karnaugh map to simplify the Boolean expression $\overline{A} \cdot \overline{B} \cdot \overline{C} + \overline{A} \cdot \overline{B} \cdot C + A \cdot B \cdot \overline{C} + A \cdot \overline{B} \cdot \overline{C} = Y$. Write the simplified Boolean expression in minterm form.

4-12. Use a Karnaugh map to simplify the Boolean expression $A \cdot \overline{B} \cdot \overline{C} \cdot \overline{D} + A \cdot \overline{B} \cdot \overline{C} \cdot D + A \cdot \overline{B} \cdot C \cdot D + A \cdot \overline{B} \cdot C \cdot \overline{D} = Y$.

4-13. From the truth table in Fig. 4-48 do the following:
   a. Write the unsimplified Boolean expression.
   b. Use a Karnaugh map to simplify the Boolean expression from a.
   c. Write the simplified minterm Boolean expression for the truth table.
   d. Draw a logic circuit from the simplified Boolean expression (use AND, OR, and NOT gates).
   e. Redraw the logic circuit from d using only NAND gates.

| INPUTS | | | OUTPUT |
|---|---|---|---|
| C | B | A | Y |
| 0 | 0 | 0 | 1 |
| 0 | 0 | 1 | 0 |
| 0 | 1 | 0 | 1 |
| 0 | 1 | 1 | 0 |
| 1 | 0 | 0 | 0 |
| 1 | 0 | 1 | 1 |
| 1 | 1 | 0 | 0 |
| 1 | 1 | 1 | 1 |

**Fig. 4-48** Truth table.

| INPUTS | | | OUTPUT |
|---|---|---|---|
| C | B | A | Y |
| 0 | 0 | 0 | 0 |
| 0 | 0 | 1 | 0 |
| 0 | 1 | 0 | 0 |
| 0 | 1 | 1 | 1 |
| 1 | 0 | 0 | 1 |
| 1 | 0 | 1 | 0 |
| 1 | 1 | 0 | 0 |
| 1 | 1 | 1 | 0 |

**Fig. 4-49** Truth table.

4-14. Use a Karnaugh map to simplify the Boolean expression $\overline{A} \cdot \overline{B} \cdot C \cdot D + A \cdot B \cdot \overline{C} \cdot \overline{D} + A \cdot B \cdot C \cdot \overline{D} + A \cdot \overline{B} \cdot C \cdot D = Y$. Write the answer as a minterm Boolean expression.

4-15. From the Boolean expression $\overline{A} \cdot \overline{B} \cdot \overline{C} \cdot D + \overline{A} \cdot \overline{B} \cdot C \cdot D + \overline{A} \cdot B \cdot \overline{C} \cdot D + A \cdot B \cdot C \cdot D + A \cdot B \cdot C \cdot \overline{D} + A \cdot \overline{B} \cdot \overline{C} \cdot \overline{D} = Y$ do the following:
   a. Draw a truth table for the expression.
   b. Use a Karnaugh map to simplify.
   c. Draw a logic circuit of the simplified Boolean expression (use AND, OR, and NOT gates).
   d. Draw a circuit to solve this problem using a 1-of-16 data selector.
   e. Draw a circuit to solve this problem using the folding technique and a 1-of-8 data selector.

4-16. From the Boolean expression $\overline{A} \cdot \overline{B} \cdot \overline{C} \cdot D \cdot E + \overline{A} \cdot B \cdot \overline{C} \cdot D \cdot E + A \cdot B \cdot \overline{C} \cdot D \cdot E + A \cdot \overline{B} \cdot \overline{C} \cdot D \cdot E + A \cdot B \cdot \overline{C} \cdot D \cdot \overline{E} + \overline{A} \cdot \overline{B} \cdot C \cdot D \cdot \overline{E} + \overline{A} \cdot \overline{B} \cdot C \cdot D \cdot E = Y$ do the following:
   a. Use a Karnaugh map to simplify.
   b. Write the simplified minterm Boolean expression.
   c. Draw a logic circuit from the simplified Boolean expression (use AND, OR, and NOT gates).

4-17. The Boolean algebra laws that allow us to convert from minterm-to-maxterm or maxterm-to-minterm forms of expressions are called _____.

4-18. Based on De Morgan's first theorem, $\overline{A + B} = $ _____.

4-19. Based on De Morgan's second theorem, $\overline{A \cdot B} = $ _____.

4-20. Using De Morgan's theorems, convert the maxterm Boolean expression $\overline{(A + \overline{B} + C) \cdot (\overline{A} + \overline{B} + \overline{C})} = Y$ to its minterm form. This will remove the long overbar.

4-21. Using De Morgan's theorems, convert the minterm Boolean expression $\overline{A \cdot \overline{B} \cdot C + A \cdot B \cdot C} = Y$ to its maxterm form. This will remove the long overbar.

4-22. Write the keyboard version of the Boolean expression $A \cdot \overline{B} + \overline{A} \cdot B = Y$.

4-23. Write the keyboard version of the Boolean expression $A \cdot \overline{B} \cdot C = Y$.

4-24. Write the keyboard version of the Boolean expression $(\overline{A} + B) (\overline{C} + D) = Y$.

4-25. Using the Logic Converter from Electronics Workbench® or MultiSIM® (a) draw the logic diagram shown in Fig. 4-50 on the Logic Converter screen, (b) generate and write its truth table, (c) generate and write it unsimplified Boolean expression, and (d) generate and copy down its simplified Boolean expression.

4-26. Using the Logic Converter from Electronics Workbench® or MultiSIM® (a) enter the truth table shown in Fig. 4-51 on the Logic Converter screen, (b) generate and write the simplified Boolean expression, and (c) generate and draw the AND-OR logic symbol diagram for the truth table.

4-27. Using the Logic Converter from Electronics Workbench® or MultiSIM® (a) enter the Boolean expression $A'C' + BC + ACD'$ on the Logic Converter screen, (b) generate and draw the 4-variable truth table, and (c) generate and draw the AND-OR logic symbol diagram that is equivalent to the Boolean expression.

4-28. Using the Logic Converter from Electronics Workbench® or MultiSIM® (a) enter the 5-variable truth table shown in Fig. 4-52 on the Logic Converter screen, (b) generate and copy the simplified Boolean expression, and (c) generate and draw the AND-OR logic symbol diagram that is equivalent to the truth table and simplified Boolean expression.

4-29. In electronic technology, PAL is a common acronym for _____.

4-30. In electronic technology, PLD is a common acronym for _____.

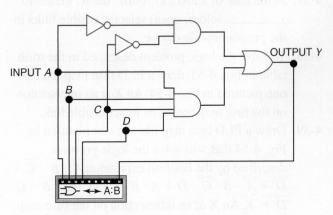

Fig. 4-50 Logic converter problem.

| INPUTS | | | | OUTPUT |
|---|---|---|---|---|
| A | B | C | D | Y |
| 0 | 0 | 0 | 0 | 0 |
| 0 | 0 | 0 | 1 | 0 |
| 0 | 0 | 1 | 0 | 0 |
| 0 | 0 | 1 | 1 | 0 |
| 0 | 1 | 0 | 0 | 0 |
| 0 | 1 | 0 | 1 | 0 |
| 0 | 1 | 1 | 0 | 1 |
| 0 | 1 | 1 | 1 | 1 |
| 1 | 0 | 0 | 0 | 1 |
| 1 | 0 | 0 | 1 | 1 |
| 1 | 0 | 1 | 0 | 1 |
| 1 | 0 | 1 | 1 | 0 |
| 1 | 1 | 0 | 0 | 0 |
| 1 | 1 | 0 | 1 | 1 |
| 1 | 1 | 1 | 0 | 1 |
| 1 | 1 | 1 | 1 | 1 |

**Fig. 4-51** Truth table.

| INPUTS | | | | | OUTPUT |
|---|---|---|---|---|---|
| A | B | C | D | D | Y |
| 0 | 0 | 0 | 0 | 0 | 0 |
| 0 | 0 | 0 | 0 | 1 | 0 |
| 0 | 0 | 0 | 1 | 0 | 0 |
| 0 | 0 | 0 | 1 | 1 | 0 |
| 0 | 0 | 1 | 0 | 0 | 1 |
| 0 | 0 | 1 | 0 | 1 | 0 |
| 0 | 0 | 1 | 1 | 0 | 0 |
| 0 | 0 | 1 | 1 | 1 | 0 |
| 0 | 1 | 0 | 0 | 0 | 0 |
| 0 | 1 | 0 | 0 | 1 | 0 |
| 0 | 1 | 0 | 1 | 0 | 0 |
| 0 | 1 | 0 | 1 | 1 | 0 |
| 0 | 1 | 1 | 0 | 0 | 1 |
| 0 | 1 | 1 | 0 | 1 | 0 |
| 0 | 1 | 1 | 1 | 0 | 0 |
| 0 | 1 | 1 | 1 | 1 | 0 |
| 1 | 0 | 0 | 0 | 0 | 0 |
| 1 | 0 | 0 | 0 | 1 | 0 |
| 1 | 0 | 0 | 1 | 0 | 0 |
| 1 | 0 | 0 | 1 | 1 | 0 |
| 1 | 0 | 1 | 0 | 0 | 1 |
| 1 | 0 | 1 | 0 | 1 | 0 |
| 1 | 0 | 1 | 1 | 0 | 0 |
| 1 | 1 | 1 | 1 | 1 | 0 |
| 1 | 1 | 0 | 0 | 0 | 1 |
| 1 | 1 | 0 | 0 | 1 | 1 |
| 1 | 1 | 0 | 1 | 0 | 1 |
| 1 | 1 | 0 | 1 | 1 | 1 |
| 1 | 1 | 1 | 0 | 0 | 1 |
| 1 | 1 | 1 | 0 | 1 | 0 |
| 1 | 1 | 1 | 1 | 0 | 0 |
| 1 | 1 | 1 | 1 | 1 | 0 |

**Fig. 4-52** Truth table.

4-31. In electronic technology, GAL is a common acronym for _____.

4-32. In electronic technology, FPGA is a common acronym for _____.

4-33. In electronic technology, CPLD is a common acronym for _____.

4-34. The _____ (PAL, CPLD) is the simpler programmable logic device and is usually used to implement combinational logic designs.

4-35. List several advantages of using PLDs to implement a logic design.

4-36. List two software development packages furnished by manufacturers to program PLDs using a personal computer and IC Burner.

4-37. In the case of a PAL, to "burn" the IC means to _____ (close, open) selected fusible links in the programmable device.

4-38. To solve the logic problem described in the truth table in Fig. 4-53, draw a PLD fuse map like the one pictured in Fig. 4-54. An $X$ at an intersection on the fuse map means an intact fusible link.

4-39. Draw a PLD fuse map like the one pictured in Fig. 4-54 that will solve the logic problem described by the Boolean expression $\overline{A} \cdot \overline{B} \cdot \overline{C} \cdot \overline{D} + \overline{A} \cdot B \cdot \overline{C} \cdot D + A \cdot B \cdot C \cdot D + A \cdot \overline{B} \cdot C \cdot \overline{D} = Y$. An X at an intersection on the fuse map means an intact fusible link.

4-40. BASIC Stamp modules are _____ (microcontroller-based, vacuum tube-based) devices that can be programmed to generate logic functions.

4-41. Refer to Fig. 4-45(a). The PBASIC code $(\sim A \& \sim B) | (A \& C)$ when used in an assignment statement would generate the logic for truth table output column _____ (Y1, Y2, Y3).

| INPUTS | | | | OUTPUT |
|---|---|---|---|---|
| A | B | C | D | Y |
| 0 | 0 | 0 | 0 | 0 |
| 0 | 0 | 0 | 1 | 1 |
| 0 | 0 | 1 | 0 | 0 |
| 0 | 0 | 1 | 1 | 0 |
| 0 | 1 | 0 | 0 | 0 |
| 0 | 1 | 0 | 1 | 0 |
| 0 | 1 | 1 | 0 | 0 |
| 0 | 1 | 1 | 1 | 1 |
| 1 | 0 | 0 | 0 | 0 |
| 1 | 0 | 0 | 1 | 0 |
| 1 | 0 | 1 | 0 | 0 |
| 1 | 0 | 1 | 1 | 1 |
| 1 | 1 | 0 | 0 | 0 |
| 1 | 1 | 0 | 1 | 1 |
| 1 | 1 | 1 | 0 | 0 |
| 1 | 1 | 1 | 1 | 0 |

**Fig. 4-53** Truth table.

4-42. Refer to Fig. 4-45(*b*). What three BASIC Stamp 2 module ports are used as inputs in this circuit?

4-43. Refer to Fig. 4-45(*b*) and the PBASIC program **'3in-3out logic problem.** What lines of code define which ports are inputs and which are outputs?

4-44. Refer to Fig. 4-45 and the **'3in-3out logic problem** listing. If only pushbuttons B and C are activated (depressed), which LED(s) will light?

4-45. Refer to the **'3in-3out logic problem** listing. Lines 15–17 are used to turn _____ (off, on) all three LEDs.

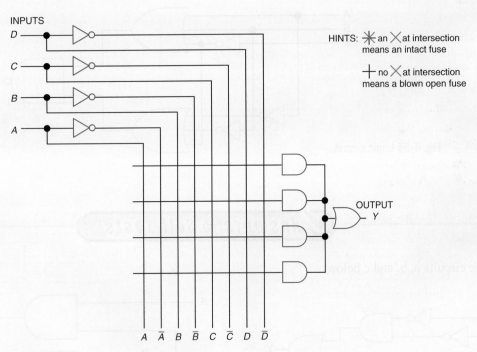

INPUTS

HINTS: ✳ an ✕ at intersection means an intact fuse

+ no ✕ at intersection means a blown open fuse

OUTPUT
Y

A  Ā  B  B̄  C  C̄  D  D̄

**Fig. 4-54** Fuse map problem.

4-1. When implemented, a minterm Boolean expression produces what pattern of logic gates?

4-2. When implemented, a maxterm Boolean expression produces what pattern of logic gates?

4-3. Simplify the Boolean expression $\overline{A} \cdot \overline{B} \cdot \overline{C} \cdot \overline{D} + \overline{A} \cdot \overline{B} \cdot \overline{C} \cdot D + \overline{A} \cdot B \cdot \overline{C} \cdot \overline{D} + \overline{A} \cdot B \cdot \overline{C} \cdot D + A \cdot \overline{B} \cdot \overline{C} \cdot \overline{D} + A \cdot \overline{B} \cdot \overline{C} \cdot D + A \cdot \overline{B} \cdot C \cdot D = Y$.

4-4. Do you think it is possible to develop a maxterm (product-of-sums) Boolean expression from a truth table?

4-5. Do you think the Karnaugh map shown in Fig. 4-18(b) can be used to simplify either minterm or maxterm Boolean expressions?

4-6. A five-variable logic problem could be solved using a 1-of-16 data selector employing the _____ technique.

4-7. A six-variable truth table would have how many combinations?

4-8. Write the keyboard version of the Boolean expression $\overline{A} \cdot \overline{B} \cdot C + A \cdot B \cdot \overline{C} + A \cdot \overline{B} \cdot C = Y$ as you may have to do when entering information into a computer circuit simulator.

4-9. Write a *maxterm* Boolean expression for the logic diagram shown in Fig. 4-55.

4-10. Using De Morgan's theorems (or circuit simulator software if available) write the *minterm* Boolean expression that would describe the logic function of the circuit in Fig. 4-55. HINT: Use the maxterm expression developed in question 4-9.

4-11. Draw a three-variable truth table that would describe the logic function of the circuit in Fig. 4-55. HINT: Work from the minterm expression developed in question 4-10.

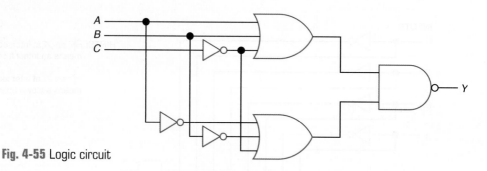

**Fig. 4-55** Logic circuit

## Answers to Self-Tests

1. See logic circuits a, b, and c below.

a.

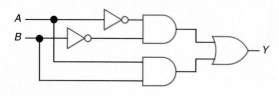

b.

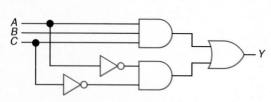

c.

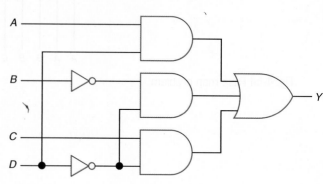

2. sum-of-products
3. product-of-sums
4. sum-of-products
5. product-of-sums
6. See logic circuits a, b, and c below.

a.

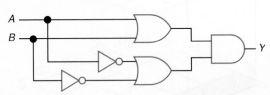

b.

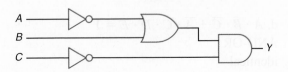

c.

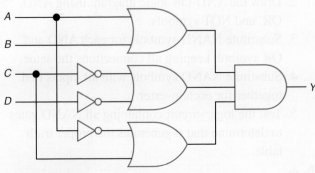

7. maxterm
8. product-of-sums
9. OR-AND
10. $C \cdot B \cdot \overline{A} + C \cdot B \cdot A = Y$
11. lines 1 and 2
12. See table below.

| INPUTS | | | OUTPUT |
|---|---|---|---|
| *C* | *B* | *A* | *Y* |
| 0 | 0 | 0 | 0 |
| 0 | 0 | 1 | 0 |
| 0 | 1 | 0 | 0 |
| 0 | 1 | 1 | 0 |
| 1 | 0 | 0 | 0 |
| 1 | 0 | 1 | 1 |
| 1 | 1 | 0 | 1 |
| 1 | 1 | 1 | 0 |

13. minterm
14. minterm
15. $C'B'A + BA' = Y$

16. T
17. T
18. F
19. $\overline{C} \cdot B \cdot \overline{A} + C \cdot \overline{B} \cdot A = Y$
20. See figure below.

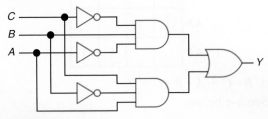

21. identical
22. Boolean, Karnaugh
23. Quine-McCluskey or tabular method
24. Maurice Karnaugh
25. 1. Start with a minterm Boolean expression.
    2. Record 1s on a Karnaugh map.
    3. Loop adjacent 1s (loops of two, four, or eight squares).
    4. Simplify by dropping terms that contain a term and its complement within a loop.
    5. OR the remaining terms (one term per loop).
    6. Write the simplified minterm Boolean expression.
26. See a–c below.

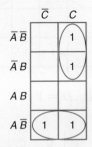

   d. $\overline{A} \cdot C + A \cdot \overline{B} = Y$
27. See a–c below.

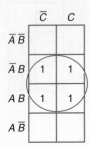

   d. $B = Y$

28. See a–c below.

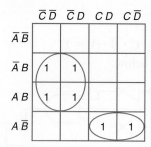

d. $B \cdot \overline{C} + A \cdot \overline{B} \cdot C = Y$

29. See a–c below.

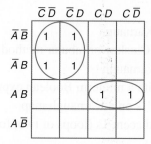

d. $\overline{A} \cdot \overline{C} + A \cdot B \cdot C = Y$

30. See a–c below.

| | $\overline{C}\,\overline{D}$ | $\overline{C}\,D$ | $C\,D$ | $C\,\overline{D}$ |
|---|---|---|---|---|
| $\overline{A}\,\overline{B}$ | | 1 | 1 | |
| $\overline{A}\,B$ | 1 | | | 1 |
| $A\,B$ | | | | |
| $A\,\overline{B}$ | | 1 | 1 | |

d. $\overline{A} \cdot B \cdot \overline{D} + \overline{B} \cdot D = Y$

31. See a–c below.

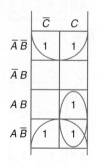

d. $\overline{B} + A \cdot C = Y$

32. See a–c below.

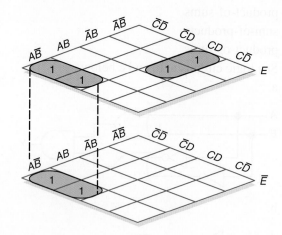

d. $A \cdot \overline{B} \cdot \overline{C} + \overline{A} \cdot C \cdot D \cdot E = Y$

33. AND-OR

34. identical

35. 1. Start with a minterm Boolean expression.
    2. Draw the AND-OR logic diagram using AND, OR, and NOT symbols.
    3. Substitute NAND symbols for each AND and OR symbol, keeping all connections the same.
    4. Substitute NAND symbols with all inputs tied together for each inverter.
    5. Test the logic circuit containing all NAND gates to determine that it generates the proper truth table.

36. a.

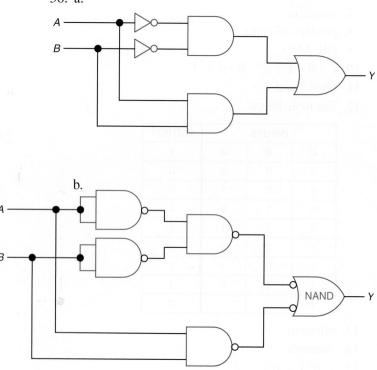

b.

37. a.

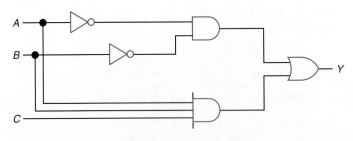

b.

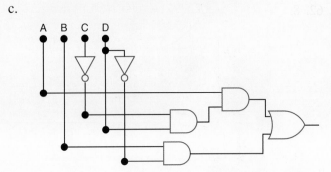

38. b.

| INPUTS | | | | OUTPUT |
|---|---|---|---|---|
| A | B | C | D | Y |
| 0 | 0 | 0 | 0 | 0 |
| 0 | 0 | 0 | 1 | 0 |
| 0 | 0 | 1 | 0 | 0 |
| 0 | 0 | 1 | 1 | 1 |
| 0 | 1 | 0 | 0 | 0 |
| 0 | 1 | 0 | 1 | 0 |
| 0 | 1 | 1 | 0 | 0 |
| 0 | 1 | 1 | 1 | 0 |
| 1 | 0 | 0 | 0 | 0 |
| 1 | 0 | 0 | 1 | 0 |
| 1 | 0 | 1 | 0 | 1 |
| 1 | 0 | 1 | 1 | 0 |
| 1 | 1 | 0 | 0 | 0 |
| 1 | 1 | 0 | 1 | 1 |
| 1 | 1 | 1 | 0 | 1 |
| 1 | 1 | 1 | 1 | 1 |

c. $A'B'CD + ACD' + ABD$

39. b.

| INPUTS | | | | OUTPUT |
|---|---|---|---|---|
| A | B | C | D | Y |
| 0 | 0 | 0 | 0 | 0 |
| 0 | 0 | 0 | 1 | 0 |
| 0 | 0 | 1 | 0 | 0 |
| 0 | 0 | 1 | 1 | 0 |
| 0 | 1 | 0 | 0 | 1 |
| 0 | 1 | 0 | 1 | 0 |
| 0 | 1 | 1 | 0 | 1 |
| 0 | 1 | 1 | 1 | 0 |
| 1 | 0 | 0 | 0 | 0 |
| 1 | 0 | 0 | 1 | 1 |
| 1 | 0 | 1 | 0 | 0 |
| 1 | 0 | 1 | 1 | 0 |
| 1 | 1 | 0 | 0 | 1 |
| 1 | 1 | 0 | 1 | 1 |
| 1 | 1 | 1 | 0 | 1 |
| 1 | 1 | 1 | 1 | 0 |

c.

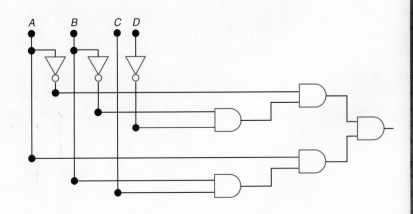

40. b. $A'B'C'D' + A'B'CD' + ABCD' + ABCD$
c. $A'B'D' + ABC$
d.

41. data selector
42. 7, W.
43. rotary
44. 15, HIGH
45. multiplexers
46. LOW
47. HIGH
48. LOW
49. ROMs, PROMs, PALs, gate arrays, programmable gate arrays
50. programmable logic device
51. programmable array logic
52. generic array logic
53. field-programmable logic array
54. combinational
55. by the local user
56. burning open selected fusible links
57. PLD burner or programmer
58. Boolean expression
59. sum-of-products
60. 12, 6, active-HIGH
61. CMOS
62. 8

63. AND
64. $\overline{A} \cdot \overline{B} + A \cdot B = Y$
65. $\overline{A + B} = \overline{A} \cdot \overline{B}$
    $\overline{A \cdot B} = \overline{A} + \overline{B}$
66. Begin. $\overline{(A + \overline{B} + \overline{C}) \cdot (\overline{A} + B + \overline{C})} = Y$

    Step 1. $\overline{(A \cdot \overline{B} \cdot \overline{C})} + \overline{(\overline{A} \cdot B \cdot \overline{C})} = Y$

    Step 2. $(\overline{A} \cdot \overline{\overline{B}} \cdot \overline{\overline{C}}) + (\overline{\overline{A}} \cdot \overline{B} \cdot \overline{\overline{C}}) = Y$

    Step 3. $(\overline{A} \cdot \overline{\overline{B}} \cdot \overline{\overline{C}}) + (\overline{\overline{A}} \cdot \overline{B} \cdot \overline{\overline{C}}) = Y$

    Step 4. Eliminate double overbars.

    End. $\overline{A} \cdot B \cdot C + A \cdot \overline{B} \cdot C = Y$

67. Begin. $\overline{A} \cdot B \cdot C + \overline{A} \cdot \overline{B} \cdot \overline{C} = Y$

    Step 1. $\overline{(\overline{A} + B + C)} \cdot \overline{(\overline{A} + \overline{B} + \overline{C})} = Y$

    Step 2. $\overline{(\overline{\overline{A}} + \overline{B} + \overline{C})} \cdot \overline{(\overline{\overline{A}} + \overline{\overline{B}} + \overline{\overline{C}})} = Y$

    Step 3. $(\overline{\overline{A}} + \overline{B} + \overline{C}) \cdot (\overline{\overline{A}} + \overline{\overline{B}} + \overline{\overline{C}}) = Y$

    Step 4. Eliminate double overbars.

    End. $(A + \overline{B} + \overline{C}) \cdot (A + B + C) = Y$

68. $(A'BC + A'B'C')'$

69.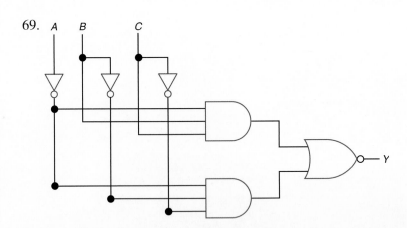

70.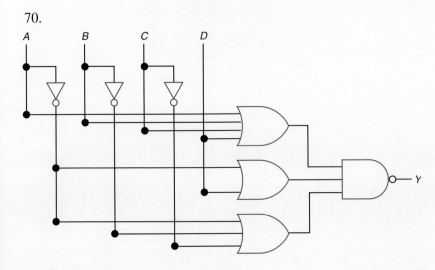

71. active-HIGH
72. yellow (Y3), red (Y1)
73. $\overline{A} \cdot \overline{B} \cdot \overline{C} + A \cdot B \cdot C = Y$
74. downloaded
75. green (Y2) and yellow (Y3)

76. $\overline{A} \cdot \overline{B} + A \cdot C = Y$
77. line 26
78. line 30
79. inputs
80. line 29

69.  A   B

70.  A   B   C   D

71. active HIGH

72. yellow (Y3), red (Y1)

73. $\overline{A} \cdot \overline{B} \cdot \overline{C} = \overline{A} \cdot B \cdot \overline{C} = Y$

74. downloaded

75. green (Y2) and yellow (Y3)

76. $A \cdot \overline{B} + \overline{A} \cdot C = Y$

77. line 26

78. line 30

79. inputs

80. line 29

# IC Specifications and Simple Interfacing

## Chapter Objectives

*This chapter will help you to:*

1. *Determine* logic levels using TTL and CMOS voltage profile diagrams.

2. *Discuss* selected TTL and CMOS IC specifications such as input and output voltages, noise margin, drive capability, fan-in, fanout, propagation delay, and power consumption.

3. *List* several safety precautions for handling and designing with CMOS ICs.

4. *Recognize* several simple switch interface and debounce circuits using both TTL and CMOS ICs.

5. *Analyze* interfacing circuits for LEDs and incandescent lamps using both TTL and CMOS ICs.

6. *Draw* TTL-to-CMOS and CMOS-to-TTL interface circuits.

7. *Describe* the operation of interface circuits for buzzers, relays, motors, and solenoids using both TTL and CMOS ICs.

8. *Analyze* interfacing circuits featuring an optoisolator.

9. *List* the primary characteristics and features of a stepper motor.

10. *Describe* the operation of stepper motor driver circuits.

11. *Use* the terms current sourcing and current sinking.

12. *Summarize* the operation of a servo motor using pulse width modulation (PWM).

13. *Characterize* the operation of a Hall-effect sensor and its application in device such as a Hall-effect switch.

14. *Demonstrate* the interfacing of an open-collector Hall-effect switch with TTL and CMOS ICs as well as LEDs.

15. *Demonstrate* simple interfacing using a BASIC Stamp microcontroller module.

16. *Troubleshoot* a simple logic circuit.

The driving force behind the increased use of digital circuits has been the availability of a variety of *logic families.* Integrated circuits within a logic family are designed to *interface* easily with one another. For instance, in the TTL logic family you may connect an output directly into the input of several other TTL inputs with no extra parts. The designer can have confidence that ICs from the same logic family will interface properly. Interfacing *between* logic families and between digital ICs and the outside world is a bit more complicated. *Interfacing* can be defined as the design of the interconnections between circuits that shift the levels of voltage and current to make them compatible. A fundamental knowledge of simple interfacing techniques is required of technicians and engineers who work with digital circuits. Most logic circuits are of no value if they are not interfaced with "real world" devices.

**Logic families: TTL and CMOS**

**Interfacing**

## 5-1 Logic Levels and Noise Margin

In any field of electronics most technicians and engineers start investigating a new device in terms of voltage, current, and resistance or impedance. In this section just the *voltage characteristics* of both TTL and CMOS ICs will be studied.

### Logic Levels

How is a logical 0 (LOW) or logical 1 (HIGH) defined? Figure 5-1 shows an inverter (such as the 7404 IC) from the bipolar TTL logic family. Manufacturers specify that for correct operation, a LOW *input* must range from

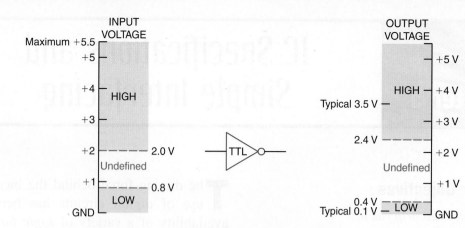

Fig. 5-1 Defining TTL input and output voltage levels.

GND to 0.8 V. Also, a HIGH *input* must be in the range from 2.0 to 5.5 V. The unshaded section from 0.8 to 2.0 V on the input side is the undefined area, or indeterminate region. Therefore, an input of 3.2 is a HIGH input. An input of 0.5 V is considered a LOW input. An input of 1.6 V is in the undefined region and should be avoided. Inputs in the undefined region yield unpredictable results at the output.

Expected outputs from the TTL inverter are shown on the right in Fig. 5-1. A typical LOW output is about 0.1 V. A typical HIGH output might be about 3.5 V. However, a HIGH output could be as low as 2.4 V according to the voltage profile diagram in Fig. 5-1. The HIGH output depends on the resistance value of the load placed at the output. The greater the load current, the lower the HIGH output voltage. The unshaded section on the output voltage side in Fig. 5-1 is the undefined region. Suspect trouble if the output voltage is in the undefined region (0.4 to 2.4 V).

The voltages given for LOW and HIGH logic levels in Fig. 5-1 are for a TTL device. These voltages are different for other logic families.

The 4000 and 74C00 series CMOS logic families of ICs operate on a wide range of power supply voltages (from +3 to +15 V). The definition of a HIGH and LOW logic level for a typical CMOS inverter is illustrated in Fig. 5-2(*a*). A 10-V power supply is being used in this voltage profile diagram.

The CMOS inverter shown in Fig. 5-2(*a*) will respond to any input voltage within

70–100 percent of $V_{DD}$ (+10 V in this example) as a HIGH. Likewise, any voltage within 0 to 30 percent of $V_{DD}$ is regarded as a LOW input to ICs in the 4000 and 74C00 series.

Typical output voltages for CMOS ICs are shown in Fig. 5-2(*a*). Output voltages are normally almost at the *voltage rails* of the power supply. In this example, a HIGH output would be about +10 V while a LOW output would be about 0 V or GND.

The 74HC00 series and the newer 74AC00 and 74ACQ00 series operate on a lower-voltage power supply (from +2 to +6 V) than the older 4000 and 74C00 series CMOS ICs. The input and output voltage characteristics are summarized in the voltage profile diagram in Fig. 5-2(*b*). The definition of HIGH and LOW for both input and output on the 74HC00, 74AC00, and 74ACQ00 series is approximately the same as for the 4000 and 74C00 series CMOS ICs. This can be seen in a comparison of the two voltage profiles in Figs. 5-2(*a*) and (*b*).

The 74HCT00 series and newer 74ACT00, 74ACTQ00, 74FCT00, and 74FCTA00 series of CMOS ICs are designed to operate on a 5-V power supply like TTL ICs. The function of the 74HCT00, 74ACT00, 74ACTQ00, 74FCT00, and 74FCTA00 series of ICs is to interface between TTL and CMOS devices. These CMOS ICs with a "T" designator can serve as direct replacements for many TTL ICs.

A voltage profile diagram for the 74HCT00, 74ACT00, 74ACTQ00, 74FCT00, and 74FCTA00 series CMOS ICs is drawn in

Data sheets for ICs at www. onsemi.com or www.ti.com.

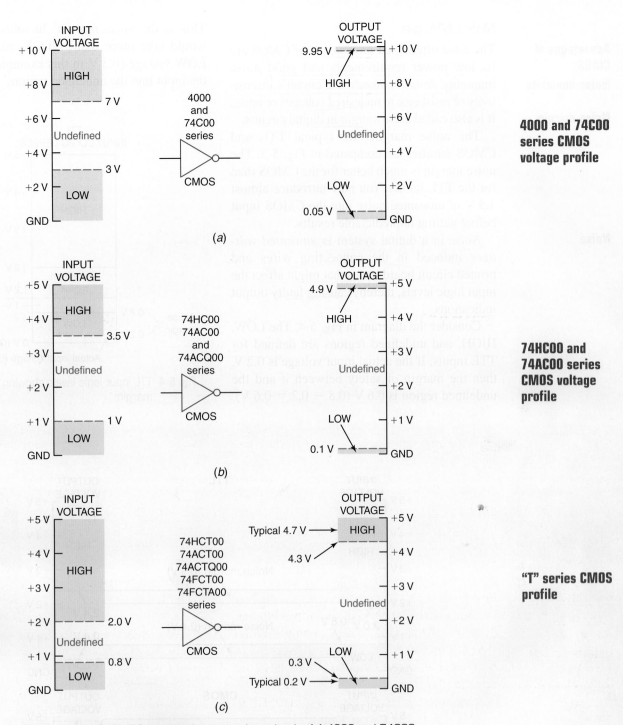

**Fig. 5-2** Defining CMOS input and output voltage levels. (a) 4000 and 74C00 series voltage profile. (b) 74HC00, 74AC00, and 74ACQ00 series voltage profile. (c) 74HCT00, 74ACT00, 74ACTQ00, 74FCT00, 74FCTA00 series voltage profile.

Fig. 5-2(c). Notice that the definition of LOW and HIGH at the *input* is the same for these "T" CMOS ICs as it is for regular bipolar TTL ICs. This can be seen in a comparison of the input side of the voltage profiles of TTL and the "T" CMOS ICs (see Figs. 5-1 and 5-2(c)). The output voltage profiles for all the CMOS ICs are similar. In summary, the "T" series CMOS ICs have typical TTL input voltage characteristics with CMOS outputs.

## Noise Margin

The most often cited *advantages of CMOS* are its low power requirements and good noise immunity. *Noise immunity* is a circuit's insensitivity or resistance to undesired voltages or noise. It is also called *noise margin* in digital circuits.

The noise margins for typical TTL and CMOS families are compared in Fig. 5-3. The noise margin is much better for the CMOS than for the TTL family. You may introduce almost 1.5 V of unwanted noise into the CMOS input before getting unpredictable results.

*Noise* in a digital system is *unwanted voltages* induced in the connecting wires and printed circuit board traces that might affect the input logic levels, thereby causing faulty output indications.

Consider the diagram in Fig. 5-4. The LOW, HIGH, and undefined regions are defined for TTL inputs. If the actual input voltage is 0.2 V, then the margin of safety between it and the undefined region is 0.6 V ($0.8 - 0.2 = 0.6$ V).

This is the *noise margin*. In other words, it would take more than +0.6 V added to the LOW voltage (0.2 V in this example) to move the input into the undefined region.

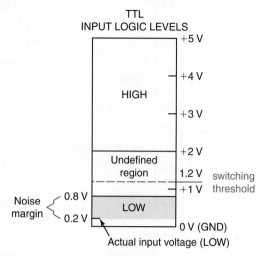

**Fig. 5-4** TTL input logic levels showing noise margin.

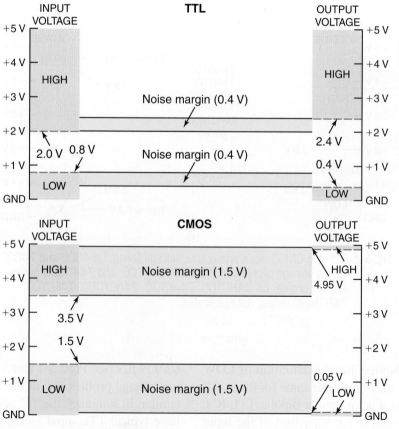

**Fig. 5-3** Defining and comparing TTL and CMOS noise margins.

In actual practice, the noise margin is even greater because the voltage must increase to the *switching threshold,* which is shown as 1.2 V in Fig. 5-4. With the actual LOW input at +0.2 V and the switching threshold at about +1.2 V, the actual noise margin is 1 V (1.2 − 0.2 = 1 V).

The switching threshold is *not* an absolute voltage. It does occur within the undefined region but varies widely because of manufacturer, temperature, and the quality of the components. However, the logic levels are guaranteed by the manufacturer.

## ✔ Self-Test

*Supply the missing word in each statement.*

1. The design of interconnections between two circuits to make them compatible is called _____.
2. An input of +3.1 V to a TTL IC would be considered _____ (HIGH, LOW, undefined).
3. An input of +0.5 V to a TTL IC would be considered _____ (HIGH, LOW, undefined).
4. An output of +2.0 V from a TTL IC would be considered _____ (HIGH, LOW, undefined).
5. An input of +6 V (10-V power supply) to a 4000 series CMOS IC would be considered _____ (HIGH, LOW, undefined).

6. A typical HIGH output (10-V power supply) from a CMOS IC would be about _____ V.
7. An input of +4 V (5-V power supply) to a 74HCT00 series CMOS IC would be considered _____ (HIGH, LOW, undefined).
8. The _____ (CMOS, TTL) family of ICs has better noise immunity.
9. The switching threshold of a digital IC is the *input* voltage at which the output logic level switches from HIGH-to-LOW or LOW-to-HIGH (T or F).
10. The 74FCT00 series of _____ (CMOS, TTL) ICs has an input voltage profile that looks like that of a TTL IC.

## 5-2 Other Digital IC Specifications

Digital logic voltage levels and noise margins were studied in the last section. In this section, other important specifications of digital ICs will be introduced. These include drive capabilities, fan-out and fan-in, propagation delay, and power dissipation.

### Drive Capabilities

A bipolar transistor has its maximum wattage and collector current ratings. These ratings determine its *drive capabilities.* One indication of output drive capability of a digital IC is called its fan-out. The *fan-out* of a digital IC is the number of "standard" inputs that can be driven by the gate's output. If the fan-out for standard TTL gates is 10, this means that the output of a single gate can drive up to 10 inputs of the gates *in the same subfamily.* A typical fan-out value for standard TTL ICs is 10. The fan-out for low-power

Schottky TTL (LS-TTL) is 20 and for the 4000 series CMOS it is considered to be about 50.

Another way to look at the current characteristics of gates is to examine their output drive and input loading parameters. The diagram in Fig. 5-5(*a*), on the next page, is a simplified view of the output drive capabilities and input load characteristics of a standard TTL gate. A standard TTL gate is capable of handling 16 mA when the output is LOW ($I_{OL}$) and 400 μA when the output is HIGH ($I_{OH}$). This seems like a mismatch until you examine the input loading profile for a standard TTL gate. The input loading (worst-case conditions) is only 40 μA with the input HIGH ($I_{IH}$) and 1.6 mA when the input is LOW ($I_{IL}$). This means that the output of a standard TTL gate can drive 10 inputs (16 mA/1.6 mA = 10). Remember, these are *worst-case conditions* and in actual bench tests under static conditions these input load currents are much less than specified.

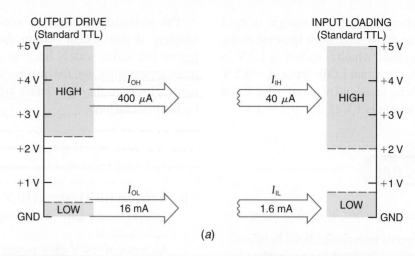

| | Device Family | Output Drive* | Input Loading |
|---|---|---|---|
| **TTL** | Standard TTL | $I_{OH} = 400\ \mu A$<br>$I_{OL} = 16\ mA$ | $I_{IH} = 40\ \mu A$<br>$I_{IL} = 1.6\ mA$ |
| | Low-Power Schottky | $I_{OH} = 400\ \mu A$<br>$I_{OL} = 8\ mA$ | $I_{IH} = 20\ \mu A$<br>$I_{IL} = 400\ \mu A$ |
| | Advanced Low-Power<br>Schottky | $I_{OH} = 400\ \mu A$<br>$I_{OL} = 8\ mA$ | $I_{IH} = 20\ \mu A$<br>$I_{IL} = 100\ \mu A$ |
| | FAST Fairchild Advanced<br>Schottky TTL | $I_{OH} = 1\ mA$<br>$I_{OL} = 20\ mA$ | $I_{IH} = 20\ \mu A$<br>$I_{IL} = 0.6\ mA$ |
| **CMOS** | 4000 Series | $I_{OH} = 400\ \mu A$<br>$I_{OL} = 400\ \mu A$ | $I_{in} = 1\ \mu A$ |
| | 74HC00 Series | $I_{OH} = 4\ mA$<br>$I_{OL} = 4\ mA$ | $I_{in} = 1\ \mu A$ |
| | FACT Fairchild Advanced<br>CMOS Technology Series<br>(AC/ACT/ACQ/ACTQ) | $I_{OH} = 24\ mA$<br>$I_{OL} = 24\ mA$ | $I_{in} = 1\ \mu A$ |
| | FACT Fairchild Advanced<br>CMOS Technology Series<br>(FCT/FCTA) | $I_{OH} = 15\ mA$<br>$I_{OL} = 64\ mA$ | $I_{in} = 1\ \mu A$ |

*Buffers and drivers may have more output drive.

(b)

**Fig. 5-5** (a) Standard TTL voltage and current profiles. (b) Output drive and input loading characteristics for selected TTL and CMOS logic families.

A summary of the *output drive* and *input loading* characteristics of several popular families of digital ICs is detailed in Fig. 5-5(b). Look over this chart of very useful information. You will need this data later when interfacing TTL and CMOS ICs.

**Fan-in**

Notice the outstanding output drive capabilities of the FACT series of CMOS ICs (see Fig. 5-5(b)). The superior drive capabilities, low power consumption, excellent speed, and great noise immunity make the FACT series

of CMOS ICs one of the preferred logic families for new designs. The newer FAST TTL logic series also has many desirable characteristics that make it suitable for new designs.

The load represented by a single gate is called the *fan-in* of that family of ICs. The input loading column in Fig. 5-5(b) can be thought of as the fan-in of these IC families. Notice that the fan-in or input loading characteristics are different for each family of ICs.

Suppose you are given the interfacing problem in Fig. 5-6(a). You are asked if the 74LS04 inverter has enough fan-out to drive the four standard TTL NAND gates on the right.

The voltage and current profiles for LS-TTL and standard TTL gates are sketched in Fig. 5-6(b). The voltage characteristics of all TTL families are compatible. The LS-TTL gate can drive 10 standard TTL gates when its output is HIGH (400 µA/40 µA = 10). However, the LS-TTL gate can drive only five standard TTL gates when it is LOW (8 mA/1.6 mA = 5). We could say that the fan-out of LS-TTL gates is only 5 when driving standard TTL gates. It is true that the LS-TTL inverter can drive four standard TTL inputs in Fig. 5-6(a).

## Propagation Delay

Speed, or quickness of response to a change at the inputs, is an important consideration in high-speed applications of digital ICs. Consider the waveforms in Fig. 5-7(a). The top waveform shows the input to an inverter going from LOW to HIGH and then from HIGH to LOW. The bottom waveform shows the output response to the changes at the input. The slight delay between the time the input changes and the time the output changes is called the *propagation delay* of the inverter. Propagation delay is measured in seconds. The propagation delay for the LOW-to-HIGH transition of the input to the inverter is different from the HIGH-to-LOW delay. Propagation delays are

**Propagation delay**

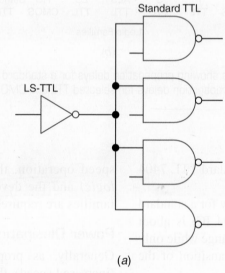

Standard TTL

LS-TTL

(a)

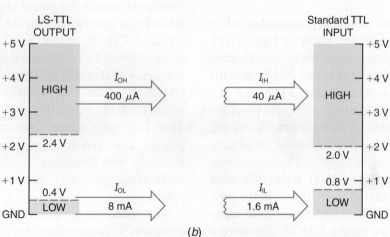

(b)

**Fig. 5-6** Interfacing LS-TTL to standard TTL problem. (*a*) Logic diagram of interfacing problem. (*b*) Voltage and current profiles for visualizing the solution to the problem.

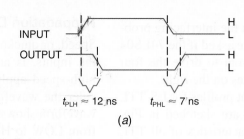

$t_{PLH} \approx 12$ ns    $t_{PHL} \approx 7$ ns

(a)

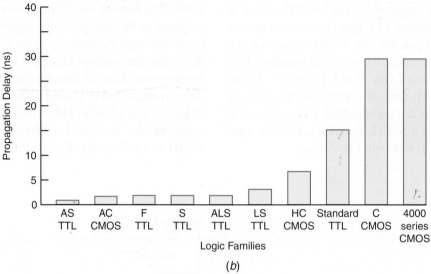

(b)

**Propagation
delays**

Fig. 5-7 (a) Waveforms showing propagation delays for a standard TTL inverter. (b) Graph of propagation delays for selected TTL and CMOS families.

**Emitter coupled
logic (ECL)**

shown in Fig. 5-7(a) for a standard TTL 7404 inverter IC.

The typical propagation delay for a standard TTL inverter (such as the 7404 IC) is about 12 ns for the LOW-to-HIGH change while only 7 ns for the HIGH-to-LOW transition of the input.

**Power
dissipation**

Representative minimum propagation delays are summarized on the graph in Fig. 5-7(b). The lower the propagation delay specification for an IC, the higher the speed. Notice that the AS-TTL (advanced Schottky TTL) and AC-CMOS are the fastest with minimum propagation delays of about 1 ns for a simple inverter. The older 4000 and 74C00 series CMOS families are the slowest families (highest propagation delays). Some 4000 series ICs have propagation delays of over 100 ns. In the past, TTL ICs were considered faster than those manufactured using the CMOS technology. Currently, however, the FACT CMOS series rival the best TTL ICs in low propagation delays (high speed). For extremely high-

**FACT series
CMOS ICs**

speed operation, the *ECL (emitter coupled logic)* and the developing gallium arsenide families are required.

## Power Dissipation

Generally, as propagation delays decrease (increased speed), the power consumption and related heat generation increase. Historically, a standard TTL IC might have a propagation delay of about 10 ns compared with a propagation delay of about 30 to 50 ns for a 4000 series CMOS IC. The 4000 CMOS IC, however, would consume only 0.001 mW, while the standard TTL gate might consume 10 mW of power. The power dissipation of CMOS increases with frequency. So at 100 kHz, the 4000 series gate may consume 0.1 mW of power.

The speed versus power table in Fig. 5-8 compares in graphic form several of the modern TTL and CMOS families. The vertical axis on the graph represents the propagation delay (speed) in nanoseconds while the horizontal

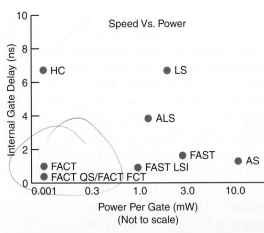

**Fig. 5-8** Speed versus power for selected TTL and CMOS families. (Courtesy of National Semiconductor Corporation.)

axis depicts the power consumption (in milliwatts) of each gate. Families with the most desirable combination of both speed and power are those near the lower left corner of the table. A few years ago many designers suggested that the ALS (advanced low-power Schottky TTL) family was the best compromise between speed and power dissipation. With the introduction of new families, it appears that the FACT (Fairchild Advanced CMOS Technology) series is now one of the best compromise logic families. Both the ALS and the FAST (Fairchild Advanced Schottky TTL) families are also excellent choices for new designs.

**Best logic families**

**Speed versus power chart**

---

## ✓ Self-Test

*Supply the missing word in each statement.*

11. The number of "standard" input loads that can be driven by an IC is called its _____ (fan-in, fan-out).
12. The _____ (4000 series CMOS, FAST TTL series) gates have more output drive capabilities.
13. Refer to Fig. 5-5(*b*). The calculated fan-out when interfacing LS-TTL to LS-TTL is _____.

14. The 4000 series CMOS gates have very low power dissipation, good noise immunity, and _____ (long, short) propagation delays.
15. Refer to Fig. 5-7(*b*). The fastest CMOS subfamily is _____.
16. All TTL subfamilies have _____ (different, the same) voltage and different output drive and input loading characteristics.

IC manufacturers refer to RoHS. RoHS stands for what?

---

---

# 5-3 MOS and CMOS ICs

## MOS ICs

The enhancement type of metal oxide semiconductor field-effect transistor (MOSFET) forms the primary component in MOS ICs. Because of their simplicity, MOS devices use less space on a silicon chip. Therefore, more functions per chip are typical in MOS devices than in bipolar ICs (such as TTL). Metal oxide semiconductor technology is widely used in large scale integration (LSI) and very large scale integration (VLSI) devices because of this packing density on the chip. Microprocessors, memory, and clock chips are typically fabricated using MOS technology. Metal oxide semiconductor circuits are typically of either the PMOS (P-channel MOS) or the newer, faster NMOS (N-channel

MOS) type. Metal oxide semiconductor chips are smaller, consume less power, and have a better noise margin and higher fan-out than bipolar ICs. The main disadvantage of MOS devices is their relative lack of speed.

## CMOS ICs

*Complementary symmetry metal oxide semiconductor (CMOS)* devices use both P-channel and N-channel MOS devices connected end to end. Complementary symmetry metal oxide semiconductor ICs are noted for their *exceptionally low power consumption*. The CMOS family of ICs also has the advantages of low cost, simplicity of design, low heat dissipation, good fan-out, wide logic swings, and good noise-margin performance. Most CMOS families of digital ICs operate on a wide range of voltages.

**Complementary symmetry metal oxide semiconductor ICs**

**PMOS**

**NMOS**

The main disadvantage of many CMOS ICs is that they are somewhat slower than bipolar digital ICs such as TTL devices. Also, extra care must be taken when handling CMOS ICs because they must be protected from static discharges. A static charge or transient voltage in a circuit can damage the very thin silicon dioxide layers inside the CMOS chip. The silicon dioxide layer acts like the dielectric in a capacitor and can be punctured by static discharge and transient voltages.

**Cautions when using CMOS**

If you do work with CMOS ICs, manufacturers suggest preventing damage from static discharge and transient voltages by:

**Conductive foam**

1. Storing CMOS ICs in special *conductive foam* or static shielding bags or containers
2. Using battery-powered soldering irons when working on CMOS chips or grounding the tips of ac-operated units
3. Changing connections or removing CMOS ICs only when the power is turned off

**4000 series**

4. Ensuring that input signals do not exceed power supply voltages
5. Always turning off input signals before circuit power is turned off

**Transmission gates**

6. Connecting *all unused input leads* to either the positive supply voltage or GND, whichever is appropriate (only unused CMOS *outputs* may be left unconnected)

**Bilateral switches**

**74C00 series**

FACT CMOS ICs are much more tolerant of static discharge.

**74HC00 series**

The extremely low power consumption of CMOS ICs makes them ideal for battery-operated portable devices. Complementary symmetry metal oxide semiconductor ICs are widely used in a variety of portable devices.

**CMOS structure**

A typical CMOS device is shown in Fig. 5-9. The top half is a P-channel MOSFET, while the

bottom half is an N-channel MOSFET. Both are enhancement-mode MOSFETs. When the input voltage ($V_{in}$) is LOW, the top MOSFET is on and the bottom unit is off. The output voltage ($V_{out}$) is then HIGH. However, if ($V_{in}$) is HIGH, the bottom device is on and the top MOSFET is off. Therefore, ($V_{out}$) is LOW. The device in Fig. 5-9 acts as an inverter.

Notice in Fig. 5-9 that the $V_{DD}$ of the CMOS unit goes to the positive supply voltage. The $V_{DD}$ lead is labeled $V_{CC}$ (as in TTL) by some manufacturers. The "D" in $V_{DD}$ stands for the *drain* supply in MOSFET. The $V_{SS}$ lead of the CMOS unit is connected to the negative of the power supply. This connection is called GND (as in TTL) by some manufacturers. The "S" in $V_{SS}$ stands for *source* supply in a MOSFET. CMOS ICs typically operate on 5-, 6-, 9-, or 12-V power supplies.

The CMOS technology is used in making several families of digital ICs. The most popular are the 4000, 74C00, 74HC00, and FACT series ICs. The *4000 series* is the oldest. This family has all the customary logic functions plus a few devices that have no equivalent in TTL families. For instance, in CMOS it is possible to produce *transmission gates* or *bilateral switches*. These gates can conduct or allow a signal to pass in either direction like relay contacts.

The 74C00 series is an older CMOS logic family that is the pin-for-pin, function-for-function equivalent of the 7400 series of TTL ICs. As an example, a 7400 TTL IC is designated as a quadruple ("quad") two-input NAND gate as is the 74C00 CMOS ICs.

The 74HC00 series CMOS logic family is designed to replace the 74C00 series and many

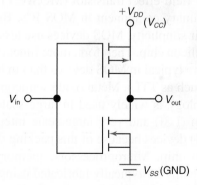

**Fig. 5-9** CMOS structure using P-channel and N-channel MOSFETs in series.

4000 series ICs. It has pin-for-pin, function-for-function equivalents for both 7400 and 4000 series ICs. It is a high-speed CMOS family with good drive capabilities. It operates on a 2- to 6-V power supply.

The FACT (Fairchild Advanced CMOS Technology) logic IC series includes the 74AC00, 74ACQ00, 74ACT00, 74ACTQ00, 74FCT00, and 74FCTA00 subfamilies. The FACT family provides pin-for-pin, function-for-function equivalents for 7400 TTL ICs. The FACT series was designed to outperform existing CMOS and most bipolar logic families. As noted before, the FACT series of CMOS ICs may be the best overall logic family currently available to the designer. It features low power consumption even at modest frequencies (0.1 mW/gate at 1 MHz). Power consumption does however increase at higher frequencies (>50 mW at 40 MHz). It has outstanding noise immunity, with the "Q" devices having patented noise-suppression circuitry. The "T" devices have TTL voltage level inputs. The propagation delays for the FACT series are outstanding (see Fig. 5-7(b)). FACT ICs show excellent resistance to static electricity. The series is also radiation-tolerant making it good in space, medical, and military applications. The output drive capabilities of the FACT family are outstanding (see Fig. 5-5(b)).

---

## ✓ Self-Test

*Supply the missing word in each statement.*

17. Large-scale integration (LSI) and very large-scale integration (VLSI) devices make extensive use of _MOS_ (bipolar, MOS) technology.
18. The letters CMOS stand for _____.
19. The most important advantage of using CMOS is its _____.
20. The $V_{SS}$ pin on a CMOS IC is connected to _____ (positive, GND) of the power supply.
21. The $V_{DD}$ pin on a CMOS IC is connected to _____ (positive, GND) of the power supply.
22. The _____ (FACT, 4000)-series of CMOS ICs are an extremely good choice for new designs because of their low power consumption, good noise immunity, excellent drive capabilities, and outstanding speed.
23. The 74FCT00 would have the same logic function and pinout as the 7400 quad 2-input NAND gate IC. (T or F)

---

## 5-4 Interfacing TTL and CMOS with Switches

One of the most common means of entering information into a digital system is the use of switches or a keyboard. Examples might be the switches on a digital clock, the keys on a calculator, or the keyboard on a microcomputer. This section will detail several methods of using a switch to enter data into either TTL or CMOS digital circuits.

Three simple switch interface circuits are depicted in Fig. 5-10. Pressing the push-button switch in Fig. 5-10(a) will drop the input of the TTL inverter to ground level or LOW. Releasing the push-button switch in Fig. 5-10(a) opens the switch. The input to the TTL inverter now is allowed to "float." In TTL, inputs usually float at a HIGH logic level.

Floating inputs on TTL are not dependable. Figure 5-10(b) is a slight refinement of the switch input circuit in Fig. 5-10(a). The 10-kΩ resistor has been added to make sure the input to the TTL inverter goes HIGH when the switch is open. The 10-kΩ resistor is called a *pull-up resistor.* Its purpose is to pull the input voltage up to +5 V when the input switch is open. Both circuits in Figs. 5-10(a) and (b) illustrate active LOW switches. They are called active LOW switches because the inputs go LOW only when the switch is activated.

An active HIGH input switch is sketched in Fig. 5-10(c). When the input switch is activated, the +5 V is connected directly to the input of the TTL inverter. When the switch is released (opened) the input is pulled LOW by the *pull-down resistor.* The value of the pull-down resistor is relatively low because the input

**Pull-up resistor**

**Pull-down resistor**

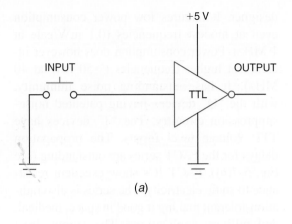

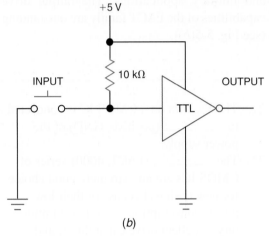

**Switch-to-CMOS interfaces**

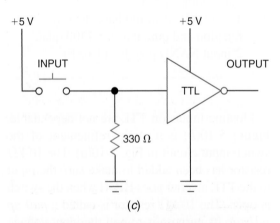

**Switch-to-TTL interfaces**

**Fig. 5-10** Switch-to-TTL interfaces. (*a*) Simple active-LOW switch interface. (*b*) Active-LOW switch interface using pull-up resistor. (*c*) Active-HIGH switch interface using pull-down resistor.

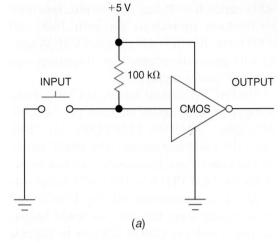

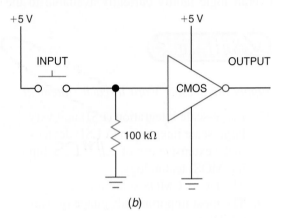

**Fig. 5-11** Switch-to-CMOS interfaces. (*a*) Active-LOW switch interface with pull-up resistor. (*b*) Active-HIGH switch interface with pull-down resistor.

illustrates an active HIGH switch feeding a CMOS inverter. The 100-kΩ pull-down resistor makes sure the input to the CMOS inverter is near ground when the input switch is open. The resistance value of the pull-up and pull-down resistors is much greater than those in TTL interface circuits. This is because the input loading currents are much greater in TTL than in CMOS. The CMOS inverter illustrated in Fig. 5-11 could be from the 4000, 74C00, 74HC00, or the FACT series of CMOS ICs.

## Switch Debouncing

The switch interface circuits in Figs. 5-10 and 5-11 work well for some applications. However, none of the switches in Figs. 5-10 and 5-11 were *debounced*. The lack of a debouncing circuit can be demonstrated by operating the counter shown in Fig. 5-12(*a*). Each press of the input switch should cause the decade (0–9)

current required by a standard TTL gate may be as high as 1.6 mA (see Fig. 5-5(*b*)).

Two switch-to-CMOS interface circuits are drawn in Fig. 5-11. An active LOW input switch is drawn in Fig. 5-11(*a*). The 100-kΩ pull-up resistor pulls the voltage to +5 V when the input switch is open. Figure 5-11(*b*)

To find out more about software switch debouncing using a BASIC stamp, visit www. stampsinclass. com.

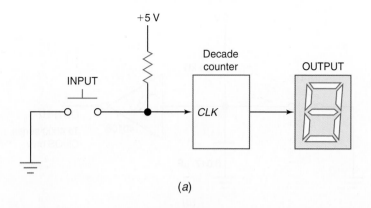

(a)

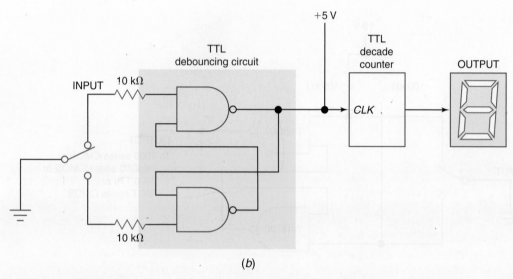

(b)

**Fig. 5-12** (a) Block diagram of switch interfaced to a decimal counter system. (b) Adding a debouncing circuit to make the decimal counter work properly.

counter to increase by 1. However, in practice each press of the switch increases the count by 1, 2, 3, or sometimes more. This means that several pulses are being fed into the clock (CLK) input of the counter each time the switch is pressed. This is caused by switch bounce.

A *switch debouncing circuit* has been added to the counting circuit in Fig. 5-12(b). The decade counter will now count each HIGH–LOW cycle of the input switch. The cross-coupled NAND gates in the debouncing circuit are sometimes called an RS *flip-flop or latch*. Flip-flops will be studied later in greater detail.

Several other switch debouncing circuits are illustrated in Fig. 5-13. The simple debouncing circuit drawn in Fig. 5-13(a) will only work on the slower 4000 series CMOS IC. The 40106 CMOS IC is a special inverter. The 40106 is a *Schmitt trigger inverter*, which means it has a "snap action" when changing to either HIGH or

LOW. A Schmitt trigger can also change a slow-rising signal (such as a sine wave) into a square wave.

The switch debouncing circuit in Fig. 5-13(b) will drive 4000, 74HC00, or FACT series CMOS or TTL ICs. Another general-purpose switch debouncing circuit is illustrated in Fig. 5-13(c). This debouncing circuit can drive either CMOS or TTL inputs. The 7403 is an *open-collector* NAND TTL IC and needs pull-up resistors as shown in Fig. 5-13(c). The external pull-up resistors make it possible to have an output voltage of just about +5 V for a HIGH. Open-collector TTL gates with external pull-up resistors are useful when driving CMOS with TTL.

Another switch debouncing circuit using the versatile 555 timer IC is sketched in Fig. 5-14. When pushbutton switch $SW_1$ is closed (see point A on output waveform) the output toggles

**Switch debouncing circuit**

**RS flip-flop or latch**

**Open-collector TTL output**

**Schmitt trigger inverter**

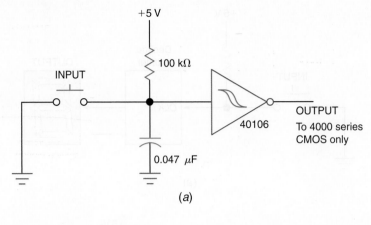

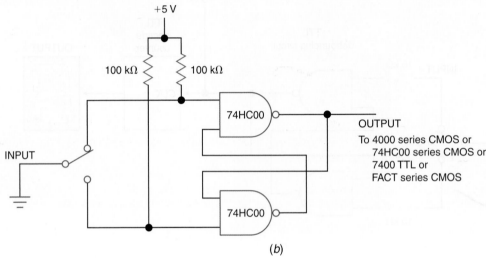

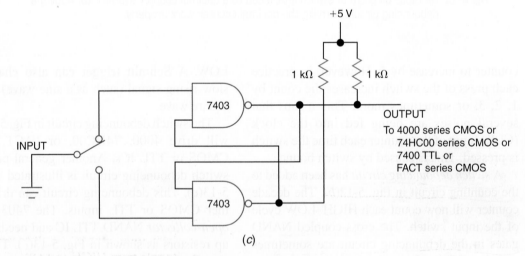

**Switch debouncing circuits**

Fig. 5-13 Switch debouncing circuits. (a) A 4000 series switch debouncing circuit. (b) General purpose switch debouncing circuit that will drive CMOS or TTL inputs. (c) Another general-purpose switch debouncing circuit that will drive CMOS or TTL inputs.

from LOW to HIGH. Later when input switch $SW_1$ is opened (see point B on waveform) the output of the 555 IC remains HIGH for a delay period. After the delay period (about 1 second

for this circuit) the output toggles from HIGH to LOW.

The delay period can be adjusted for the switch debouncing circuit shown in Fig. 5-14.

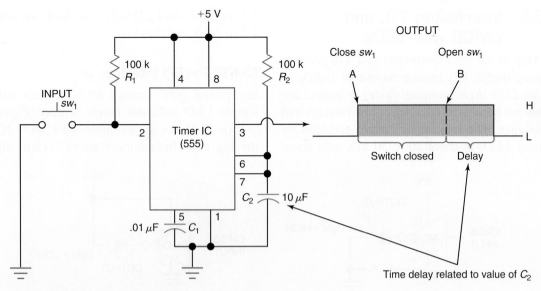

**Fig. 5-14** Switch debouncing circuit using the 555 timer IC.

One method of adjusting the time delay is to change the capacitance value of $C_2$. Decreasing the value of $C_2$ will decrease the delay time at the output of the 555 IC. Increasing the capacitance value of $C_2$ will increase the delay time.

**Self-Test**

*Answer the following questions.*

24. Refer to Fig. 5-10(*a*). The input to the TTL inverter goes _____ (HIGH, LOW) when the switch is pressed (closed) but _____ (floats HIGH, goes LOW) when the input push button is open.

25. Refer to Fig. 5-10(*b*). The 10-k$\Omega$ resistor, which assures the input of the TTL inverter, will go HIGH when the switch that is open is called a _____ (filter, pull-up) resistor.

26. Refer to Fig. 5-12(*b*). The cross-coupled NAND gates that function as a debouncing circuit are sometimes called a(n) _____ or latch.

27. Refer to Fig. 5-10(*c*). Pressing the switch causes the input of the inverter to go _____ (HIGH, LOW) while the output goes _____ (HIGH, LOW).

28. Refer to Fig. 5-11. The inverters and associated resistors form switch debouncing circuits (T or F).

29. Refer to Fig. 5-12(*a*). This decade counter circuit lacks what circuit?

30. Refer to Fig. 5-13(*c*). The 7403 is a TTL inverter with a(n) _____ output.
    a. Open collector
    b. Totem pole
    c. Tri state

31. Refer to Fig. 5-14. Pressing (closing) input switch $SW_1$ caused the output of the 555 IC to toggle from _____ (HIGH-to-LOW, LOW-to-HIGH).

32. Refer to Fig. 5-14. Releasing (opening) input switch SW1 (see point B on output waveform) causes the output of the 555 IC to _____.
    a. Immediately toggle from HIGH to LOW
    b. Toggle from LOW-to-HIGH after a delay time of about 1 millisecond
    c. Toggle from HIGH-to-LOW after a delay time of about 1 second

33. Refer to Fig. 5-14. The time delay at the output of the 555 IC can be decreased by _____ (decreasing, increasing) the capacitance value of $C_2$.

## 5-5 Interfacing TTL and CMOS with LEDs

Many of the lab experiments you will perform using digital ICs require an output indicator. The *LED (light-emitting diode)* is perfect for this job because it operates at low currents and voltages. The maximum current required by many LEDs is about 20 to 30 mA with about 2 V applied. An LED will light dimly on only 1.7 to 1.8 V and 2 mA.

### CMOS-To-LED Interfacing

Interfacing 4000 series CMOS devices with simple LED indicator lamps is easy. Figure 5-15(*a–f*) shows six examples of CMOS ICs driving LED indicators. Figures 5-15(*a*) and

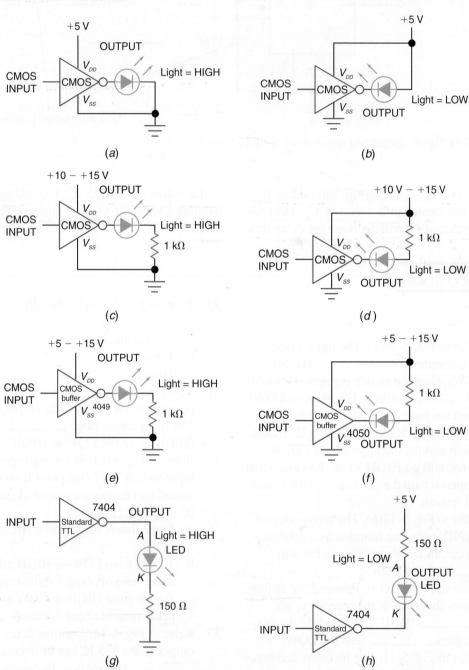

(a)

(b)

(c)

(d)

(e)

(f)

(g)

(h)

**Fig. 5-15** Simple CMOS- and TTL-to-LED interfacing. (*a*) CMOS active HIGH. (*b*) CMOS active-LOW. (*c*) CMOS active HIGH, supply voltage = 10 to 15 V. (*d*) CMOS active LOW, supply voltage = 10 to 15 V. (*e*) CMOS inverting buffer to LED interfacing. (*f*) CMOS noninverting buffer to LED interfacing. (*g*) TTL active-HIGH. (*h*) TTL active-LOW.

**Light-emitting diode (LED)**

**Simple CMOS-to-LED interfacing**

**Simple TTL-to-LED interfacing**

(b) show the CMOS supply voltage at +5 V. At this low voltage, no limiting resistors are needed in series with the LEDs. In Fig. 5-15(a), when the output of the CMOS inverter goes HIGH, the LED output indicator lights. The opposite is true in Fig. 5-15(b): when the CMOS output goes LOW, the LED indicator lights.

Figures 5-15(c) and (d) show the 4000 series CMOS ICs being operated on a higher supply voltage (+10 to +15 V). Because of the higher voltage, a 1-kΩ limiting resistor is placed in series with the LED output indicator lights. When the output of the CMOS inverter in Fig. 5-15(c) goes HIGH, the LED output indicator lights. In Fig. 5-14(d), however, the LED indicator is activated by a LOW at the CMOS output.

Figures 5-15(e) and (f) show CMOS buffers being used to drive LED indicators. The circuits may operate on voltages from +5 to +15 V. Figure 5-15(e) shows the use of an inverting CMOS buffer (like the 4049 IC), while Fig. 5-15(f) uses the noninverting buffer (like the 4050 IC). In both cases, a 1-kΩ limiting resistor must be used in series with the LED output indicator.

## TTL-To-LED Interfacing

Standard TTL gates are sometimes used to drive LEDs directly. Two examples are illustrated in Figs. 5-15(g) and (h). When the output of the inverter in Fig. 5-15(g) goes HIGH, current will flow through the LED causing it to light. The indicator light in Fig. 5-15(h) only lights when the output of the 7404 inverter goes LOW. The circuits in Fig. 5-15 are not recommended for critical uses because they exceed the output current ratings of the ICs. However, the circuits in Fig. 5-15 have been tested and work properly as simple output indicators.

## Current Sourcing and Current Sinking

When reading technical literature or listening to technical discussions you may encounter terms such as *current sourcing* and *current sinking*. The idea behind these terms is illustrated in Fig. 5-16 using TTL ICs to drive LEDs.

In Fig. 5-16(a) the output of the TTL AND gate is HIGH. This HIGH at the output of the AND gate lights the LED. In this example, we talk of the IC as being the source of current (conventional current flow from + to −). The *sourcing current* is sketched on the schematic diagram in Fig. 5-16(a). The source current appears to "flow from the IC" through the external circuit (LED and limiting resistor) to ground.

**Current sourcing**

**Conventional current flow**

In Fig. 5-16(b) the output of the TTL NAND gate is LOW. This LOW at the output of the NAND gate lights the LED. In this example, we refer to the IC as sinking the current. The *sinking current* is sketched on the schematic diagram in Fig. 5-16(b). The sinking current appears to start with +5V above the external circuit (limiting resistor and LED) and "sink to ground" through the external circuit (limiting resistor and LED) and the output pin of the NAND IC.

**Current sinking**

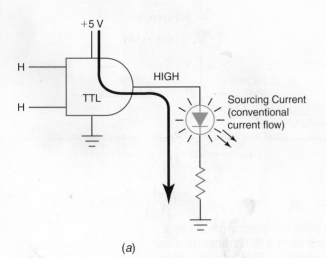

(a)

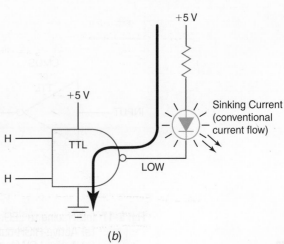

(b)

**Fig. 5-16** (a) Current sourcing. (b) Current sinking.

## Improved LED Output Indicators

Three improved LED output indicator designs are diagramed in Fig. 5-17. Each of the circuits uses transistor drivers and can be used with either CMOS or TTL. The LED in Fig. 5-17(a) lights when the output of the inverter goes HIGH. The LED in Fig. 5-17(b) lights when the output of the inverter goes LOW. Notice that the indicator in Fig. 5-17(b) uses a PNP instead of an NPN transistor.

The LED indicator circuits in Figs. 5-17(a) and (b) are combined in Fig. 5-17(c). The red light ($LED_1$) will light when the inverter's out- put is HIGH. During this time $LED_2$ will be off. When the output of the inverter goes LOW, transistor $Q_1$ turns off while $Q_2$ turns on. The green light ($LED_2$) lights when the output of the inverter is LOW.

The circuit in Fig. 5-17(c) is a very basic logic probe. However, its accuracy is less than most logic probes.

The indicator light shown in Fig. 5-18 uses an incandescent lamp. When the output of the inverter goes HIGH, the transistor is turned on and the lamp lights. When the inverter's out- put is LOW, the lamp does not light.

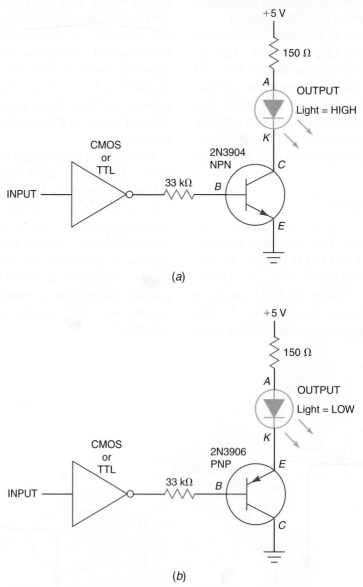

(a)

(b)

**Fig. 5-17** Interfacing to LEDs using a transistor driver circuit. (a) Active-HIGH output using a NPN transistor driver. (b) Active-LOW output using a PNP transistor driver (simplified logic probe).

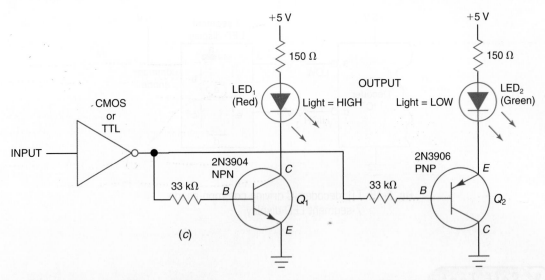

Fig. 5-17 (cont.) Interfacing to LEDs using a transistor driver circuit. (c) HIGH-LOW indicator circuit (simplified logic probe).

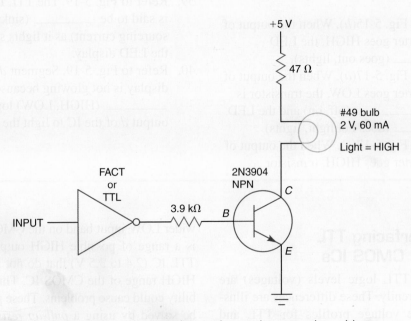

Fig. 5-18 Interfacing to an incandescent lamp using a transistor driver circuit.

Interfacing to an incandescent lamp

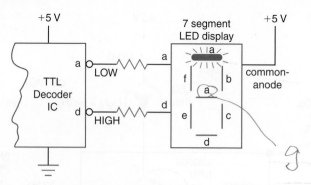

**Fig. 5-19** TTL decoder IC driving common-anode 7-segment LED display.

## Self-Test

*Supply the missing word(s) in each statment.*

34. Refer to Fig. 5-15(*a–f*). The _____ (4000, FAST) series CMOS ICs are being used to drive the LEDs in these circuits.

35. Refer to Fig. 5-15(*h*). When the output of the inverter goes HIGH, the LED _____ (goes out, lights).

36. Refer to Fig. 5-17(*a*). When the output of the inverter goes LOW, the transistor is turned _____ (off, on) and the LED _____ (does not light, lights).

37. Refer to Fig. 5-17(*c*). When the output of the inverter goes HIGH, transistor _____ ($Q_1$, $Q_2$) is turned on and the _____ (green, red) LED lights.

38. Refer to Fig. 5-19. The TTL Decoder IC has _____ (active-HIGH, active-LOW) outputs.

39. Refer to Fig. 5-19. The TTL Decoder IC is said to be _____ (sinking current, sourcing current) as it lights segment *a* of the LED display.

40. Refer to Fig. 5-19. Segment *d* on the display is not glowing because it takes a _____ (HIGH, LOW) logic level at output *d* of the IC to light the LED.

---

**Interfacing TTL and CMOS ICs**

## 5-6 Interfacing TTL and CMOS ICs

CMOS and TTL logic levels (voltages) are defined differently. These differences are illustrated in the voltage profiles for TTL and CMOS shown in Fig. 5-20(*a*). Because of the differences in voltage levels, CMOS and TTL ICs usually cannot simply be connected together. Just as important, current requirements for CMOS and TTL ICs are different.

**Pull-up resistor**

Look at the voltage and current profile in Fig. 5-20(*a*). Note that the output drive currents for the standard TTL are more than adequate to drive CMOS inputs. However, the voltage profiles do not match. The LOW outputs from the TTL are compatible because they fit within the

**TTL-CMOS interfacing**

**CMOS-to-TTL interfacing**

wider LOW input band on the CMOS IC. There is a range of possible HIGH outputs from the TTL IC (2.4 to 3.5 V) that do not fit within the HIGH range of the CMOS IC. This incompatibility could cause problems. These problems can be solved by using a *pull-up resistor* between gates to pull the HIGH output of the standard TTL up closer to +5 V. A completed circuit for interfacing standard TTL to CMOS is shown in Fig. 5-20(*b*). Note the use of the 1-kΩ pull-up resistor. This circuit works for driving either 4000 series, 74HC00, or FACT series CMOS ICs.

Several other examples of TTL-to-CMOS and CMOS-to-TTL interfacing using a common 5-V power supply are detailed in Fig. 5-21. Figure 5-21(*a*) shows the popular LS-TTL driving any CMOS gate. Notice the use of a

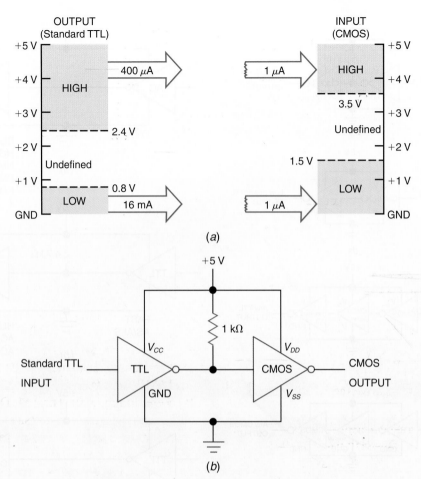

**Fig. 5-20** TTL-to-CMOS interfacing. (*a*) TTL output and CMOS input profiles for visualizing compatibility. (*b*) TTL-to-CMOS interfacing using a pull-up resistor.

**TTL-to-CMOS interfacing**

2.2-kΩ pull-up resistor. The pull-up resistor is being used to pull the TTL HIGH up near +5 V so that it will be compatible with the input voltage characteristics of CMOS ICs.

In Fig. 5-21(*b*), a CMOS inverter (any series) is driving an LS-TTL inverter *directly*. Complementary symmetry metal oxide semiconductor ICs can drive LS-TTL and ALS-TTL (advanced low-power Schottky) inputs: most CMOS ICs cannot drive standard TTL inputs without special interfacing.

Manufacturers have made interfacing easier by designing special buffers and other interface chips for designers. One example is the use of the 4050 noninverting buffer in Fig. 5-21(*c*). The 4050 buffer allows the CMOS inverter to have enough drive current to operate up to two standard TTL inputs.

The problem of voltage incompatibility from TTL (or NMOS) to CMOS was solved in Fig. 5-20 using a pull-up resistor. Another

method of solving this problem is illustrated in Fig. 5-21(*d*). The 74HCT00 series of CMOS ICs is specifically designed as a convenient interface between TTL (or NMOS) and CMOS. Such an interface is implemented in Fig. 5-21(*d*) using the 74HCT34 noninverting IC.

The *74HCT00 series of CMOS ICs* is widely used when interfacing between NMOS devices and CMOS. The NMOS output characteristics are almost the same as for LS-TTL.

The modern *FACT series of CMOS ICs* has excellent output drive capabilities. For this reason FACT series chips can drive TTL, CMOS, NMOS, or PMOS ICs directly as illustrated in Fig. 5-22(*a*). The output voltage characteristics of TTL do not match the input voltage profile of 74HC00, 74AC00, and 74ACQ00 series CMOS ICs. For this reason, a pull-up resistor is used in Fig. 5-22(*b*) to make sure the HIGH output voltage of the TTL gate is pulled up near the +5-V rail of the power supply. Manufacturers

**74HCT00 series of CMOS ICs**

**FACT series of CMOS ICs**

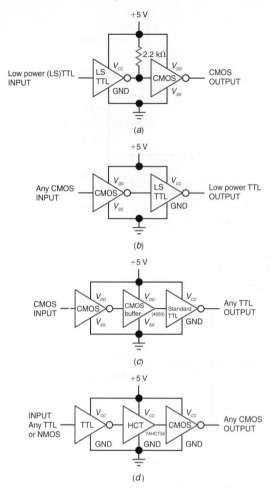

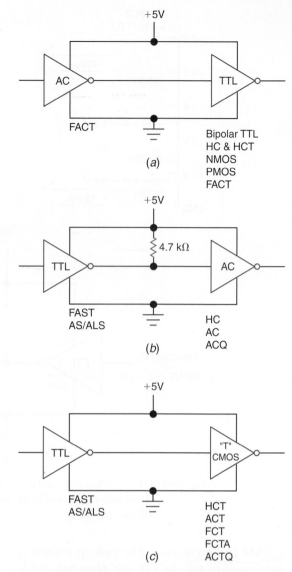

**Fig. 5-21** Interfacing TTL and CMOS when both use a common +5-V power supply. (a) Low-power Schottky TTL to CMOS interfacing using a pull-up resistor. (b) CMOS to low-power Schottky TTL interfacing. (c) CMOS to standard TTL interfacing using a CMOS buffer IC. (d) TTL to CMOS interfacing using a 74HCT00 series IC.

**Fig. 5-22** Interfacing FACT with other families. (a) FACT driving most other TTL and CMOS families. (b) TTL-to-FACT interfacing using a pull-up resistor. (c) TTL-to-"T" CMOS ICs.

produce "T"-type CMOS gates that have the input voltage profile of a TTL IC. TTL gates can directly drive any 74HCT00, 74ACT00, 74FCT00, 74FCTA00, or 74ACTQ00 series CMOS IC, as summarized in Fig. 5-22(c).

Interfacing CMOS devices with TTL devices takes some added components when each operates on a *different voltage power supply*. Figure 5-23 shows three examples of TTL-to-CMOS and CMOS-to-TTL interfacing. Figure 5-23(a) shows the TTL inverter driving a general-purpose NPN transistor. The transistor and associated resistors translate the lower-voltage TTL outputs to the higher-voltage inputs needed to operate the CMOS inverter. The CMOS out-

**CMOS buffer**

put has a voltage swing from about 0 to almost +10 V. Figure 5-23(b) shows an open-collector TTL buffer and a 10-kΩ pull-up resistor being used to translate the lower TTL to the higher CMOS voltages. The 7406 and 7416 TTL ICs are two inverting, open-collector (OC) buffers.

Interfacing between a higher-voltage CMOS inverter and a lower-voltage TTL inverter is shown in Fig. 5-23(c). The 4049 *CMOS buffer* is used between the higher-voltage CMOS inverter and the lower-voltage TTL IC. Note that the CMOS buffer is powered by the lower-voltage (+5 V) power supply in Fig. 5-23(c).

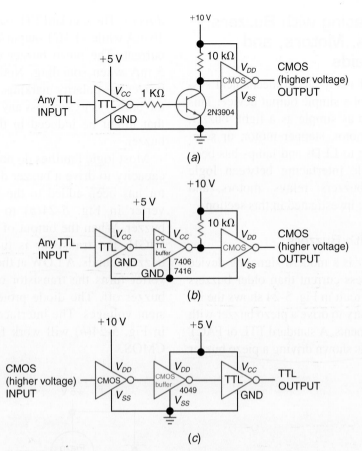

**Fig. 5-23** Interfacing TTL and CMOS when each use a different power supply voltage. (*a*) TTL-to-CMOS interfacing using a driver transistor. (*b*) TTL-to-CMOS interfacing using an open collector TTL buffer IC. (*c*) CMOS-to-TTL interfacing using a CMOS buffer IC.

Looking at the voltage and current profiles (such as in Fig. 5-20(*a*)) is a good starting point when learning about or designing an interface. Manufacturers' manuals are also very helpful.

Several techniques are used to interface between different logic families. These include the use of pull-up resistors and special interface ICs. Sometimes no extra parts are needed.

**✔Self-Test**

*Supply the missing word in each statement.*

41. Refer to Fig. 5-20(*a*). According to this profile of TTL output and CMOS input characteristics, the logic devices _____ (are, are not) voltage compatible.
42. Refer to Fig. 5-21(*a*). The 2.2-kΩ resistor in this circuit is called a _____ resistor.

43. Refer to Fig. 5-21(*c*). The 4050 buffer is a special interface IC that solves the _____ (current drive, voltage) incompatibility between the logic families.
44. Refer to Fig. 5-23(*a*). The _____ (NMOS IC, transistor) translates the TTL logic levels to the higher-voltage CMOS logic levels.

## 5-7 Interfacing with Buzzers, Relays, Motors, and Solenoids

The objective of many electromechanical systems is to control a simple output device. This device might be as simple as a light, buzzer, relay, electric motor, stepper motor, or solenoid. Interfacing to LEDs and lamps has been explored. Simple interfacing between logic elements and buzzers, relays, motors, and solenoids will be investigated in this section.

### Interfacing with Buzzers

**Piezo buzzer**

The *piezo buzzer* is a modern signaling device drawing much less current than older buzzers and bells. The circuit in Fig. 5-24 shows the interfacing necessary to drive a piezo buzzer with digital logic elements. A standard TTL or FACT CMOS inverter is shown driving a piezo buzzer *directly.* The standard TTL output can sink up to 16 mA while a FACT output has 24 mA of drive current. The piezo buzzer draws about 3 to 5 mA when sounding. Notice that the piezo buzzer has polarity markings. The diode across the buzzer is to suppress any transient voltages that might be induced in the system by the buzzer.

Most logic families do not have the current capacity to drive a buzzer directly. A transistor has been added to the output of the inverter in Fig. 5-24(b) to drive the piezo buzzer. When the output of the inverter goes HIGH, the transistor is turned on and the buzzer sounds. A LOW at the output of the inverter turns the transistor off, switching the buzzer off. The diode protects against transient voltages. The interface circuit sketched in Fig. 5-24(b) will work for both TTL and CMOS.

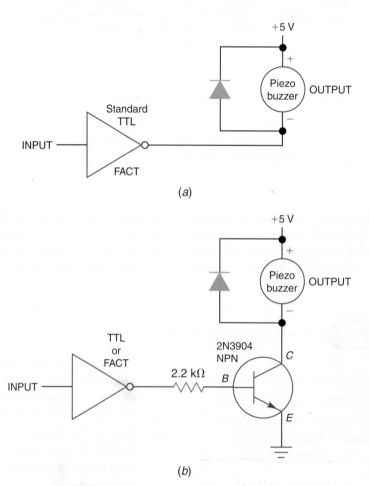

(a)

(b)

**Fig. 5-24** Logic device to buzzer interfacing. (a) Standard TTL or FACT CMOS inverter driving a piezo buzzer directly. (b) TTL or CMOS interfaced with buzzer using a transistor driver.

## Interfacing Using Relays

A *relay* is an excellent method of isolating a logic device from a high-voltage circuit. Figure 5-25 shows how a TTL or CMOS inverter could be interfaced with a relay. When the output of the inverter goes HIGH, the transistor is turned on and the relay is activated. When activated, the normally open (NO) contacts of the relay close as the armature clicks downward. When the output of the inverter in Fig. 5-25 goes LOW, the transistor stops conducting and the relay is deactivated. The armature springs upward to its normally closed (NC) position. The *clamp diode* across the relay coil prevents voltage spikes which might be induced in the system.

The circuit in Fig. 5-26(a) uses a relay to isolate an electric motor from the logic devices. Notice that the logic circuit and dc motor have separate power supplies. When the output of

the inverter goes HIGH, the transistor is turned on and the NO contacts of the relay snap closed. The dc motor operates. When the output of the inverter goes LOW, the transistor stops conducting and the relay contacts spring back to their NC position. This turns off the motor. The electric motor in Fig. 5-26(a) produces rotary motion. A solenoid is an electrical device that can produce linear motion. A solenoid is being driven by a logic gate in Fig. 5-26(b). Note the separate power supplies. This circuit works the same as the motor interface circuit in Fig. 5-26(a).

In summary, voltage and current characteristics of most buzzers, relays, electric motors, and solenoids are radically different from those of logic circuits. Most of these electrical devices need special interfacing circuits to drive and isolate the devices from the logic circuits.

**Relay**

**Clamp diode**

**Logic device to relay interface**

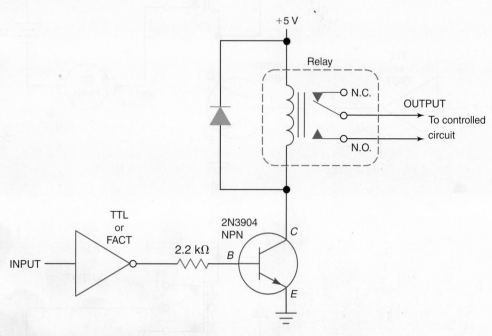

**Fig. 5-25** TTL or CMOS interfaced with a relay using a transistor driver circuit.

---

### ✓ Self-Test

*Supply the missing word(s) in each statement.*

45. Refer to Fig. 5-24(a). If the piezo buzzer draws only 6 mA, it _____ (is, is not) possible for a 4000 series CMOS IC to drive the buzzer directly (see Fig. 5-5(b) for 4000 series data).

46. Refer to Fig. 5-24(b). When the input to the inverter goes LOW, the transistor turns _____ (off, on) and the buzzer _____ (does not sound, sounds).

47. Refer to Fig. 5-25. The purpose of the diode across the coil of the relay is to suppress _____ (sound, transient voltages) induced in the circuit.

48. Refer to Fig. 5-26(a). The dc motor will run only when a _____ (HIGH, LOW) appears at the output of the inverter.
49. If an electric motor produces rotary motion then a solenoid produces _____ (linear, circular) motion.
50. The main purpose of the relay in Fig. 5-26 is to _____ (combine, isolate) the logic circuitry from the higher voltage/current motor or solenoid.

51. Refer to Fig. 5-26(a). If the input to the inverter is LOW its output goes HIGH which _____ (turns on, turns off) the NPN transistor.
52. Refer to Fig. 5-26(a). When the transistor is turned on current flows through the coil of the relay and the armature snaps from the _____ (NC to the NO, NO to the NC) position which activates the motor circuit.

**Logic device to motor interfacing**

**Logic device to solenoid interfacing**

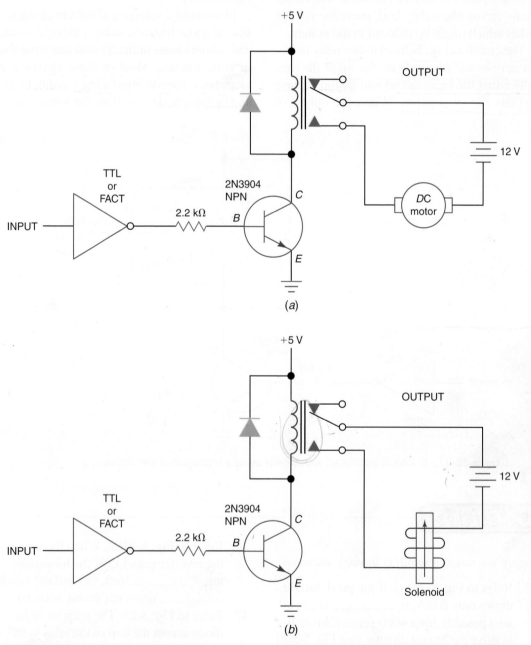

**Fig. 5-26** Using a relay to isolate higher voltage/current circuits from digital circuits. (*a*) Interfacing TTL or CMOS with an electric motor. (*b*) Interfacing TTL or CMOS with a solenoid.

## 5-8 Optoisolators

The relay featured in Fig. 5-26 isolated the lower voltage digital circuitry from the high voltage/current devices such as a solenoid and electric motor. *Electromechanical relays* are relatively large and expensive but are a widely used method of control and isolation. Electromechanical relays can cause unwanted voltage spikes and noise due to the coil windings and opening and closing of contact points. A useful alternative to an electromagnetic relay when interfacing with digital circuits is the *optoisolator* or *optocoupler.* One close relative of the optoisolator is the *solid-state relay.*

One economical optoisolator is featured in Fig. 5-27. The *4N25 optoisolator* consists of a *gallium arsenide infrared-emitting diode* optically coupled to a silicon *phototransistor detector* enclosed in a six-pin dual in-line package (DIP). Figure 5-27(*a*) details the pin diagram for the 4N25 optoisolator with the names of the pins. On the input side, the LED is typically activated with a current of about 10 to 30 mA. When the input LED is activated, the light activates (turns on) the phototransistor. With no current through the LED the output phototransistor of the optoisolator is turned off (high resistance from emitter-to-collector).

A simple test circuit using the 4N25 optoisolator is shown in Fig. 5-27(*b*). The digital signal from the output of a TTL or FACT inverter directly drives the infrared-emitting diode. The circuit is designed so the LED is activated when the output of the inverter goes LOW, which allows the inverter to sink the 10 to 20 mA LED current to ground. When the LED is activated, infrared light shines (inside the package) activating the phototransistor. The transistor is turned on (low resistance from emitter-to-collector) dropping the voltage at the collector (OUTPUT) to near 0 V. If the output of the

inverter goes HIGH, the LED does not light and the NPN phototransistor turns off (high resistance from emitter-to-collector). The output (at the collector) is pulled to about +12 V (HIGH) by the 10-kΩ pull-up resistor. In this example, notice that the input side of the circuit operates on +5 V while the output side in this example uses a separate +12 V power supply. In summary, the input and output sides of the circuit are isolated from one another. When pin 2 of the optoisolator goes LOW, the output at the collector of the transistor goes LOW. The grounds of the separate power supplies should not be connected to complete the isolation between the low- and high-voltage sides of the optoisolator.

**Optoisolator**

A simple application of the optoisolator being used to interface between TTL circuitry and a piezo buzzer is diagramed in Fig. 5-27(*c*). In this example the pull-up resistor is removed because we are using the NPN phototransistor in the optoisolator to sink the 2 to 4 mA of current when the transistor is activated. A LOW at the output of the inverter (pin 2 of optoisolator) activates the LED, which in turn activates the phototransistor.

To control heavier loads using the optoisolator we could attach a power transistor to the output as is done in Fig. 5-27(*d*). In this example, if the LED is activated it activates the phototransistor. The output of the optoisolator (pin 5) drops LOW, which turns off the power transistor. The emitter-to-collector resistance of the power transistor is high, turning off the dc motor. When the output of the TTL inverter goes HIGH it turns off the LED and the phototransistor in the optoisolator. The voltage at output pin 5 goes positive, which turns on the power transistor and operates the dc motor.

If the power transistor (or other power-handling device such as a triac) in Fig. 5-27(*d*) were housed in the isolation unit the entire device is sometimes a *solid-state relay.* Solid-state

**Solid-state relay**

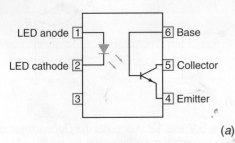

(*a*)

**Fig. 5-27** (*a*) The 4N25 optoisolator pin out and six-pin DIP.

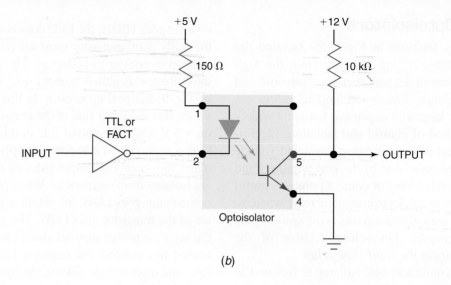

(b)

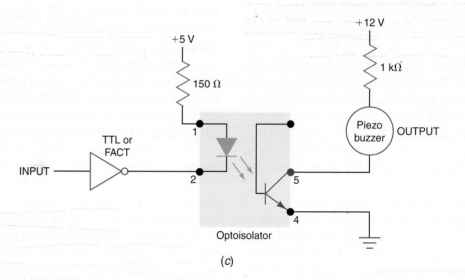

(c)

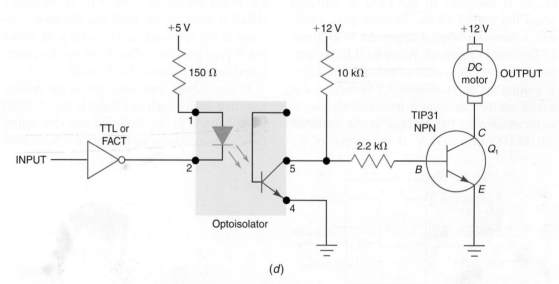

(d)

**Fig. 5-27 (cont.)** (b) Basic optoisolator circuit separates 5 V and 12 V circuits. (c) Optoisolator driving piezo buzzer. (d) Optoisolator isolating low-voltage digital circuit from high voltage/current motor circuit.

(a)

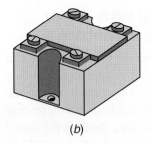

(b)

**Fig. 5-28** (a) Solid-state relay—small PC-mounted package.
(b) Solid-state relay—heavy duty package.

relays can be purchased to handle a variety of outputs included in either ac or dc loads. The output circuitry in a solid-state relay may be more complicated than that shown in Fig. 5-27(d).

Several examples of solid-state relay packages are shown in Fig. 5-28. The unit in Fig. 5-28(a) is a smaller PC-mounted unit. The larger bolted-on solid-state relay has screw terminals and can handle greater ac currents and voltages.

In summary, it is common to isolate digital circuitry from some devices because of high operating voltages and currents or because of dangerous feedback in the form of voltage

spikes and noise. Traditionally, electromagnetic relays have been used for isolation, but optoisolators and solid-state relays are an inexpensive and effective alternative when interfacing with digital circuits. A typical optoisolator, shown in Fig. 5-27(a), contains an infrared-emitting diode that activates a phototransistor. If you are building an interface project using the parallel port from an IBM-compatible or PC, you will want to use optoisolators between your circuits and the computer. The PC parallel-port outputs and inputs operate with TTL level signals. Good isolation protects your computer from voltage spikes and noise.

**✓ Self-Test**

*Supply the missing word(s) in each statement.*

53. Refer to Fig. 5-26(a). The _____ (relay, transistor) isolates the digital circuitry from the higher voltage and noisy dc motor circuit.

54. The 4N25 optoisolator device contains an infrared-emitting diode optically coupled to a _____ (phototransistor, triac) detector enclosed in a six-pin DIP.

55. Refer to Fig. 5-27(b). If the output of the TTL inverter goes LOW, the infrared LED _____ (does not light, lights), which _____ (activates, deactivates) the phototransistor and the voltage at pin 5 (output) goes _____ (HIGH, LOW).

56. Refer to Fig. 5-27(b). The 10-kΩ resistor connecting the collector of the phototransistor to +12 V is called a(n) _____ resistor.

57. Refer to Fig. 5-27(c). If the output of the TTL inverter goes HIGH the LED

_____ (does not light, lights), which _____ (activates, deactivates) the phototransistor and the voltage at pin 5 (output) goes (HIGH, LOW) and the buzzer _____ (does not sound, sounds).

58. Refer to Fig. 5-27(d). If the output of the TTL inverter goes HIGH the LED does not light, which deactivates (turns off) the phototransistor and the voltage at pin 5 (output) goes more positive. This positive-going voltage at the base of the power transistor _____ (turns on, turns off) $Q_1$ and the dc motor _____ (does not run, runs).

59. The _____ (electromagnetic, solid-state) relay is a close relative of an optoisolator.

60. Refer to Fig. 5-27(c). If the input to the inverter is HIGH then the piezo buzzer _____ (will not sound, will sound).

## 5-9 Interfacing with Servo and Stepper Motors

The dc motor mentioned previously in this chapter is a device that rotates continuously when power is applied. The control over the dc motor is limited to ON-OFF, or if you reverse the direction of current flow through the motor the direction of rotation reverses. A simple dc motor does not facilitate good speed control and it will not rotate a given number of degrees to stop for angular positioning. Where precision positioning or exacting speed are required, a regular dc motor does not do the job.

### Servo Motor

Both the servo and the stepper motor can rotate to a given position and stop and also reverse direction. The word "servo" is short for *servo motor.* "Servo" is a general term for a motor in which either the angular position or speed can be controlled precisely by a servo loop which uses feedback from the output back to input for control. The most common servos are the inexpensive units used in model aircraft, model cars, and some educational robot kits. These servos are geared-down dc motors with built-in electronics that respond to different pulse widths. These servos use feedback to ensure the device rotates to and stays at the current angular position. These servos are popular in remote-control models and toys. They commonly have three wires (one wire for input and two wires for power) and are not commonly used for continuous rotation.

The position of a hobby servo's output shaft is determined by the width or duration of the control pulse. The width of the control pulse commonly varies from about 1 to 2 msec. The concept of controlling the hobby servo motor using a control pulse is sketched in Fig. 5-29. The *pulse generator* emits a constant frequency of about 50 Hz. The *pulse width* (or *pulse duration*) can be changed by the operator using an input device such as a potentiometer or joystick. The internally geared motor and feedback-and-control circuitry inside the servo motor responds to the continuous stream of pulses by rotating to a new angular position. As an example, if the pulse width is 1.5 msec the shaft moves to the middle of its range as illustrated in Fig. 5-29(*a*). If the pulse width decreases to 1 msec the output shaft takes a new position rotating about 90° clockwise as shown in Fig. 5-29(*b*). Finally if the

pulse width increases to 2 msec as in Fig. 5-19(*c*), the output shaft moves counterclockwise to its new position.

The changing of the pulse duration is called *pulse-width modulation* (PWM). In the example shown in Fig. 5-29 the pulse generator outputs a constant frequency of 50 Hz but the pulse width can be adjusted.

A sketch of the internal functions of a servo motor is shown in Fig. 5-29(*d*). The servo contains a dc motor and speed-reducing gear. The last gear drives the output shaft and is also connected to a potentiometer. The potentiometer senses the angular position of the output. The varying resistance of the potentiometer is fed back to the control circuitry and repeatedly *compares* the pulse-width of the external (input) pulse with an internally generated pulse from a one-shot inside the control circuit. The internal pulse width is varied based on the feedback from the potentiometer.

For the servo motor in Fig. 5-29, suppose the *external pulse width is 1.5-msec* and the internal pulse width is 1.0-msec. After comparing the pulses the control circuitry would start to rotate the output shaft in a CCW direction. After each external pulse (50 times per second) the control circuitry would make a small CCW shaft adjustment until the external and internal pulse widths are both 1.5-msec. At this point the shaft would stop in the position shown in Fig. 5-29(*a*).

Next for the servo motor in Fig. 5-29, suppose the *external pulse width changes to 1-msec* and the internal pulse width, based on the feedback from the potentiometer, is at 1.5-msec. After comparing the pulses the control circuitry would start to rotate the output shaft in a CW direction. After each external pulse (50 times per second) the control circuitry would make a small CW shaft adjustment until the external and internal pulse width are both 1.0-msec. At this point the shaft would stop in the position shown in Fig. 5-29(*b*).

When both external and internal pulse widths are equal for the servo motor in Fig. 5-29, the control circuitry stops the dc motor. For instance, if both the external and internal pulse widths are 2.0-msec, then the output shaft would freeze in the position shown in Fig. 5-29(*c*).

Some hobby servo motors may have opposite rotational characteristics from the unit featured in Fig. 5-29. Some servos are internally wired so that a narrow pulse (1-msec) would cause full CCW rotation instead of CW rotation shown in Fig. 5-29(*b*). Likewise, a wide

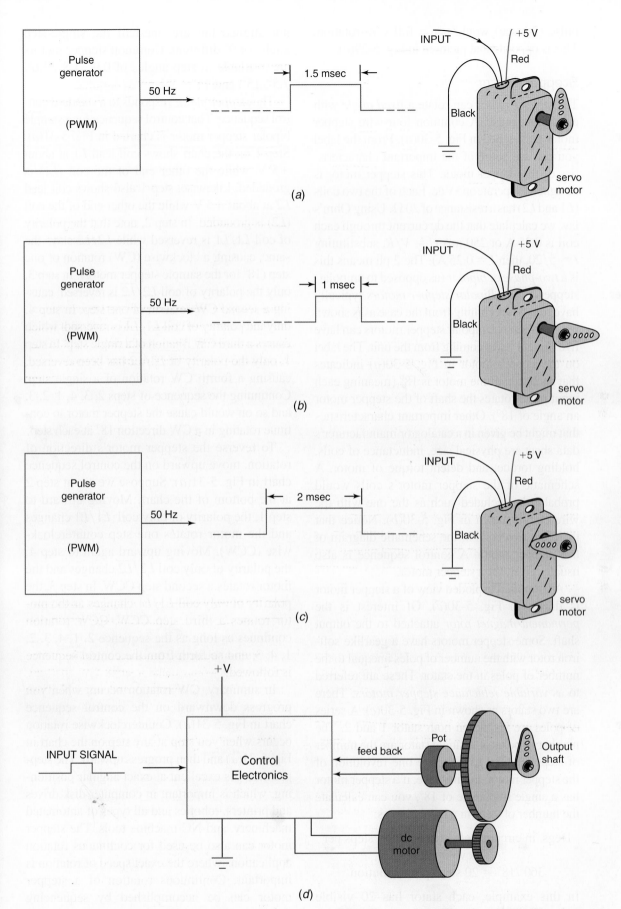

**Fig. 5-29** Controlling the angular position of a hobby servo motor using pulse-width modulation (PWM). (NOTE: Some hobby servos rotate in the opposite direction as the pulse width increases.)

pulse (2-msec) would cause full CW rotation. This is opposite that pictured in Fig. 5-29(c).

## Stepper Motor

Control sequence for stepper motor

The *stepper motor* can rotate a *fixed angle* with each input pulse. A common four-wire stepper motor is sketched in Fig. 5-30(a). From the label you can see some of the important characteristics of the stepper motor. This stepper motor is designed to operate on 5 V dc. Each of the two coils (L1 and L2) has a resistance of 20 $\Omega$. Using Ohm's law we calculate that the dc current through each coil is 0.25 A or 250 mA ($I = V/R$, substituting $I = 5/20$, then $I = 0.25$ A). The 2 ph means this is a *two-phase* or *bipolar* (as opposed to unipolar) stepper motor. *Bipolar stepper motors* typically have four wires coming from the case as is shown in Fig. 5-30(a). Unipolar stepper motors can have five to eight wires coming from the unit. The label on the stepper motor in Fig. 5-30(a) indicates that each step of the motor is 18° (meaning each input pulse rotates the shaft of the stepper motor an angle of 18°). Other important characteristics that might be given in a catalog or manufacturer's data sheet are physical size, inductance of coils, holding torque, and detent torque of motor. A schematic of the stepper motor's coils would probably be included such as the one with the wire colors shown in Fig. 5-30(b). Notice that there are two coils in the schematic diagram of this stepper motor. A control sequence is also usually given for a stepper motor.

Bipolar stepper motor

A simplified exploded view of a stepper motor is drawn in Fig. 5-30(c). Of interest is the *permanent magnet rotor* attached to the output shaft. Some stepper motors have a gearlike soft-iron rotor with the number of poles unequal to the number of poles in the stator. These are referred to as *variable reluctance stepper motors*. There are two stators as shown in Fig. 5-30(c). A series of poles are visible on both stator 1 and 2. The number of poles on a single stator are the number of steps required to complete one revolution of the stepper motor. For instance, if a stepper motor has a single step angle of 18°, you can calculate the number of steps in a revolution as:

Variable reluctance stepper motor

Degs. in circle/single-step angle = steps per revolution
$$360°/18° = 20 \text{ steps per revolution}$$

In this example, each stator has 20 visible poles. Notice that the poles of stator 1 and 2 are not aligned but are one-half the single-step angle, or 9° different. Common stepper motors are available in step angles of 0.9°, 1.8°, 3.6°, 7.5°, 15°, and 18°.

The *stepper motor* responds to a standard control sequence. That control sequence for a sample bipolar stepper motor is charted in Fig. 5-31(a). Step 1 on the chart shows coil lead L1 at about +5 V, while the other end of the coil ($\overline{L1}$) is grounded. Likewise, step 1 also shows coil lead L2 at about +5 V while the other end of the coil ($\overline{L2}$) is grounded. In step 2, note that the polarity of coil $L1/\overline{L1}$ is reversed while $L2/\overline{L2}$ stays the same, causing a clockwise (CW) rotation of one step (18° for the sample stepper motor). In step 3, only the polarity of coil $L2/\overline{L2}$ is reversed, causing a second CW rotation of one step. In step 4, only the polarity of coil $L1/\overline{L1}$ is reversed, which causes a third CW rotation of a single step. In step 1, only the polarity of $L2/\overline{L2}$ has been reversed, causing a fourth CW rotation of a single step. Continuing the sequence of steps 2, 3, 4, 1, 2, 3, and so on would cause the stepper motor to continue rotating in a CW direction 18° at each step.

To reverse the stepper motor's direction of rotation, move upward on the control sequence chart in Fig. 5-31(a). Suppose we are at step 2 at the bottom of the chart. Moving upward to step 1, the polarity of only coil $L1/\overline{L1}$ changes and the motor rotates one step counterclockwise (CCW). Moving upward again to step 4, the polarity of only coil $L2/\overline{L2}$ changes and the motor rotates a second step CCW. In step 3, the polarity of only coil $L1/\overline{L1}$ changes as the motor rotates a third step CCW. CCW rotation continues as long as the sequence 2, 1, 4, 3, 2, 1, 4, 3, and so forth from the control sequence is followed.

In summary, CW rotation occurs when you progress downward on the control sequence chart in Fig. 5-31(a). Counterclockwise rotation occurs when you stop at any step on the chart in Fig. 5-31(a) and then progress upward. The stepper motor is excellent at exact angular positioning, which is important in computer disk drives and printers, robotics and all types of automated machinery, and NC machine tools. The stepper motor can also be used for continuous rotation applications where the exact speed of rotation is important. Continuous rotation of a stepper motor can be accomplished by sequencing through the control sequence quickly. For

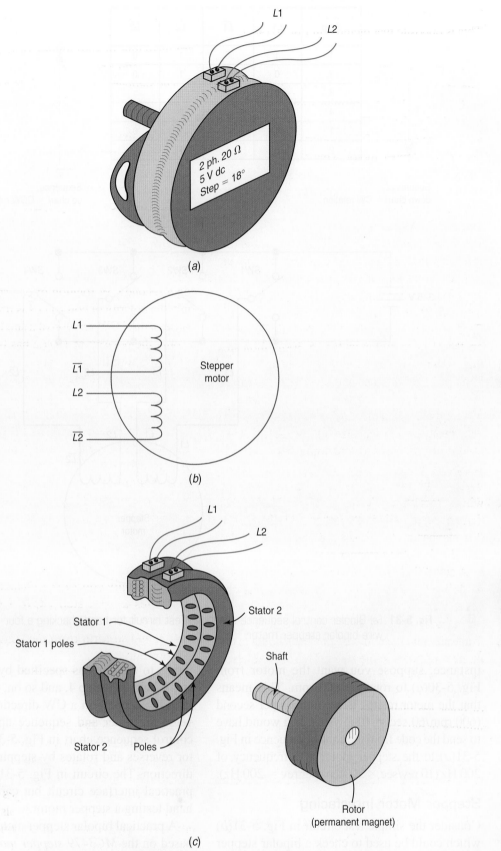

2 ph. 20 Ω
5 V dc
Step = 18°

(a)

L1
$\overline{L1}$
L2
$\overline{L2}$

Stepper
motor

(b)

Stator 1
Stator 1 poles

Stator 2

Shaft

Stator 2
Poles

Rotor
(permanent magnet)

(c)

**Fig. 5-30** (a) Typical four-wire stepper motor. (b) Schematic of four-wire
bipolar stepper motor. (c) Simple exploded view of typical
permanent magnet type stepper motor.

Internet
Connection

Simulations of
several types of
motors including
a stepper motor
can be found at
Motorola's web-
site.

| Step | L1 | $\overline{L1}$ | L2 | $\overline{L2}$ |
|------|----|----|----|----|
| 1 | 1 | 0 | 1 | 0 |
| 2 | 0 | 1 | 1 | 0 |
| 3 | 0 | 1 | 0 | 1 |
| 4 | 1 | 0 | 0 | 1 |
| 1 | 1 | 0 | 1 | 0 |
| 2 | 0 | 1 | 1 | 0 |

Sequence:
down chart = CW rotation

Sequence:
up chart = CCW rotation

(a)

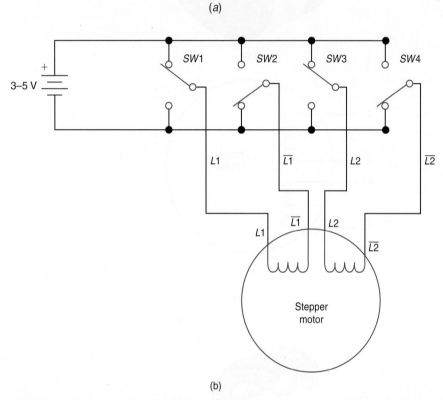

(b)

**Fig. 5-31** (*a*) Bipolar control sequence chart. (*b*) Test circuit for hand checking a four-wire bipolar stepper motor.

instance, suppose you want the motor from Fig. 5-30(*a*) to rotate at 600 rpm. This means that the motor rotates 10 revolutions per second (600 rpm/60 sec = 10 rev/sec). You would have to send the code from the control sequence in Fig. 5-31(*a*) to the stepper motor at a frequency of 200 Hz (10 rev/sec × 20 steps per rev = 200 Hz).

### Stepper Motor Interfacing

Consider the simple test circuit in Fig. 5-31(*b*) which could be used to check a bipolar stepper motor. The single-pole double-throw (SPDT) switches are currently set to deliver the voltages defined by step 1 on the control sequence chart in Fig. 5-31(*a*). As you change the voltage

inputs to the coils as specified by step 2, then step 3, and then step 4, and so on, the motor rotates by stepping in a CW direction. If you reverse the order and sequence upward on the control sequence chart in Fig. 5-31(*a*) the motor reverses and rotates by stepping in a CCW direction. The circuit in Fig. 5-31(*b*) is an impractical interface circuit but can be used for hand testing a stepper motor.

A practical bipolar stepper motor interface is based on the *MC3479 stepper motor driver IC* from Motorola. The schematic diagram in Fig. 5-32(*a*) details how you might wire the MC3479 driver IC to a bipolar stepper motor. The MC3479 IC has a logic section that generates

INPUT

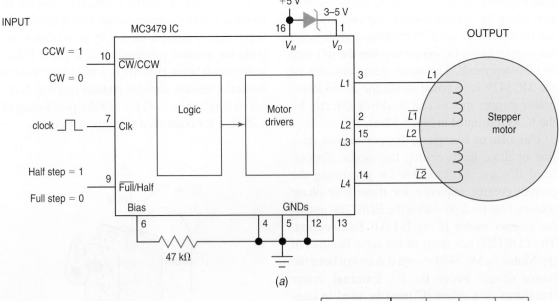

(a)

| Step | L1 | L2 | L3 | L4 |
|------|----|----|----|----|
| 1 | 1 | 0 | 1 | 0 |
| 2 | 0 | 1 | 1 | 0 |
| 3 | 0 | 1 | 0 | 1 |
| 4 | 1 | 0 | 0 | 1 |
| 1 | 1 | 0 | 1 | 0 |
| 2 | 0 | 1 | 1 | 0 |

(b)

| Step | L1 | L2 | L3 | L4 |
|------|----|----|----|----|
| 1 | 1 | 0 | 1 | 0 |
| 2 | 1 | 1 | 1 | 0 |
| 3 | 0 | 1 | 1 | 0 |
| 4 | 0 | 1 | 1 | 1 |
| 5 | 0 | 1 | 0 | 1 |
| 6 | 1 | 1 | 0 | 1 |
| 7 | 1 | 0 | 0 | 1 |
| 8 | 1 | 0 | 1 | 1 |
| 1 | 1 | 0 | 1 | 0 |
| 2 | 1 | 1 | 1 | 0 |
| 3 | 0 | 1 | 1 | 0 |
| 4 | 0 | 1 | 1 | 1 |

(c)

**Fig. 5-32** (a) Using the MC3479 Stepper Motor Driver IC to interface with a bipolar stepper motor. (b) Control sequence of the MC3479 IC in the full-step mode. (c) Control sequence for the MC3479 IC in the half-step mode.

the proper control sequence to drive a bipolar stepper motor. The motor driver section has a drive capability of 350 mA per coil. Each step of the motor is triggered by a single positive-going clock pulse entering the CLK input (pin 7) of the IC. One input control sets the direction of rotation of the stepper motor. A logic 0 at the CW/CCW input to the MC3479 allows CW rotation while a logic 1 input at pin 10 changes to CCW rotation of the stepper motor.

The MC3479 IC also has a Full/Half input (pin 9) which can change the operation of the IC from stepping by full steps or half steps. In the *full-step mode,* the stepper motor featured

in Fig. 5-30 rotates 18° for each clock pulse (each single step). In the *half-step mode,* the stepper motor rotates half of a regular step or only 9° per clock pulse. The *control sequence* used by the MC3479 IC in the full-step mode is shown in chart form in Fig. 5-32(b). Note that this is the same control sequence used in Fig. 5-31(a). The control sequence used by the MC3479 IC in the half-step mode is detailed in chart form in Fig. 5-32(c). These control sequences are standard for bipolar or two-phase stepper motors and are built into the logic block of the *MC3479 stepper motor driver IC.* Specialized ICs such as the MC3479 stepper motor driver are usually the simplest and

**MC 3479 Stepper motor driver IC**

least expensive method of solving the problem of generating the correct control sequences, allowing for either CW or CCW rotation, and allowing the stepper motor to operate in either the full-step or half-step mode. The motor driver circuitry of the MC3479 is included inside the IC so lower power stepper motors can be driven directly by the IC as illustrated in Fig. 5-32(*a*).

Unipolar or four-phase stepper motors have five or more leads exiting the motor. Specialized ICs are also available for generating the correct control sequence for these four-phase motors. One such product is the *EDE1200 unipolar stepper motor IC* by E-LAB Engineering. The EDE1200 has many of the same features of the Motorola MC3479 except it does not have the motor drivers inside the IC. External driver transistors or a driver IC must be used in conjunction with the EDE1200 unipolar stepper motor IC. The control sequence for four-phase (unipolar) and two-phase (bipolar) stepper motors is different.

In summary, a simple permanent magnet dc motor is good for continuous rotation applications. Servo motors (such as the hobby servo motor) are good for angular positioning of a shaft. Pulse-width modulation (PWM) is a technique used to rotate the servo to an exact angular position. Stepper motors can be used for angular positioning of a shaft or for controlled continuous rotation.

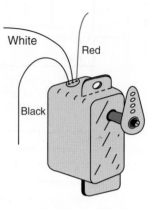

White
Red
Black

**Fig. 5-33** Art for self-test questions 64, 65, and 66.

✓ **Self-Test**

*Answer the following questions.*

61. The _____ (dc motor, servo motor) is a good choice for continuous rotation applications not requiring speed control.

62. The _____ (dc motor, stepper motor) is a good choice for applications that require exact angular positioning of a shaft.

63. Both the servo and stepper motors can be used in applications that require exact angular positioning (T or F).

64. Refer to Fig. 5-33. This device, which might be found in a radio-controlled airplane or car, is called a _____ (servo motor, stepper motor).

65. Refer to Fig. 5-33. The red lead is connected to + of the power supply, the black lead to ground, and the white lead to the servo is the _____ (input, output) lead.

66. Refer to Fig. 5-33. This hobby servo motor is controlled by inputs from a pulse generator using _____ (pulse-amplitude, pulse-width) modulation.

67. The device featured in Fig. 5-30 is a _____ (bipolar, unipolar) stepper motor.

68. The chart in Fig. 5-31(*a*) shows the _____ sequence for a _____ (bipolar, unipolar) stepper motor.

69. Refer to Fig. 5-31(*a*). If we are at step 4 and progress upward on the control sequence chart to step 3, the stepper motor rotates in a _____ (CCW, CW) direction.

70. Refer to Fig. 5-32(*a*). The _____ (logic, motor drive) block inside the MC3479 IC assures that the control sequence for driving a bipolar stepper motor is followed.

71. Refer to Fig. 5-32(*a*). The maximum drive current for each coil available using the MC3479 is _____ (10, 350) mA, which allows it to drive many smaller stepper motors directly.

72. Refer to Fig. 5-32(*a*) and assume input pins 9 and 10 are HIGH. When a clock pulse enters pin 7, the attached stepper motor rotates _____ (CCW, CW) a _____ (full step, half step).

## 5-10 Using Hall-Effect Sensors

The *hall-effect sensor* is often used to solve difficult switching applications. Hall-effect sensors are *magnetically-activated* sensors or switches. Hall-effect sensors are immune to environmental contaminants and are suitable for use under severe conditions. Hall-effect sensors operate reliably under oily and dirty, hot or cold, bright or dark, and wet or dry conditions.

Several examples of where Hall-effect sensors and switches might be used in a modern automobile are graphically summarized in Fig. 5-34. Hall-effect sensors and switches are also used in other applications such as ignition systems, security systems, mechanical limit switches, computers, printers, disk drives, keyboards, machine tools, position detectors, and brushless dc motor commutators.

Many of the advances in automotive technology revolve around accurate reliable sensors sending data to the central computer. The central computer gathers the sensor data and controls many functions of the engine and other systems of the automobile. The computer also gathers and stores data from the sensors to be used by the *On-Board Diagnostics* system (OBD I or the newer OBD II). Only some of the many sensors in an automobile are Hall-effect devices.

### Basic Hall-Effect Sensor

The basic Hall-effect sensor is a semiconductor material represented in Fig. 5-35(a). A source voltage (bias voltage) will cause a constant bias current to flow through the Hall-effect sensor. As demonstrated in Fig. 5-35(a), when a magnetic field is present a voltage is generated by the Hall-effect sensor. The Hall-voltage is proportional to the strength of the magnetic field. As an example, if no magnetic field is present then the sensor will produce no Hall-effect voltage at the output. As the magnetic field increases the Hall voltage increases proportionally. In summary, if a biased Hall sensor is placed in a magnetic field the voltage output will be directly proportional to the strength of the magnetic field. The Hall effect was discovered by E. F. Hall in 1879.

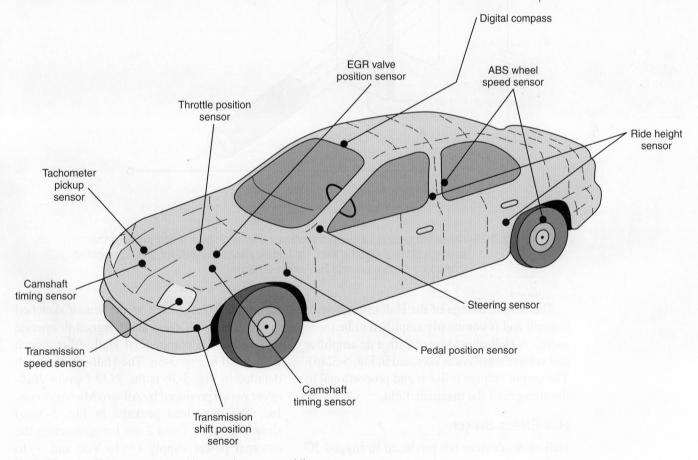

**Fig. 5-34** Hall-effect sensors used in a modern automobile.

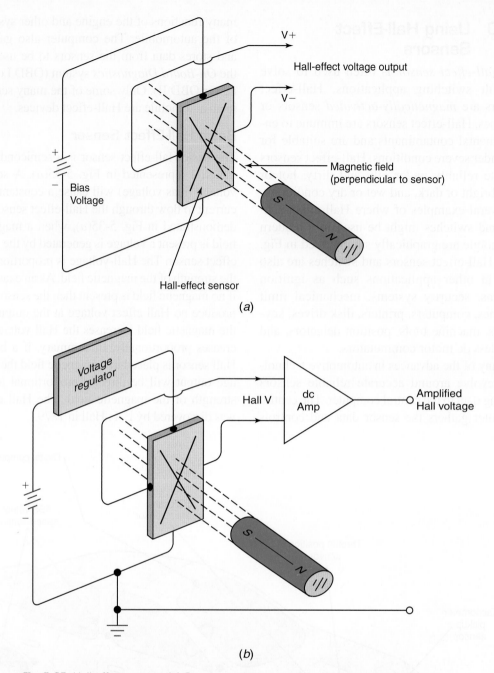

V+

Hall-effect voltage output

V−

Magnetic field
(perpendicular to sensor)

Bias
Voltage

S
N

Hall-effect sensor

*(a)*

Voltage
regulator

Hall V

dc
Amp

Amplified
Hall voltage

S
N

*(b)*

**Fig. 5-35** Hall-effect sensor. *(a)* Sensor produces a small voltage proportional to the strength of the magnetic field. *(b)* Adding a voltage regulator and dc amplifier to produce a more useable Hall-effect sensor.

The output voltage of the Hall-effect sensor is small and is commonly amplified to be more useful. A Hall-effect sensor with a dc amplifier and voltage regulator is sketched in Fig. 5-35(*b*). The output voltage is linear and proportional to the strength of the magnetic field.

### Hall-Effect Switch

Hall-effect devices are produced in rugged IC packages. Some are designed to generate a lin-

ear output voltage such as the sensor sketched in Fig. 5-35(*b*). Others are designed to operate as switches. A commercial Hall-effect switch is featured in Fig. 5-36. The Hall-effect switch detailed in Fig. 5-36 is the *3132 bipolar Hall-effect switch* produced by Allegro Microsystems, Inc. The three-lead package in Fig. 5-36(*a*) shows that pins 1 and 2 are for connecting the external power supply (+ to Vcc and − to ground). Pin 3 is the output of this bouncefree

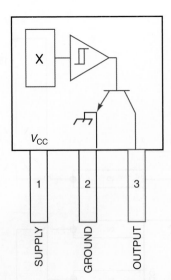

Pinning is shown viewed from branded side.

(a)

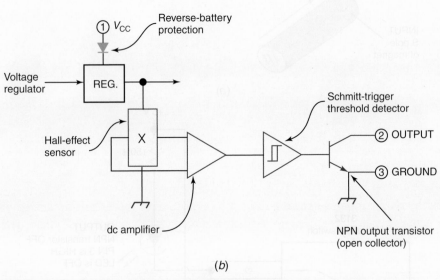

(b)

**Fig. 5-36** Allegro Microsystem's 3132 Bipolar Hall-effect switch. (a) Pin diagram. (b) Functional block diagram.

For more information on Hall-effect sensors and switches, visit www.allegromicro.com.

switch. The pinout in Fig. 5-36(a) is correct when viewing the 3132 Hall-effect switch from the printed side (branded side) of the IC package. A functional block diagram of Allegro Microsystem's 3132 Hall-effect switch is drawn in Fig. 5-36(b). Notice that the symbol for the Hall-effect sensor is a rectangle with an X inside. Added to the sensor are several sections that convert the analog Hall-effect device into a digital switch. The *Schmitt-trigger threshold detector* produces a snap-action bouncefree output needed for digital switching. Its output is either HIGH or LOW. The open-collector output transistor is included so the IC can drive a load up to 25 mA continuously.

The two most important characteristics of a magnetic field are its *strength* and its *polarity* (South or North poles of magnet). Both of these characteristics are used in the operation of the bipolar 3132 Hall-effect switch. To demonstrate the operation of the bipolar Hall-effect switch study the circuit in Fig. 5-37(a) including the 3132 IC. An output indicator LED with 150-Ω limiting resistor has been added at the output of the IC.

In Fig. 5-37(a), the *south pole* of the magnet approaches the branded side (side with printing) of the IC causing the internal NPN transistor turn to ON. This causes pin 3 of the IC to drop LOW causing the LED to light.

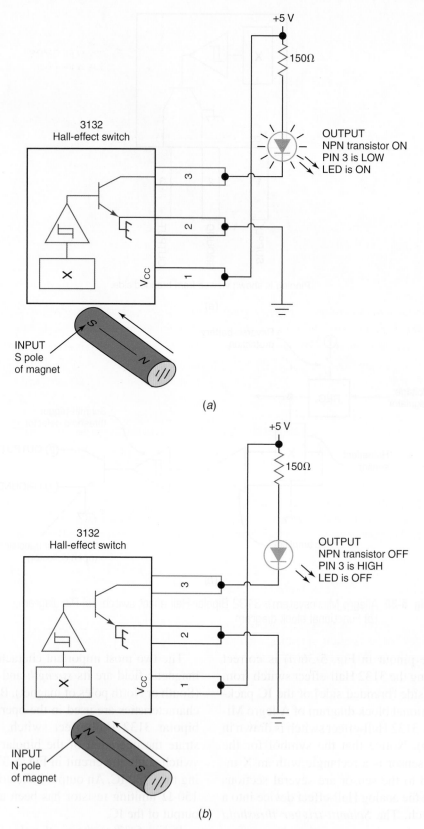

**Fig. 5-37** Controlling 3132 Hall-effect switch with opposite poles of a magnet (a) Turning ON switch with S pole. (b) Turning OFF switch with N pole.

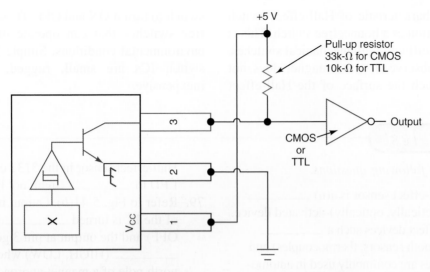

**Fig. 5-38** Interfacing Hall-effect switch IC with either TTL or CMOS.

In Fig. 5-37(*b*), the *north pole* of the magnet approaches the branded side (side with printing) of the IC causing the internal NPN transistor turn OFF. This causes pin 3 of the IC to go HIGH and the LED does not light.

The 3132 Hall-effect switch was *bipolar* because it required a S pole and then a N pole to make it toggle between ON and OFF. Unipolar Hall-effect switches are also available which turn ON and OFF by just increasing (switch ON) and decreasing (switch OFF) the magnetic field strength and not changing polarity. One such unipolar Hall-effect switch is the 3144 by Allegro Microsystems, Inc. The *3144 unipolar Hall-effect switch* is a close relative the the 3132 bipolar Hall-effect switch you have already studied. The unipolar 3144 IC shares the same pinout diagram [Fig. 5-36(*a*)] and functional block diagram [Fig. 5-36(*b*)] as the bipolar 3132 Hall-effect switch. The 3144 Hall-effect switch features a snap-action digital output. The 3144 IC also features an NPN output transistor that will sink 25mA.

The Hall-effect switch IC drawn in Fig. 5-38 has an NPN driver transistor with an open collector. Interfacing a Hall-effect switch IC with digital ICs (TTL or CMOS) requires the use of a pull-up resistor as shown in Fig. 5-38. Typical values for the pull-up are shown as 33k-Ω for CMOS and 10k-Ω for TTL. The Hall-effect switch shown in Fig. 5-38 could be either the 3132 or 3144 ICs.

## Gear-Tooth Sensing

Other common Hall-effect switching devices include gear-tooth sensing ICs. Gear-tooth sensing ICs contain one or more Hall-effect sensors and

a built-in permanent magnet. A sketch of typical gear-tooth sensing IC and gear is shown in Fig. 5-39. The South pole of the permanent magnet produces a magnetic field that varies with the position of the gear. When a gear-tooth moves into position to shorten the air gap, the field gets stronger and the Hall-effect sensor switches. Gear-tooth sensors are commonly used in mechanical systems including automobiles to count the position, rotation, and speed of gears.

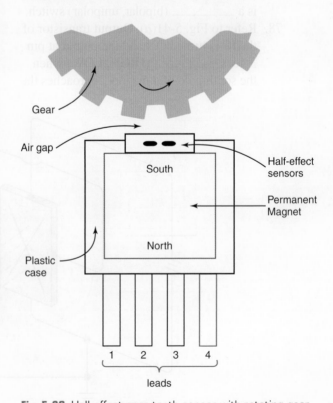

**Fig. 5-39** Hall-effect gear-tooth sensor with rotating gear for triggering.

It is a characteristic of Hall-effect switch ICs to function as a bouncefree switch. This is sometimes difficult with mechanical switches. You will observe that the magnet does not have to touch the surface of the Hall-effect switch to turn it ON and OFF. These are touch free switches that can operate under serve environmental conditions. Simple Hall-effect switch ICs are small, rugged, and very inexpensive.

## ✓ Self-Test

*Answer the following questions.*

73. A Hall-effect sensor is a(n) _____ (magnetically, optically)-activated device.

74. Hall-effect devices such a _____ (gear-tooth sensors, thermocouples) and switches are commonly used in automobiles because they are reliable, inexpensive, and can operate under severe conditions.

75. Refer to Fig. 5-40. The semiconductor material shown with the X on it is called the _____ (electromagnet, Hall-effect sensor).

76. Refer to Fig. 5-40. Moving the permanent magnet closer to the Hall-effect sensor increases the magnetic field causing the output voltage to _____ (decrease, increase).

77. Refer to Fig. 5-41. The 3132 Hall-effect IC is a _____ (bipolar, unipolar) switch.

78. Refer to Fig. 5-41(*a*). Output transistor of the IC is turned ON and the output at pin 3 goes _____ (HIGH, LOW) when the south pole of a magnet approaches the Hall-effect sensor in the 3132 causing the LED to _____ (light, not light).

79. Refer to Fig. 5-41(*b*). Output transistor of the IC is turned _____ (ON, OFF) and the output at pin 3 goes _____ (HIGH, LOW) when the north pole of a magnet approaches the Hall-effect sensor in the 3132 causing the LED to not light.

80. Refer to Fig. 5-36. The snap-action causing a digital output (either HIGH or LOW) from the IC is caused by the _____ (dc amplifier, Schmitt-trigger) section of the IC.

81. Hall-effect switches are small, bounce-free, rugged, and _____ (inexpensive, very expensive).

82. The open-collector of the NPN driver transistor used in both the 3132 and 3144 Hall-effect switches requires the use a _____ (pull-up, transition) resistor when sending digital signals to either CMOS or TTL logic devices.

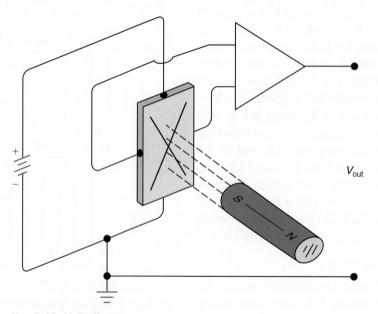

**Fig. 5-40** Hall-effect sensor.

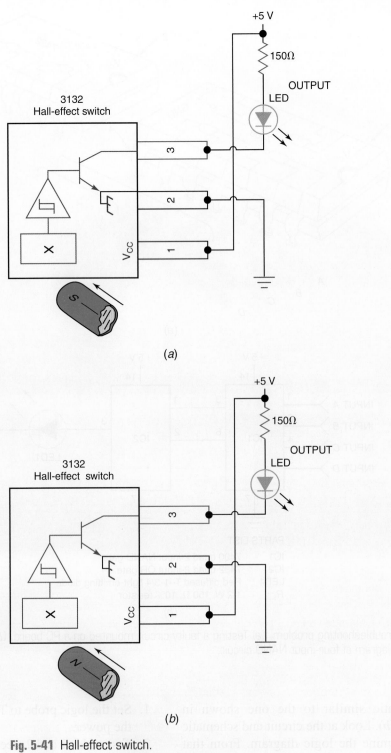

(a)

(b)

**Fig. 5-41** Hall-effect switch.

# 5-11 Troubleshooting Simple Logic Circuits

One test equipment manufacturer suggests that about three-quarters of all faults in digital circuits occur because of open input or output circuits. Many of these faults can be isolated in a logic circuit using a logic probe.

Consider the combinational logic circuit mounted on a printed circuit board in Fig. 5-42(a). The equipment manual might include

**Troubleshooting using a logic probe**

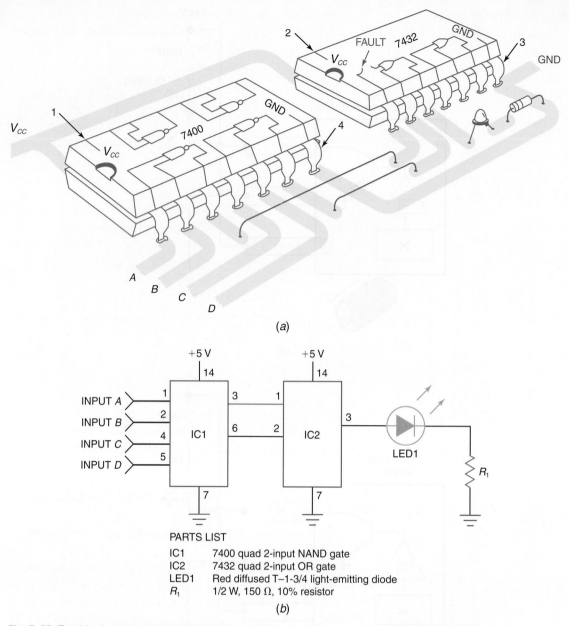

**Fig. 5-42** Troubleshooting problem. (*a*) Testing a faulty circuit mounted on a PC board. (*b*) Schematic diagram of four-input NAND circuit.

a schematic similar to the one shown in Fig. 5-42(*b*). Look at the circuit and schematic and determine the logic diagram. From that you can determine the Boolean expression and truth table. You will find that in this example, two NAND gates are feeding an OR gate. This is equivalent to the four-input NAND function.

The fault in the circuit in Fig. 5-42(*a*) is shown as an open circuit in the input to the OR gate. Now let's troubleshoot the circuit to see how we find this fault.

1. Set the logic probe to TTL and connect the power.
2. Test nodes 1 and 2 (see Fig. 5-42(*a*)). *Result:* Both are HIGH.
3. Test nodes 3 and 4. *Result:* Both are LOW. *Conclusion:* Both ICs have power.
4. Test the four-input NAND circuit's unique state (inputs *A, B, C,* and *D* are all HIGH). Test at pins 1, 2, 4, and 5 of the 7400 IC. *Results:* All inputs are HIGH but the LED still glows and indicates a

HIGH output. *Conclusion:* The unique state of the four-input NAND circuit is faulty.

5. Test the outputs of the NAND gates at pins 3 and 6 of the 7400 IC. *Results:* Both outputs are LOW. *Conclusions:* The NAND gates are working.

6. Test the inputs to the OR gate at pins 1 and 2 of the 7432 IC. *Results:* Both inputs are LOW. *Conclusions:* The OR gate inputs at pins 1 and 2 are correct but the output is still incorrect. Therefore the OR gate is faulty and the 7432 IC needs to be replaced.

*Supply the missing word in each statement.*

83. Most faults in digital circuits occur because of _____ (open, short) circuits in the inputs and outputs.
84. A simple piece of test equipment, such as a(n) _____, can be used for checking a digital logic circuit for open circuits in the inputs and outputs.
85. Refer to Fig. 5-42. With inputs *A, B, C,* and *D* all HIGH, the output (pin 3 of IC2) should be _____ (HIGH, LOW).

## 5-12 Interfacing the Servo (BASIC Stamp Module)

Programmable devices are very common in modern digital electronics. This section will explore interfacing of the BASIC Stamp 2 microcontroller module with a simple servo.

Review section 5-9 on servo motors. Hobby servo motor operation is summarized in Fig. 5-29. Notice the use of pulse-width modulation (PWM) to control the angular position of the servo motor. In this section, you will program a BASIC Stamp 2 microcontroller module to act as the *PWM pulse generator* sketched in Figs. 5-29(*a*), (*b*), and (*c*). Notice in Fig. 5-29 that the positive pulse widths are 2 msec for fully CCW rotation, 1 msec for fully CW rotation, or 1.5 msec for centering of the servo's output shaft.

Consider the hobby servo motor connected to the BASIC Stamp 2 module in Fig. 5-43. This is a test circuit to rotate the servo (1) fully CCW, (2) fully CW, and (3) to finally center the output shaft.

The procedure for solving the logic problem with the use of the BASIC Stamp 2 module is detailed below. The steps in wiring and programming the BASIC Stamp 2 module are:

1. Refer to Fig. 5-43. Wire the hobby servo motor to port P14 of the BASIC Stamp 2 module. Note the color coding (red = $V_{dd}$

and black = $V_{ss}$ or GND) of the power wires.
2. Load the PBASIC text editor program (version for the BS2 IC) into the PC. Type your PBASIC program describing the **'Servo Test 1**. A PBASIC program titled **'Servo Test 1** is listed in the shaded box.
3. Attach a serial cable between the PC and the BASIC Stamp 2 development board (such as the Board of Education by Parallax, Inc.).
4. With the BASIC Stamp 2 module turned on, download your PBASIC program from the PC to BS2 module using the RUN command.
5. Disconnect the serial cable from the BS2 module.
6. Observe the rotation of the servo output shaft. The PBASIC program stored in EEPROM program memory in the BASIC Stamp 2 module will restart each time the BS2 IC is turned on.

### PBASIC Program—ServoTest 1

Consider the PBASIC program titled **'ServoTest 1**. Lines 1 and 2 both begin with an apostrophe (') meaning these are *remark statements*. Remark statements are used to clarify the program and are not executed by the microcontroller. Line 3 is a

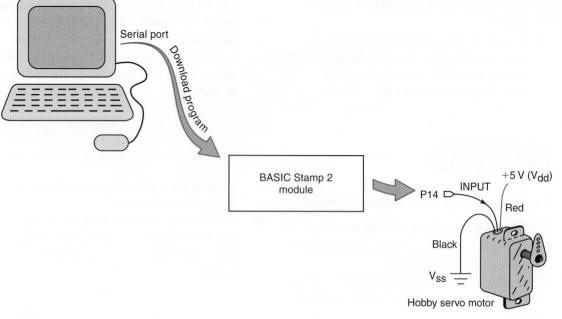

**Fig. 5-43** Hobby servo motor connected to a BASIC Stamp 2 module for testing.

line of code to *declare a variable* that will be used later in the program As an example, L3 reads **C VAR Word**. This tells the microcontroller that

**C** is a variable name that will hold a word length value (16 bits). The 16-bit variable **C** can hold a range of values from 0 to 65535 in decimal.

```
'ServoTest 1                    'Title of program (Fig. 5-43)                    L1
'Test servo in 3 different positions, CCW, CW and centered                       L2

C      VAR  Word                'Declare C as variable, 16 bit length            L3

FOR C = 1 TO 75                 'Begin counting loop, C = 1 thru 75              L4
PULSOUT 14, 1000                'Pulse output (HIGH) at pin 14 for 2 ms          L5
PAUSE 20                        'Pause for 20 ms, output LOW                     L6
NEXT                            'Back to FOR if C < 75                           L7

FOR C = 1 TO 75                 'Begin counting loop, C = 1 thru 75              L8
PULSOUT 14, 500                 'Pulse output (HIGH) at pin 14 for 1 ms          L9
PAUSE 20                        'Pause for 20 ms, output LOW                     L10
NEXT                            'Back to FOR if C < 75                           L11

FOR C = 1 TO 75                 'Begin counting loop, C = 1 thru 75              L12
PULSOUT 14, 750                 'Pulse output (HIGH) at pin 14 for 1.5 ms        L13
PAUSE 20                        'Pause for 20 ms, output LOW                     L14
NEXT                            'Back to FOR if C < 75                           L15

END                                                                              L16
```

Lines 4–7 produce the full CCW rotation of the servo motor shaft. The FOR-NEXT loop will be executed 75 times (C = 1 to 75). The **PULSOUT 14, 1000** code (L5) generates a HIGH pulse at pin 14 for 2 milliseconds (2 μseconds × 1000 = 2000 μs = 2 ms). Pin 14 then drops LOW after the 2 ms positive pulse. The **PAUSE 20** code (L6) allows pin 14 to remain LOW for 20 ms. This first FOR-NEXT loop (L4–7) will cause the hobby servo motor to turn fully CCW.

Lines 8–11 produce the Full CW rotation of the servo motor shaft. The FOR-NEXT loop will be executed 75 times (C = 1 to 75). The **PULSOUT 14, 500** code (L9) generates a HIGH pulse at pin 14 for 1 milliseconds (2 μseconds × 500 = 1000 μs = 1 ms). Pin 14 then drops LOW after the 1 ms positive pulse. The **PAUSE 20** code (L10) allows pin 14 to remain LOW for 20 ms. This second FOR-NEXT loop (L8–11) will cause the hobby servo motor to turn fully CW.

Lines 12–15 cause the servo motor shaft to center itself. The FOR-NEXT loop will be executed 75 times (C = 1 to 75). The **PULSOUT 14, 750** code (L13) generates a HIGH pulse at pin 14 for 1.5 milliseconds (2 μseconds × 750 = 1500 μs = 1.5 ms). Pin 14 then drops LOW after the 1.5 ms positive pulse. The **PAUSE 20** code (L14) allows pin 14 to remain LOW for 20 ms. This last FOR-NEXT loop (L12–15) will cause the hobby servo motor shaft to move to the center of its range. The **END** statement (L16) causes the program to stop executing.

The PBASIC program 'ServoTest 1 will run once while the BS2 BASIC Stamp 2 module is powered. The PBASIC program is held in EEPROM program memory for future use. Turning the BS2 OFF and then ON again will restart the program. Downloading a different PBASIC program to the BASIC Stamp module will erase the old program and start execution of the new listing.

## ✔ Self-Test

*Answer the following questions.*

86. The angular position of a hobby servo motor is controlled by a technique called _____ (amplitude, pulse-width) modulation.
87. Refer to the PBASIC program 'ServoTest 1. Variable C may hold only a single bit of data. (T or F)
88. Refer to the PBASIC program 'ServoTest 1. Each of the three FOR-NEXT loops will be repeated _____ (20, 75) times.

89. Refer to the PBASIC program 'ServoTest 1. The purpose of the **PAUSE 20** statement is to permit the microcontroller to cool off for 20 minutes. (T or F)
90. Refer to the PBASIC program 'ServoTest 1. The **PULSOUT 14, 750** output a positive pulse to pin 14 with a time duration of _____ milliseconds.
91. Refer to the PBASIC program 'ServoTest 1 and Fig. 5-43. What is the effect on the servo's output shaft when the FOR-NEXT loop in lines 12–15 is totally executed (75 times through the loop)?

# Chapter 5 Summary and Review

1. Interfacing is the design of circuitry between devices that shifts voltage and current levels to make them compatible.

2. Interfacing between members of the same logic family is usually as simple as connecting one gate's output to the next logic gate's input, etc.

3. In interfacing between logic families or between logic devices and the "outside world," the voltage and current characteristics are very important factors.

4. Noise margin is the amount of unwanted induced voltage that can be tolerated by a logic family. Complementary symmetry metal oxide semiconductor ICs have better noise margins than TTL families.

5. The fan-out and fan-in characteristics of a digital IC are determined by its output drive and input loading specifications.

6. Propagation delay (or speed) and power dissipation are important IC family characteristics.

7. The ALS-TTL, FAST (Fairchild Advanced Schottky TTL), and FACT (Fairchild Advanced CMOS Technology) logic families are very popular owing to a combination of low power consumption, high speed, and good drive capabilities. Earlier TTL and CMOS families are still in wide use in existing equipment.

8. Most CMOS ICs are sensitive to static electricity and must be stored and handled properly. Other precautions to be observed include turning off an input signal before circuit power and connecting all unused inputs.

9. Simple switches can drive logic circuits using pull-up and pull-down resistors. Switch debouncing is usually accomplished using latch circuits.

10. Driving LEDs and incandescent lamps with logic devices usually requires a driver transistor.

11. Most TTL-to-CMOS and CMOS-to-TTL interfacing requires some additional circuitry. This can take the form of a simple pull-up resistor, special interface IC, or transistor driver.

12. Interfacing digital logic devices with buzzers and relays usually requires a transistor driver circuit. Electric motors and solenoids can be controlled by logic elements using a relay to isolate them from the logic circuit.

13. Optoisolators are also called optocouplers. Solid-state relays are a variation of the optoisolator. Optoisolators are used to electrically isolate digital circuitry from circuits that contain motors or other high voltage/current devices that might cause voltage spikes and noise.

14. Hobby servo motors are used for angular positioning of an output shaft. A pulse generator employing pulse-width modulation (PWM) is used to drive a these inexpensive servo motors.

15. Hobby servo motors can be driven by programmable devices such as the BASIC Stamp 2 microcontroller module.

16. Stepper motors operate on dc and are useful in applications where precise angular positioning or speed of an output shaft is important.

17. Stepper motors are classified as either bipolar (two-phase) or unipolar (four-phase). Other important characteristics are step angle, voltage, current, coil resistance, and torque.

18. Specialized ICs are useful for interfacing and driving stepper motors. The logic section of the IC generates the correct control sequence to step the motor.

19. A Hall-effect sensor is a magnetically-activated device used in Hall-effect switches. Hall-effect switches are classified as either bipolar (need S and N poles of magnet to activate) or unipolar (need S pole or no magnetic field to activate).

20. External magnetic fields are commonly used to activate a Hall-effect sensor or switch. Gear-tooth sensors have Hall-effect sensors and a permanent

magnet encapsulated in the IC. Hall-effect gear-tooth sensors are triggered by ferrous metals (such as steel gear teeth) passing near the IC.

21. Each logic family has its own definition of logical HIGH and LOW. Logic probes test for these levels.

*Answer the following questions.*

5-1. Applying 3.1 V to a TTL input is interpreted by the IC as a(n) _____ (HIGH, LOW, undefined) logic level (5-V power source).

5-2. A TTL output of 2.0 V is considered a(n) _____ (HIGH, LOW, undefined) output (5-V power source).

5-3. Applying 2.4 V to a CMOS input (10-V power supply) is interpreted by the IC as a(n) _____ (HIGH, LOW, undefined) logic level.

5-4. Applying 3.0 V to a 74HC00 series CMOS input (5-V power supply) is interpreted by the IC as a(n) _____ (HIGH, LOW, undefined) logic level.

5-5. A "typical" HIGH output voltage for a TTL gate would be about _____ (0.1, 0.8, 3.5) V.

5-6. A "typical" LOW output voltage for a TTL gate would be about _____ (0.1, 0.8, 3.5) V.

5-7. A "typical" HIGH output voltage for a CMOS gate (10-V power supply) would be about _____ V.

5-8. A "typical" LOW output voltage for a CMOS gate (10-V power supply) would be about _____ V.

5-9. Applying 3.0 V to a 74HCT00 series CMOS input (5-V power supply) is interpreted by the IC as a(n) _____ (HIGH, LOW, undefined) logic level.

5-10. Applying 1.0 V to a 74HCT00 series CMOS input (5-V power supply) is interpreted by the IC as a(n) _____ (HIGH, LOW, undefined) logic level.

5-11. The _____ (CMOS, TTL) logic family has better noise immunity.

5-12. Refer to Fig. 5-3. The noise margin for the TTL family is about _____ V.

5-13. Refer to Fig. 5-3. The noise margin for the CMOS family is about _____ V.

5-14. Refer to Fig. 5-4. The *switching threshold* for TTL is always exactly 1.4 V (T or F).

5-15. The fan-out for standard TTL is said to be _____ (10, 100) when driving other standard TTL gates.

5-16. Refer to Fig. 5-5(*b*). A single ALS-TTL output will drive _____ (5, 50) standard TTL inputs.

5-17. Refer to Fig. 5-5(*b*). A single 74HC00 series CMOS output has the capacity to drive at least _____ (10, 50) LS-TTL inputs.

5-18. Refer to Fig. 5-44. If both family *A* and *B* are TTL, the inverter _____ (can, may not be able to) drive the AND gates.

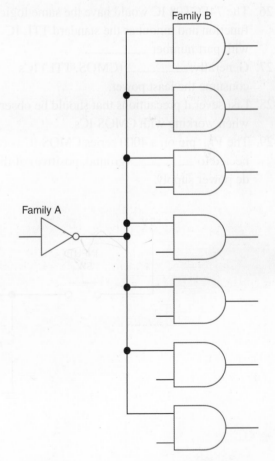

**Fig. 5-44** Interfacing problem.

5-19. Refer to Fig. 5-44. If family *A* is ALS-TTL and family *B* is standard TTL, the inverter _____ (can, may not be able to) drive the AND gates.

5-20. Refer to Fig. 5-44. If both families *A* and *B* are ALS-TTL, the inverter _____ (can, may not be able to) drive the AND gates.

5-21. The _____ (4000, 74AC00) series CMOS ICs have greater output drive capabilities.

5-22. Refer to Fig. 5-7(*b*). The _____ logic family has the lowest propagation delays and is considered the _____ (fastest, slowest).

5-23. Refer to Fig. 5-7(*b*). The _____ logic family has the highest propagation delays and is considered the _____ (fastest, slowest).

5-24. Refer to Fig. 5-7(*b*). The _____ is the fastest CMOS family.

5-25. Refer to Fig. 5-7. Which CMOS family of ICs would be recommended for a new design because of its high speed, low power consumption, good noise immunity, and excellent drive capabilities?

5-26. The 74FCT08 IC would have the same logic function and pinout as the standard TTL IC with part number _____.

5-27. Generally, _____ (CMOS, TTL) ICs consume the least power.

5-28. List several precautions that should be observed when working with CMOS ICs.

5-29. The $V_{DD}$ pin on a 4000 series CMOS IC is connected to _____ (ground, positive) of the dc power supply.

5-30. Refer to Fig. 5-10(*b*). With the switch open, the inverter's input is _____ (HIGH, LOW) while the output is _____ (HIGH, LOW).

5-31. Refer to Fig. 5-11(*a*). When the switch is open, the _____ resistor causes the input of the CMOS inverter to be pulled HIGH.

5-32. Refer to Fig. 5-45. Component $R_1$ is called a _____ resistor.

5-33. Refer to Fig. 5-45. Closing $SW_1$ causes the input to the inverter to go _____ (HIGH, LOW) and the LED _____ (goes out, lights).

5-34. Refer to Fig. 5-45. With $SW_1$ open, a _____ (HIGH, LOW) appears at the input of the inverter causing the output LED to _____ (go out, light).

5-35. The common switch debouncing circuits in Figs. 5-13(*b*) and (*c*) are called RS flip-flops or _____.

5-36. Refer to Fig. 5-14. Closing input switch $SW_1$ causes the output of the 555 IC to toggle from _____ (HIGH-to-LOW, LOW-to-HIGH).

5-37. Refer to Fig. 5-14. Opening input switch $SW_1$ causes the output of the 555 IC to toggle from HIGH-to-LOW _____.
  a. immediately
  b. after a delay of about 1 second
  c. after a delay of about 1 microsecond

5-38. A TTL output can drive a regular CMOS input with the addition of a(n) _____ resistor.

5-39. Any CMOS gate can drive at least one LS-TTL input (T or F).

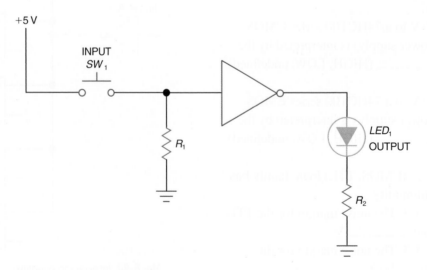

**Fig. 5-45** Interfacing problem.

5-40. A 4000 series CMOS output can drive a standard TTL input with the addition of a(n) _____.

5-41. Open-collector TTL gates require the use of _____ resistors at the outputs.

5-42. Refer to Fig. 5-24(*b*). The transistor functions as a(n) _____ (AND gate, driver) in this circuit.

5-43. Refer to Fig. 5-24(*b*). When the input to the inverter goes LOW, its output goes _____ (HIGH, LOW) which _____ (turns off, turns on) the transistor allowing current to flow through the transistor and piezo buzzer to sound the buzzer.

5-44. Refer to Fig. 5-26(*a*). When the input to the inverter goes LOW, its output goes HIGH which _____ (turns off, turns on) the NPN transistor; the coil of the relay is _____ (activated, deactivated), the relay armature clicks downward, and the dc motor _____ (rotates, will not rotate).

5-45. Refer to Fig. 5-26(*b*). When the input to the inverter goes HIGH, its output goes LOW which _____ (turns off, turns on) the NPN transistor; the coil of the relay is _____ (activated, deactivated), the armature of the relay _____ (clicks, will not click) downward, and the solenoid _____ (is, will not be) activated.

5-46. Refer to Fig. 5-27. The 4N25 optoisolator contains a gallium arsenide _____ (infrared-emitting diode, incandescent lamp) optically coupled to a phototransistor output.

5-47. Refer to Fig. 5-27(*b*). If the input to the inverter goes HIGH, its output goes LOW which _____ (activates, deactivates) the LED, the phototransistor is _____ (turned off, turned on), and the output voltage goes _____ (HIGH, LOW).

5-48. Refer to Fig. 5-27(*c*). The piezo buzzer sounds when the input to the inverter goes _____ (HIGH, LOW).

5-49. Refer to Fig. 5-27(*d*). This is an example of good design practice by using an optoisolator to isolate the low-voltage digital circuit from the higher-voltage noisy motor circuit (T or F).

5-50. Refer to Fig. 5-27(*d*). The dc motor turns on when a _____ (HIGH, LOW) logic level appears at the input of the inverter.

5-51. A solid-state relay is a close relative of the optoisolator (T or F).

5-52. The electromagnetic device well suited to continuous rotation in either direction is the _____ (dc motor, hobby servo motor).

5-53. Refer to Fig. 5-46. The pulse generator will vary the _____ causing the servo motor to adjust the angular position of the output shaft.
   a. Frequency from about 30- to 100-Hz
   b. Pulse width from about 1- to 2-msec
   c. Pulse amplitude from about 1- to 5-V

5-54. A _____ (dc motor, stepper motor) should be used when the application calls for exact angular positioning of a shaft (as in a robot wrist).

5-55. The stepper motor sketched in Fig. 5-30(*a*) is classified as a unipolar or four-phase unit (T or F).

5-56. The device featured in Fig. 5-30 is a _____ (permanent magnet, variable reluctance) type stepper motor.

5-57. The step angle for the stepper motor in Fig. 5-30(*a*) is _____ degrees.

5-58. The control sequence shown in Fig. 5-31(*a*) is for a _____ (bipolar, unipolar) stepper motor.

5-59. Refer to Fig. 5-32(*a*). How is the MC3479 IC described by its manufacturer?

5-60. Refer to Fig. 5-32(*a*) and assume pins 9 and 10 of the MC3479 IC are LOW. When a single clock pulse enters the CLK input (pin 7) the stepper motor rotates a _____ (full step, half step) in the _____ (CCW, CW) direction.

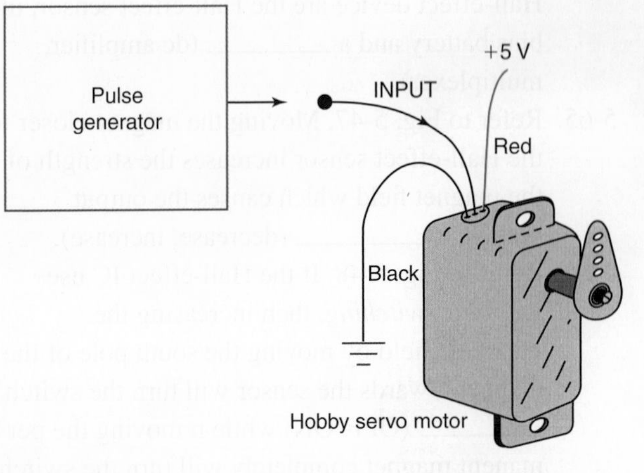

**Fig. 5-46** Driving a servo motor.

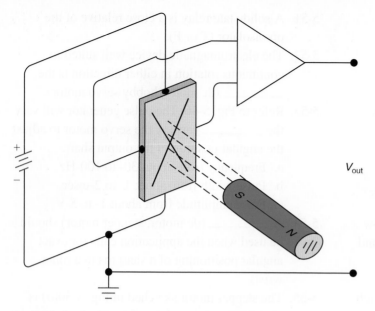

Fig. 5-47 Art for chapter review questions 5-64, 5-65, and 5-69.

$V_{out}$

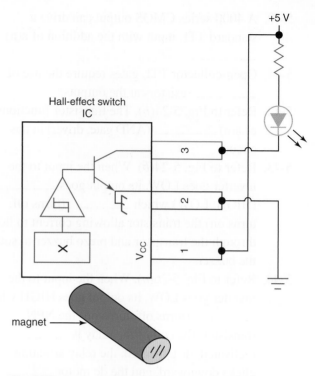

Hall-effect switch
IC

magnet

Fig. 5-48 Art for chapter review questions 5-66, 5-67, 5-68, and 5-70.

5-61. Refer to Fig. 5-32(a) and assume pins 9 and 10 of the MC3479 IC are HIGH and the stepper motor has a step angle of 18°. Under these conditions, how many clock pulses must enter the CLK input to cause the stepper motor to rotate one revolution?

5-62. The Hall-effect sensor is a _____ (magnetically-, pressure-) activated device.

5-63. Hall-effect devices such as gear-tooth sensors and switches are commonly used in automobiles because they are rugged, reliable, operate under severe conditions, and are inexpensive. (T or F)

5-64. Refer to Fig. 5-47. The sections of this Hall-effect device are the Hall-effect sensor, the bias battery and a _____ (dc amplifier, multiplexer).

5-65. Refer to Fig. 5-47. Moving the magnet closer to the Hall-effect sensor increases the strength of the magnet field which causes the output voltage to _____ (decrease, increase).

5-66. Refer to Fig. 5-48. If the Hall-effect IC uses *unipolar switching*, then increasing the magnetic field by moving the south pole of the magnet towards the sensor will turn the switch _____ (OFF, ON) while removing the permanent magnet completely will turn the switch _____ (OFF, ON).

5-67. Refer to Fig. 5-48. If the IC is the bipolar 3132 Hall-effect switch then the _____ (N, S) pole of the magnet will turn the device ON while the _____ (N, S) pole will turn the output transistor OFF.

5-68. Refer to Fig. 5-48. If the IC is the bipolar 3132 Hall-effect switch then moving the north pole of the magnet near the sensor will turn _____ (OFF, ON) the switch, the voltage at pin 3 will _____ (drop LOW, raise HIGH), and the LED will _____ (light, not light).

5-69. Refer to Fig. 5-47. The output of this device is _____ (analog, digital) in nature.

5-70. Refer to Fig. 5-48. The output of this IC is _____ (analog, digital) in nature.

5-71. Refer to Fig. 5-43. The BASIC Stamp 2 _____ (audio-amplifier, microcontroller) module substitutes as a PWM generator to rotate the servo motor.

5-72. In Parallax's PBASIC language, the statement **PULSOUT 14, 750** generates 14 negative pulses each 750 μs long. (T or F)

5-1. How would you define *interfacing*?

5-2. How do you define *noise* in a digital system?

5-3. What is the propagation delay of a logic gate?

5-4. List several advantages of CMOS logic elements.

5-5. Refer to Fig. 5-5(*b*). A single 4000 series CMOS output has the capacity to drive at least _____ LS-TTL input(s).

5-6. Refer to Fig. 5-43. If family *A* is standard TTL and family *B* is ACT-CMOS, the inverter _____ (can, may not be able to) drive the AND gates.

5-7. Refer to Fig. 5-17(*c*). Explain the operation of this HIGH-LOW indicator circuit.

5-8. What is the purpose of "T"-type CMOS ICs (HCT, ACT, etc.)?

5-9. What electromechanical device could be used to isolate higher voltage equipment (such as motors or solenoids) from a logic circuit?

5-10. An electric motor converts electric energy into _____ motion.

5-11. A(n) _____ is an electromechanical device that converts electric energy into linear motion.

5-12. Why is the FACT-CMOS series considered by many engineers to be one of the best logic families for new designs?

5-13. Refer to Fig. 5-25. Explain the circuit action when the inverter input is LOW.

5-14. Refer to Fig. 5-26(*a*). Explain the circuit action when the inverter input is HIGH.

5-15. An optoisolator prevents the transmission of _____ (the signal, unwanted noise) from one electronic system to another that operates on a different voltage.

5-16. Refer to Fig. 5-27(*d*). Explain the circuit action when the inverter input is LOW.

5-17. If the coil resistance of a 12-V stepper motor is $40\Omega$, what is the current draw for the coil?

5-18. If a stepper motor is designed with a step angle of 3.6°, how many steps are required for one revolution of the motor?

5-19. Why are Hall-effect devices such as switches and gear-tooth sensors so widely used in modern automobiles?

5-20. Explain what we mean by current sinking.

5-21. What is PWM and how is it used to drive a hobby servo motor?

5-22. Refer to Fig. 5-32(*c*). What do you notice as you progress down the control sequence for a stepper motor (*Hint:* direction of current flow through windings)?

5-23. Refer to Fig. 5-36(*b*). What is the purpose of the Schmitt-trigger in the Hall-effect switch?

5-24. Refer to Fig. 5-36(*b*). The output of the driver transistor in this IC is the _____ (open-collector, totem-pole) type.

5-25. Describe the difference between the operation of a bipolar and a unipolar Hall-effect switch.

# Answers to Self-Tests

1. interfacing
2. HIGH
3. LOW
4. undefined
5. undefined
6. +10
7. HIGH
8. CMOS
9. T
10. CMOS
11. fan-out
12. FAST TLL series
13. 20 (8 mA/400 $\mu$A = 20)
14. long
15. FACT series CMOS
16. the same
17. MOS
18. complementary symmetry metal oxide semiconductor
19. low-power consumption
20. GND
21. positive
22. FACT
23. T
24. LOW, floats HIGH
25. pull-up
26. RS flip-flop
27. HIGH, LOW
28. F
29. switch debouncing circuit
30. open collector
31. LOW-to-HIGH
32. C
33. decreasing
34. 4000
35. goes out
36. off, does not light
37. $Q_1$, red
38. active-LOW
39. sinking current
40. LOW
41. are not
42. pull-up
43. current drive

44. transistor
45. is not
46. on, sounds
47. transient voltages
48. HIGH
49. linear
50. isolate
51. turns on
52. NC to the NO
53. relay
54. phototransistor
55. lights, activates, LOW
56. pull-up

57. does not light, deactivates, HIGH, does not sound
58. turns on, runs
59. solid-state
60. will sound
61. dc motor
62. stepper motor
63. T
64. servo motor
65. input
66. pulse-width
67. bipolar

68. control, bipolar
69. CCW
70. logic
71. 350
72. CCW, half step
73. magnetically
74. gear-tooth sensors
75. Hall-effect sensor
76. increase
77. bipolar
78. LOW, light
79. OFF, HIGH
80. Schmitt-trigger

81. inexpensive
82. pull-up
83. open
84. logic probe or voltmeter
85. LOW
86. pulse-width
87. F
88. 75
89. F
90. 1.5
91. rotates to the center of its range

## Chapter 6

# Encoding, Decoding, and Seven-Segment Displays

## Chapter Objectives

*This chapter will help you to:*

1. *Identify* the characteristics and applications of several commonly used codes.

2. *Convert* decimal numbers to BCD code and BCD to decimal numbers.

3. *Compare* decimal numbers with excess-3 code, Gray code, and 8421 BCD code.

4. *Convert* ASCII code to letters and numbers, and characters to ASCII code.

5. *Demonstrate* the coding of a seven-segment display.

6. *Describe* the construction and important characteristics of LCD, LED, and vacuum fluorescent (VF) seven-segment displays.

7. *Demonstrate* the operation of several TTL and CMOS BCD-to-seven segment decoder/driver ICs used for driving LED, LCD, and VF seven-segment displays.

8. *Troubleshoot* a faulty decoder/driver seven-segment display circuit.

We use the decimal code to represent numbers. Digital electronic circuits use various forms of binary. Many special codes are used in digital electronics to represent numbers, letters, punctuation marks, and control characters. This chapter covers several common codes used in digital electronic equipment. Electronic translators, which convert from one code to another, are widely used in digital electronics. This chapter introduces you to several common encoders and decoders used for translating from code to code.

In modern electronic systems, the encoding and decoding may be performed by hardware or by computer programs or software. In computer jargon, to *encrypt* means to encode. So an *encoder* is an electronic device that translates from decimal to an encrypted code (such as binary) which is not as easy to interpret. In general, to encode means to convert input information to a code useful to digital circuitry.

In general, to *decode* means to translate from one code to another. In common use, a *decoder* would be a logic device that translates from an encrypted code into a code that is more understandable. An example of decoding would be to translate from binary to decimal.

## 6-1 The 8421 BCD Code

How would you represent the decimal number 926 in binary form? In other words, how would you convert 926 to the binary number 1110011110? The decimal-to-binary conversion would be done using the repeated divide-by-2 method illustrated in Fig. 6-1.

Following the repeated divide-by-2 process shown in Fig. 6-1, recall that the decimal number 926 is first divided by 2 yielding a quotient of 463 with a remainder of 0. The remainder of 0 becomes the least-significant bit (LSB is the 1s place) in the binary number. Next the first quotient is divided-by-2 yielding 231 ($463/2 =$ 231) with a remainder of 1. This remainder of 1 holds the 2s place in the binary number. This process is continuous until the quotient becomes 1. When the quotient becomes 0 the process is complete. Studying Fig. 6-1 will help refresh your memory about the repeated divide-by-2 process used to convert a decimal number to its binary equivalent.

The binary number 1110011110 does not make much sense to most of us. A code that uses binary in a different way than in the preceding example is called the *8421 binary-coded decimal code*. This code is frequently referred to as just the *BCD code*.

**Converting from decimal to 8421 BCD code**

**Converting from BCD to decimal**

**BCD code**

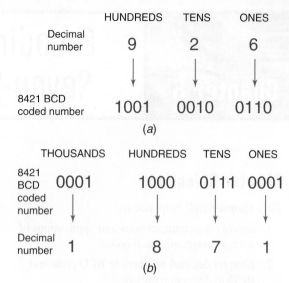

**Fig. 6-2** (a) Converting from decimal to 8421 BCD code. (b) Converting from BCD to decimal.

The decimal number 926 is converted to the BCD (8421) code in Fig. 6-2(a). The result is that the decimal number 926 equals 1001 0010 0110 in the 8421 BCD code. Notice from Fig. 6-2(a) that each group of four binary digits represents a decimal digit. The right group (0110) represents the 1s place value in the decimal number. The middle group (0010) represents the 10s place value in the decimal number. The left group (1001) represents the 100s place value in the decimal number.

Suppose you are given the 8421 BCD number 0001 1000 0111 0001. What decimal number does this represent? Figure 6-2(b) shows how you translate from the BCD code to a decimal number. We find that the BCD number 0001 1000 0111 0001 is equal to the decimal number 1871. The 8421 BCD code does not use the numbers 1010 1011 1100 1101 1110 1111. These are considered invalid numbers.

The 8421 BCD code is very widely used in digital systems. As pointed out, it is common practice to substitute the term "BCD code" to mean the 8421 BCD code. A word of caution, however: some BCD codes do have different weightings of the place values, such as the 4221 code and the excess-3 code. If seven-segment displays need to show decimal digits 0 through 9, the BCD code is a good choice.

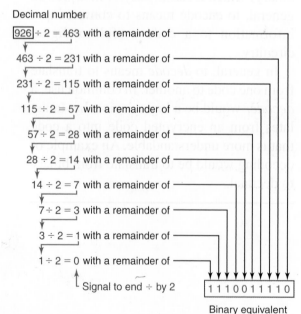

**Fig. 6-1** Converting decimal to binary numbers using the repeated divide-by-2 method.

**Converting decimal to binary numbers**

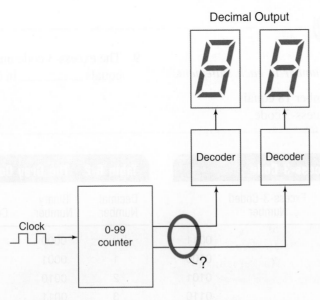

Decimal Output

Fig. 6-3 Two-digit counter with decimal output.

*Supply the missing number in each statement.*

1. The decimal number 29 is the same as _____ in binary.
2. The decimal number 29 is the same as _____ in the 8421 BCD code.
3. The 8421 BCD number 1000 0111 0110 0101 equals _____ in decimal.
4. The _____ (ASCII, 8421 BCD) code would be the preferred output from the counter shown in Fig. 6-3.

5. Refer to Fig. 6-3. If the output from the counter is 0111 1001 $_{\text{BCD}}$ entering the decoders what will the seven-segment displays read?
6. Refer to Fig. 6-3. If the seven-segment displays read decimal 85, then the BCD code between the counter and decoders is _____.
7. Refer to Fig. 6-3. If the seven-segment displays read decimal 81, then the BCD code between the counter and decoders will be 0101 0001. (T or F)

## 6-2 The Excess-3 Code

The term "BCD" is a general term, usually referring to an 8421 code. Another code that is really a BCD code is the *excess-3 code*. To convert a decimal number to the excess-3 form we *add 3 to each digit of the decimal number* and convert to binary form. Figure 6-4 shows how the decimal number 4 is converted to the excess-3 code number 0111. Some decimal numbers are converted to excess-3 code in Table 6-1. You probably have noticed that the excess-3 code for decimal numbers is rather difficult to figure out. This is because the binary digits are not weighted as they are in regular binary numbers and in the 8421 BCD code.

The excess-3 code is used in some arithmetic circuits because it is self-complementing.

The 8421 and excess-3 codes are but two of many BCD codes used in digital electronics. The 8421 code is by far the most widely used BCD code.

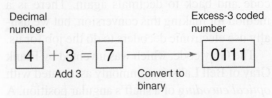

Fig. 6-4 Converting a decimal number to the excess-3 code.

**Excess-3 code**

**Converting a decimal number to the excess-3 code**

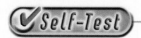 

*Supply the missing number in each statement.*

8. The decimal number 18 equals _____ in excess-3 code.

9. The excess-3 code number 1001 0011 equals _____ in decimal.

| Table 6-1 | The Excess-3 Code | | |
|---|---|---|---|
| Decimal Number | | Excess-3-Coded Number | |
| 0 | | | 0011 |
| 1 | | | 0100 |
| 2 | | | 0101 |
| 3 | | | 0110 |
| 4 | | | 0111 |
| 5 | | | 1000 |
| 6 | | | 1001 |
| 7 | | | 1010 |
| 8 | | | 1011 |
| 9 | | | 1100 |
| 14 | | 0100 | 0111 |
| 27 | | 0101 | 1010 |
| 38 | | 0110 | 1011 |
| 459 | 0111 | 1000 | 1100 |
| 606 | 1001 | 0011 | 1001 |
| | Hundreds | Tens | Ones |

| Table 6-2 | The Gray Code | | |
|---|---|---|---|
| Decimal Number | Binary Number | 8421 BCD Coded Number | Gray-Code Number |
| 0 | 0000 | 0000 | 0000 |
| 1 | 0001 | 0001 | 0001 |
| 2 | 0010 | 0010 | 0011 |
| 3 | 0011 | 0011 | 0010 |
| 4 | 0100 | 0100 | 0110 |
| 5 | 0101 | 0101 | 0111 |
| 6 | 0110 | 0110 | 0101 |
| 7 | 0111 | 0111 | 0100 |
| 8 | 1000 | 1000 | 1100 |
| 9 | 1001 | 1001 | 1101 |
| 10 | 1010 | 0001 0000 | 1111 |
| 11 | 1011 | 0001 0001 | 1110 |
| 12 | 1100 | 0001 0010 | 1010 |
| 13 | 1101 | 0001 0011 | 1011 |
| 14 | 1110 | 0001 0100 | 1001 |
| 15 | 1111 | 0001 0101 | 1000 |
| 16 | 10000 | 0001 0110 | 11000 |
| 17 | 10001 | 0001 0111 | 11001 |

## 6-3 The Gray Code

**Gray code**

Table 6-2 compares the *Gray code* with some codes you already know. The important characteristic of the Gray code is that only *one bit changes* as you *count* from top to bottom, as shown in Table 6-2. The Gray code cannot be used in arithmetic circuits. The Gray code is used for input and output devices in digital systems. You can see from Table 6-2 that the Gray code is not classed as one of the many BCD codes. Also notice that it is quite difficult to translate from decimal numbers to the Gray code and back to decimals again. There is a method for making this conversion, but we usually use electronic decoders to do the job for us.

The Gray code, which was invented by Frank Gray of Bell Labs, is commonly associated with *optical encoding* of a shaft's angular position. A simple example of this idea is sketched in Fig. 6-5. The encoder disk is attached to a shaft. The lighter areas of the disk represent transparent

**Optical encoding**

areas; the darker areas are opaque. A light source (usually infrared) shines from above the disk and light detectors are positioned below. The disk is free to rotate while the light sources and detectors stay in their position.

In the example shown in Fig. 6-5, light passes through all three transparent areas activating all three light detectors. In this example the detectors send the Gray code 111 to the Gray code to binary decoder. The decoder translates Gray code 111 to binary 101. As this is only a 3-bit shaft position encoder disk, the resolution is only 1 of 8. It can only detect a change in angular shaft position each 45° (360°/8 = 45°). The encoder disk in Fig. 6-5 is not very practical but serves to show how shaft positioning might be accomplished using the Gray code.

The Gray and excess-3 codes are not used extensively today. The purpose of mentioning

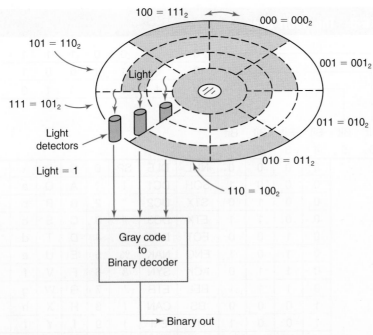

$100 = 111_2$ ↔ $000 = 000_2$

$101 = 110_2$

Light

$001 = 001_2$

$111 = 101_2$

$011 = 010_2$

Light detectors

Light = 1

$010 = 011_2$

$110 = 100_2$

Gray code
to
Binary decoder

→ Binary out

**Fig. 6-5** Gray code used on shaft encoder disk to determine angular position.

 **Self-Test**

*Answer the following questions.*

10. The Gray code _____ (is, is not) a BCD-type code.
11. What characteristic is most important about the Gray code?

12. The inventor of the Gray code was _____ of Bell Labs.
13. The Gray code is most commonly associated with _____ of a shaft's angular position using an encoder disk.

them briefly is to make you aware that many codes exist in digital equipment. The codes you will probably encounter most often are binary, BCD (8421), and ASCII.

## 6-4 The ASCII Code

The ASCII code is widely used to send information to and from microcomputers. The standard ASCII code is a 7-bit code used in transferring coded information from keyboards and to computer displays and printers. The abbreviation ASCII (pronounced "ask-ee") stands for the *American Standard Code for Information Interchange.*

Table 6-3 is a summary of the ASCII code. The ASCII code is used to represent numbers, letters, punctuation marks, as well as control characters. For instance, the 7-bit ASCII code 111 1111 stands for DEL from the top chart. From the bottom chart we see that DEL means delete.

What is the coding for "A" in ASCII? Locate A on the top chart in Table 6-3. Assembling the 7-bit code gives 100 0001 = A. This is the code you would expect to be sent to a microcomputer's CPU if you pressed the A key on the keyboard.

Some care must be used in applying Table 6-3 to specific equipment. Be aware that the shaded control characters may have other meanings on specific computers or other equipment. However, common control characters such as BEL (bell), BS (backspace), LF (line feed), CR (carriage return), DEL (delete), and SP (space) are used on most computers. The exact meaning of the ASCII control codes should be looked up in your equipment manual.

The ASCII code is an *alphanumeric code.* It can represent both letters and numbers. Several other alphanumeric codes are *EBCDIC* (extended binary-coded decimal interchange code), *Baudot,* and *Hollerith.*

**American Standard Code for Information Interchange**

**Alphanumeric code**

**EBCDIC**

**Baudot**

**Hollerith**

## Table 6-3    The ASCII Code

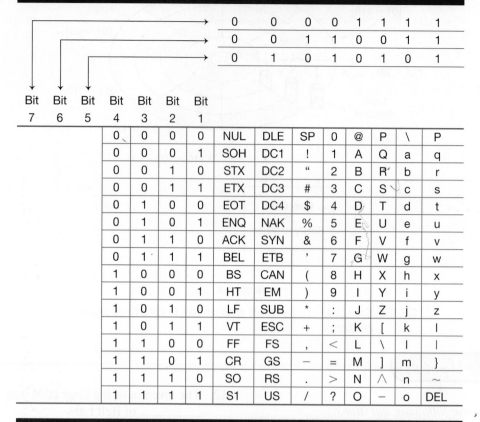

| Bit 7 | Bit 6 | Bit 5 | Bit 4 | Bit 3 | Bit 2 | Bit 1 | 000 | 001 | 010 | 011 | 100 | 101 | 110 | 111 |
|---|---|---|---|---|---|---|---|---|---|---|---|---|---|---|
|  |  |  | 0 | 0 | 0 | 0 | NUL | DLE | SP | 0 | @ | P | \ | P |
|  |  |  | 0 | 0 | 0 | 1 | SOH | DC1 | ! | 1 | A | Q | a | q |
|  |  |  | 0 | 0 | 1 | 0 | STX | DC2 | " | 2 | B | R | b | r |
|  |  |  | 0 | 0 | 1 | 1 | ETX | DC3 | # | 3 | C | S | c | s |
|  |  |  | 0 | 1 | 0 | 0 | EOT | DC4 | $ | 4 | D | T | d | t |
|  |  |  | 0 | 1 | 0 | 1 | ENQ | NAK | % | 5 | E | U | e | u |
|  |  |  | 0 | 1 | 1 | 0 | ACK | SYN | & | 6 | F | V | f | v |
|  |  |  | 0 | 1 | 1 | 1 | BEL | ETB | ' | 7 | G | W | g | w |
|  |  |  | 1 | 0 | 0 | 0 | BS | CAN | ( | 8 | H | X | h | x |
|  |  |  | 1 | 0 | 0 | 1 | HT | EM | ) | 9 | I | Y | i | y |
|  |  |  | 1 | 0 | 1 | 0 | LF | SUB | * | : | J | Z | j | z |
|  |  |  | 1 | 0 | 1 | 1 | VT | ESC | + | ; | K | [ | k | l |
|  |  |  | 1 | 1 | 0 | 0 | FF | FS | , | < | L | \ | l | \| |
|  |  |  | 1 | 1 | 0 | 1 | CR | GS | – | = | M | ] | m | } |
|  |  |  | 1 | 1 | 1 | 0 | SO | RS | . | > | N | ∧ | n | ~ |
|  |  |  | 1 | 1 | 1 | 1 | S1 | US | / | ? | O | – | o | DEL |

## Control Functions

| | | | |
|---|---|---|---|
| NUL | Null | DLE | Data link escape |
| SOH | Start of heading | DC1 | Device control 1 |
| STX | Start of text | DC2 | Device control 2 |
| ETX | End of text | DC3 | Device control 3 |
| EOT | End of transmission | DC4 | Device control 4 |
| ENQ | Enquiry | NAK | Negative acknowledge |
| ACK | Acknowledge | SYN | Synchronous idle |
| BEL | Bell | ETB | End of transmission block |
| BS | Backspace | CAN | Cancel |
| HT | Horizontal tabulation (skip) | EM | End of medium |
| LF | Line feed | SUB | Substitute |
| VT | Vertical tabulation (skip) | ESC | Escape |
| FF | Form feed | FS | File separator |
| CR | Carriage return | GS | Group separator |
| SO | Shift out | RS | Record separator |
| SI | Shift in | US | Unit separator |
| DEL | Delete | SP | Space |

*Answer the following questions.*

14. ASCII is classified as a(n) _____ code because it can represent both numbers and letters.

15. The letters ASCII stand for _____.

16. The letter R is represented by the 7-bit ASCII code _____.

17. The ASCII code 010 0100 represents what character?

## 6-5 Encoders

A digital system using an *encoder* is shown in Fig. 6-6. The encoder in this system must translate the decimal input from the keyboard to an 8421 BCD code. This encoder is called a *10-line-to-4-line priority encoder* by the manufacturer. Figure 6-7(a) is a block diagram of this encoder. If the decimal input 3 on the encoder is activated, then the logic circuit inside the unit outputs the BCD number 0011 as shown.

A more accurate description of a 10-line-to-4-line priority encoder is shown in Fig. 6-7(b). This is a connection diagram for the 74147 10-line-to-4-line priority encoder. Note the bubbles at both the inputs (1 to 9) and the outputs (A to D). The bubbles mean that the 74147 priority encoder has both *active low inputs* and *active low outputs*. A truth table is given for the 74147 priority encoder in Fig. 6-7(c). Note that

only low logic levels (L on the truth table) activate the appropriate input. The active state for the outputs on this IC are also LOW. Notice that in the last line of the truth table in Fig. 6-7(c) the L (logical 0) at input 1 activates only the A output (the least significant bit of the four-bit group).

The 74147 TTL IC in Fig. 6-7(c) is packaged in a 16-pin DIP. Internally, the IC consists of circuitry equivalent to about 30 logic gates.

The 74147 encoder in Fig. 6-7 has a *priority* feature. This means that if two inputs are activated at the same time, only the larger number will be encoded. For instance, if both the 9 and the 4 inputs were activated (LOW), then the output would be LHHL, representing decimal 9. Note that the outputs need to be complemented (inverted) to form the true binary number of 1001.

**10-line-to-4-line priority encoder**

**Active low inputs**

**Active low outputs**

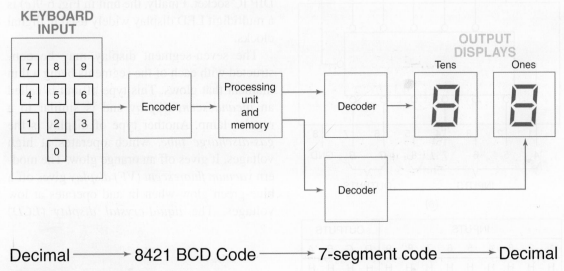

Decimal ⟶ 8421 BCD Code ⟶ 7-segment code ⟶ Decimal

**Fig. 6-6** A digital system.

---

### Self-Test

*Answer the following questions.*

18. Refer to Fig. 6-7. The 74147 encoder IC has active _____ (HIGH, LOW) inputs and active _____ (HIGH, LOW) outputs.

19. Refer to Fig. 6-7. If only input 7 of the 74147 encoder is LOW, what is the logic state at each of the four outputs?

20. Refer to Fig. 6-7(b). What is the meaning of a bubble on the logic symbol at input 4 (pin 1 on the 74147 IC)?

21. Refer to Fig. 6-7. If both inputs 2 and 8 go LOW on the 74147 encoder, what is the logic state at each of the four outputs?

## DECIMAL INPUTS

## BCD OUTPUT

**Seven-segment display**

0 0 1 1

(D) (C) (B) (A)

Activate →

10-line-to-4-line encoder

1
2
3
4
5
6
7
8
9

D
C
B
A

*(a)*

OUTPUT          INPUTS          OUTPUT

$V_{CC}$  NC  D  3  2  1  9  A

16  15  14  13  12  11  10  9

74147

1  2  3  4  5  6  7  8

4  5  6  7  8  C  B  GND

INPUTS          OUTPUTS

*(b)*

| INPUTS | | | | | | | | | OUTPUTS | | | |
|---|---|---|---|---|---|---|---|---|---|---|---|---|
| 1 | 2 | 3 | 4 | 5 | 6 | 7 | 8 | 9 | D | C | B | A |
| H | H | H | H | H | H | H | H | H | H | H | H | H |
| X | X | X | X | X | X | X | X | L | L | H | H | L |
| X | X | X | X | X | X | X | L | H | L | H | H | H |
| X | X | X | X | X | X | L | H | H | H | L | L | L |
| X | X | X | X | X | L | H | H | H | H | L | L | H |
| X | X | X | X | L | H | H | H | H | H | L | H | L |
| X | X | X | L | H | H | H | H | H | H | L | H | H |
| X | X | L | H | H | H | H | H | H | H | H | L | L |
| X | L | H | H | H | H | H | H | H | H | H | L | H |
| L | H | H | H | H | H | H | H | H | H | H | H | L |

H = HIGH logic level, L = LOW logic level, X = Don't care

*(c)*

**Incandescent display**

**Gas-discharge tube**

**Vacuum fluorescent (VF) display**

**Liquid-crystal display (LCD)**

**74147 encoder IC**

**Fig. 6-7**  *(a)* 10-line-to-4-line encoder.
*(b)* Pin diagram for 74147 encoder IC.
*(c)* Truth table for 74147 encoder.

# 6-6 Seven-Segment LED Displays

The common task of decoding from machine language to decimal numbers is suggested in the system in Fig. 6-6. A very common output device used to display decimal numbers is the *seven-segment display.* The seven segments of the display are labeled *a* through *g* in Fig. 6-8(*a*). The displays representing decimal digits 0 through 9 are shown in Fig. 6-8(*b*). For instance, if segments *a*, *b*, and *c* are lit, a decimal 7 is displayed. If, however, all segments *a* through *g* are lit, a decimal 8 is displayed.

Several common seven-segment display packages are shown in Fig. 6-9. The seven-segment LED display in Fig. 6-9(*a*) fits a regular 14-pin DIP IC socket. Another single-digit seven-segment LED display is shown in Fig. 6-9(*b*). This display fits crosswise into a wider DIP IC socket. Finally, the unit in Fig. 6-9(*c*) is a multidigit LED display widely used in digital clocks.

The seven-segment display may be constructed with each of the segments being a thin filament that glows. This type of unit is called an *incandescent display* and is similar to a regular lamp. Another type of display is the *gas-discharge tube,* which operates at high voltages. It gives off an orange glow. The modern *vacuum fluorescent (VF) display* gives off a blue-green glow when lit and operates at low voltages. The *liquid-crystal display (LCD)*

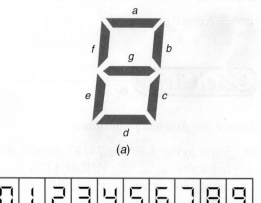

*(a)*

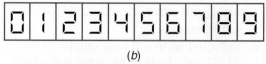

*(b)*

**Fig. 6-8**  *(a)* Segment identification. *(b)* Decimal numbers on typical seven-segment display.

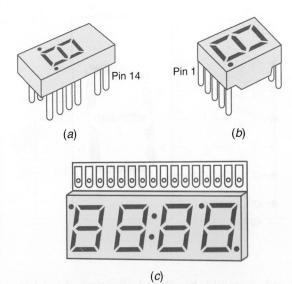

(a)                              (b)

(c)

**Fig. 6-9** (a) DIP seven-segment LED display. (b) A common 10-pin single-digit package. Note the location of pin 1. Pins are numbered counterclockwise from pin 1 when viewed from the top of the display. (c) Multidigit package.

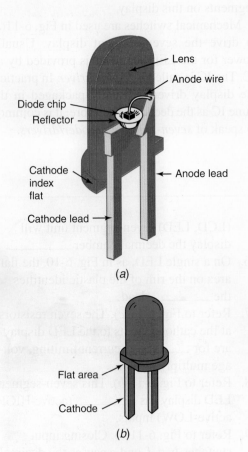

Lens
Anode wire
Diode chip
Reflector
Cathode index flat
Anode lead
Cathode lead

(a)

Flat area
Cathode

(b)

**Fig. 6-10** (a) Cutaway view of standard light-emitting diode. (b) Identifying the cathode lead of the LED.

creates numbers in a black or silvery color. The common LED display gives off a characteristic reddish glow when lit.

A basic single *LED (light-emitting diode)* is illustrated in Fig. 6-10. The cutaway view of the LED in Fig. 6-10(a) shows the small exposed diode chip with a reflector to project the light upward toward the plastic lens.

Of importance during use is the index area on the rim of the LED shown in Fig. 6-10(b). The flat area shows the cathode side of the LED. The LED is basically a *PN-junction diode*. When the diode is forward-biased, current flows through the PN junction and the LED lights and is focused by the plastic lens. Many LEDs are fabricated from *gallium arsenide* (GaAs) and several related materials. LEDs come in several colors including red, green, orange, blue, and amber.

A single LED is being tested in Fig. 6-11(a). When the switch (SW1) is closed, current flows from the 5-V power supply through the LED, causing it to light. The series resistor limits current to about 20 mA. Without the limiting resistor the LED would burn out. Typically, LEDs can accept only about 1.7 to 2.1 V across their terminals when lit. Being a diode, the LED is sensitive to polarity. Hence, the *cathode (K)* must be toward the negative (GND) terminal while the *anode (A)* must be toward the positive terminal of the power supply.

A *seven-segment LED display* is shown in Fig. 6-11(b). Each segment (a through g) contains an LED, as shown by the seven symbols. The display shown has all the anodes tied together and coming out the right side as a single connection (common anode). The inputs on the left go to the various segments of the display. The device in Fig. 6-11(b) is referred to as a *common-anode seven-segment LED display*. These units can also be purchased in *common-cathode* form.

To understand how segments on the display are activated and lit, consider the circuit in Fig. 6-11(c). If switch b is closed, current flows from GND through the limiting resistor to the b-segment LED and out the common-anode connection to the power supply. Only segment b will light.

Suppose you wanted the decimal 7 to light on the display in Fig. 6-11(c). Switches a, b,

**PN-junction diode**

**Gallium arsenide (GaAs)**

More on LEDs
www.howstuff
works.com

**Seven-segment LED display**

**Common-anode**
**Common-cathode**

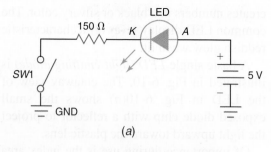

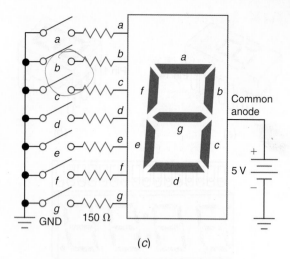

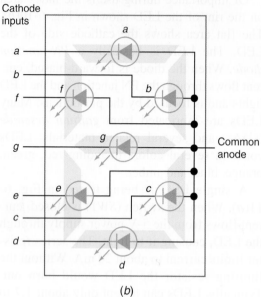

Fig. 6-11 (a) Operation of a simple LED. (b) Wiring a common-anode seven-segment LED display. (c) Driving a seven-segment LED display with switches.

**Display driver**

**Seven-segment decoder/drivers**

and $c$ would be closed, lighting the LED segments $a$, $b$, and $c$. The decimal 7 would light on the display. Likewise, if the decimal 5 were to be lit, switches $a$, $c$, $d$, $f$, and $g$ would be closed. These five switches would ground the correct segments, and a decimal 5 would appear on the display. Note that it takes a GND voltage (LOW logic level) to activate the LED segments on this display.

Mechanical switches are used in Fig. 6-11(c) to drive the seven-segment display. Usually power for the LED segments is provided by an IC. The IC is called a *display driver*. In practice, the display driver is usually packaged in the same IC as the decoder. Therefore, it is common to speak of *seven-segment decoder/drivers*.

### Self-Test

*Supply the missing word or words in each statement.*

22. Refer to Fig. 6-8(a). If segments $a$, $c$, $d$, $f$, and $g$ are lit, the decimal number _____ will appear on the seven-segment display.
23. The seven-segment unit that gives off a blue-green glow is a(n) _____ (vacuum fluorescent, incandescent, LCD, LED) display.
24. The letters "LED" stand for _____.
25. Refer to Fig. 6-11(c). If switches $b$ and $c$ are closed, segments _____ and _____ will light. This _____

(LCD, LED) seven-segment unit will display the decimal number _____.
26. On a single LED, as in Fig. 6-10, the flat area on the rim of the plastic identifies the _____ lead.
27. Refer to Fig. 6-11(c). The seven resistors at the cathode inputs to the LED display are for _____ (current limiting, voltage multiplying).
28. Refer to Fig. 6-11(b). This seven-segment LED display has _____ (active-HIGH, active-LOW) inputs.
29. Refer to Fig. 6-11(c). Closing input switches $b$, $c$, $f$, and $g$ causes the decimal number _____ to be displayed.

## 6-7 Decoders

A *decoder*, like an encoder, is a code translator. Figure 6-6 shows two decoders being used in the system. The decoders are translating the 8421 BCD code to a seven-segment display code that lights the proper segments on the displays. The display will be a decimal number. Figure 6-12 shows the BCD number 0101 at the input of the *BCD-to-seven-segment decoder/driver*. The decoder activates outputs *a, c, d, f,* and *g* to light the segments shown in Fig. 6-12. The decimal number 5 lights up on the display.

Decoders come in several varieties, such as the ones illustrated in Fig. 6-13. Notice in Fig. 6-13 that the same block diagram is used for the 8421 BCD, the excess-3, and the Gray decoder.

Other decoders are available such as BCD converters, BCD-to-binary converters, 4-to-16-line decoders, and 2-to-4-line decoders. Other encoders available are a decimal-to-octal and 8-to-3 line priority encoder.

Decoders, like encoders, are *combinational logic circuits* with several inputs and outputs.

Most decoders contain from 20 to 50 gates. Most decoders and encoders are packaged in single IC packages. Specialized encoders and decoders can be fabricated using programmable logic devices (PLDs).

Decoding can also be achieved using flexible programmable devices such as BASIC Stamp modules. These modules by Parallax contain a microcontroller and associated EEPROM memory.

**Decoder**

**Combinational logic circuits**

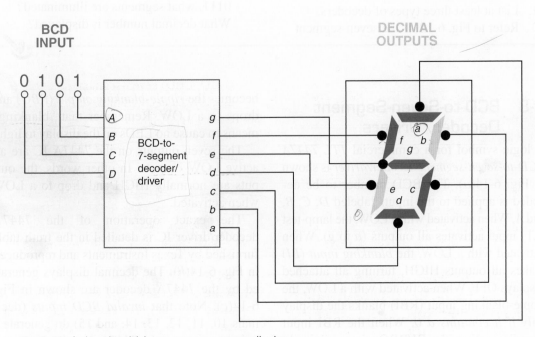

**BCD INPUT**

**DECIMAL OUTPUTS**

0 1 0 1

A
B
C
D

BCD-to-7-segment decoder/driver

g
f
e
d
c
b
a

**Fig. 6-12** A decoder driving a seven-segment display.

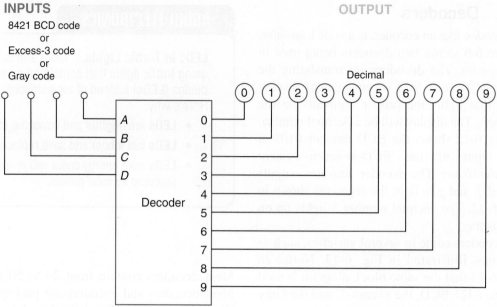

Fig. 6-13 A typical decoder block diagram. Note that inputs may be 8421 BCD, excess-3, or Gray code.

30. Refer to Fig. 6-12. If the BCD input to the decoder/driver is 1000, which segments on the display will light? The seven-segment LED display will read what decimal number?
31. List at least three types of decoders.
32. Refer to Fig. 6-12. If the seven-segment display were an LED-type, seven _____ (limiting resistors, switches) would need to be added between the decoder and the display.
33. Refer to Fig. 6-12. If the display reads decimal 3, the BCD input is _____.
34. Refer to Fig. 6-12. If the BCD input is 0111, what segments are illuminated? What decimal number is displayed?

## 6-8 BCD-to-Seven-Segment Decoder/Drivers

**7447A BCD-to-seven-segment decoder/driver**

**Blanking input (BI)**

**Invalid BCD inputs**

A logic symbol for a commercial *TTL 7447A BCD-to-seven-segment decoder/driver* is shown in Fig. 6-14(a). The BCD number to be decoded is applied to the inputs labeled *D, C, B,* and *A*. When activated with a LOW, the lamp-test (LT) input activates all outputs (*a* to *g*). When activated with a LOW, the *blanking input (BI)* makes all outputs HIGH, turning all attached displays OFF. When activated with a LOW, the ripple-blanking input (RBI) blanks the display *only if it contains a 0*. When the RBI input becomes active, the BI/RBO pin temporarily becomes the *ripple-blanking output (RBO)* and drops to a LOW. Remember that "blanking" means to cause no LEDs on the display to light.

The seven outputs on the 7447A IC are all active LOW outputs. In other words, the outputs are normally HIGH and drop to a LOW when activated.

The exact operation of the 7447A decoder/driver IC is detailed in the truth table furnished by Texas Instruments and reproduced in Fig. 6-14(b). The decimal displays generated by the 7447A decoder are shown in Fig. 6-14(c). Note that *invalid BCD inputs* (decimals 10, 11, 12, 13, 14, and 15) do generate a unique output on the 7447A decoder.

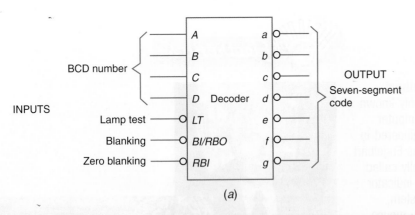

(a)

| Decimal or function | INPUTS | | | | | | BI/BRO | OUTPUTS | | | | | | | Note |
|---|---|---|---|---|---|---|---|---|---|---|---|---|---|---|---|
| | LT | RBI | D | C | B | A | | a | b | c | d | e | f | g | |
| 0 | H | H | L | L | L | L | H | ON | ON | ON | ON | ON | ON | OFF | |
| 1 | H | X | L | L | L | H | H | OFF | ON | ON | OFF | OFF | OFF | OFF | |
| 2 | H | X | L | L | H | L | H | ON | ON | OFF | ON | ON | OFF | ON | |
| 3 | H | X | L | L | H | H | H | ON | ON | ON | ON | OFF | OFF | ON | |
| 4 | H | X | L | H | L | L | H | OFF | ON | ON | OFF | OFF | ON | ON | |
| 5 | H | X | L | H | L | H | H | ON | OFF | ON | ON | OFF | ON | ON | |
| 6 | H | X | L | H | H | L | H | OFF | OFF | ON | ON | ON | ON | ON | |
| 7 | H | X | L | H | H | H | H | ON | ON | ON | OFF | OFF | OFF | OFF | |
| 8 | H | X | H | L | L | L | H | ON | ON | ON | ON | ON | ON | ON | 1 |
| 9 | H | X | H | L | L | H | H | ON | ON | ON | OFF | OFF | ON | ON | |
| 10 | H | X | H | L | H | L | H | OFF | OFF | OFF | ON | ON | OFF | ON | |
| 11 | H | X | H | L | H | H | H | OFF | OFF | ON | ON | OFF | OFF | ON | |
| 12 | H | X | H | H | L | L | H | OFF | ON | OFF | OFF | OFF | ON | ON | |
| 13 | H | X | H | H | L | H | H | ON | OFF | OFF | ON | OFF | ON | ON | |
| 14 | H | X | H | H | H | L | H | OFF | OFF | OFF | ON | ON | ON | ON | |
| 15 | H | X | H | H | H | H | H | OFF | OFF | OFF | OFF | OFF | OFF | OFF | |
| BI | X | X | X | X | X | X | L | OFF | OFF | OFF | OFF | OFF | OFF | OFF | 2 |
| RBI | H | L | L | L | L | L | L | OFF | OFF | OFF | OFF | OFF | OFF | OFF | 3 |
| LT | L | X | X | X | X | X | H | ON | ON | ON | ON | ON | ON | ON | 4 |

H = HIGH level, L = LOW level, X = Irrelevant

Notes:

1. The blanking input (BI) must be open or held at a HIGH logic level when output functions 0 through 15 are desired. The ripple-blanking input (RBI) must be open or HIGH if blanking of a decimal zero is not desired.
2. When a LOW logic level is applied directly to the blanking input (BI), all segment outputs are OFF regardless of the level of any other input.
3. When ripple-blanking input (RBI) and inputs A, B, C, and D are at a LOW level with the lamp test (LT) input HIGH, all segment outputs go OFF and the ripple-blanking output (RBO) goes to a LOW level (response condition).
4. When the blanking input/ripple-blanking output (BI/RBO) is open or held HIGH and a LOW is applied to the lamp test (LT) input, all segment outputs are ON.

(b)

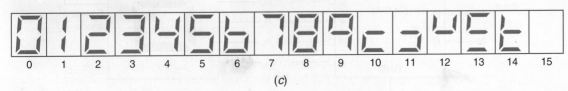

(c)

**Fig. 6-14** (a) Logic symbol for 7447A TTL decoder IC. (b) Truth table for 7447A decoder. (*Courtesy of Texas Instruments, Inc.*) (c) Format of readouts on seven-segment display using the 7447A decoder IC.

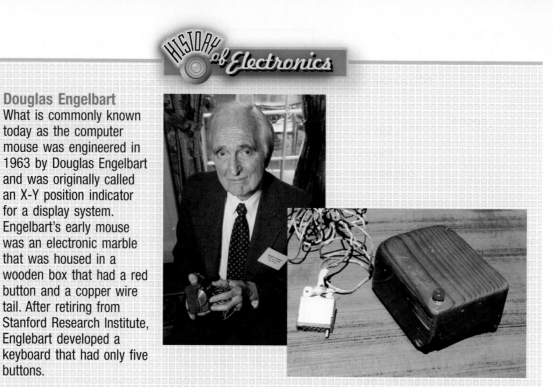
**Good design practice**

**"Floating" inputs**

**Blanking leading zeros**

The 7447A decoder/driver IC is typically connected to a common-anode seven-segment LED display. Such a circuit is shown in Fig. 6-15. It is especially important that the seven 150-$\Omega$ limiting resistors be wired between the 7447A IC and the seven-segment display.

Assume that the BCD input to the 7447A decoder/driver in Fig. 6-15 is 0001 (LLLH). This is equal to line 2 of the truth table in Fig. 6-14(*b*). This input combination causes segments *b* and *c* on the seven-segment display to light (outputs *b* and *c* drop to LOW). Decimal

1 is displayed. The LT and two BIs are not shown in Fig. 6-15. When not connected, they are assumed to be "floating" HIGH and therefore disabled in this circuit. *Good design practice* suggests that these "*floating*" inputs should be connected to +5 V to make sure they stay HIGH.

Many applications, such as a calculator or cash register, require that the *leading zeros be blanked*. The illustration in Fig. 6-16 shows the use of the 7447A decoder/drivers operating a group of displays as in a cash register. The

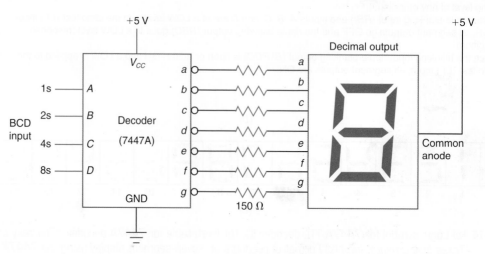

**Fig. 6-15** Wiring a 7447A decoder and seven-segment LED display.

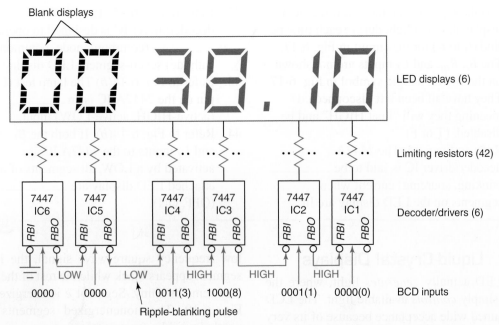

**Fig. 6-16** Using the ripple-blanking input (RBI) of the 7447A decoder/driver to blank leading zeros in a multidigit display.

six-digit display example details how the blanking of leading 0s would be accomplished using the 7447A IC driving LED displays.

The current inputs to the six decoders are shown across the bottom of the drawing in Fig. 6-16. The current BCD input is 0000 0000 0011 1000 0001 0000 (003810 in decimal). The two left 0s should be blanked, so the display reads 38.10. The blanking of the leading 0s is handled by wiring the RBI and RBO pins to each 7447A decoder IC together as shown in Fig. 6-16.

Working from left to right in Fig. 6-16, notice that the RBI input of IC6 is grounded. From the 7447A decoder's truth table in Fig. 6-14(b) it can be determined that when RBI is LOW and when all BCD inputs are LOW, then all segments of the display are blanked or off.

Also the RBO is forced LOW. This LOW is passed to the RBI of IC5.

Continuing in Fig. 6-16 with the BCD input to IC5 as 0000 and the RBI at LOW, the display is also blanked. The RBO of IC5 is forced LOW and is passed to the RBI input of IC4. Even with the RBI LOW, IC4 *does not* blank the display because the BCD input is 0011. The RBO of IC4 remains HIGH, which is sent to IC3.

A question arises about the right LED display in Fig. 6-16. The BCD input to IC1 is $0000_{BCD}$ and a zero (0) appears on the display. The zero on the IC1 display is not blanked because the RBI input is not activated (RBI = HIGH). The first row of the truth table in Fig. 6-14(b) shows that the 7447A decoder/driver will display the zero when the RBI is HIGH.

![Self-Test]

*Check your understanding by answering the following questions.*

35. Refer to Fig. 6-14. The 7447A decoder/driver IC has active _____ (HIGH, LOW) BCD inputs and active _____ (HIGH, LOW) outputs.

36. Refer to Fig. 6-14. The lamp test, blanking, and zero-blanking inputs of the 7447A are active _____ (HIGH, LOW) inputs.

37. The RBI and RBO inputs of the 7447A are commonly used for blanking _____ on calculator and cash register multidigit displays.

38. What will the seven-segment display read during each time period ($t_1$ to $t_7$) for the circuit shown in Fig. 6-17?

39. List the segments on the seven-segment display that will light during each time period ($t_1$ to $t_7$) for the circuit in Fig. 6-17.
40. The $B_I$, $R_{BI}$, and $L_T$ inputs are not shown on the 7447A's logic symbol in Fig. 6-17. They have all been left disconnected meaning they will "float HIGH" and be disabled. (T or F)
41. Refer to Fig 6-15. The 7447A decoder/driver IC is said to be _____ (sinking, sourcing) current when segments on the LED display are lit.

42. Refer to Fig. 6-15. The 7447A decoder/driver IC is designed to operate a _____ (common-anode, common-cathode) seven-segment LED display.
43. Refer to Fig. 6-14(b) The lamp test (LT) pin on the 7447A IC is an _____ (active-HIGH, active-LOW) input.
44. Refer to Fig. 6-14(b). If both the $B_I$ and $L_T$ inputs to the 7447A IC are activated by a LOW, all segments of an attached LED display are _____ (OFF, ON).

---

**Liquid-crystal displays**

## 6-9 Liquid-Crystal Displays

The LED actually *generates* light, where the LCD simply *controls* available light. The LCD has gained wide acceptance because of its very low power consumption. The LCD is also well suited for use in sunlight or in other brightly lit areas. The DMM (digital multimeter) in Fig. 6-18 uses a modern LCD.

**Nematic fluid**

The LCD is also suited for more complex displays than just seven-segment decimal. The LCD display in Fig. 6-18 contains an analog scale across the bottom as well as the larger digital readout. In practice you will find that the DMM LCD has several other symbols, which you can also observe on the DMM in Fig. 6-18.

### Monochrome LCD

**Field-effect LCD**

The construction of a common LCD unit is shown in Fig. 6-19. This unit is called the *field-effect LCD*. When a segment is energized by a

low frequency square-wave signal, the LCD segment appears black while the rest of the surface remains shiny. Segment *e* is energized in Fig. 6-19. The nonenergized segments are nearly invisible.

The key to LCD operation is the liquid crystal, or *nematic fluid*. This nematic fluid is sandwiched between two glass plates. An ac voltage is applied across the nematic fluid, from the top metalized segments to the metalized backplane. When affected by the field of the ac voltage, the nematic fluid transmits light differently and the energized segment appears as black on a silvery background.

The twisted-nematic field-effect LCD uses a polarizing filter on the top and bottom of the display as shown in Fig. 6-19. The backplane and segments are internally wired to contacts on the edge of the LCD. Only two of the many contacts are shown in Fig 6-19.

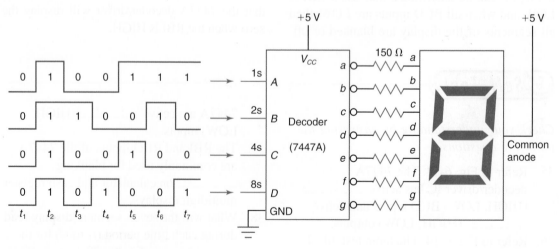

**Fig. 6-17** Decoder-LED display circuit.

**Fig. 6-18** Digital multimeter (DMM) using a liquid-crystal display.

## Driving the LCD

Decimal 7 is displayed on the LCD shown in Fig 6-20. The BCD-to-seven-segment decoder on the left is receiving a BCD input of 0111. This input activates the *a*, *b*, and *c* outputs of the decoder (*a*, *b*, and *c* are HIGH in this example). The remaining outputs of the decoder are LOW (*d*, *e*, *f*, and *g* = LOW). The 100-Hz square-wave input is always applied to the backplane of the display. This signal is also applied to each of the CMOS XOR gates used to drive the LCD. Note that the XOR gates produce an inverted waveform when activated (*a*, *b*, and *c* XOR gates are activated). The 180° out-of-phase signals to the backplane and segments *a*, *b*, and *c* cause these areas of the LCD to turn black. The in-phase signals from XOR gates *d*, *e*, *f*, and *g* do not cause these segments to be activated. Therefore, these segments remain nearly invisible.

The XOR gates used as LCD drivers in Fig. 6-20 are CMOS units. TTL XOR gates are not used because they cause a small dc offset voltage to be developed across the LCD's nematic fluid. The *dc voltage* across the nematic fluid *will destroy the LCD* in a short time.

In actual practice, the decoder and XOR LCD display drivers pictured in Fig. 6-20 are usually packaged in a single CMOS IC. The 100-Hz square-wave signal frequency is not critical and may range from 30 to 200 Hz. Liquid-crystal displays are sensitive to low temperatures. At below-zero temperatures the LCD display's turn-on and turn-off times become very slow. However, the long lifetime and extremely low power consumption make them ideal for battery or solar cell operation.

**Digital multimeter**

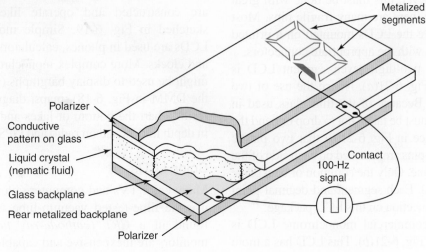

**Fig. 6-19** Construction of a field-effect LCD.

**Construction of a field-effect LCD**

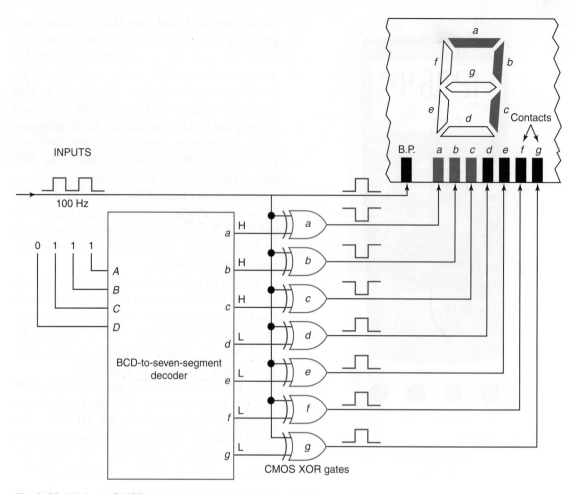

INPUTS

100 Hz

0 1 1 1

A
B
C
D

BCD-to-seven-segment
decoder

H  a
H  b
H  c
L  d
L  e
L  f
L  g

a
b
c
d
e
f
g

CMOS XOR gates

B.P.  a  b  c  d  e  f  g

Contacts

**Fig. 6-20** Wiring a CMOS decoder/driver system to an LCD.

## Commercial LCDs

Figure 6-21 illustrates two examples of typical commercial monochrome LCD devices. Note that both have pins which can be soldered into a printed circuit board. In the lab, these LCD displays may be plugged into solderless mounting boards. However, this must be done with great care because there are many fragile pins. Most labs will have the LCDs mounted on a printed circuit board with the appropriate connectors.

A simple two-digit seven-segment LCD is sketched in Fig. 6-21(a). Notice the use of two glass plates. Because of the thin glass used in LCDs, care must be taken not to drop or bend the display. Notice in Fig. 6-21(a) that two plastic headers with pins are fastened on each side of the glass backplane. Only the common or backplane pin is labeled. Each segment and decimal point has a pin connection on this LCD package.

Another commercial monochrome LCD is illustrated in Fig. 6-21(b). This LCD has a more complex display, including symbols. This unit comes in a 40-pin package. All segments, decimal points, and symbols are assigned a pin number. Only the backplane or common pin is noted on the drawing. Manufacturers' data sheets must be consulted for actual pin numbers.

Inexpensive monochrome LCDs use the *twisted-nematic field-effect technology*. These are constructed and operate like the one sketched in Fig. 6-19. Simple monochrome LCDs are used in phones, calculators, watches, and clocks. More complex monochrome LCDs might be used to display bargraphs (such as on the DMM in Fig. 6-18), maps, diagram waveforms, chart the bottom of lakes and show fish in depth finders and on radio and GPS receivers.

## Color LCDs

Many color TVs and computer color monitors use the time-tested vacuum-tube technology found in a *CRT (cathode-ray tube)*. CRT monitors are inexpensive and capable of excellent performance. CRTs display bright colors in

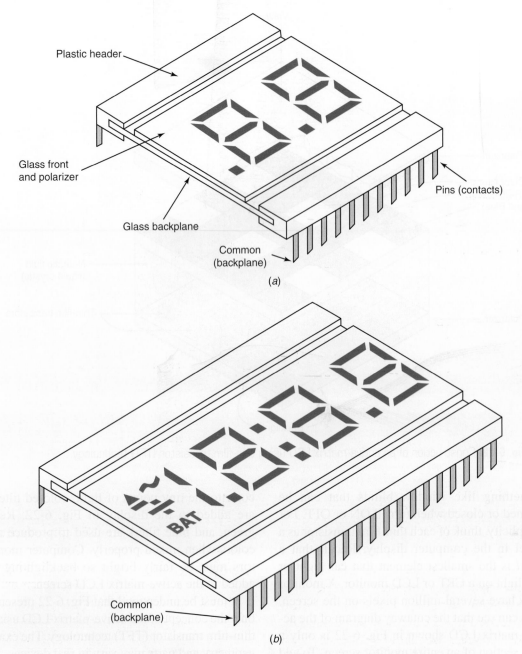

Plastic header

Glass front
and polarizer

Glass backplane

Common
(backplane)

Pins (contacts)

*(a)*

Common
(backplane)

*(b)*

**Fig. 6-21** Commercial liquid-crystal displays. (*a*) Two-digit LCD. (*b*) LCD with 3 1/2 digits and symbols.

Commercial liquid-crystal displays

Thin-film transistors

Active-matrix LCDs (AMLCD)

high resolutions. CRTs do have the disadvantages of being large and heavy and use a lot of power.

Color LCDs are commonly used in lightweight battery-power laptop computers. Color LCDs are appearing on computer desktops in place of the bulky CRT monitors. Color LCDs are generally classified as either *passive-matrix LCDs* or the newer more *expensive active-matrix LCDs (AMLCD)*. The future favors the active-matrix LCD because it is faster, brighter, and has a wider viewing angle than

the passive-matrix types. However, active-matrix LCDs are more expensive than passive-matrix type.

A simplified sketch of the construction of an active-matrix LCD is shown in Fig. 6-22. Like the monochrome LCD it has polarizers top and bottom. The active-matrix LCD also contains nematic fluid (liquid crystal) sandwiched in the sealed display much like the monochrome type display. Below the nematic fluid are *thin-film transistors* that can be turned ON or OFF individually. The thin-film transistors are

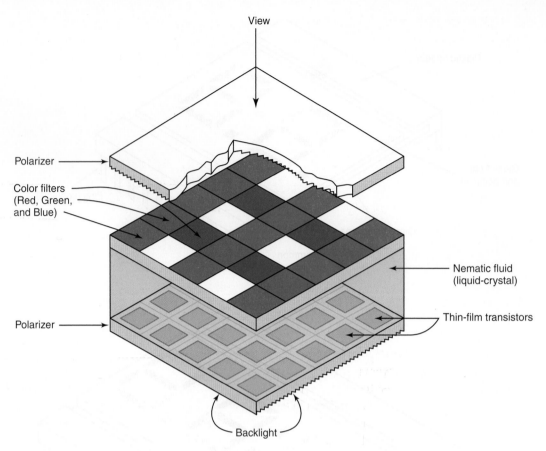

Fig. 6-22 Construction of an active-matrix LCD using thin-film transistor (TFT) technology.

**Pixel**

something like window blinds that can be opened or closed when turned ON or OFF. For simplicity, think of each thin-film transistor as a pixel in the computer display. Recall that a *pixel* is the smallest element that can light or not light on a CRT or LCD monitor. A monitor may have several million pixels on the screen. You can see that the cutaway diagram of the active-matrix LCD shown in Fig. 6-22 is only a tiny section of an entire monitor screen. To add

color to the tiny points of light, colored filters are added to the display in Fig. 6-22. Red, green, and blue filters are used to produce all colors when mixed properly. Computer monitors must be fairly bright so backlighting is added to the active-matrix LCD screen.

It must be understood that Fig. 6-22 presents only the concept of the active-matrix LCD using thin-film transistor (TFT) technology. The exact geometry and parts may vary in real devices.

## Self-Test

*Supply the missing word or words in each statement.*

45. Digits appear _____ (black, silver) on a _____ (black, silver) background on the field-effect LCD.
46. The LCD uses a liquid crystal, or _____ fluid, which transmits light differently when affected by a magnetic field from an ac voltage.

47. A(n) _____ (ac, dc) voltage applied to an LCD will damage the unit.
48. The LCD unit consumes a _____ (large amount, moderate amount, very small amount) of power.
49. Refer to Fig. 6-23. The XOR gates used to drive the LCD display are _____ (CMOS, TTL) devices.
50. Refer to Fig. 6-23. With an input of $0101_{BCD}$ to the decoder, the output display

with form the decimal _____ on the monochrome LCD.

51. Refer to Fig. 6-23. With the BCD input to the decoder of 0101, what segments of the monochrome LCD are activated and show up as dark on a lighter background?

52. Refer to Fig. 6-23. With the BCD input to the decoder of 0101, the outputs of XOR gates a, c, d, f, and g are _____ (in-phase, 180° out-of-phase) with the B.P. (backplane) signal.

## 6-10 Using CMOS to Drive an LCD Display

A block diagram of an LCD decoder-driver system is sketched in Fig. 6-24(a). The input is 8421 BCD. The latch is a temporary memory to hold the BCD data. The BCD-to-seven-segment decoder operates somewhat like the 7447A decoder that was studied earlier. Note that the output from the decoder in Fig. 6-24(a) is in seven-segment code. The last block before the display is the LCD driver. This consists of XOR gates as in Fig. 6-20. The drivers and backplane (common) of the display must be driven with a 100-Hz square-wave signal. In actual practice the latch, decoder, and LCD driver are all available in a single CMOS package. The 74HC4543 and 4543 ICs described by the manufacturer as a *BCD-to-seven-segment latch/decoder/driver for LCDs* are such packages.

A wiring diagram for a single LCD driver circuit is shown in Fig. 6-24(b). The 74HC4543 decoder/driver CMOS IC is being employed. The 8421 BCD input is 0011 (decimal 3). The $0011_{BCD}$ is decoded into seven-segment code. A separate 100-Hz clock feeds the signal to both the LCD backplane (common) and the Ph (phase) input of the 74HC4543 IC. The driving signals in this example are sketched for each segment of the LCD. Note that *only out-of-phase signals will activate a segment*. In-phase signals (such as segments e and f in this example) do not activate LCD segments.

A pin diagram for the *74HC4543 BCD-to-seven-segment latch/decoder/driver CMOS IC* is reproduced in Fig. 6-25(a). Detailed information on the operation of the 74HC4543 IC is contained in the truth table in Fig. 6-25(b). On the output side of the truth table, an "H" means the segment is on while an "L" means the

**74HC4543 BCD-to-seven-segment latch/decoder/driver CMOS IC**

**BCD-to-seven-segment latch/decoder/driver for LCD**

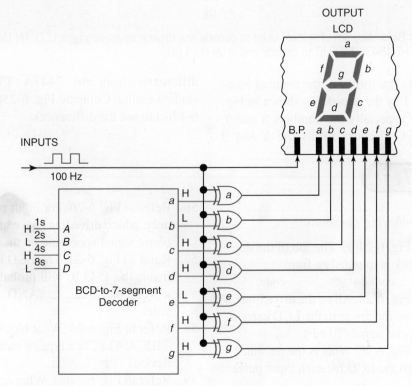

**Fig. 6-23** Driving a liquid-crystal display.

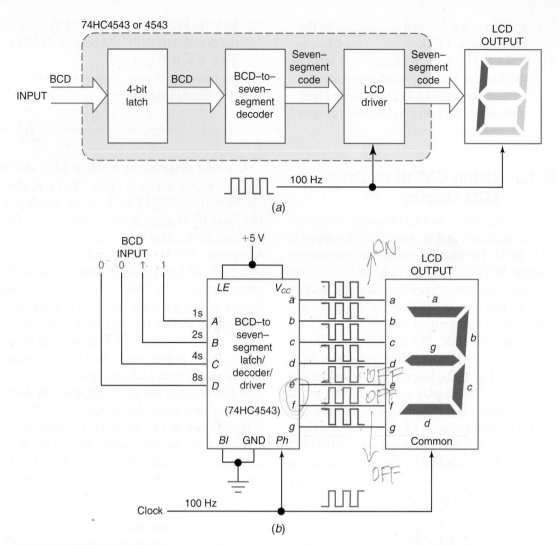

**Fig. 6-24** (*a*) Block diagram of system used to decode and drive a seven-segment LCD. (*b*) Using the 74HC4543 CMOS IC to decode and drive the LCD.

segment is off. The format of the decimal numbers generated by the decoder is shown in Fig. 6-25(*c*). Note especially the numbers 6 and 9. The 74HC4543 decoder forms the 6 and 9

differently from the 7447A TTL decoder studied earlier. Compare Fig. 6-25(*c*) with Fig. 6-14(*c*) to see the differences.

✓ *Self-Test*

*Answer the following questions.*

53. Refer to Fig. 6-24(*a*). The job of the decoder block is to translate from _____ code to _____ code.
54. Refer to Fig. 6-24. All of the drive lines going from the driver to the LCD carry a square-wave signal (T or F).
55. Refer to Fig. 6-26. What is the decimal reading on the LCD for each input pulse ($t_1$ to $t_5$)?

56. Refer to Fig. 6-26. For input pulse $t_5$ only, which drive lines have an *out-of-phase* signal appearing on them?
57. Refer to Fig. 6-24. The LCD driver block inside the 4543 IC will probably contain a group of _____ (AND, XOR) gates.
58. Refer to Fig. 6-24. What block in the 74HC4543 IC is a type of memory device?
59. Refer to Fig. 6-24(*b*). What activates segments *a*, *b*, *c*, *d*, and *g* on the LCD?

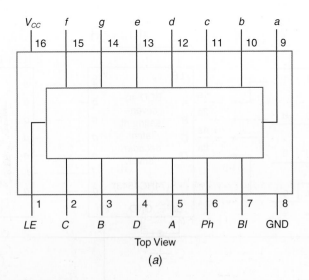

Top View

(a)

Truth Table

| INPUTS | | | | | | | OUTPUTS | | | | | | | |
|---|---|---|---|---|---|---|---|---|---|---|---|---|---|---|
| LE | BI | Ph* | D | C | B | A | a | b | c | d | e | f | g | Display |
| X | H | L | X | X | X | X | L | L | L | L | L | L | L | Blank |
| H | L | L | L | L | L | L | H | H | H | H | H | H | L | 0 |
| H | L | L | L | L | L | H | L | H | H | L | L | L | L | 1 |
| H | L | L | L | L | H | L | H | H | L | H | H | L | H | 2 |
| H | L | L | L | L | H | H | H | H | H | H | L | L | H | 3 |
| H | L | L | L | H | L | L | L | H | H | L | L | H | H | 4 |
| H | L | L | L | H | L | H | H | L | H | H | L | H | H | 5 |
| H | L | L | L | H | H | L | H | L | H | H | H | H | H | 6 |
| H | L | L | L | H | H | H | H | H | H | L | L | L | L | 7 |
| H | L | L | H | L | L | L | H | H | H | H | H | H | H | 8 |
| H | L | L | H | L | L | H | H | H | H | H | L | H | H | 9 |
| H | L | L | H | L | H | L | L | L | L | L | L | L | L | Blank |
| H | L | L | H | L | H | H | L | L | L | L | L | L | L | Blank |
| H | L | L | H | H | L | L | L | L | L | L | L | L | L | Blank |
| H | L | L | H | H | L | H | L | L | L | L | L | L | L | Blank |
| H | L | L | H | H | H | L | L | L | L | L | L | L | L | Blank |
| H | L | L | H | H | H | H | L | L | L | L | L | L | L | Blank |
| L | L | L | X | X | X | X | ** | | | | | | | ** |
| † | † | H | † | | | | Inverse of Output Combinations Above | | | | | | | Display as above |

X = don't care
† = same as above combinations
* = for liquid crystal readouts, apply a square wave to Ph
** = depends upon the BCD code previously applied when LE = H

(b)

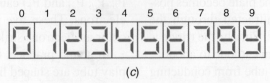

(c)

**Fig. 6-25** The 74HC4543 BCD-to-seven-segment latch/decoder/driver CMOS IC. (a) Pin diagram. (b) Truth table. (c) Format of digits formed by the 74HC4543 decoder IC.

74HC4543
BCD-to-seven-segment
latch/decoder/
driver CMOS IC

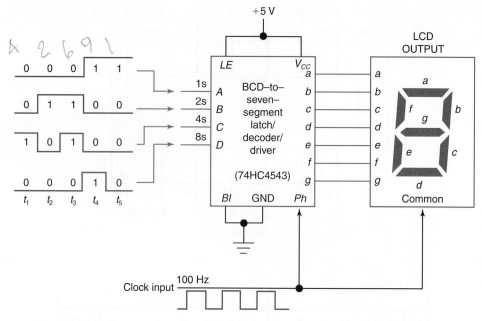

**Fig. 6-26** Decoder-LCD circuit for problems 55 and 56.

## 6-11 Vacuum Fluorescent Displays

**Vacuum fluorescent (VF) display**

**Plate (P)**

**Grid (G)**

**Cathode (K)**

**Filament**

**Heater**

**Zinc-oxide fluorescent material**

The *vacuum fluorescent (VF) display* is a modern relative of the triode vacuum tube. A schematic symbol for a triode vacuum tube is sketched in Fig. 6-27(a). The three parts of the triode tube are called the *plate (P), grid (G),* and *cathode (K).* The cathode is also called the *filament* or the *heater.* The plate is also called the *anode.*

The cathode/heater is a fine tungsten wire coated with a material such as barium oxide. The cathode gives off electrons when heated. The grid is a stainless steel screen. The plate can be thought of as the "collector of electrons" in the triode tube.

Assume that the cathode (K) of the triode tube in Fig. 6-27(a) is heated and has "boiled off" some electrons into the vacuum surrounding the cathode. Next, assume that the grid (G) becomes positive. The electrons are attracted to the grid. Next, assume that the plate (P) becomes positive. When the plate becomes positive, electrons will be attracted through the screenlike grid to the plate. Finally, the triode is conducting electricity from cathode to anode.

You can stop the triode tube from conducting in one of two ways. First, make the grid slightly negative (leave the plate positive). This will repel the electrons and they will not pass through the

grid to the plate. Second, leave the grid positive and drop the plate voltage to 0. The plate will not attract electrons and the triode tube does not conduct electricity from cathode to anode.

The schematic symbol in Fig. 6-27(b) represents a single digit of a VF display. Notice the single cathode (K), single grid (G), and seven plates ($P_a$ to $P_g$). Each of the seven plates has been coated with a *zinc oxide fluorescent material.* Electrons striking the fluorescent material on the plate will cause it to glow a blue-green color. The seven plates in the schematic in Fig. 6-27(b) represent the seven segments of a normal numeric display. Note that the entire unit is held in a glass enclosure containing a vacuum.

A single-digit seven-segment display is being operated in Fig. 6-27(c). The cathode (heater) is being powered by direct current in this example, with +12 V applied to the grid (G). Two plates ($P_c$ and $P_f$) are grounded. Each of the remaining five plates has +12 V applied. The high positive voltage on the five plates ($P_a$, $P_b$, $P_d$, $P_e$, and $P_g$) causes these plates to attract electrons. These five plates glow a blue-green color as electrons strike their surface.

In actual practice, the plates of the VF display tube are shaped like segments of a number or other shapes. Figure 6-28(a) is a view of the physical arrangement of cathode, grid, and plates. Notice that the plates are arranged in

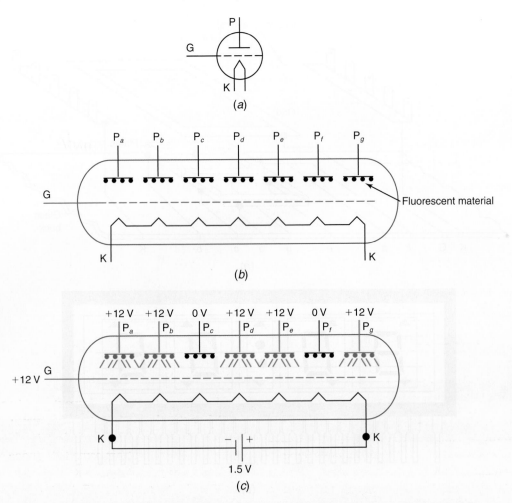

**Fig. 6-27** (*a*) Schematic symbol for a triode vacuum tube. (*b*) Schematic symbol for a single digit of a VF display. (*c*) Lighting plates on the VF display.

**Triode vacuum tube**

**VF display**

seven-segment format in this display unit. The screen above the segments is the grid. Above the grid are the cathodes (filaments or heaters). Each segment, grid, or cathode lead comes out the side of the sealed glass vacuum tube. The VF display shown in Fig. 6-28(*a*) would be viewed from above the unit looking downward. The fine wire cathodes and grid would be almost invisible. Lighted segments (plates) show through the mesh (grid).

A commercial vacuum fluorescent display is sketched in Fig. 6-28(*b*). This VF display contains 4 seven-segment numeric displays, a colon, and 10 triangle-shaped symbols. The internal parts of most VF displays are visible through the sealed glass package. Visible in the display are the cathodes (filaments or heaters) stretched horizontally across the display. These are very fine wires and are barely visible on a commercial display. Next, the grids are shown in five sections. Each of the five grids can be activated individually. Finally, the fluorescent-coated plates form the numeric segments, colons, and other symbols.

Vacuum fluorescent displays are based on an older technology but they have gained great favor in recent years. This is because they can operate at relatively low voltages and power and have an extremely long life and fast response. They can display in various colors (with filters), have good reliability, and low cost. Vacuum fluorescent displays are compatible with the popular 4000 series CMOS family of ICs. They are widely used in readouts found in automobiles, VCRs, TVs, household appliances, and digital clocks.

For information on solid state displays, visit the website for Lumex, Inc.

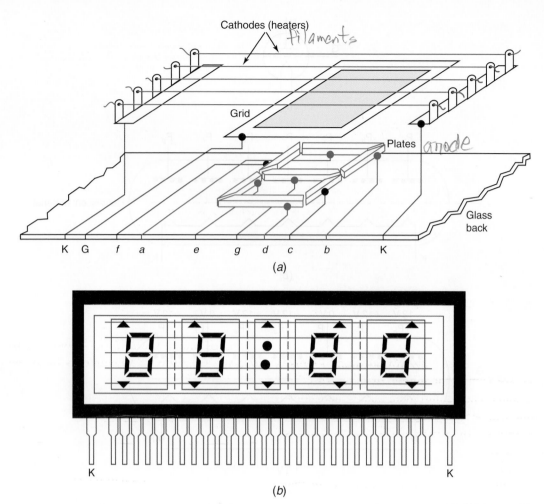

(a)

(b)

Fig. 6-28 (a) Typical construction of VF display. (b) Commercial four-digit VF display.

✔ *Self-Test*

*Answer the following questions.*

60. A VF display glows with a(n) _____ color when activated.

61. Refer to Fig. 6-29. Which plates of this VF display will glow?

62. Name the parts of the vacuum fluorescent display labeled *A*, *B*, and *C* in Fig. 6-30.

63. Refer to Fig. 6-30. What segments of the VF display glow, and what decimal number is lit?

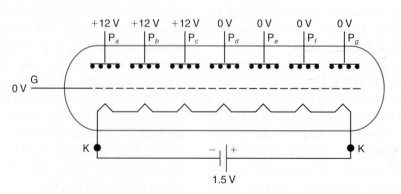

Fig. 6-29 VF display with no positive grid voltage.

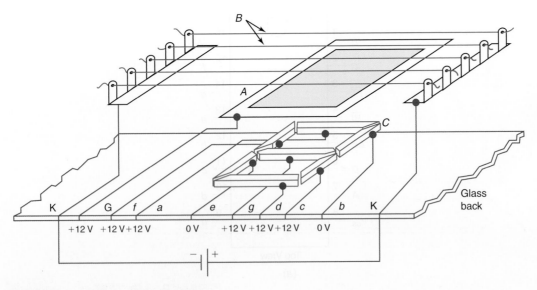

**Fig. 6-30** Single-digit VF display problem.

## 6-12 Driving a VF Display

The voltage requirements for operating VF displays are somewhat higher than for LED or LCD units. This requirement makes them compatible with 4000 series CMOS ICs. Recall that 4000 series CMOS ICs can operate on voltages up to 18 V.

A wiring diagram of a simple BCD decoder/driver circuit is detailed in Fig. 6-31. In this example, $1001_{BCD}$ is translated into the decimal 9 on the VF display. The circuit uses the 4511 *BCD-to-seven-segment latch/decoder/driver CMOS IC.*

In this example, the *a*, *b*, *c*, *f*, and *g* output lines are HIGH ($+12$ V) with only the *d* and *e* lines LOW.

The $+12$-V supply is connected directly to the grid in Fig. 6-31. The cathode (filament or heater) circuit contains a resistor ($R_1$) to limit the current through the heaters to a safe level. The $+12$ V is also used to supply power for the 4511 decoder/driver CMOS IC. Note the labels on the power connections to the 4511 IC. The $V_{DD}$ pin connects to $+12$ V while $V_{ss}$ goes to ground (GND).

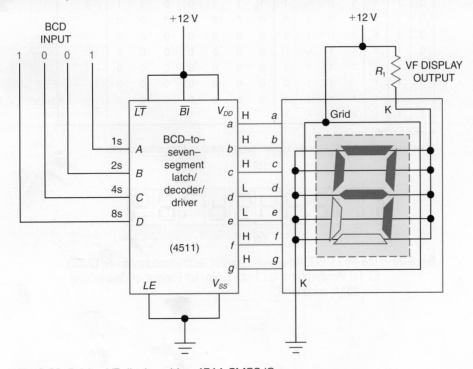

**Fig. 6-31** Driving VF display with a 4511 CMOS IC.

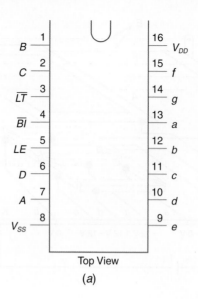

Top View

(a)

Truth Table

| INPUTS | | | | | | | OUTPUTS | | | | | | | |
|---|---|---|---|---|---|---|---|---|---|---|---|---|---|---|
| LE | $\overline{BI}$ | $\overline{LT}$ | D | C | B | A | a | b | c | d | e | f | g | Display |
| X | X | 0 | X | X | X | X | 1 | 1 | 1 | 1 | 1 | 1 | 1 | 8 |
| X | 0 | 1 | X | X | X | X | 0 | 0 | 0 | 0 | 0 | 0 | 0 | |
| 0 | 1 | 1 | 0 | 0 | 0 | 0 | 1 | 1 | 1 | 1 | 1 | 1 | 0 | 0 |
| 0 | 1 | 1 | 0 | 0 | 0 | 1 | 0 | 1 | 1 | 0 | 0 | 0 | 0 | 1 |
| 0 | 1 | 1 | 0 | 0 | 1 | 0 | 1 | 1 | 0 | 1 | 1 | 0 | 1 | 2 |
| 0 | 1 | 1 | 0 | 0 | 1 | 1 | 1 | 1 | 1 | 1 | 0 | 0 | 1 | 3 |
| 0 | 1 | 1 | 0 | 1 | 0 | 0 | 0 | 1 | 1 | 0 | 0 | 1 | 1 | 4 |
| 0 | 1 | 1 | 0 | 1 | 0 | 1 | 1 | 0 | 1 | 1 | 0 | 1 | 1 | 5 |
| 0 | 1 | 1 | 0 | 1 | 1 | 0 | 0 | 0 | 1 | 1 | 1 | 1 | 1 | 6 |
| 0 | 1 | 1 | 0 | 1 | 1 | 1 | 1 | 1 | 1 | 0 | 0 | 0 | 0 | 7 |
| 0 | 1 | 1 | 1 | 0 | 0 | 0 | 1 | 1 | 1 | 1 | 1 | 1 | 1 | 8 |
| 0 | 1 | 1 | 1 | 0 | 0 | 1 | 1 | 1 | 1 | 0 | 0 | 1 | 1 | 9 |
| 0 | 1 | 1 | 1 | 0 | 1 | 0 | 0 | 0 | 0 | 0 | 0 | 0 | 0 | |
| 0 | 1 | 1 | 1 | 0 | 1 | 1 | 0 | 0 | 0 | 0 | 0 | 0 | 0 | |
| 0 | 1 | 1 | 1 | 1 | 0 | 0 | 0 | 0 | 0 | 0 | 0 | 0 | 0 | |
| 0 | 1 | 1 | 1 | 1 | 0 | 1 | 0 | 0 | 0 | 0 | 0 | 0 | 0 | |
| 0 | 1 | 1 | 1 | 1 | 1 | 0 | 0 | 0 | 0 | 0 | 0 | 0 | 0 | |
| 0 | 1 | 1 | 1 | 1 | 1 | 1 | 0 | 0 | 0 | 0 | 0 | 0 | 0 | |
| 1 | 1 | 1 | X | X | X | X | | | | * | | | | * |

X = Don't care

* Depends upon the BCD code appiled during the 0 to 1 transition of LE.

(b)

(c)

**4511 BCD-to-seven-segment latch/decoder/driver CMOS IC**

Fig. 6-32   The 4511 BCD-to-seven-segment latch/decoder/driver CMOS IC. (a) Pin diagram. (b) Truth table. (c) Format of digits using the 4511 decoder IC.

The pin diagram, truth table, and number formats for the 4511 CMOS IC are shown in Fig. 6-32. The 4511 BCD-to-seven-segment latch/-decoder/driver IC's pin diagram is shown in Fig. 6-32(*a*). This is the top view of this 16-pin DIP CMOS IC. Internally, the 4511 IC is organized like the 74HC4543 unit. The latch, decoder, and driver sections are illustrated in the shaded section of the block diagram in Fig. 6-24(*a*).

The truth table in Fig. 6-32(*b*) shows seven inputs to the 4511 decoder/driver IC. The BCD data inputs are labeled *D*, *C*, *B*, and *A*. The $\overline{LT}$ input stands for lamp test. When activated by a LOW (row 1 on the truth table), all outputs go HIGH and light all segments of an attached display. The $\overline{BI}$ input stands for blanking input. When $\overline{BI}$ is activated with a LOW, all outputs go LOW and all

segments of an attached display are blanked. The *LE* (latch enable) input can be used like a memory to hold data on display while the BCD input data changes. If *LE* = 0, then data passes through the 4511 IC. However, if *LE* = 1 then the last data present at the data inputs (*D*, *C*, *B*, *A*) is latched and held on the display. The *LE*, $\overline{BI}$, and $\overline{LT}$ inputs are all disabled in the circuit in Fig. 6-31.

Next, refer to the output side of the truth table in Fig. 6-32. On the 4511 IC, a HIGH or 1 is an active output. In other words, an output of 1 turns on a segment on the attached display. Therefore, a 0 output means the display's segment is turned off.

The format of the digits generated by the 4511 BCD-to-seven-segment decoder IC are illustrated in Fig. 6-32(*c*). Especially note the formation of the decimal numbers 6 and 9.

## ✔ Self-Test

*Answer the following questions.*

64. Refer to Fig. 6-31. The +12-V power supply is being used because the _____ (CMOS, TTL) 4511 decoder/driver IC and the _____ (LCD, VF) display operate properly at this voltage.

65. Refer to Fig. 6-31. What is the purpose of resistor $R_1$ in this circuit?

66. Refer to Fig. 6-33. What is the decimal reading on the vacuum fluorescent display for each input pulse ($t_1$ to $t_4$)?

67. Refer to Fig. 6-33. During pulse ($t_4$) what voltages are applied to the seven plates (segments) of the VF display?

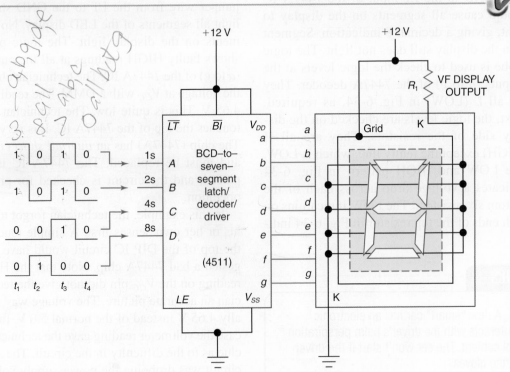

**Fig. 6-33** Decoder–VF display pulse-train problem.

Internet
Connection

DLP (Digital
Light Processing)
en.wikipedia.
org or
www.dlp.com

**Troubleshooting
faulty
decoder/LED
display circuit**

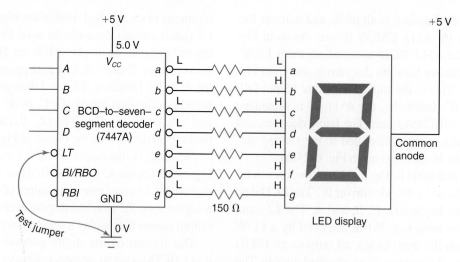

**Fig. 6-34** Troubleshooting a faulty decoder/LED display circuit.

## 6-13 Troubleshooting a Decoding Circuit

Consider the BCD-to-seven-segment decoder circuit in Fig. 6-34. The problem is that segment *a* of the display does not light. The technician first checks the circuit visually. Then the IC is checked for signs of excessive heat. The $V_{CC}$ and GND voltages are checked with a voltmeter or logic probe. In this example, the results of these tests did not locate the problem. Next, the temporary jumper wire from GND to the LT input of the 7447A IC should cause all segments on the display to light, giving a decimal 8 indication. Segment *a* on the display still does not light. The logic probe is used to check the logic levels at the outputs (*a* to *g*) of the 7447A decoder. They are all L (LOW) in Fig. 6-34, as required. Next, the logic levels are checked on the display side of the resistors. They are all H (HIGH) except the faulty line, which is LOW. The LOW and HIGH pattern in Fig. 6-34 indicates a voltage drop across each of the bottom six resistors. The LOW indications on both ends of the top resistor in Fig. 6-34 indi-

**Internal short
circuit**

cate an open circuit in the segment *a* section of the seven-segment display. Segment *a* of the display must be faulty. The entire seven-segment LED display is replaced. The replacement must have the same pin diagram and be a common-anode LED display. After replacement, the circuit is checked for proper operation.

The circuit shown in Fig. 6-35 produces no display. The hurried technician checks the $V_{CC}$ and GND pins with a logic probe. The readings as shown in Fig. 6-35 seem all right. The test jumper wire from the LT to the GND should light all segments of the LED display. No segments on the display light. The logic probe shows faulty HIGH readings at all the outputs (*a* to *g*) of the 7447A IC. The technician checks the voltage at $V_{CC}$ with a DMM. The reading is 4.65 V. This is quite low. The technician now touches the top of the 7447A IC. It is very hot. The chip (7447A) has an *internal short circuit* and must be replaced. The 7447A IC is replaced, and the circuit is checked for proper operation.

In this example, the technician forgot to use his or her own senses first. A simple touch of the top of the DIP IC circuit would have suggested a bad 7447A chip. Note that the HIGH reading on the $V_{CC}$ pin did not give the technician an accurate picture. The voltage was actually 4.65 V instead of the normal 5.0 V. In this case the voltmeter reading gave the technician a clue as to the difficulty in the circuit. The short circuit was dropping the power supply voltage to 4.65 V.

**ABOUT ELECTRONICS**

**Sobering Sensors.** A new "smart" car has an electronic steering wheel that interacts with the driver's palm perspiration and measures alcohol content. The car won't start if the driver is intoxicated or wearing gloves.

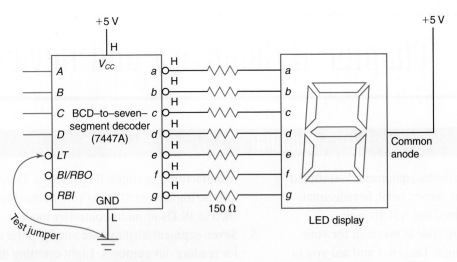

**Fig. 6-35** Troubleshooting a faulty decoder circuit with blank LED display.

*Answer the following questions.*

68. What is the first step in troubleshooting a digital logic circuit?

69. An internal _____ (open, short) circuit in a TTL IC will many times cause the IC to become excessively hot.

# Chapter 6 Summary and Review

## Summary

1. Many codes are used in digital equipment. You should be familiar with decimal, binary, octal, hexadecimal, 8421 BCD, excess-3, Gray, and ASCII codes.

2. Converting from code to code is essential for your work in digital electronics. Table 6-4 will aid you in converting from several of the codes.

3. The most popular alphanumeric code is the 7-bit ASCII code. The ASCII code is widely used in microcomputer keyboard and display interfacing. Extended ASCII uses 8-bits.

4. Electronic translators are called encoders and decoders. The complicated logic circuits are manufactured in single IC packages. Decoding can also be implemented using programmable devices such as PLDs or microcontroller modules.

5. Seven-segment displays are very popular devices for reading out numbers. Light-emitting diode (LED), liquid-crystal display (LCD), and vacuum fluorescent (VF) types are popular displays.

6. The BCD-to-seven-segment decoder/driver is a common decoding device. It translates from BCD machine language to decimal numbers. The decimal numbers appear on seven-segment LED, LCD, or VF displays.

**Table 6-4**

| Decimal Number | Binary Number | BCD Codes 8421 | | BCD Codes Excess-3 | | Gray code |
|---|---|---|---|---|---|---|
| 0 | 0000 | | 0000 | | 0011 | 0000 |
| 1 | 0001 | | 0001 | | 0100 | 0001 |
| 2 | 0010 | | 0010 | | 0101 | 0011 |
| 3 | 0011 | | 0011 | | 0110 | 0010 |
| 4 | 0100 | | 0100 | | 0111 | 0110 |
| 5 | 0101 | | 0101 | | 1000 | 0111 |
| 6 | 0110 | | 0110 | | 1001 | 0101 |
| 7 | 0111 | | 0111 | | 1010 | 0100 |
| 8 | 1000 | | 1000 | | 1011 | 1100 |
| 9 | 1001 | | 1001 | | 1100 | 1101 |
| 10 | 1010 | 0001 | 0000 | 0100 | 0011 | 1111 |
| 11 | 1011 | 0001 | 0001 | 0100 | 0100 | 1110 |
| 12 | 1100 | 0001 | 0010 | 0100 | 0101 | 1010 |
| 13 | 1101 | 0001 | 0011 | 0100 | 0110 | 1011 |
| 14 | 1110 | 0001 | 0100 | 0100 | 0111 | 1001 |
| 15 | 1111 | 0001 | 0101 | 0100 | 1000 | 1000 |
| 16 | 10000 | 0001 | 0110 | 0100 | 1001 | 11000 |
| 17 | 10001 | 0001 | 0111 | 0100 | 1010 | 11010 |
| 18 | 10010 | 0001 | 1000 | 0100 | 1011 | 11011 |
| 19 | 10011 | 0001 | 1001 | 0100 | 1100 | 11010 |
| 20 | 10100 | 0010 | 0000 | 0101 | 0011 | 11110 |

*Answer the following questions.*

6-1. Write the binary numbers for the decimal numbers in *a* to *f*:
  a. 17      d. 75
  b. 31      e. 150
  c. 42      f. 300

6-2. Write the 8421 BCD numbers for the decimal numbers in *a* to *f*:
  a. 17      d. 1632
  b. 31      e. 47,899
  c. 150     f. 103,926

6-3. Write the decimal numbers for the 8421 BCD numbers in *a* to *f*:
  a. 0010
  b. 1111
  c. 0011 0000
  d. 0111 0001 0110 0000
  e. 0001 0001 0000 0000 0000
  f. 0101 1001 1000 1000 0101

6-4. Write the excess-3 code numbers for the decimal numbers in *a* to *d*:
  a. 7       c. 59
  b. 27      d. 318

6-5. Why is the excess-3 code used in some arithmetic circuits?

6-6. List two codes you learned about that are classified as BCD codes.

6-7. Write the Gray code numbers for the decimal numbers in *a* to *f*.
  a. 1       d. 4
  b. 2       e. 5
  c. 3       f. 6

6-8. The _____ (Gray, XS3) code is commonly associated with optical encoding of a shaft's angular position.

6-9. The important characteristic of the Gray code is that only one digit changes as you decrement or increment the count (T or F).

6-10. Refer to Table 6-3. The 7-bit ASCII code for the capital letter S is _____.

6-11. The letters "ASCII" stand for _____.

6-12. Standard ASCII is a(n) _____ -bit _____ (alphanumeric, BCD) code that is used to represent number, letters, punctuation marks, and control characters.

6-13. List two general names for code translators, or electronic code converters.

6-14. A(n) _____ (decoder, encoder) is the electronic device used to convert the decimal input of a calculator keypad to the BCD code used by the central processing unit.

6-15. A(n) _____ (decoder, encoder) is the electronic device used to convert the BCD of the central processing unit of a calculator to the decimal display output.

6-16. Which segments of the seven-segment display *will light* when the following decimal numbers appear? Use the letters *a, b, c, d, e, f,* and *g* as answers.
  a. 0       f. 5
  b. 1       g. 6
  c. 2       h. 7
  d. 3       i. 8
  e. 4       j. 9

6-17. The seven-segment displays that you will use emit light (usually red) and are of the _____ (LCD, LED) type.

6-18. The _____ (LCD, LED) seven-segment display is used where battery operation demands low power consumption.

6-19. The _____ (LCD, LED) display is used where the unit must be read in bright light.

6-20. Refer to Fig. 6-36. All the outputs from the 7447A decoder are _____ (HIGH, LOW). This is _____ (correct, not correct) for this circuit.

6-21. Both a voltmeter and a _____ are used to troubleshoot the circuit in Fig. 6-36.

6-22. Refer to Fig. 6-36. Segment *b* of the LED display appears to be _____ (open, partially short circuited). The display should be replaced with a common- _____ LED display having the same pin diagram as the one in the circuit.

6-23. Refer to Fig. 6-37. With the BCD input shown, the six-digit display reads _____.

6-24. The front and back panels of a _____ (LCD, LED) seven-segment display are made of glass and can be broken by rough handling.

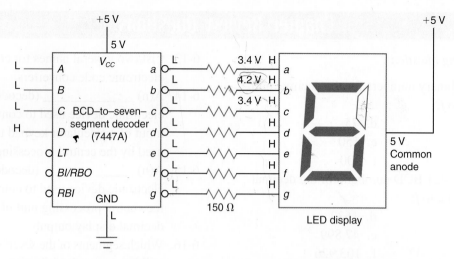

**Fig. 6-36** Troubleshooting problems. Logic levels and voltages given on faulty decoder/LED display circuit.

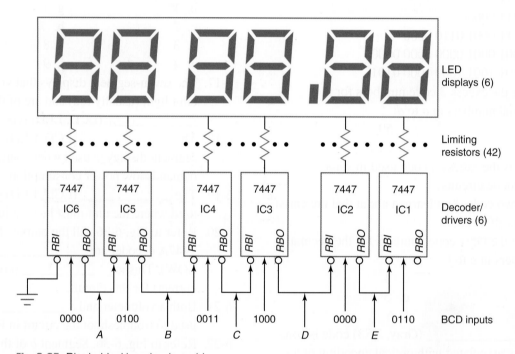

**Fig. 6-37** Ripple-blanking circuit problem.

6-25. Refer to Fig. 6-38. With the driving signals shown, the LCD will display the decimal number _____. The input must be the BCD number _____.

6-26. Vacuum fluorescent displays can operate on 12 V, which makes them very compatible with _____ (CMOS, TTL) ICs and automotive applications.

6-27. Refer to Fig. 6-39. What will the VF seven-segment display read for each input pulse?

6-28. Refer to Fig. 6-39. List the approximate voltages at each of the seven plates and the grid of the VF display during pulse $t_4$.

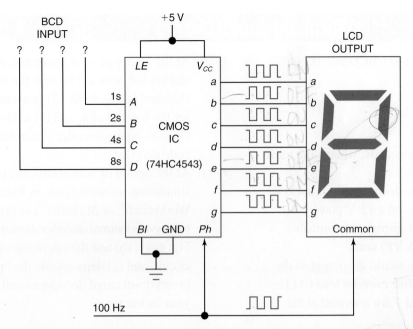

**Fig. 6-38** Decoder/LCD circuit problem.

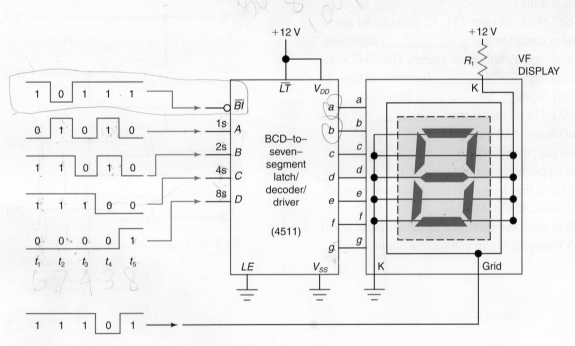

**Fig. 6-39** Decoder/VF display pulse-train problem.

6-1. Convert the following 8421 BCD numbers to binary.
   a. 0011 0101
   b. 1001 0110
   c. 0111 0100

6-2. As you count in the Gray code, what is the most important characteristic?

6-3. Refer to Fig. 6-6. If the decoder chip is a 4511 and the circuit operates on a 12-V power supply, then the output displays are probably _____ (LED, VF) units.

6-4. Refer to Fig. 6-7. Why would the output of the 74147 10-line to four-line encoder read 0111 when *both* inputs 2 and 7 are activated at the same time?

6-5. What is the purpose of the 7447A TTL IC, and with which type of seven-segment display is it compatible?

6-6. The 7447A decoder TTL IC contains 44 gates and is considered a _____ (combinational, sequential) logic circuit. The 7447A decoder has _____ (number) active-LOW inputs, _____ (number) active-HIGH inputs, and _____ (number) active-LOW outputs.

6-7. List the condition (HIGH or LOW) of each of the ripple blanking lines *A* to *E* in Fig. 6-37.

6-8. Refer to Fig. 6-38. List the three functions of the 74HC4543 CMOS IC.

6-9. What are some reasons a designer might select a VF display for an automotive application?

6-10. At the option of your instructor, use circuit simulation software to (1) draw the logic circuit sketched in Fig. 6-40, (2) generate a truth table for the logic circuit, and (3) determine if it is a Gray-to-binary decoder or a binary-to-Gray code decoder.

6-11. At the option of your instructor, use circuit simulation software (such as Electronics Workbench® or MultiSim®) to (a) draw the binary-to-decimal decoder circuit sketched in Fig. 6-41, (b) test the operation of the decoder circuit, and (c) demonstrate the operation of the binary-to-decimal decoder circuit simulation to your instructor.

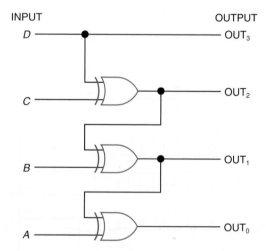

Fig. 6-40 Logic circuit.

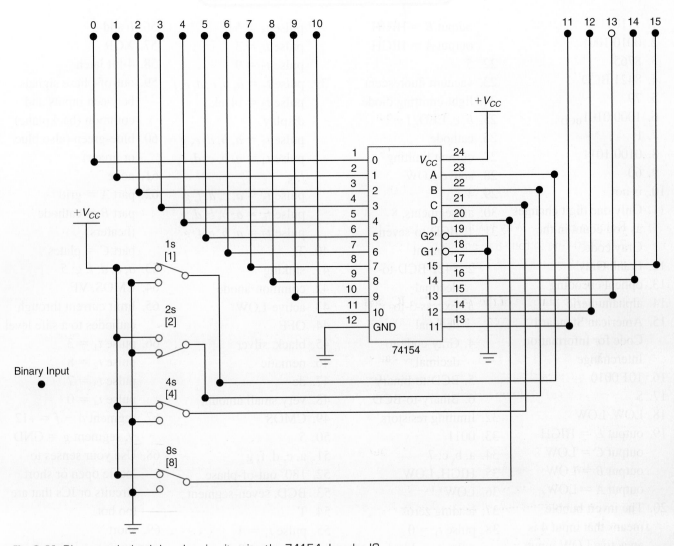

**Fig. 6-41** Binary-to-decimal decoder circuit using the 74154 decoder IC.

1. 11101
2. 0010 1001
3. 8765
4. 8421 BCD
5. 79
6. $1000\ 0101_{BCD}$
7. F
8. 0100 1011
9. 60
10. is not
11. Only one digit changes as you count in the Gray code.
12. Frank Gray
13. optical encoding
14. alphanumeric
15. American Standard Code for Information Interchange
16. 101 0010
17. $
18. LOW, LOW
19. output $D$ = HIGH
    output $C$ = LOW
    output $B$ = LOW
    output $A$ = LOW
20. The invert bubble means that input 4 is an active LOW input; it is activated by a logical 0.
21. output $D$ = LOW
    output $C$ = HIGH
    output $B$ = HIGH
    output $A$ = HIGH
22. 5
23. vacuum fluorescent
24. light-emitting diode
25. $b, c$, LED, 1
26. cathode
27. current limiting
28. active-LOW
29. 4
30. all segments, 8
31. 1. BCD-to-seven-segment
    2. 8421-BCD-to-decimal
    3. Excess-3-to-decimal
    4. Gray-code-to-decimal
    5. BCD-to-binary
    6. Binary-to-BCD
32. limiting resistors
33. 0011
34. a, b, c; 7
35. HIGH, LOW
36. LOW
37. leading zeros
38. pulse $t_1$ = 0
    pulse $t_2$ = blank display (not a BCD number)
    pulse $t_3$ = 2
    pulse $t_4$ = 8

pulse $t_5$ = 5
pulse $t_6$ = 3
pulse $t_7$ = 9
39. pulse $t_1$ = $a, b, c, d, e, f$
    pulse $t_2$ = blank display
    pulse $t_3$ = $a, b, d, e, g$
    pulse $t_4$ = $a, b, c, d, e, f, g$
    pulse $t_5$ = $a, c, d, f, g$
    pulse $t_6$ = $a, b, c, d, g$
    pulse $t_7$ = $a, b, c, f, g$
40. T
41. sinking
42. common-anode
43. active-LOW
44. OFF
45. black, silver
46. nematic
47. dc
48. very small amount
49. CMOS
50. 5
51. a, c, d, f, g
52. 180° out-of-phase
53. BCD, seven-segment
54. T
55. pulse $t_1$ = 4
    pulse $t_2$ = 2
    pulse $t_3$ = 6
    pulse $t_4$ = 9
    pulse $t_5$ = 1

56. $b$ and $c$
57. XOR
58. 4-bit latch
59. out-of-phase signals between inputs and common (backplane)
60. blue-green (also blue or green)
61. none
62. part $A$ = grid
    part $B$ = cathode (heaters)
    part $C$ = plates
63. $a, c, d, f, g$; 5
64. CMOS, VF
65. limit current through cathodes to a safe level
66. pulse $t_1$ = 3
    pulse $t_2$ = 8
    pulse $t_3$ = 7
    pulse $t_4$ = 0
67. segment $a - f$ = +12 V, segment $g$ = GND
68. Use your senses to locate open or short circuits or ICs that are too hot.
69. short

# Flip-Flops

## Chapter Objectives

*This chapter will help you to:*

1. *Memorize* the block diagram and explain the function of each input and output on several types of flip-flops.

2. *Use* truth tables to determine the mode of operation and outputs of a flip-flop.

3. *Interpret* flip-flop waveform diagrams to determine the mode of operation, outputs, and the type of triggering.

4. *Discuss* the organization and use of a 4-bit latch and *predict* the operation of the IC.

5. *Classify* flip-flops as synchronous or asynchronous and *compare* the triggering of the synchronous units.

6. *Describe* the operation of Schmitt trigger devices and *cite* their applications.

7. *Compare* traditional with newer IEEE/ANSI flip-flop symbols.

**E**ngineers classify logic circuits into two groups. We have already worked with *combination logic circuits* using AND, OR, and NOT gates. The other group of circuits is classified as *sequential logic circuits*. Sequential circuits involve timing and memory devices. The basic building block for combinational logic circuits is the logic gate. The basic building block for sequential logic circuits is the *flip-flop (FF)*. This chapter covers several types of flip-flop circuits. In later chapters you will wire flip-flops together. Flip-flops are wired to form *counters, shift registers,* and various *memory devices*.

**Combination logic circuits**

**Sequential logic circuits**

**Counters**

**Shift registers**

**Memory devices**

## 7-1 The R-S Flip-Flop

The logic symbol for the *R-S flip-flop* is drawn in Fig. 7-1. Notice that the R-S flip-flop has two inputs, labeled *S* and *R*. The two outputs are labeled $Q$ and $\overline{Q}$ (say "not Q" or "Q not"). In flip-flops the outputs are always opposite, or *complementary*. In other words, if output $Q = 1$, then output $\overline{Q} = 0$, and so on. The R-S flip-flop symbol from Fig. 7-1 labels the outputs as *normal* and *complementary*. The letters *S* and *R* at the left of the R-S flip-flop symbol are often referred to as the *set* and *reset* inputs.

The R-S flip-flop may also be referred to as an *R-S latch*. The term "latch" refers to its use as a temporary memory device. A latch such as the R-S flip-flop in Fig. 7-1 can hold one bit of information.

**R-S flip-flop**

**Complementary**

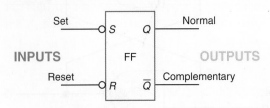

**Fig. 7-1** Logic symbol for an R-S flip-flop.

## Table 7-1　Truth table for R-S flip-flop

| Mode of operation | Inputs | | Outputs | | Effect on output $Q$ |
|---|---|---|---|---|---|
| | $S$ | $R$ | $Q$ | $\overline{Q}$ | |
| Prohibited | 0 | 0 | 1 | 1 | Prohibited-Do not use |
| Set | 0 | 1 | 1 | 0 | For setting $Q$ to 1 |
| Reset | 1 | 0 | 0 | 1 | For resetting $Q$ to 0 |
| Hold | 1 | 1 | $Q$ | $\overline{Q}$ | Depends on previous state |

**Timing diagrams**

**Waveform diagram**

**Set and reset inputs**

**Set condition**

**Reset condition**

**Hold condition**

**R-S latch**

The truth table in Table 7-1 details the operation of the R-S flip-flop. When the $S$ and $R$ inputs are both 0, both outputs go to a logical 1. This is called a *prohibited state* for the flip-flop and is not used. The second line of the truth table shows that when input $S$ is 0 and $R$ is 1, the $Q$ output is set to logical 1. This is called the *set condition*. The third line shows that when input $R$ is 0 and $S$ is 1, output $Q$ is reset (cleared) to 0. This is called the *reset condition*. Line 4 in the truth table shows both inputs ($R$ and $S$) at 1. This is the idle or at rest condition and leaves $Q$ and $\overline{Q}$ in their previous complementary states. This is called the *hold condition*.

From Table 7-1, it may be observed that it takes a logical 0 to activate the set (set $Q$ to 1). It also takes a logical 0 to activate the reset, or clear (clear $Q$ to 0). Because it takes a logical 0 to enable, or activate, the flip-flop, the logic symbol in Fig. 7-1 has invert bubbles at the $R$ and $S$ inputs. These invert bubbles indicate that the set and reset inputs are activated by a logical 0.

R-S flip-flops can be purchased in an IC package, or they can be wired from logic gates, as shown in Fig. 7-2. The NAND gates in Fig. 7-2 form an R-S flip-flop. This NAND-gate R-S flip-flop operates according to the truth table in Table 7-1.

Many times *timing diagrams,* or *waveforms,* are given for sequential logic circuits. These diagrams show the voltage level and timing between inputs and outputs and are similar to what you would observe on an oscilloscope. The horizontal distance is *time,* and the vertical distance is *voltage.* Figure 7-3 shows the input waveforms ($R$, $S$) and the output waveforms ($Q$, $\overline{Q}$) for the R-S flip-flop. The bottom of the diagram lists the lines of the truth table from Table 7-1. The $Q$ waveform shows the set and reset conditions of the output; the logic levels (0, 1) are on the right side of the waveforms. Waveform diagrams of the type shown in Fig. 7-3 are very common when dealing with sequential logic circuits. Study this diagram to see what it tells you. The waveform diagram is really a type of truth table.

Recall that there are three types of multivibrators (MVs). They are the monostable MV, the astable MV, and the bistable multivibrator. The R-S flip-flop is one of several bistable MVs. The R-S flip-flop is most commonly known as a *latch* and is listed under this heading in IC catalogs. A latch is a fundamental binary memory device for holding data. Latches are commonly used at the output of a digital device to hold the data until the next device is ready to receive the input. Latches are commonly organized into groups of 4-bits, 8-bits or more into *registers*. An 8-bit register would be a group of eight latches holding a byte of information. You may recall that R-S flip-flops were also used for switch debouncing.

Commercial versions of the R-S flip-flop are available. One example is the *74LS279 Quad S-R Latch IC*. It contains four latches like the one you studied from Fig. 7-2. Later in this chapter you will study the *7475/74LS75/74HC75 4-Bit Latch* in detail.

Do you know the logic symbol and truth table for the R-S flip-flop? Do you know the four modes of operation for the R-S flip-flop?

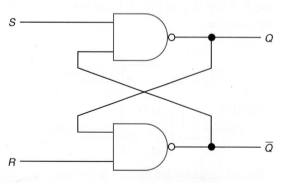

**Fig. 7-2** Wiring an R-S flip-flop using NAND gates.

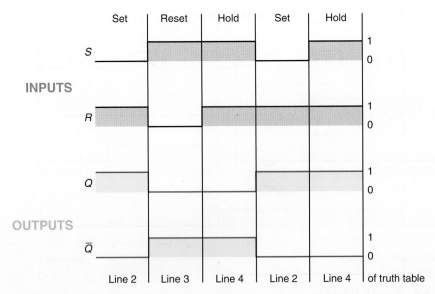

**Fig. 7-3** Waveform diagram for an R-S flip-flop.

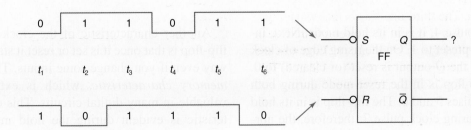

**Fig. 7-4** R-S flip-flop pulse-train problem.

---

✓ **Self-Test**

*Answer the following questions.*

1. The R-S flip-flop in Fig. 7-1 has active _____ (HIGH, LOW) inputs.
2. List the mode of operation of the R-S flip-flop for each input pulse shown in

Fig. 7-4. Answer with the terms "set" and "reset," "hold," and "prohibited."

3. List the binary output at the normal output (*Q*) of the R-S flip-flop for each of the pulses shown in Fig. 7-4.

---

## 7-2 The Clocked R-S Flip-Flop

The logic symbol for a *clocked R-S flip-flop* is shown in Fig. 7-5. Observe that it looks almost like an R-S flip-flop except that it has one extra input labeled *CLK* (for clock). Figure 7-6 diagrams the operation of the clocked R-S flip-flop. The *CLK input* is at the top of the diagram. Notice that the clock pulse (1) has no effect on output $Q$ with inputs $S$ and $R$ in the 0 position. The flip-flop is in the *idle,* or *hold,* mode during clock pulse 1. At the preset $S$ position, the $S$ (set) input is moved to 1, but output $Q$ is not yet set to 1. The rising

edge of clock pulse 2 permits $Q$ to go to 1. Pulses 3 and 4 have no effect on output $Q$. During pulse 3, the flip-flop is in its set mode, and

**Clocked R-S flip-flop**

**CLK input**

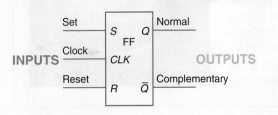

**Fig. 7-5** Logic symbol for a clocked R-S flip-flop.

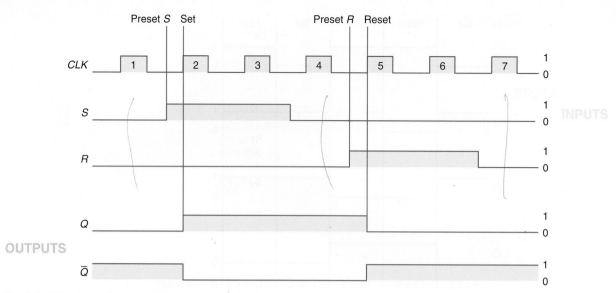

**Fig. 7-6** Waveform diagram for a clocked R-S flip-flop.

Waveform
diagram for a
clocked R-S
flip-flop

Memory
characteristic

during pulse 4, it is in its hold mode. Next, input $R$ is preset to 1. On the rising edge of clock pulse 5, the $Q$ output is reset (or cleared) to 0. The flip-flop is in the reset mode during both clock pulses 5 and 6. The flip-flop is in its hold mode during clock pulse 7; therefore, the normal output ($Q$) remains at 0.

Notice that the outputs of the clocked R-S flip-flop *change only on a clock pulse.* We say that this flip-flop operates *synchronously;* it operates *in step with* the clock. *Synchronous operation* is very important in most digital circuits, where each step must happen in an exact order.

Another characteristic of the clocked R-S flip-flop is that once it is set or reset it stays that way even if you change some inputs. This is a *memory characteristic,* which is extremely valuable in many digital circuits. This characteristic is evident during the hold mode of operation. In the waveform diagram in Fig. 7-6, this flip-flop is in the hold mode during clock pulses 1, 4, and 7.

Figure 7-7(a) shows a truth table for the clocked R-S flip-flop. Notice that only the top three lines of the truth table are usable; the bottom line is prohibited and not used. Observe that the $R$ and $S$ inputs to the clocked R-S flip-flop are active HIGH inputs. That is, it takes a HIGH on input $S$ while $R = 0$ to cause output $Q$ to be set to 1.

Figure 7-7(b) shows a wiring diagram of a clocked R-S flip-flop. Notice that two NAND gates have been added to the inputs of the R-S flip-flop to add the clocked feature.

It is important to remember that the memory characteristics exhibited by flip-flops are among the fundamental reasons why digital technology has become so widely used in modern electronics products. It is strongly suggested that you actually experiment with R-S and clocked R-S flip-flops either on a circuit simulator or with actual ICs on a solderless breadboard. Operating flip-flops in the lab will help you better understand their operation.

## ABOUT ELECTRONICS

**Heart of Glass.** *Preforms* are the pieces made to begin the construction of light guides used in fiber optics. In a preform, you can see the concentric rings of glass, each bonding with the next. At the start, they are about $1/2$ inch in diameter. Then the layered core and its surrounding rings are stretched into a fiber no thicker than a hair and many kilometers long.

| Mode of operation | INPUTS | | | OUTPUTS | | |
|---|---|---|---|---|---|---|
| | CLK | S | R | Q | $\overline{Q}$ | Effect on output Q |
| Hold | ⊓ | 0 | 0 | No change | No change | |
| Reset | ⊓ | 0 | 1 | 0 | 1 | Reset or cleared to 0 |
| Set | ⊓ | 1 | 0 | 1 | 0 | Set to 1 |
| Prohibited | ⊓ | 1 | 1 | 1 | 1 | Prohibited— do not use |

(a)

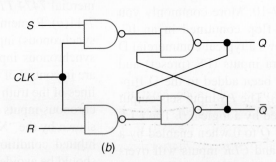

(b)

**Fig. 7-7** (a) Truth table for a clocked R-S flip-flop. (b) Wiring a clocked R-S flip-flop using NAND gates.

Truth table for a clocked R-S flip-flop

Wiring a clocked R-S flip-flop using NAND gates

 **Self-Test**

*Answer the following questions.*

4. The set and reset inputs (S, R) of the clocked R-S flip-flop in Fig. 7-5 are active _____ (HIGH, LOW) inputs.

5. List the mode of operation of the clocked R-S flip-flop for each input pulse shown in Fig. 7-8. Answer with the terms "set," "reset," "hold," and "prohibited."

6. List the binary output at the normal output (Q) of the clocked R-S flip-flop for each of the pulses shown in Fig. 7-8.

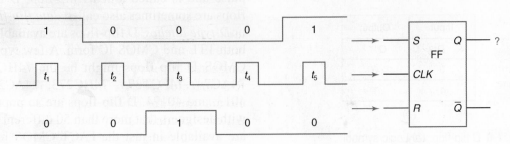

**Fig. 7-8** Clocked R-S flip-flop pulse-train problem.

## 7-3 The D Flip-Flop

**D flip-flop**

The logic symbol for the *D flip-flop* is shown in Fig. 7-9(*a*). It has only one *data input* (*D*) and a clock input (*CLK*). The outputs are labeled *Q* and $\overline{Q}$. The D flip-flop is often called a *delay flip-flop*. The word "delay" describes what happens to the data, or information, at input *D*. The data (a 0 or 1) at input *D* is *delayed one clock pulse* from getting to output *Q*. A simplified truth table for the D flip-flop is shown in Fig. 7-9(*b*). Notice that output *Q* follows input *D after one clock pulse* (see $Q^{n+1}$ column).

**Delay flip-flop**

A D flip-flop may be formed from a clocked R-S flip-flop by adding an inverter, as shown in Fig. 7-10. More commonly you will use a D flip-flop contained in an IC. Figure 7-11(*a*) shows a typical commercial D flip-flop. Two extra inputs (PS (preset) and CLR (clear)) have been added to the D flip-flop in Fig. 7-11(*a*). The *PS* input sets output *Q* to 1 when enabled by a logical 0. The *CLR* input clears output *Q* to 0 when enabled by a logical 0. The PS and *CLR* inputs will override the *D* and *CLK* inputs. The *D* and *CLK* inputs operate as they did in the D flip-flops in Fig. 7-9.

**7474 TTL D flip-flop**

Note the addition of a small triangle on the CLK input of the IC symbol in Fig. 7-11(*a*). This small triangle inside the 7474 IC symbol in Fig. 7-11(*a*) means the flip-flop is *edge triggered*. During synchronous operation, edge

**Edge triggering**

**Synchronous operation**

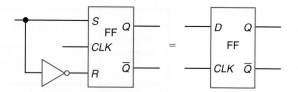

**Fig. 7-10** Wiring a D flip-flop.

triggering means the bit of data present at input D is transferred to output Q on the positive edge (LOW-to-HIGH transition) of the clock pulse. The 7474 D-flip-flop IC is said to be *positive-edge triggered.*

A more detailed truth table for the commercial *7474 TTL D flip-flop* is shown in Fig. 7-11(*b*). Remember that the asynchronous (not synchronous) inputs (*PS* and *CLR*) override the synchronous inputs. The asynchronous inputs are in control of the D flip-flop in the first three lines of the truth table in Fig. 7-11(*b*). The synchronous inputs (*D* and *CLK*) are irrelevant as shown by the "X"s on the truth table. The prohibited condition, line 3 on the truth table, should be avoided.

With both asynchronous inputs disabled (*PS* = 1 and *CLR* = 1), the D flip-flop can be set and reset using the *D* and *CLK* inputs. The last two lines of truth table use a clock pulse to transfer data from input *D* to output *Q* of the flip-flop. Being in step with the clock, this is called *synchronous operation.* Note that this flip-flop uses the LOW-to-HIGH transition of the clock pulse to transfer data from input *D* to output *Q*.

D flip-flops are sequential logic devices which are widely used temporary memory devices. D flip-flops are wired together to form *shift registers* and *storage registers.* These registers are commonly used in digital systems. Remember that the D flip-flop *delays* data from reaching output *Q* one clock pulse and is called a *delay flip-flop.* D flip-flops are sometimes also called *data flip-flops* or *D-type latches.* D flip-flops are available in both TTL and CMOS IC form. A few typical CMOS D flip-flops might be the 74HC74, 74AC74, 74FCT374, 74HC273, 74AC273, 4013, and 40174. D flip-flops are so popular with designers that more than 50 different ICs are available in just the FACT CMOS logic series.

**Shift registers**

**Storage registers**

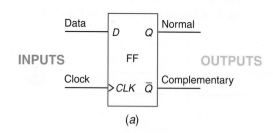

(a)

| Input | Output |
|-------|--------|
| *D* | $Q^{n+1}$ |
| 0 | 0 |
| 1 | 1 |

(b)

**Fig. 7-9** D flip-flop. (*a*) Logic symbol. (*b*) Simplified truth table.

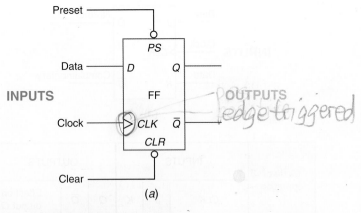

INPUTS

OUTPUTS *edge-triggered*

| Mode of operation | INPUTS | | | | OUTPUTS | |
|---|---|---|---|---|---|---|
| | Asynchronous | | Synchronous | | | |
| | PS | CLR | CLK | D | Q | $\bar{Q}$ |
| Asynchronous set | 0 | 1 | X | X | 1 | 0 |
| Asynchronous reset | 1 | 0 | X | X | 0 | 1 |
| Prohibited | 0 | 0 | X | X | 1 | 1 |
| Set | 1 | 1 | ↑ | 1 | 1 | 0 |
| Reset | 1 | 1 | ↑ | 0 | 0 | 1 |

0 = LOW
1 = HIGH
X = Irrelevant
↑ = LOW-to-HIGH transition of clock pulse

(b)

**Fig. 7-11** (a) Logic symbol for commercial D flip-flop. (b) Truth table for 7474 D flip-flop.

 Self-Test

*Answer the following questions.*

7. List the mode of operation of the 7474 D flip-flop for each input pulse shown in Fig. 7-12. Answer with the terms

"asynchronous set," "asynchronous reset," "prohibited," "set," and "reset."

8. List the binary output at the normal output (Q) of the D flip-flop for each of the pulses shown in Fig. 7-12.

**7474 D flip-flop**

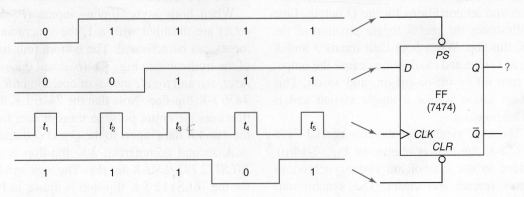

**Fig. 7-12** D flip-flop problem for test items 7 and 8.

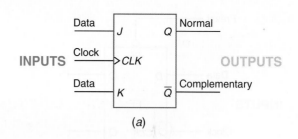

(a)

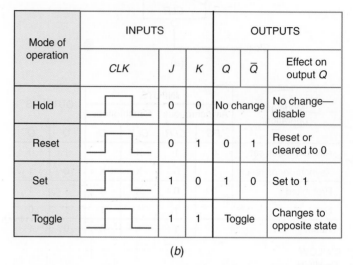

| Mode of operation | INPUTS | | | OUTPUTS | | |
|---|---|---|---|---|---|---|
| | CLK | J | K | Q | $\overline{Q}$ | Effect on output Q |
| Hold | ⎍ | 0 | 0 | No change | No change—disable | |
| Reset | ⎍ | 0 | 1 | 0 | 1 | Reset or cleared to 0 |
| Set | ⎍ | 1 | 0 | 1 | 0 | Set to 1 |
| Toggle | ⎍ | 1 | 1 | Toggle | | Changes to opposite state |

(b)

**Fig. 7-13** J-K flip-flop. (a) Logic symbol. (b) Truth table.

## 7-4 The J-K Flip-Flop

**J-K flip-flop**

The *J-K flip-flop* has the features of all the other types of flip-flops. The logic symbol for the J-K flip-flop is illustrated in Fig. 7-13(a). The inputs labeled *J* and *K* are the data inputs. The input labeled *CLK* is the clock input. Outputs *Q* and *$\overline{Q}$* are the usual normal and complementary outputs on a flip-flop. A truth table for the J-K flip-flop is shown in Fig. 7-13(b). When the *J* and *K* inputs are both 0, the flip-flop is in the *hold* mode. In the hold mode, the data inputs have no effect on the outputs. The outputs "hold" the last data present.

Lines 2 and 3 of the truth table show the reset and set conditions for the *Q* output. Line 4 illustrates the useful *toggle* position of the J-K flip-flop. When both data inputs *J* and *K* are at 1, repeated clock pulses cause the output to turn off-on-off-on-off-on, and so on. This off-on action is like a toggle switch and is called *toggling*.

**Toggling**

The logic symbol for the commercial *7476 TTL J-K flip-flop* is shown in Fig. 7-14(a). Added to the symbol are two asynchronous inputs (preset and clear). The synchronous inputs are the *J* and *K* data and clock inputs. The

customary normal (*Q*) and complementary (*$\overline{Q}$*) outputs are also shown. A detailed truth table for the commercial *7476 J-K flip-flop* is drawn in Fig. 7-14(b). Recall that asynchronous inputs (such as *PS* and *CLR*) override synchronous inputs. The asynchronous inputs are activated in the first three lines of the truth table. The synchronous inputs are irrelevant (overridden) in the first three lines in Fig. 7-14(b); therefore, an "X" is placed under the *J*, *K*, and *CLK* inputs for these rows. The prohibited state occurs when both asynchronous inputs are activated at the same time. The prohibited state is not useful and should be avoided.

When both asynchronous inputs (*PS* and *CLR*) are disabled with a 1, the synchronous inputs can be activated. The bottom four lines of the truth table in Fig. 7-14(b) detail the *hold, reset, set,* and *toggle* modes of operation for the 7476 J-K flip-flop. Note that the 7476 J-K flip-flop uses the entire pulse to transfer data from the *J* and *K* data inputs to the *Q* and *$\overline{Q}$* outputs.

A second commercial J-K flip-flop is the *74LS112 TTL-LS J-K flip-flop*. The logic symbol for the 74LS112 J-K flip-flop is drawn in Fig. 7-15(a). The 74LS112 flip-flop features two

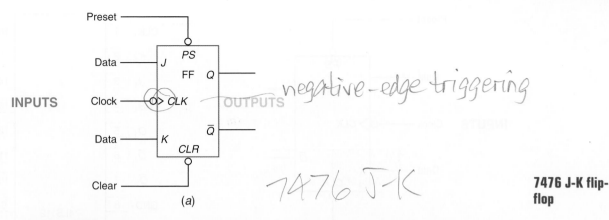

Preset

Data — J — PS

FF — Q

INPUTS

Clock — ◯> CLK

OUTPUTS

Data — K — Q̄

CLR

Clear

(a)

*negative-edge triggering*

*7476 J-K*

| Mode of operation | INPUTS | | | | | OUTPUTS | |
|---|---|---|---|---|---|---|---|
| | Asynchronous | | Synchronous | | | | |
| | PS | CLR | CLK | J | K | Q | Q̄ |
| Asynchronous set | 0 | 1 | X | X | X | 1 | 0 |
| Asynchronous reset | 1 | 0 | X | X | X | 0 | 1 |
| Prohibited | 0 | 0 | X | X | X | 1 | 1 |
| Hold | 1 | 1 | ⊓ | 0 | 0 | No change | |
| Reset | 1 | 1 | ⊓ | 0 | 1 | 0 | 1 |
| Set | 1 | 1 | ⊓ | 1 | 0 | 1 | 0 |
| Toggle | 1 | 1 | ⊓ | 1 | 1 | Opposite state | |

0 = LOW
1 = HIGH
X = Irrelevant
⊓ = Positive clock pulse

(b)

**Fig. 7-14** (a) Logic symbol for commercial J-K flip-flop. (b) Truth table for 7476 J-K flip-flop.

active-LOW asynchronous inputs (preset and clear). The two data inputs are labeled $J$ and $K$. The clock ($CLK$) input has a bubble with the $>$ symbol inside the block. This means that the 74LS112 flip-flop uses *negative-edge triggering*. In other words the flip-flop is activated on the HIGH-to-LOW transition of the input clock pulse. The 74LS112 J-K flip-flop has the customary normal ($Q$) and complementary ($\overline{Q}$) outputs.

The pin diagram for a 16-pin DIP 74LS112 IC is sketched in Fig. 7-15(b). Notice that the 74LS112 IC contains two J-K flip-flops both with asynchronous inputs ($PS$ and $CLR$) and complementary outputs ($Q$ and $\overline{Q}$). The 74LS112 IC is also available in other IC packages.

The truth table for the 74LS112 J-K flip-flop is drawn in Fig. 7-15(c). The 74LS112 flip-flop has the same modes of operation as the 7476. The first three lines of the truth table show the

asynchronous inputs ($PS$ and $CLR$) overriding the synchronous inputs ($J$, $K$, and $CLK$). Notice that the asynchronous pins are active-LOW inputs. The last four lines of the truth table detail the hold, reset, set, and toggle modes of operation. The $CLK$ inputs triggers the flip-flop on the HIGH-to-LOW transition of the clock pulse. This is called negative *edge-triggering*. The last line of the truth table in Fig. 7-15(c) is the useful toggle mode. With the asynchronous inputs disabled ($PS = 1$, $CLR = 1$) and the data inputs both HIGH ($J = 1$, $K = 1$), each clock pulse will cause the outputs to toggle to their opposite state. For instance, output $Q$ might go from HIGH, LOW, HIGH, LOW, on repeated clock pulses. This is a useful feature when building circuits such as counters.

J-K flip-flops are used in many digital circuits. You will use the J-K flip-flop especially in

**negative-edge triggering**

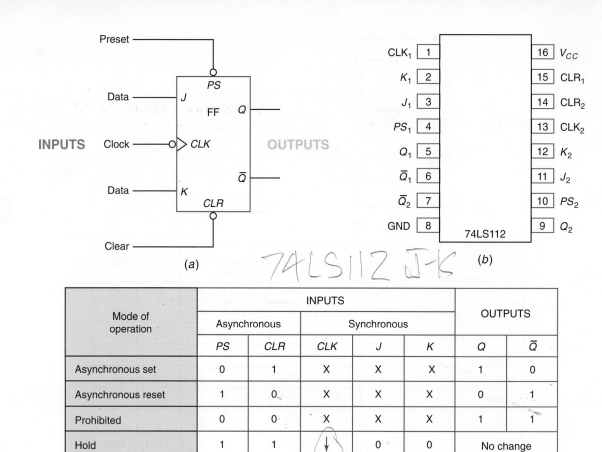

74LS112 J-K

| Mode of operation | INPUTS | | | | | OUTPUTS | |
| --- | --- | --- | --- | --- | --- | --- | --- |
| | Asynchronous | | Synchronous | | | | |
| | $PS$ | $CLR$ | $CLK$ | $J$ | $K$ | $Q$ | $\bar{Q}$ |
| Asynchronous set | 0 | 1 | X | X | X | 1 | 0 |
| Asynchronous reset | 1 | 0 | X | X | X | 0 | 1 |
| Prohibited | 0 | 0 | X | X | X | 1 | 1 |
| Hold | 1 | 1 | ↓ | 0 | 0 | No change | |
| Reset | 1 | 1 | ↓ | 0 | 1 | 0 | 1 |
| Set | 1 | 1 | ↓ | 1 | 0 | 1 | 0 |
| Toggle | 1 | 1 | ↓ | 1 | 1 | Opposite state | |

0 = LOW
1 = HIGH
X = Irrelevant
↓ = HIGH-to-LOW clock transition

negative-edge triggering

(c)

**Fig. 7-15** 74LS112 J-K flip-flop IC. (a) Logic symbol, (b) pin diagram, (c) truth table.

**Counters**

*counters.* Counters are found in almost every digital system.

In summary, the J-K flip-flop is considered the "universal" flip-flop. Its unique feature is the toggle mode of operation so useful in designing counters. When the J-K flip-flop is

**T flip-flop**

wired for use only in the toggle mode, it is commonly called a *T flip-flop.* J-K flip-flops are available in both TTL and CMOS IC form. Typical CMOS J-K flip-flops are the 74HC76, 74AC109, and 4027 ICs.

*Answer the following questions.*

9. List the mode of operation of the 7476 J-K flip-flop after each input pulse shown in Fig. 7-16. Answer with the terms "asynchronous set," "asynchronous

reset," "prohibited," "hold," "reset," "set," and "toggle."

10. List the binary output at the normal output ($Q$) of the J-K flip-flop after each of the pulses shown in Fig. 7-16.

11. Refer to Fig. 7-17. Both J-K flip-flops are in the _____ (reset, set, toggle) mode of operation in this circuit.

12. Refer to Fig. 7-17. The 74LS112 flip-flops used in this circuit are triggered on the _____ (HIGH-to-LOW, LOW-to-HIGH) transition of a clock pulse.

13. List the 2-bit binary output of the J-K flip-flops after each of the pulses shown in Fig. 7-17.

14. Refer to Fig. 7-17. The circuit using J-K flip-flops seems to operate as a 2-bit _____ (adder, counter).

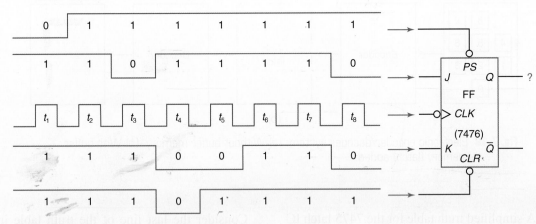

**Fig. 7-16** J-K flip-flop problem for test items 9 and 10.

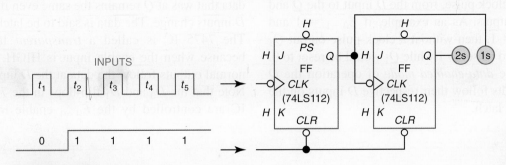

**Fig. 7-17** J-K flip-flop problem for test items 11–14.

## 7-5 IC Latches

Consider the block diagram of the digital system in Fig. 7-18(*a*). Press and hold the decimal number 7 on the keyboard. A 7 will be observed on the seven-segment display. Release the 7 on the keyboard and the 7 disappears from the display. It is obvious that a *memory device* is needed to hold the BCD code for 7 at the inputs to the decoder. A device that serves as a temporary buffer memory is called a *latch*. A 4-bit latch has been added to the system in Fig. 7-18(*b*). Now when the decimal number 7 on the keyboard is *pressed and released,* the seven-segment display continues to show a 7.

The term "latch" refers to a digital storage device. The *D flip-flop* is a good example of a device used to latch data. However, other types of flip-flops are also used for the latching function.

Manufacturers have developed many latches in IC form. The logic diagram for the *7475 TTL 4-bit transparent latch* is shown in Fig. 7-19(*a*). This unit has four D flip-flops enclosed in a single IC package. The $D_0$ data input and the normal $Q_0$ and complementary $\overline{Q}_0$ outputs form the first D flip-flop. The enable input ($E_{0-1}$) is similar to the clock input on the D flip-flop. When $E_{0-1}$ is enabled, data at both $D_0$ and $D_1$ are transferred to their outputs.

**D flip-flop**

**7475 TTL 4-bit transparent latch**

**Memory device**

**Latch**

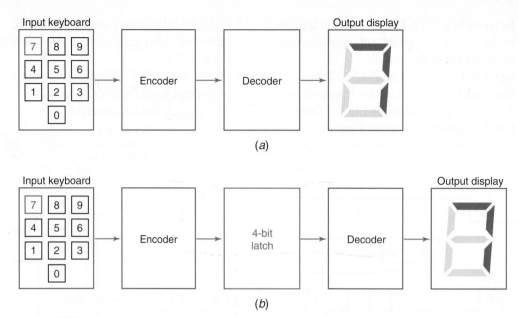

Fig. 7-18 Electronic encoder/decoder system. (*a*) Without buffer memory. (*b*) With buffer memory (latch) added.

**Electronic encoder/decoder system**

A simplified truth table for the 7475 latch IC is shown in Fig. 7-19(*b*). If the enable input is at a logical 1, data is transferred, without a separate clock pulse, from the $D$ input to the $Q$ and $\overline{Q}$ outputs. As an example, if $E_{0-1} = 1$ and $D_1 = 1$, then without a clock pulse output $Q_1$ would be set to 1 while $\overline{Q}_1$ would be reset to 0. In the *data-enabled mode* of operation the $Q$ outputs follow their respective $D$ inputs on the 7475 latch.

**Data-latched mode**

**Data-enabled mode**

Consider the last line of the truth table in Fig. 7-19(*b*). When the enable input drops to 0, the 7475 IC enters the *data-latched mode*. The data that was at $Q$ remains the same even if the $D$ inputs change. The data is said to be latched. The 7475 IC is called a *transparent* latch because when the enable input is HIGH, the normal outputs follow the data at the $D$ inputs. Note that the $D_0$ and $D_1$ flip-flops in the 7475 IC are controlled by the $E_{0-1}$ enable input

## ABOUT ELECTRONICS

**Onboard Car Electronics.** Alpine Electronics has developed a DVD-based navigation system, which uses global positioning systems, to help drivers navigate. Using a DVD drive paired with a dashboard monitor, the system displays driving directions through map visuals, combined with voice-guided, turn-by-turn instructions. You can use this system to find the shortest route, find the route using the fewest toll roads, or even find the nearest ATM.

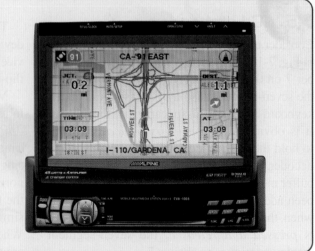

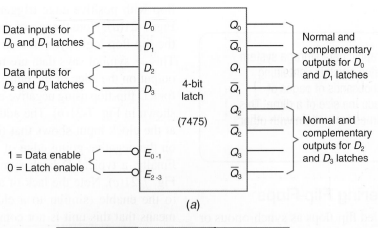

Data inputs for $D_0$ and $D_1$ latches $\begin{cases} \\ \\ \end{cases}$ $D_0$
$D_1$

Data inputs for $D_2$ and $D_3$ latches $\begin{cases} \\ \\ \end{cases}$ $D_2$
$D_3$

4-bit latch

(7475)

$1$ = Data enable $0$ = Latch enable $\begin{cases} \\ \\ \end{cases}$ $E_{0\text{-}1}$
$E_{2\text{-}3}$

$Q_0$
$\overline{Q}_0$ $\Big\}$ Normal and complementary outputs for $D_0$ and $D_1$ latches

$Q_1$
$\overline{Q}_1$

$Q_2$
$\overline{Q}_2$ $\Big\}$ Normal and complementary outputs for $D_2$ and $D_3$ latches

$Q_3$
$\overline{Q}_3$

(a)

| Mode of operation | INPUTS | | OUTPUTS | |
|---|---|---|---|---|
| | $E$ | $D$ | $Q$ | $\overline{Q}$ |
| Data enabled | 1 | 0 | 0 | 1 |
| | 1 | 1 | 1 | 0 |
| Data latched | 0 | X | No change | |

0 = LOW
1 = HIGH
X = Irrelevant

(b)

**Fig. 7-19** (a) Logic symbol for commercial 7475 4-bit transparent latch. (b) Truth table for 7475 D latch.

whereas the $E_{2-3}$ input controls the $D_2$ and $D_3$ pair of flip-flops.

One use of a flip-flop is to hold, or latch, data. When used for this purpose, the flip-flop is called a latch. Flip-flops have many other uses, including *counters, shift registers, delay units,* and *frequency dividers.*

Latches are available in all logic families. Several typical CMOS latches are the 4042, 4099, 74HC75, and 74HC373 ICs. Latches are sometimes built into other ICs such as the 4511 and 4543 BCD-to-seven segment latch/decoder/driver chips you may have already studied.

One of the primary advantages of digital over analog circuitry is the availability of easy-to-use memory devices. The latch is the most fundamental memory device used in digital electronics. Almost all digital equipment contains simple memory devices called latches.

Counters
Shift registers
Delay unit
Frequency dividers

**Self-Test**

*Supply the missing word in each statement.*

15. When the 7475 latch IC is in its data-enabled mode of operation, the _____ outputs follow their respective $D$ inputs.

16. A _____ (HIGH, LOW) at the enable inputs places the 7475 latch IC in the data-latched mode of operation.

17. In the data-latched mode, a change at any of the $D$ inputs to the 7475 latch IC has _____ (an immediate effect on their respective outputs; no effect on the outputs).

18. When a flip-flop is used to temporarily hold data, it is sometimes called _____.

## 7-6 Triggering Flip-Flops

**Triggering flip-flops**

**Synchronous flip-flops**

**J-K master/slave flip-flop**

**Positive-edge-triggered flip-flop**

**Negative-edge-triggered flip-flop**

We have classified flip-flops as synchronous or asynchronous in their operation. *Synchronous flip-flops* are all those that have a clock input. We found that the clocked R-S, the D, and the J-K flip-flop operate in step with the clock.

When using manufacturers' data manuals, you will notice that many synchronous flip-flops are also classified as either *edge-triggered* or *master/slave*. Figure 7-20 shows two edge-triggered flip-flops in the toggle position. On clock pulse 1, the positive edge (positive-going edge) of the pulse is identified. The second waveform shows how the *positive-edge-triggered flip-flop* toggles each time a positive-going pulse comes along (see pulses 1 to 4). On pulse 1 in Fig. 7-20, the negative edge (negative-going edge) of the pulse is also labeled. The bottom waveform shows how the *negative-edge-triggered flip-flop* toggles. Notice that it changes state, or toggles, each time a negative-going pulse comes along (see pulses 1 to 4). Especially notice the difference in timing between the positive- and negative-edge-triggered flip-flops. This triggering time difference is quite important for some applications.

It is common to show the type of triggering on the flip-flop. The logic symbol for a D flip-flop with positive-edge triggering is shown in Fig. 7-21(*a*). Note the use of the small > inside the flip-flop logic symbol near the clock input. This > symbol says data are transferred to the output on the edge of the pulse. A logic symbol for a D flip-flop using negative-edge triggering is shown in Fig. 7-21(*b*). The added invert bubble at the clock input shows that triggering occurs on the negative-going edge of the clock pulse. Finally, a typical D latch symbol is shown in Fig. 7-21(*c*). Note the lack of a > symbol next to the enable (similar to a clock) input. This means that this unit is not considered an edge-triggered unit. Like the R-S flip-flop, the D latch is considered asynchronous. Recall that the D latch normal output ($Q$) follows its input ($D$) when the enable ($E$) input is HIGH. The data are latched when the enable input drops to LOW. Several manufacturers label the enable input with a "G" on the D latch.

Another class of flip-flop triggering is the master/slave type. The *J-K master/slave flip-flop* uses the entire pulse (positive edge and negative edge) to trigger the flip-flop. Figure 7-22 shows the triggering of a master/slave flip-flop. Pulse 1 shows four positions (*a* to *d*) on the waveform. The following sequences of operation takes place in the master/slave flip-flop at each point on the clock pulse:

- Point *a*: leading edge—isolate input from output
- Point *b*: leading edge—enter information from *J* and *K* inputs
- Point *c*: trailing edge—disable *J* and *K* inputs
- Point *d*: trailing edge—transfer information from input to output

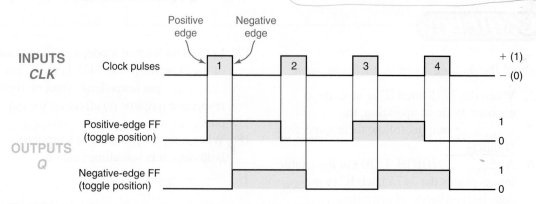

**Fig. 7-20** Waveforms for positive- and negative-edge-triggered flip-flops.

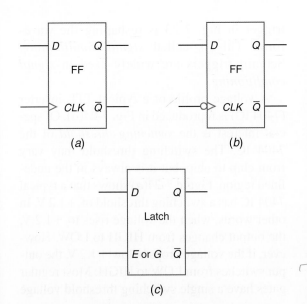

Fig. 7-21 (a) Logic symbol for positive-edge-triggered D flip-flop. (b) Logic symbol for negative-edge-triggered D flip-flop. (c) Logic symbol for D latch.

A very interesting characteristic of the master/slave flip-flop is shown during pulse 2, Fig. 7-22. Notice that at the beginning of pulse 2 the outputs are disabled. For a very brief moment, the J and K inputs are moved to the toggle positions (see point e) and then disabled. The J-K master/slave flip-flop "remembers" that the J and K inputs were in toggle positions, and it toggles at point f on the waveform diagram. This memory characteristic happens only while the clock pulse is high (at logical 1).

Master/slave triggering has become obsolete with the newer edge-triggered flip-flops. For instance, the master/slave 7476 flip-flop is replaced by the 74LS76 device. It has the same pin diagrams and functions but the newer 74LS76 IC uses negative-edge triggering.

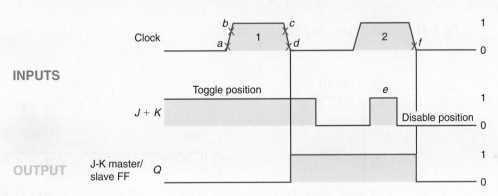

Fig. 7-22 Triggering the J-K master/slave flip-flop.

Triggering the J-K master/slave flip-flop

**Self-Test**

*Supply the missing word in each statement.*

19. A positive-edge-triggered flip-flop changes state on the _____ (HIGH-to-LOW, LOW-to-HIGH) transition of the clock pulse.
20. A negative-edge-triggered flip-flop changes state on the _____ (HIGH-to-LOW, LOW-to-HIGH) transition of the clock pulse.
21. The ">" near the clock input inside a flip-flop logic symbol means _____.

22. The 74LS112 J-K flip-flop detailed in Fig. 7-15 uses _____ (positive-edge triggering, negative-edge triggering).
23. The 7474 D flip-flop detailed in Fig. 7-11 uses _____ (positive-edge triggering, negative-edge triggering).
24. J-K master/slave flip-flops (such as the 7476 IC) using the entire pulse to trigger the unit have become obsolete replaced by the newer edge-triggered flip-flops. (T or F)

# 7-7 Schmitt Trigger

**Rise and fall times**

**Signal conditioning**

**Switching threshold**

**Schmitt trigger inverter**

Digital circuits prefer waveforms with fast *rise and fall times*. The waveform on the right side of the inverter symbol in Fig. 7-23 is an example of a good digital signal. The square wave's L-to-H and H-to-L edges are vertical. This means that the rise and fall times are very fast (almost instantaneous).

The waveform to the left of the inverter symbol in Fig. 7-23 has very slow rise and fall times. The poor waveform on the left in Fig. 7-23 might lead to unreliable operation if fed directly into counters, gates, or other digital circuitry. In this example, a *Schmitt trigger inverter* is being used to "square up" the input signal and make it more useful. The Schmitt

trigger in Fig. 7-23 is reshaping the waveform. This is called *signal conditioning*. Schmitt triggers are widely used in *signal conditioning*.

A voltage profile of a typical TTL inverter (7404 IC) is reproduced in Fig. 7-24(a). Of special interest is the *switching threshold* of the 7404 IC. The switching threshold may vary from chip to chip, but it is always in the undefined region. Figure 7-24(a) shows that a typical 7404 IC has a switching threshold of +1.2 V. In other words, when the voltage rises to +1.2 V, the output changes from HIGH to LOW. However, if the voltage drops below +1.2 V, the output switches from LOW to HIGH. Most regular gates have a single switching threshold voltage

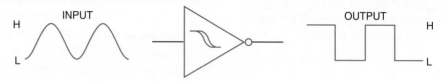

**Fig. 7-23** Schmitt trigger used for wave shaping.

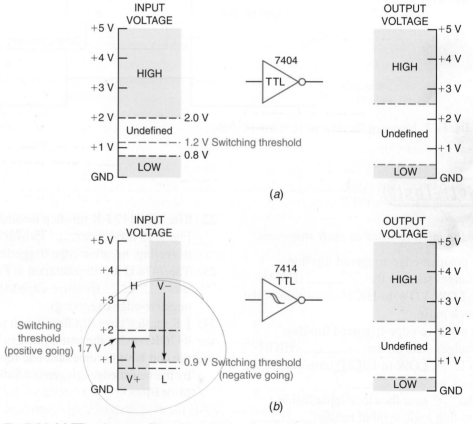

(a)

(b)

**Fig. 7-24** (a) TTL voltage profiles with switching threshold. (b) Voltage profiles for 7414 TTL Schmitt trigger IC showing switching thresholds.

whether the input voltage is rising (L to H) or falling (H to L).

A voltage profile for a *7414 Schmitt trigger inverter TTL IC* is sketched in Fig. 7-24(*b*). Note that the *switching threshold is different* for positive-going (V +) and negative-going (V −) voltages. The voltage profile for the 7414 IC shows that the switching threshold is 1.7 V for a positive-going (V +) input voltage. However, the switching threshold is 0.9 V for a negative-going (V −) input voltage. The difference between these switching thresholds (1.7 V and 0.9 V) is called *hysteresis*. Hysteresis provides

for excellent noise immunity and helps the Schmitt trigger square up waveforms with slow rise and fall times.

Schmitt triggers are also available in CMOS. These include the 40106, 4093, 74HC14, and 74AC14 ICs.

One of the characteristics of a bistable multivibrator (or flip-flop) is that its outputs are either HIGH or LOW. When changing states (H to L or L to H), they do so rapidly without the outputs being in the undefined region. This "snap action" of the output is also characteristic of Schmitt triggers.

---

## Self-Test

*Answer the following questions.*

25. The _____ is a good device for squaring up a waveform with slow rise and fall times.
26. Draw the schematic symbol for a Schmitt trigger inverter.

27. A Schmitt trigger is said to have _____ because its switching thresholds are different for positive-going and negative-going inputs.
28. Schmitt triggers are commonly used for _____ (memory, signal conditioning).

---

## 7-8  IEEE Logic Symbols

The flip-flop logic symbols you have learned are the traditional ones recognized by most workers in the electronics industry. Manufacturers' data manuals usually include traditional symbols and are recently including new IEEE standard logic symbols.

The table in Fig. 7-25 shows the traditional flip-flop and latch logic symbols learned in this chapter along with their IEEE counterparts. All IEEE logic symbols are rectangular and include the number of the IC directly above the symbol. Smaller rectangles show the number of duplicate devices in the package. Notice that all inputs are on the left of the IEEE symbol and outputs are on the right.

The IEEE 7474 D flip-flop symbol shows four inputs labeled "*S*" (for "set"), ">*C*1" (for "positive-edge-triggered clock"), "1*D*" (for "data"), and "*R*" (for "reset"). The triangles at the *S* and *R* inputs on the IEEE 7474 symbol identify them as active-LOW inputs. The 7474 outputs are on the right of the IEEE symbol with no internal identifying markings. The $\overline{Q}$ outputs

have triangles suggesting active-LOW outputs. The markings inside the IEEE logic symbols are standard while the outside markings vary from manufacturer to manufacturer.

Consider the IEEE logic symbol for the 7476 dual master/slave J-K flip-flop in Fig. 7-25. The internal inputs are marked "*S*" ("set"), "1*J*" ("*J* data"), "*C*1" ("Clock"), "1*K*" ("*K* data"), and "*R*" ("reset"). The "7476" above the symbol identifies the specific TTL IC. The markings near the *Q* and $\overline{Q}$ outputs are the unique IEEE symbols for pulse-triggering. The IEEE logic symbol for the 7476 IC shows that there are two active-LOW inputs (*S* and *R*) and a single active-LOW output ($\overline{Q}$) on each J-K flip-flop. The active-LOW inputs and outputs are marked with a small right triangle. The symbol is repeated below to indicate that the 7476 IC package contains two identical J-K flip-flops.

The IEEE standard logic symbol for the 7475 4-bit transparent latch is reproduced in Fig. 7-25. Note the four rectangles to represent the four D-type latches in the 7475 IC package. The four $\overline{Q}$ output leads are marked with small triangles.

| Flip-Flop/ Latch IC | Traditional Logic Symbol | IEEE Logic Symbol* |
|---|---|---|
| 7474 TTL Dual D Flip-flop | Preset — Data — D PS Q FF (7474) Clock — CLK Q̄ CLR Clear — | 7474<br>1 Preset — S<br>1 Clock — >C1<br>1 Data — 1D<br>1 Clear — R — 1Q / 1Q̄<br>2 Preset / 2 Clock / 2 Data / 2 Clear — 2Q / 2Q̄ |
| 7476 TTL Dual Master/ Slave J-K Flip-flop | Preset — PS Data — J Q FF Clock — CLK (7476) Data — K Q̄ CLR Clear — | 7476<br>1 Preset — S<br>1 J Data — 1J<br>1 Clock — >C1<br>1 K Data — 1K<br>1 Clear — R — 1Q / 1Q̄<br>2 Preset / 2 J Data / 2 Clock / 2 K Data / 2 Clear — 2Q / 2Q̄ |
| 7475 TTL 4-bit Transparent Latch | Data Inputs { D₀ Q₀ / D₁ Q̄₀ / D₂ Q₁ / D₃ Q̄₁ } Enable Inputs { E₀₋₁ Q₂ / Q̄₂ / E₂₋₃ Q₃ / Q̄₃ } | 7475<br>1 Data — 1D — 1Q / 1Q̄<br>Enable — C1 / C2<br>2 Data — 2D — 2Q / 2Q̄<br>3 Data — 3D — 3Q / 3Q̄<br>Enable — C3 / C4<br>4 Data — 4D — 4Q / 4Q̄ |

*IEEE Standard 91-1984

**Fig. 7-25** Comparing traditional and IEEE symbols for several flip-flops.

*Answer the following questions.*

29. The "C" marking inside the IEEE symbol stands for the control input or the _____ inputs on the flip-flops.

30. The complementary ($\overline{Q}$) outputs of the flip-flops and latches on an IEEE symbol are designated with the _____ symbol.

31. The asynchronous clear on the 7474 and 7476 flip-flop are active- _____ inputs and are marked with the letter "R," which stands for _____.

# Chapter 7 Summary and Review

1. Logic circuits are classified as combinational or sequential. Combinational logic circuits use AND, OR, and NOT gates and do not have a memory characteristic. Sequential logic circuits use flip-flops and involve a memory characteristic.

2. Flip-flops are wired together to form counters, registers, and memory devices.

3. Flip-flop outputs are opposite, or complementary.

4. The table in Fig. 7-26 summarizes some basic flip-flops.

5. Waveform (timing) diagrams are used to describe the operation of sequential devices.

6. Flip-flops can be edge-triggered or master/slave types. Flip-flops can be pulse-or edge-triggered.

7. Special flip-flops called latches are widely used in most digital circuits as temporary buffer memories.

8. Schmitt triggers are special devices that are used for signal conditioning.

9. Figure 7-25 compares traditional flip-flop/latch symbols with the newer IEEE logic symbols.

| Circuit | Logic Symbol | Truth Table | | | | Remarks |
|---------|-------------|------|------|------|------|---------|
| R-S flip-flop | $-\circ\, S \quad Q-$ <br> $\quad FF$ <br> $-\circ\, R \quad \bar{Q}-$ <br> *inverted data* | $S$ | $R$ | $Q$ | | R-S latch <br> Set-reset flip-flop |
| | | 0 | 0 | prohibited | | |
| | | 0 | 1 | 1 | set | |
| | | 1 | 0 | 0 | reset | |
| | | 1 | 1 | | hold | (asynchronous) |
| Clocked R-S flip-flop | $-S \quad Q-$ <br> $\quad FF$ <br> $-CLK$ <br> $-R \quad \bar{Q}-$ | $CLK$ | $S$ $R$ | $Q$ | | |
| | | ⊓ | 0 0 | hold | | |
| | | ⊓ | 0 1 | 0 reset | | |
| | | ⊓ | 1 0 | 1 set | | |
| | | ⊓ | 1 1 | prohibited | | (synchronous) |
| D flip-flop | $-D \quad Q-$ <br> $\quad FF$ <br> $-\!\!\triangleright CLK \quad \bar{Q}-$ | $CLK$ | $D$ | $Q$ | | Delay flip-flop <br> Data flip-flop |
| | | ↑ | 0 | 0 | | |
| | | ↑ | 1 | 1 | | (synchronous) |
| | | ↑ = L-to-H transition of clock | | | | |
| J-K flip-flop | $-J \quad Q-$ <br> $\quad FF$ <br> $-\circ\!\!\triangleright CLK$ <br> $-K \quad \bar{Q}-$ | $CLK$ | $J$ $K$ | $Q$ | | Most universal FF |
| | | ↓ | 0 0 | hold | | |
| | | ↓ | 0 1 | 0 reset | | |
| | | ↓ | 1 0 | 1 set | | |
| | | ↓ | 1 1 | toggle | | (synchronous) |
| | | ↓ = H-to-L transition of clock | | | | |

**Fig. 7-26** Summary of basic flip-flops.

*Answer the following questions.*

7-1. Logic _____ are the basic building blocks of combinational logic circuits; the basic building blocks of sequential circuits are devices called _____.

7-2. List one type of asynchronous and three types of synchronous flip-flops.

7-3. Draw a traditional logic symbol for the following flip-flops:
   a. J-K            c. Clocked R-S
   b. D              d. R-S

7-4. Draw a truth table for the following flip-flops:
   a. J-K (with negative-edge triggering)
   b. D (with positive-edge triggering)
   c. Clocked R-S
   d. R-S

7-5. If both synchronous and the asynchronous inputs on a J-K flip-flop are activated, which input will control the output?

7-6. When we say the flip-flop is in the set condition, we mean output _____ is at a logical _____.

7-7. When we say the flip-flop is in the reset, or clear, condition, we mean output _____ is at a logical _____.

7-8. On a timing, or waveform, diagram the horizontal distance stands for _____ and the vertical distance stands for _____.

7-9. Refer to Fig. 7-6. This waveform diagram is for a(n) _____ flip-flop. This flip-flop is _____-edge triggered.

7-10. List two types of edge-triggered flip-flops.

7-11. The "D" in "flip-flop" stands for _____, or data.

7-12. D flip-flops are widely used as temporary memories called _____.

7-13. If a flip-flop is in its toggle mode of operation, what will the output act like upon repeated clock pulses?

7-14. Identify these acronyms used on traditional flip-flop logic symbols:
   a. *CLK*          e. *PS*
   b. *CLR*          f. *R*
   c. *D*            g. *S*
   d. FF

7-15. Give a descriptive name for the following TTL ICs:
   a. 7474
   b. 7475
   c. 74LS112

7-16. The 7474 IC is a(n) _____-edge-triggered unit.

7-17. List the modes of operation of the 7474 IC.

7-18. List the mode of operation of the 7476 J-K flip-flop for each input pulse shown in Fig. 7-27.

7-19. List the binary outputs at the normal output (*Q*) of the J-K flip-flop after each time period ($t_1$–$t_7$) shown in Fig. 7-27.

7-20. List the mode of operation of the 7475 4-bit latch for each time period ($t_1$ through $t_7$) shown in Fig. 7-28.

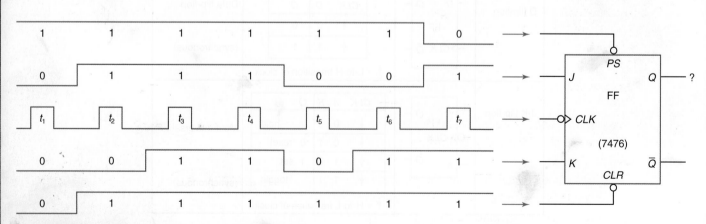

**Fig. 7-27** Pulse-train problem.

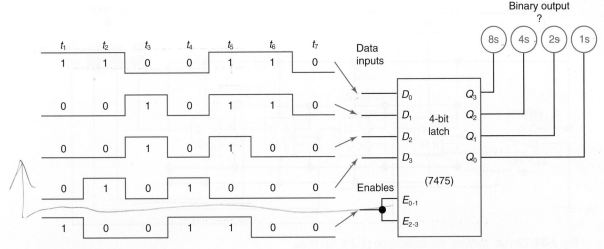

Fig. 7-28 Pulse-train problem.

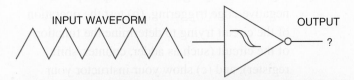

Fig. 7-29 Sample problem.

7-21. List the binary output (4-bit) at the output indicators of the 7475 4-bit latch for each time period ($t_1$ through $t_7$) shown in Fig. 7-28.

7-22. Refer to Fig. 7-29. The output waveform on the right of the logic symbol will be a _____ (sine, square) wave.

7-23. The inverter in Fig. 7-29 is being used as a signal _____ (conditioner, multiplexer) in this circuit.

7-24. The logic symbol in Fig. 7-29 is for a symbol _____ (two words) inverter IC.

7-25. Identify these markings found inside and on the leads of the IEEE flip-flop/latch logic symbols.

a.  C          e.  J
b.  S          f.  K
c.  R          g.  ⌐
d.  D          h.  >C

## Critical Thinking Questions

7-1. List two other names sometimes given for an R-S flip-flop.

7-2. Explain the difference between asynchronous and synchronous devices.

7-3. Draw the traditional and IEEE logic symbols for a D flip-flop (7474 IC) and a J-K flip-flop (7476 IC).

7-4. Refer to Fig. 7-3. Notice that line 4 is listed two times across the bottom. Why does output $Q = 0$ in the first case and then 1 in the second case when inputs $R$ and $S$ are both 1 in each case?

7-5. Explain how the 74LS112 J-K flip-flop is triggered.

7-6. What is the fundamental difference between a combinational logic and a sequential logic circuit?

7-7. List several devices that are built using J-K flip-flops.

7-8. Explain why Schmitt trigger devices tend to "square up" inputs with slow rise times.

7-9. At the option of your instructor, use circuit simulation software to (1) draw the flip-flop circuit sketched in Fig. 7-30; (2) test the operation of the flip-flop circuit; (3) make a truth table for

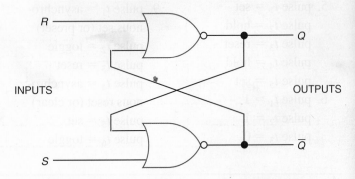

Fig. 7-30 Flip-flop circuit.

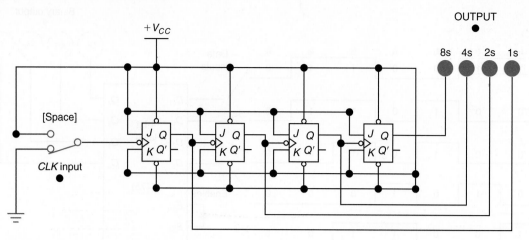

Fig. 7-31 Circuit showing the application of J-K flip-flops.

the flip-flop (something like Table 7-1) listing the modes of operation as "set," "reset," "hold," and "prohibited"; and (4) determine if it acts more like an R-S or a J-K flip-flop.

7-10. At the option of your instructor, use circuit simulation software to (a) draw the circuit shown in Fig. 7-31 using a generic J-K flip-flop with negative-edge triggering, (b) test the operation of the circuit trying to determine the function of the circuit (such as adder, counter, shift register), and (c) show your instructor your circuit simulation.

## Answers to Self-Tests

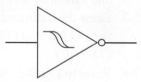

1. LOW
2. pulse $t_1$ = set
   pulse $t_2$ = reset
   pulse $t_3$ = hold
   pulse $t_4$ = set
   pulse $t_5$ = hold
   pulse $t_6$ = reset
3. pulse $t_1$ = 1
   pulse $t_2$ = 0
   pulse $t_3$ = 0
   pulse $t_4$ = 1
   pulse $t_5$ = 1
   pulse $t_6$ = 0
4. HIGH
5. pulse $t_1$ = set
   pulse $t_2$ = hold
   pulse $t_3$ = reset
   pulse $t_4$ = hold
   pulse $t_5$ = set
6. pulse $t_1$ = 1
   pulse $t_2$ = 1
   pulse $t_3$ = 0

pulse $t_4$ = 0
pulse $t_5$ = 1
7. pulse $t_1$ = asynchronous set (or preset)
   pulse $t_2$ = reset
   pulse $t_3$ = set
   pulse $t_4$ = asynchronous reset (or clear)
   pulse $t_5$ = set
8. pulse $t_1$ = 1
   pulse $t_2$ = 0
   pulse $t_3$ = 1
   pulse $t_4$ = 0
   pulse $t_5$ = 1
9. pulse $t_1$ = asynchronous set (or preset)
   pulse $t_2$ = toggle
   pulse $t_3$ = reset
   pulse $t_4$ = asynchronous reset (or clear)
   pulse $t_5$ = set
   pulse $t_6$ = toggle

pulse $t_7$ = toggle
pulse $t_8$ = hold
10. pulse $t_1$ = 1
    pulse $t_2$ = 0
    pulse $t_3$ = 0
    pulse $t_4$ = 0
    pulse $t_5$ = 1
    pulse $t_6$ = 0
    pulse $t_7$ = 1
    pulse $t_8$ = 1
11. toggle
12. HIGH-to-LOW
13. pulse $t_1$ = 00
    pulse $t_2$ = 01
    pulse $t_3$ = 10
    pulse $t_4$ = 11
    pulse $t_5$ = 00
14. counter
15. $Q$ (normal)
16. LOW
17. no effect on the outputs

18. latch
19. LOW-to-HIGH
20. HIGH-to-LOW
21. edge-triggering
22. negative-edge triggering
23. positive-edge triggering
24. T
25. Schmitt trigger
26. See figure below

27. hysteresis
28. signal conditioning
29. clock
30. triangle
31. LOW, reset

# Counters

## Chapter Objectives

*This chapter will help you to:*

1. *Draw* a circuit diagram of a ripple counter using J-K flip-flops.

2. *Analyze* the circuit action of any mod-3 through mod-8 synchronous counter.

3. *Understand* the operation of and *draw* a block diagram of a frequency divider circuit.

4. *Interpret* data sheets for several commercial TTL and CMOS counter ICs.

5. *Predict* the operation of a 4-bit magnitude comparator IC from its truth table.

6. *Analyze* the operation of an electronic "guess the number" game containing a clock, a counter, and a 4-bit magnitude comparator.

7. *Determine* the output for a variety of counters based on a series of inputs.

8. *Understand* and *explain* the details of a counting system driven by an optical sensor.

9. *Interpret* the data sheet for a 3-digit BCD counter IC featuring internal latches and display multiplexing.

10. *Analyze* the use of a 3-digit BCD counter IC to count revolutions of a motor using a Hall-effect sensor input.

11. *Adapt* the Hall-effect based BCD counter to *design* an experimental tachometer.

12. *Troubleshoot* a faulty ripple counter circuit.

Almost any complex digital system contains several *counters*. A counter's job is the obvious one of counting events or periods of time or putting events into sequence. Counters also do some not so obvious jobs: dividing frequency, addressing, and serving as memory units. This chapter discusses several types of counters and their uses. Flip-flops are wired together to form circuits that count. Because of the wide use of counters, manufacturers also make self-contained counters in IC form. Many counters are available in all TTL and CMOS families. Some counter ICs contain other devices such as signal conditioning circuitry, latches, and display multiplexers.

## 8-1 Ripple Counters

Counting in binary and decimal is illustrated in Fig. 8-1. With four binary places (*D, C, B,* and *A*), we can count from 0000 to 1111 (0 to 15 in decimal). Notice that column *A* is the 1s binary place, or least significant digit (LSD). The term "least significant bit" (*LSB*) is usually used. Column *D* is the 8s binary place, or most significant digit (MSD). The term "most significant bit" (*MSB*) is usually used. Notice that the 1s column changes state the most often. If we design a counter to count from binary 0000 to 1111, we need a device that has 16 different output states: a *modulo* (mod)-*16 counter.* The *modulus of a counter* is the number of different states the counter must go through to complete its counting cycle.

A mod-16 counter using four J-K flip-flops is diagrammed in Fig. 8-2(*a*). Each J-K flip-flop is in its toggle position (*J* and *K* both at 1). Assume the outputs are cleared to 0000. As clock pulse 1 arrives at the clock (*CLK*) input of flip-flop 1 (FF 1), it toggles (on the

**LSB**

**MSB**

**Modulo-16 counter**

**Modulus of a counter**

| BINARY COUNTING | | | | DECIMAL COUNTING |
|---|---|---|---|---|
| D | C | B | A | |
| 8s | 4s | 2s | 1s | |
| 0 | 0 | 0 | 0 | 0 |
| 0 | 0 | 0 | 1 | 1 |
| 0 | 0 | 1 | 0 | 2 |
| 0 | 0 | 1 | 1 | 3 |
| 0 | 1 | 0 | 0 | 4 |
| 0 | 1 | 0 | 1 | 5 |
| 0 | 1 | 1 | 0 | 6 |
| 0 | 1 | 1 | 1 | 7 |
| 1 | 0 | 0 | 0 | 8 |
| 1 | 0 | 0 | 1 | 9 |
| 1 | 0 | 1 | 0 | 10 |
| 1 | 0 | 1 | 1 | 11 |
| 1 | 1 | 0 | 0 | 12 |
| 1 | 1 | 0 | 1 | 13 |
| 1 | 1 | 1 | 0 | 14 |
| 1 | 1 | 1 | 1 | 15 |

**Binary count**

**Fig. 8-1** Counting sequence for a 4-bit electronic counter.

negative edge) and the display shows 0001. Clock pulse 2 causes FF 1 to toggle again, returning output $Q$ to 0, which causes FF 2 to toggle to 1. The count on the display now reads 0010. The counting continues, with each flip-flop output triggering the next flip-flop on its negative-going pulse. Look back at Fig. 8-1 and see that column $A$ (1s column) must change state on every count. This means that FF 1 in Fig. 8-2(a) must toggle for each pulse. FF 2 must toggle only half as often as FF 1, as seen from column $B$ in Fig. 8-1. Each more significant bit in Fig. 8-1 toggles less often.

The counting of the mod-16 counter is shown up to a count of decimal 10 (binary 1010) by waveforms in Fig. 8-2(b). The *CLK* input is shown on the top line. The state of each flip-flop (FF 1, FF 2, FF 3, FF 4) is shown on the waveforms below. The *binary count* is shown across the bottom of the diagram. Especially note the vertical lines on

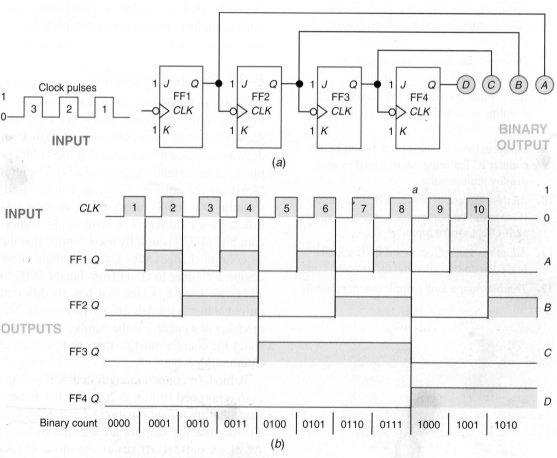

(a)

(b)

**Fig. 8-2** Mod-16 counter. (a) Logic diagram. (b) Waveform diagram.

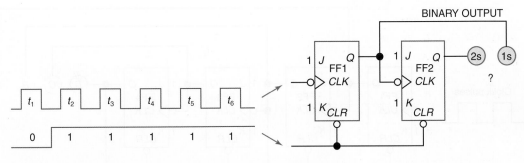

BINARY OUTPUT

**Fig. 8-3** Counter problem for test items 1 through 4.

Fig. 8-2(*b*); these lines show that the clock triggers only FF 1. FF 1 triggers FF 2, FF 2 triggers FF 3, and so on. Because one flip-flop affects the next one, it takes some time to toggle all the flip-flops. For instance, at point *a* on pulse 8, Fig. 8-2(*b*), notice that the clock triggers FF 1, causing it to go to 0. This in turn causes FF 2 to toggle from 1 to 0. This in turn causes FF 3 to toggle from 1 to 0. As output *Q* of FF 3 reaches 0, it triggers FF 4, which toggles from 0 to 1. We see that the changing of states is a chain reaction that

*ripples* through the counter. For this reason this counter is called a *ripple counter*.

The counter we studied in Fig. 8-2 could be described as a ripple counter, a mod-16 counter, a 4-bit counter, or an asynchronous counter. All these names describe something about the counter. The ripple and *asynchronous* labels mean that all the flip-flops do not trigger at one time. The mod-16 description comes from the number of states the counter cycles through. The *4-bit* label tells how many binary places there are at the output of the counter.

**Ripple counter**

**Asynchronous counter**

**4-bit counter**

**Self-Test**

*Answer the following questions.*

1. The unit in Fig. 8-3 is a(n) _____ -bit ripple counter.
2. The unit in Fig. 8-3 is a mod- _____ counter.

3. Each J-K flip-flop in Fig. 8-3 is in the _____ (hold, reset, set, toggle) mode because inputs *J* and *K* are both HIGH.
4. List the binary output after each of the six input pulses shown in Fig. 8-3.

## 8-2  Mod-10 Ripple Counters

The counting sequence for a mod-10 counter is from 0000 to 1001 (0 to 9 in decimal). This is down to the heavy line in Fig. 8-1. This *mod-10 counter,* then, has four place values: 8s, 4s, 2s, and 1s. This takes four flip-flops connected as a ripple counter in Fig. 8-4. We must *add* a NAND gate to the ripple counter to clear all the flip-flops back to zero *immediately after* the 1001 (9) count. The *trick* is to look at Fig. 8-1 and determine what the *next count will be after 1001.* You will find it is 1010 (decimal 10). You must feed the two 1s in the 1010 into a NAND gate as shown in

Fig. 8-4. The NAND gate then clears the flip-flop back to 0000. The counter then starts its count from 0000 up to 1001 again. We say we are using the NAND gate to reset the counter to 0000. By using a NAND gate in this manner, we can make several other modulous counters. Fig. 8-4 illustrates a mod-10 ripple counter. This type of counter might also be called a *decade* (meaning 10) *counter.*

Ripple counters can be constructed from individual flip-flops. Manufacturers also produce ICs with all four flip-flops inside a single package. Some IC counters even contain the reset NAND gate, such as the one you used in Fig. 8-4.

**Mod-10 ripple counter**

**Decade counter**

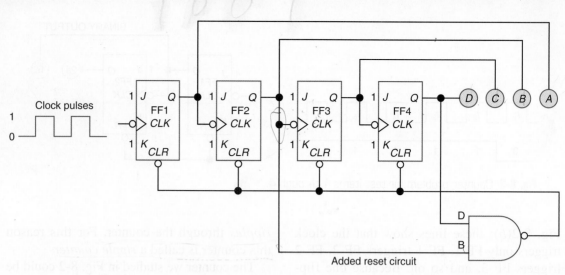

Fig. 8-4 Logic diagram for a mod-10 ripple counter.

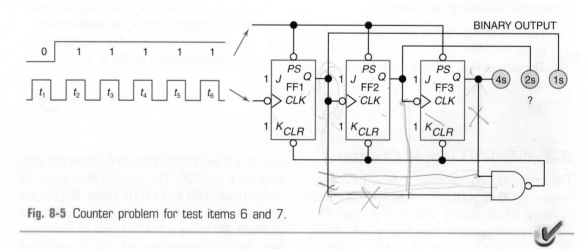

Fig. 8-5 Counter problem for test items 6 and 7.

## Self-Test

*Answer the following questions.*

5. Refer to Fig. 8-4. This is the logic diagram for a mod-10 _____ (ripple, synchronous) counter. Because it has 10 states (counts from 0 through 9), it is also called a(n) _____ counter.

6. The circuit in Fig. 8-5 is a _____ (ripple, synchronous) mod-_____ counter.

7. List the binary output after each of the six input pulses shown in Fig. 8-5.

---

## 8-3  Synchronous Counters

The ripple counters we have studied are asynchronous counters. Each flip-flop does not trigger exactly in step with the clock pulse. For some high-frequency operations, it is necessary to have all stages of the counter trigger together. There is such a counter: *a synchronous counter*.

A synchronous counter is shown in Fig. 8-6(*a*). This logic diagram is for a 3-bit (mod-8) counter. First notice the *CLK* connections. The clock is connected directly to the *CLK* input of

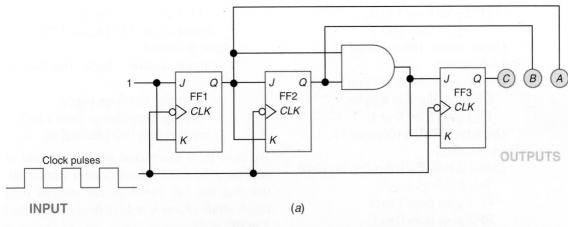

Clock pulses

INPUT

(a)

OUTPUTS

| ROW | NUMBER OF CLOCK PULSES | BINARY COUNTING SEQUENCE | | | DECIMAL COUNT |
|-----|------------------------|---------|---|---|---------------|
|     |                        | C | B | A |               |
| 1   | 0                      | 0 | 0 | 0 | 0             |
| 2   | 1                      | 0 | 0 | 1 | 1             |
| 3   | 2                      | 0 | 1 | 0 | 2             |
| 4   | 3                      | 0 | 1 | 1 | 3             |
| 5   | 4                      | 1 | 0 | 0 | 4             |
| 6   | 5                      | 1 | 0 | 1 | 5             |
| 7   | 6                      | 1 | 1 | 0 | 6             |
| 8   | 7                      | 1 | 1 | 1 | 7             |
| 9   | 8                      | 0 | 0 | 0 | 0             |

(b)

**Fig. 8-6** A 3-bit synchronous counter. (a) Logic diagram. (b) Counting sequence.

each flip-flop. We say that the *CLK* inputs are connected in *parallel*. Figure 8-6(b) gives the counting sequence of this counter. Column *A* is the binary 1s column, and FF 1 does the counting for this column. Column *B* is the binary 2s column, and FF 2 counts this column. Column *C* is the binary 4s column, and FF 3 counts this column.

Let us study the counting sequence of this mod-8 counter by referring to Fig. 8-6(a) and (b):

Pulse 1—row 2
   *Circuit action:* Each flip-flop is pulsed by the clock.
      Only FF 1 can toggle because it is the only one with 1s applied to both *J* and *K* inputs.
      FF 1 goes from 0 to 1.
   *Output result:* 001 (decimal 1).

Pulse 2—row 3
   *Circuit action:* Each flip-flop is pulsed.
      Two flip-flops toggle because they have 1s applied to both *J* and *K* inputs.
      FF 1 and FF 2 both toggle.
      FF 1 goes from 1 to 0.
      FF 2 goes from 0 to 1.
   *Output result:* 010 (decimal 2).

Pulse 3—row 4
   *Circuit action:* Each flip-flop is pulsed.
      Only one flip-flop toggles.
      FF 1 toggles from 0 to 1.
   *Output result:* 011 (decimal 3).

Pulse 4—row 5
   *Circuit action:* Each flip-flop is pulsed.
      All flip-flops toggle to opposite state.
      FF 1 goes from 1 to 0.

FF 2 goes from 1 to 0.
FF 3 goes from 0 to 1.
*Output result:* 100 (decimal 4).
Pulse 5—row 6
*Circuit action:* Each flip-flop is pulsed.
Only one flip-flop toggles.
FF 1 goes from 0 to 1.
*Output result:* 101 (decimal 5).
Pulse 6—row 7
*Circuit action:* Each flip-flop is pulsed.
Two flip-flops toggle.
FF 1 goes from 1 to 0.
FF 2 goes from 0 to 1.
*Output result:* 110 (decimal 6).
Pulse 7—row 8
*Circuit action:* Each flip-flop is pulsed.
Only one flip-flop toggles.

FF 1 goes from 0 to 1.
*Output result:* 111 (decimal 7).
Pulse 8—row 9
*Circuit action:* Each flip-flop is pulsed.
All three flip-flops toggle.
All flip-flops change from 1 to 0.
*Output result:* 000 (decimal 0).

We now have completed the explanation of how the *3-bit synchronous counter* works. Notice that the J-K flip-flops are used in their *toggle mode* (*J* and *K* at 1) or *hold mode* (*J* and *K* at 0).

Synchronous counters are most often purchased in IC form. Synchronous counters are available in both TTL and CMOS.

## Self-Test

*Supply the missing word in each statement.*

8. A counter that triggers all the flip-flops at the same instant is called a _____ (ripple, synchronous) counter.
9. Clock inputs are connected in _____ (parallel, series) on a synchronous counter.
10. Refer to Fig. 8-6(*a*). FF 1 is always in the _____ (hold, reset, set, toggle) mode in this circuit.

11. Refer to Fig. 8-6. On clock pulse 4, _____ (only FF 1 toggles; both FF 1 and FF 2 toggle; only FF 3 toggles; all the flip-flops toggle), producing a binary count of 100 at the outputs of the counter.
12. Refer to Fig. 8-6. The purpose of the AND gate is to place _____ (FF1, FF2, FF3) in the toggle mode two times during the counting cycle [rows 4 and 8 from Fig. 8-6(*b*)] while this flip-flop stays in the hold mode during other times.

## 8-4 Down Counters

Up to now we have used counters that count upward (0, 1, 2, 3, 4, . . .). Sometimes, however, we must count downward (9, 8, 7, 6, . . .) in digital systems. A counter that counts from higher to lower numbers is called a *down counter.*

A logic diagram of a *mod-8 asynchronous down counter* is shown in Fig. 8-7(*a*); the counting sequence for this counter is listed in Fig. 8-7(*b*). Note how much the down counter in Fig. 8-7(*a*) looks like the up counter in Fig.

8-2(*a*). The only difference is in the "carry" from FF 1 to FF 2 and the carry from FF 2 to FF 3. The up counter carries from *Q* to the *CLK* input of the next flip-flop. The down counter carries from $\overline{Q}$ (not *Q*) to the *CLK* input of the next flip-flop. Notice that the down counter has a preset (*PS*) control to preset the counter to 111 (decimal 7) to start the downward count. FF 1 is the binary 1s place (column *A*) counter. FF 2 is the 2s place (column *B*) counter. FF 3 is the 4s place (column *C*) counter.

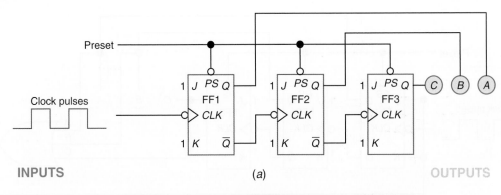

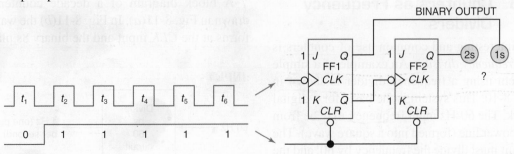

| NUMBER OF CLOCK PULSES | BINARY COUNTING SEQUENCE | | | DECIMAL COUNT |
|---|---|---|---|---|
| | C | B | A | |
| 0 | 1 | 1 | 1 | 7 |
| 1 | 1 | 1 | 0 | 6 |
| 2 | 1 | 0 | 1 | 5 |
| 3 | 1 | 0 | 0 | 4 |
| 4 | 0 | 1 | 1 | 3 |
| 5 | 0 | 1 | 0 | 2 |
| 6 | 0 | 0 | 1 | 1 |
| 7 | 0 | 0 | 0 | 0 |
| 8 | 1 | 1 | 1 | 7 |
| 9 | 1 | 0 | 0 | 6 |

(b)

**Fig. 8-7** A 3-bit ripple down counter. (a) Logic diagram. (b) Counting sequence.

## Self-Test

*Answer the following questions.*

13. Refer to Fig. 8-7(a). All flip-flops are in the _____ (hold, reset, set, toggle) mode in this counter.

14. Refer to Fig. 8-7(a). It takes a (HIGH-to-LOW, LOW-to-HIGH) transition of the clock pulse to trigger these J-K flip-flops.

15. Refer to Fig. 8-7. On clock pulse 1, _____ (only FF 1 toggles; both FF 1 and FF 2 toggle; only FF 3 toggles; all the flip-flops toggle) producing a binary count of 110 at the outputs of the counter.

16. List the counter's binary output for each of the six input pulses shown in Fig. 8-8.

**Fig. 8-8** Counter problem for test item 16.

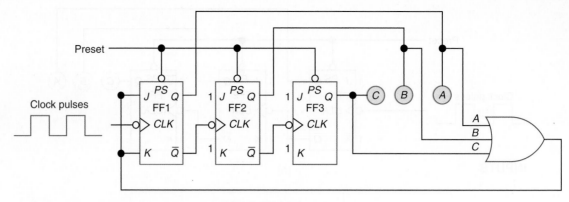

Fig. 8-9 A 3-bit down counter with self-stopping feature.

## 8-5 Self-Stopping Counters

The down counter shown in Fig. 8-7(*a*) *recirculates*. That is, when it gets to 000 it starts at 111, then 110, and so forth. However, sometimes you want a counter to *stop* when a sequence is finished. Figure 8-9 illustrates how you could stop the down counter in Fig. 8-7 at the 000 count. The counting sequence is shown in Fig. 8-7(*b*). In Fig. 8-9 we add an OR gate to place a logical 0 on the *J* and *K* inputs of FF 1 when the count at outputs *C*, *B*,

and *A* reaches 000. The preset must be enabled (*PS* to 0) again to start the sequence at 111 (decimal 7).

Up or down counters can be stopped after any sequence of counts by using a logic gate or combination of gates. The output of the gate is fed back to the *J* and *K* inputs of the first flip-flop in a ripple counter. The logical 0s fed back to the *J* and *K* inputs of FF 1 in Fig. 8-9 place it in the hold mode. This stops FF 1 from toggling, thereby stopping the count at 000.

✔ **Self-Test**

*Supply the missing word or words in each statement.*

17. Refer to Fig. 8-9. This is the logic diagram for a self-stopping 3-bit _____ (down, up) counter.

18. Refer to Fig. 8-9. With an output count of 000, the OR gate outputs a _____

(HIGH, LOW). This places FF 1 in the _____ (hold, toggle) mode.

19. Refer to Fig. 8-9. With an output count of 111, the OR gate outputs a _____ (HIGH, LOW). This places FF 1 in the _____ (hold, toggle) mode.

## 8-6 Counters as Frequency Dividers

An interesting and common use of counters is for *frequency division*. An example of a simple system using a frequency divider is shown in Fig. 8-10. This system is the basis for a digital clock. The 60-Hz input frequency may be from the power line (formed into a square wave). The circuit must divide the frequency by 60, and the output will be one pulse per second (1 Hz). This is a seconds timer.

A block diagram of a decade counter is drawn in Fig. 8-11(*a*). In Fig. 8-11(*b*) the waveforms at the *CLK* input and the binary 8s place

INPUT  OUTPUT

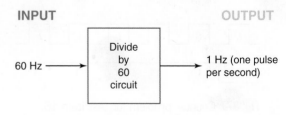

Fig. 8-10 A 1-second timer system.

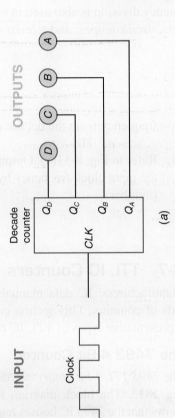

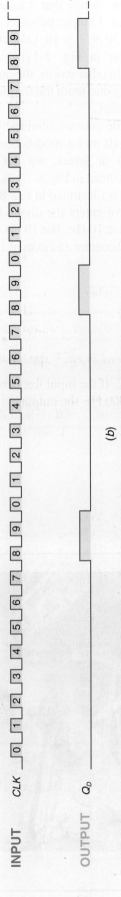

INPUT
Clock

OUTPUTS

D C B A

Decade
counter

$Q_D$ $Q_C$ $Q_B$ $Q_A$

CLK

(a)

INPUT CLK

OUTPUT $Q_D$

(b)

**Fig. 8-11** Decade counter used as a divide-by-10 counter. (a) Logic diagram. (b) Waveform diagram.

## 8-7 TTL IC Counters

Manufacturers' IC data manuals contain long lists of counters. This section covers only two representative types of TTL IC counters.

### The 7493 4-Bit Counter

(output $Q_D$) are shown. Notice that it takes 30 input pulses to produce 3 output pulses. Using division, we find that $30 \div 3 = 10$. Output $Q_D$ of the decade counter in Fig. 8-11(*a*) is a *divide-by-10* counter. In other words, the output frequency at $Q_D$ is only one-tenth the frequency at the input of the counter.

If we use the decade counter (divide-by-10 counter) from Fig. 8-10 and a mod-6 counter (divide-by-6 counter) in series, we get the *divide-by-60* circuit we need in Fig. 8-10. A diagram of such a system is illustrated in Fig. 8-12. The 60-Hz square wave enters the divide-by-6 counter and comes out at 10 Hz. The 10 Hz then enters the divide-by-10 counter and exits at 1 Hz.

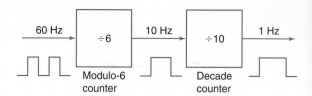

**Fig. 8-12** Practical divide-by-60 circuit used as a 1-second timer.

You are already aware that counters are used as frequency dividers in *digital* timepieces, such as electronic *digital clocks*, automobile digital clocks, and digital wristwatches. Frequency division is also used in *frequency counters, oscilloscopes,* and *television receivers.*

**Divide-by-60 circuit**

**Digital clock**

**Frequency counter**

**Oscilloscopes**

---

### ✔ Self-Test

*Supply the missing word in each statement.*

20. Refer to Fig. 8-12. If the input frequency on the left is 60,000 Hz, the output frequency from the decade counter is _____ Hz.

21. Refer to Fig. 8-11(*a*). Output A divides the input clock frequency by _____ (number).

---

**7493 TTL 4-bit binary counter**

## 8-7  TTL IC Counters

Manufacturers' IC data manuals contain long lists of counters. This section covers only two representative types of TTL IC counters.

### The 7493 4-Bit Counter

The *7493 TTL 4-bit binary counter* is detailed in Fig. 8-13. The block diagram in Fig. 8-13(*a*) shows that the 7493 IC houses four J-K flip-flops wired as a ripple counter. If you look carefully at Fig. 8-13(*a*), you will notice that the bottom three J-K flip-flops are prewired internally as a 3-bit ripple counter with output $Q_B$ connected to the clock input of the next lower J-K flip-flop and output $Q_C$ connected internally to the clock input of the bottom J-K flip-flop. Importantly, the top J-K flip-flop *does not* have its $Q_A$ output internally connected to the next lower flip-flop. To use the 7493 IC as a 4-bit ripple counter (mod-16), you have to *externally connect output $Q_A$ to input B,* which is the *CLK* input of the second flip-flop. A counting sequence for the 7493 IC wired as a 4-bit ripple counter is reproduced in Fig. 8-13(*c*). Consider the *J* and *K* inputs to each flip-flop in

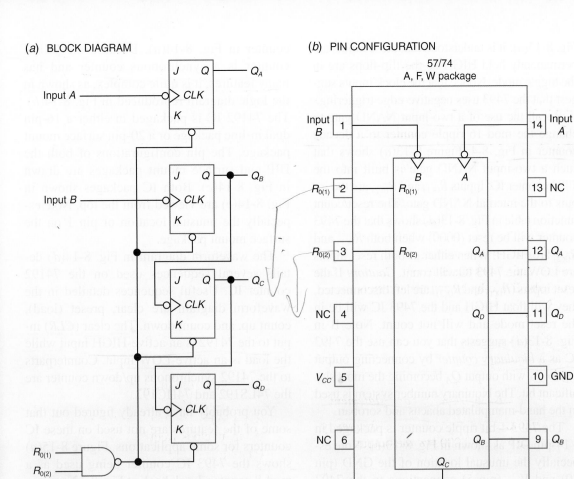

(b) PIN CONFIGURATION

57/74
A, F, W package

| | Pin | | | Pin | |
|---|---|---|---|---|---|
| Input B | 1 | | | 14 | Input A |
| $R_{0(1)}$ | 2 | $R_{0(1)}$ | | 13 | NC |
| $R_{0(2)}$ | 3 | $R_{0(2)}$ | $Q_A$ | 12 | $Q_A$ |
| NC | 4 | | $Q_D$ | 11 | $Q_D$ |
| $V_{CC}$ | 5 | | | 10 | GND |
| NC | 6 | | $Q_B$ | 9 | $Q_B$ |
| NC | 7 | $Q_C$ | | 8 | $Q_C$ |

The J and K inputs shown without connection for reference only and are functionally at a high level.

(c) COUNT SEQUENCE

| COUNT | OUTPUT | | | |
|---|---|---|---|---|
| | $Q_D$ | $Q_C$ | $Q_B$ | $Q_A$ |
| 0 | L | L | L | L |
| 1 | L | L | L | H |
| 2 | L | L | H | L |
| 3 | L | L | H | H |
| 4 | L | H | L | L |
| 5 | L | H | L | H |
| 6 | L | H | H | L |
| 7 | L | H | H | H |
| 8 | H | L | L | L |
| 9 | H | L | L | H |
| 10 | H | L | H | L |
| 11 | H | L | H | H |
| 12 | H | H | L | L |
| 13 | H | H | L | H |
| 14 | H | H | H | L |
| 15 | H | H | H | H |

Output $Q_A$ is connected to input B.

(d) RESET/COUNT FUNCTION TABLE

| RESET INPUTS | | OUTPUT | | | |
|---|---|---|---|---|---|
| $R_0 (1)$ | $R_0 (2)$ | $Q_D$ | $Q_C$ | $Q_B$ | $Q_A$ |
| H | H | L | L | L | L |
| L | X | Count | | | |
| X | L | Count | | | |

NOTES:
A. Output $Q_A$ is connected to input B for BCD count (or binary count).
B. Output $Q_D$ is connected to input A for biquinary count.
C. H = high level, L = low level, X = irrelevant.

Fig. 8-13 A 4-bit binary counter IC (7493). (a) Block diagram. (b) Pin configuration. (c) Count sequence. (d) Reset/count function table.

4-bit binary counter IC (7493)

Fig. 8-13(*a*): it is understood that these inputs are permanently held HIGH so the flip-flops are in the toggle mode. Notice that the clock inputs suggest that the 7493 uses negative-edge triggering.

Recall the use of a two-input NAND gate to change the mod-16 ripple counter to a decade counter in Fig. 8-4. Figure 8-13(*a*) shows that such a two-input NAND gate is built into the 7493 counter IC. Inputs $R_{0(1)}$ and $R_{0(2)}$ are the inputs to the internal NAND gate. The reset/count function table in Fig. 8-13(*d*) shows that the 7493 counter will be reset (0000) when both $R_{0(1)}$ and $R_{0(2)}$ are HIGH. When either or both reset inputs are LOW, the 7493 IC will count. *Caution:* If the reset inputs ($R_{0(1)}$ and $R_{0(2)}$) are left disconnected, they will float HIGH and the 7493 IC will be in the reset mode and will not count. Note B in Fig. 8-13(*d*) suggests that you can use the 7493 IC as a *biquinary counter* by connecting output $Q_D$ to $Q_A$ with output $Q_A$ becoming the most significant bit. The biquinary number system is used in the hand-manipulated abacus and soroban.

The 7493 4-bit ripple counter is packaged in a 14-pin DIP as shown in Fig. 8-13(*b*). Note especially the unusual location of the GND (pin 10) and $V_{CC}$ (pin 5) connections to the 7493 counter, which are commonly on the corners of the many ICs.

### The 74192 Up/Down Decade Counter

A second TTL IC counter is detailed in Fig. 8-14. It is the *74192 up/down decade counter IC*. Read the manufacturer's description of the IC counter in Fig. 8-14(*a*). Because the 74192 counter is a synchronous counter and has many features, it is quite complex, as shown in the logic diagram reproduced in Fig. 8-14(*b*). The 74192 IC is packaged in either a 16-pin dual in-line package or a 20-pin surface mount package. The pin configurations of both the DIP and surface mount packages are drawn in Fig. 8-14(*c*). Both IC packages shown in Fig. 8-14(*c*) are viewed from the top. Note especially the unusual location of pin 1 on the surface mount package.

The waveform diagram in Fig. 8-14(*d*) details several sequences used on the 74192 counter IC. Useful sequences detailed in the waveform diagram are clear, preset (load), count up, and count down. The clear (*CLR*) input to the 74192 is an active-HIGH input while the load is an active-LOW input. Counterparts to the 74192 synchronous up/down counter are the 74LS192 and 74HC192.

You probably have already figured out that some of the features are not used on these IC counters for some applications. Figure 8-15(*a*) shows the 7493 IC counter being used as a mod-8 counter. Look back at Fig. 8-13 and notice that several inputs and an output are not being used. Figure 8-15(*b*) shows the 74192 counter being used as a decade down counter. Six inputs and two outputs are not being used in this circuit. Simplified logic diagrams similar to those in Fig. 8-15 are more common than the complicated diagrams in Figs. 8-13(*a*) and 8-14(*b*).

**Biquinary counter**

**74192 up/down decade counter IC**

---

![Self-Test]

*Answer the following questions.*

22. Refer to Fig. 8-13. If both inputs to the NAND gate (pins 2 and 3 on the 7493 IC) are HIGH, the output from the 7493 counter will be _____ (4 bits).

23. Refer to Fig. 8-13. The 7493 IC is a(n) _____ -bit _____ (down, up) counter.

24. Refer to Fig. 8-14. The 74192 IC is a _____ (decade, mod-16) up/down _____ (ripple, synchronous) counter.

25. Refer to Fig. 8-14. The clock input to the 74192 for counting upward is pin _____ (number) on the DIP IC.

26. Refer to Fig. 8-14. The 74192 IC has an active _____ (HIGH, LOW) clear input.

27. List the output frequency at points *B, C,* and *D* in Fig. 8-16.

28. The 7493 IC is a ripple divide-by-2, divide-by-4, and divide-by- _____ unit in Fig. 8-16.

## (a) DESCRIPTION

This monolithic circuit is a synchronous reversible (up/down) counter having a complexity of 55 equivalent gates. Synchronous operation is provided by having all flip-flops clocked simultaneously so that the outputs change coincidentally with each other when so instructed by the steering logic. This mode of operation eliminates the output counting spikes which are normally associated with asynchronous (ripple-clock) counters.

The outputs of the four master-slave flip-flops are triggered by a low-to-high-level transition of either count (clock) input. The direction of counting is determined by which count input is pulsed while the other count input is high.

All four counters are fully programmable; that is, each output may be preset to either level by entering the desired data at the data inputs while the load input is low. The output will change to agree with the data inputs independently of the count pulses. This feature allows the counters to be used as modulo-N dividers by simply modifying the count length with the preset inputs.

A clear input has been provided which forces all outputs to the low level when a high level is applied. The clear function is independent of the count and load inputs. The clear, count, and load inputs are buffered to lower the drive requirements. This reduces the number of clock drivers, etc., required for long words.

These counters were designed to be cascaded without the need for external circuitry. Both borrow and carry outputs are available to cascade both the up- and down-counting functions. The borrow output produces a pulse equal to the count-down input when the counter underflows. Similarly, the carry output produces a pulse equal in width to the count-up input when an overflow condition exists. The counters can then be easily cascaded by feeding the borrow and carry outputs to the count-down and count-up inputs respectively of the succeeding counter.

## (b) LOGIC DIAGRAM

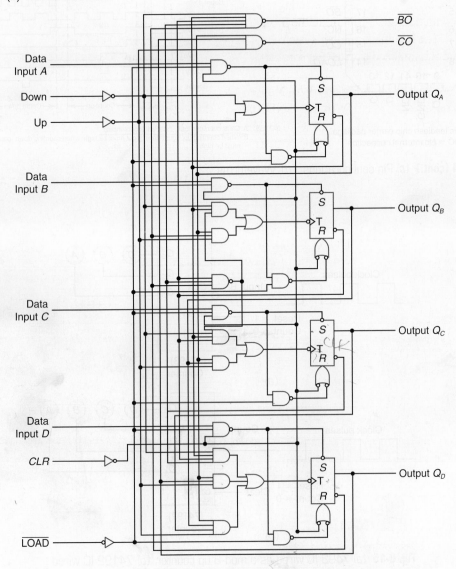

**Fig. 8-14** Synchronous decade up/down counter IC (74192). (a) Description. (b) Logic diagram.

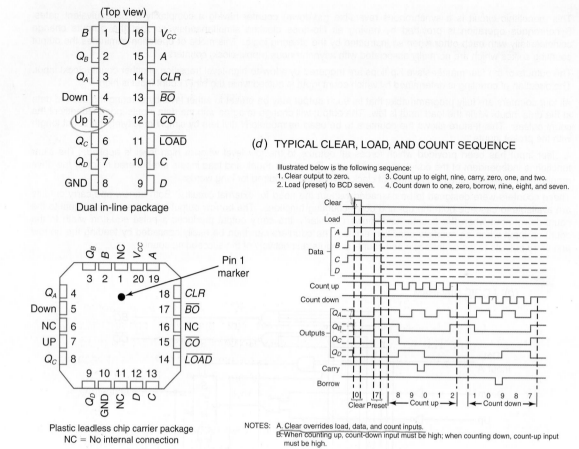

(d) TYPICAL CLEAR, LOAD, AND COUNT SEQUENCE

Illustrated below is the following sequence:
1. Clear output to zero.
2. Load (preset) to BCD seven.
3. Count up to eight, nine, carry, zero, one, and two.
4. Count down to one, zero, borrow, nine, eight, and seven.

NOTES: A. Clear overrides load, data, and count inputs.
B. When counting up, count-down input must be high; when counting down, count-up input must be high.

Dual in-line package

Plastic leadless chip carrier package
NC = No internal connection

**Fig. 8-14 (cont.)** (c) Pin configurations. (d) Waveforms.

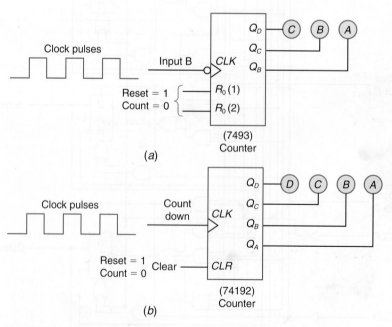

**Mod-8 counter**

**Decade down counter**

**Fig. 8-15** (a) 7493 IC wired as a mod-8 up counter. (b) 74192 IC wired as a decade down counter.

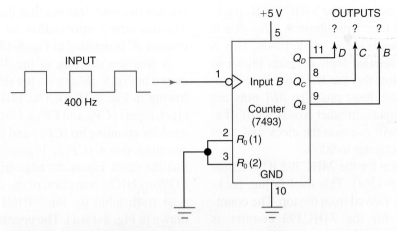

**Fig. 8-16** Counter problem for test items 27 and 28.

## 8-8 CMOS IC Counters

Manufacturers of CMOS chips offer a variety of counters in IC form. This section covers only two types of CMOS counters.

### The 74HC393 4-Bit Binary Counter

The diagrams in Fig. 8-17 detail a *74HC393 dual 4-bit binary ripple counter*. A *function diagram* (something like a logic diagram) of the 74HC393 counter IC is shown in Fig. 8-17(*a*). Note that the IC contains two 4-bit binary ripple counters. The table in Fig. 8-17(*b*) gives the names and functions of each input and output pin on the 74HC393 IC. Note that the clock inputs are labeled with the letters $\overline{CP}$ instead of *CLK*, as used earlier. Pin labels vary from manufacturer to manufacturer. For this reason, you must learn to use manufacturer's data sheets for exact information.

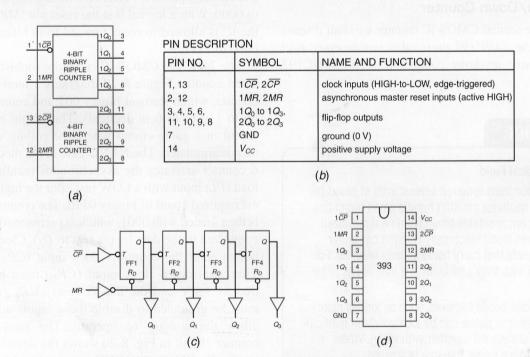

**Fig. 8-17** CMOS dual 4-bit binary counter IC (74HC393). (*a*) Function diagram. (*b*) Pin descriptions. (*c*) Detailed logic diagram. (*d*) Pin diagram.

### PIN DESCRIPTION

| PIN NO. | SYMBOL | NAME AND FUNCTION |
|---|---|---|
| 1, 13 | $1\overline{CP}$, $2\overline{CP}$ | clock inputs (HIGH-to-LOW, edge-triggered) |
| 2, 12 | 1MR, 2MR | asynchronous master reset inputs (active HIGH) |
| 3, 4, 5, 6, 11, 10, 9, 8 | $1Q_0$ to $1Q_3$, $2Q_0$ to $2Q_3$ | flip-flop outputs |
| 7 | GND | ground (0 V) |
| 14 | $V_{CC}$ | positive supply voltage |

(*b*)

_stopped Here_

**T flip-flop**

Each 4-bit counter in the 74HC393 IC package consists of four T flip-flops. A *T flip-flop* is any flip-flop that is in the toggle mode. This is shown in the detailed logic diagram drawn in Fig. 8-17(c). Note that the *MR* input is an asynchronous master reset pin. The *MR* pins are active HIGH inputs. In other words, a HIGH at the *MR* input will override the clock and reset the individual counter to 0000.

A pin diagram for the 74HC393 IC is reproduced in Fig. 8-17(d). This dual in-line package IC is being viewed from the top. The counting sequence for the 74HC393 counter is binary 0000 through 1111 (0 to 15 in decimal).

**Modes of operation for 74HC193 counter**

The functional diagram in Fig. 8-17(a) and logic diagram in Fig. 8-17(c) both suggest that the counters are triggered on the HIGH-to-LOW transition of the clock pulse. The outputs ($Q_0$, $Q_1$, $Q_2$, $Q_3$) of the ripple counter are asynchronous (not exactly in step with the clock). As with all ripple counters, there is a slight delay in outputs because the first flip-flop triggers the second, the second the third, and so forth. Note that the > symbol at the clock ($\overline{CP}$ inputs) has been omitted by this manufacturer. Again, many variations occur in both labels and logic diagrams from manufacturer to manufacturer.

**74HC193 presettable synchronous 4-bit binary up/down counter IC**

## The 74HC193 4-Bit Binary Up/Down Counter

The second CMOS IC counter we shall discuss is the *74HC193 presettable synchronous 4-bit binary up/down counter IC.* The 74HC193

counter has more features than the 74HC393 IC. Manufacturer's information on the 74HC193 counter IC is detailed in Fig. 8-18.

A function diagram of the 74HC193 IC is drawn in Fig. 8-18(a) with pin descriptions following in Fig. 8-18(b). The 74HC193 has two clock inputs ($CP_U$ and $CP_D$). One clock input is used for counting up ($CP_U$) and the other when counting down ($CP_D$). Figure 8-18(b) notes that the clock inputs are edge-triggered on the LOW-to-HIGH transition of the clock pulse.

A truth table for the 74HC193 counter is drawn in Fig. 8-18(d). The operating modes for the counter on the left give an overview of the many functions of the 74HC193 counter. Its modes of operation are *reset, parallel load, count up,* and *count down.* The truth table in Fig. 8-18(d) also makes it clear which pins are inputs and which are outputs.

Typical clear (reset), preset (parallel load), count up, and count down sequences are shown in Fig. 8-18(e). Waveforms are useful when investigating an IC's typical operations or timing.

Figures 8-19 and 8-20 show two possible applications for the CMOS counter ICs studied in this section. Figure 8-19 shows a logic diagram for a 74HC393 IC wired as a simple 4-bit binary counter. The *MR* (master reset) pin must be tied to either 0 or 1. The *MR* input is an active HIGH input so a 1 clears the binary outputs to 0000. With a logical 0 at the reset pin (*MR*), the IC is allowed to count upward from binary 0000 to 1111.

The 74HC193 CMOS IC is a more sophisticated counter. Figure 8-20 diagrams a mod-6 counter, which starts at binary 001 and counts up to 110 (1 to 6 in decimal). This might be useful in a game circuit where the rolling of dice is simulated. The NAND gate in the mod-6 counter activates the asynchronous parallel load ($\overline{PL}$) input with a LOW just after the highest required count of binary 0110. The counter is then loaded with 0001, which is permanently connected to the data inputs ($D_0$ to $D_3$). Clock pulses enter the count-up clock input ($CP_U$). The count-down clock input ($CP_D$) must be tied to +5 V and the master reset (*MR*) pin must be grounded to disable these inputs and allow the counter to operate. The mod-6 counter circuit in Fig. 8-20 shows the flexibility of the CMOS 74HC193 presettable 4-bit up/down counter IC.

---

**ABOUT ELECTRONICS**

**Devices in the Medical Field**

- In the past, blood tests required several vials of blood because testing machines couldn't handle small quantities. A new method encapsulates blood within a dime-sized "vial" and moves blood electrically within a computer chip with channels that carry liquid instead of wires. For this procedure, less than a billionth of a liter needs to be sampled.

- Deeply embedded bodily ailments, such as kidney disorders, that cause scar tissue can be located when medical personnel use ultrasound together with touch. When organs do not move freely, the area is scarred.

---

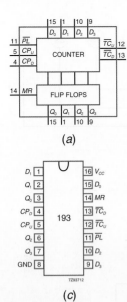

(a)

(c)

### PIN DESCRIPTION

| PIN NO. | SYMBOL | NAME AND FUNCTION |
|---|---|---|
| 3, 2, 6, 7 | $Q_0$ to $Q_3$ | flip-flop outputs |
| 4 | $CP_D$ | count down clock input* |
| 5 | $CP_U$ | count up clock input* |
| 8 | GND | ground (0 V) |
| 11 | $\overline{PL}$ | asynchronous parallel load input (active LOW) |
| 12 | $\overline{TC}_U$ | terminal count up (carry) output (active LOW) |
| 13 | $\overline{TC}_D$ | terminal count down (borrow) output (active LOW) |
| 14 | MR | asynchronous master reset input (active HIGH) |
| 15, 1,10, 9 | $D_0$ to $D_3$ | data inputs |
| 16 | $V_{CC}$ | positive supply voltage |

\* LOW-to-HIGH, edge triggered

(b)

| OPERATING MODE | INPUTS | | | | | | | | OUTPUTS | | | | | |
|---|---|---|---|---|---|---|---|---|---|---|---|---|---|---|
| | MR | $\overline{PL}$ | $CP_U$ | $CP_D$ | $D_0$ | $D_1$ | $D_2$ | $D_3$ | $Q_0$ | $Q_1$ | $Q_2$ | $Q_3$ | $\overline{TC}_U$ | $\overline{TC}_D$ |
| reset (clear) | H | X | X | L | X | X | X | X | L | L | L | L | H | L |
| | H | X | X | H | X | X | X | X | L | L | L | L | H | H |
| parallel load | L | L | X | L | L | L | L | L | L | L | L | L | H | L |
| | L | L | X | H | L | L | L | L | L | L | L | L | H | H |
| | L | L | L | X | H | H | H | H | H | H | H | H | L | H |
| | L | L | H | X | H | H | H | H | H | H | H | H | H | H |
| count up | L | H | ↑ | H | X | X | X | X | count up | | | | H* | H |
| count down | L | H | H | ↑ | X | X | X | X | count down | | | | H | H** |

\* $\overline{TC}_U$ = $CP_U$ at terminal count up (HHHH)  
\*\* $\overline{TC}_D$ = $CP_D$ at terminal count down (LLLL)

H = HIGH voltage level  
L = LOW voltage level  
X = don't care  
↑ = LOW-to-HIGH clock transition

(d)

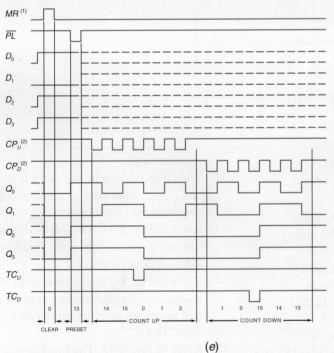

(1) Clear overrides load, data and count inputs.  
(2) When counting up the count down clock input ($CP_D$) must be HIGH, when counting down the count up clock input ($CP_U$) must be HIGH.

**Sequence**  
Clear (reset outputs to zero);  
load (preset) to binary thirteen;  
count up to fourteen, fifteen,  
   terminal count up, zero, one  
   and two;  
count down to one, zero,  
   terminal count down, fifteen,  
   fourteen and thirteen.

(e)

**Fig. 8-18** CMOS presettable 4-bit synchronous up/down counter IC (74HC193). (a) Function diagram. (b) Pin descriptions. (c) Pin diagram. (d) Truth table. (e) Typical clear, preset, and count sequence.

CMOS presettable 4-bit synchronous up/down counter IC (74HC193)

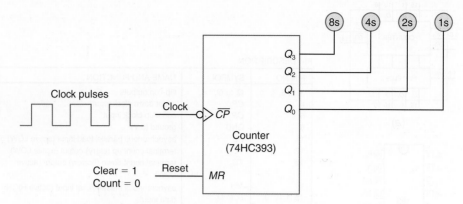

**4-bit binary counter**

Fig. 8-19 A 74HC393 IC wired as a 4-bit binary counter.

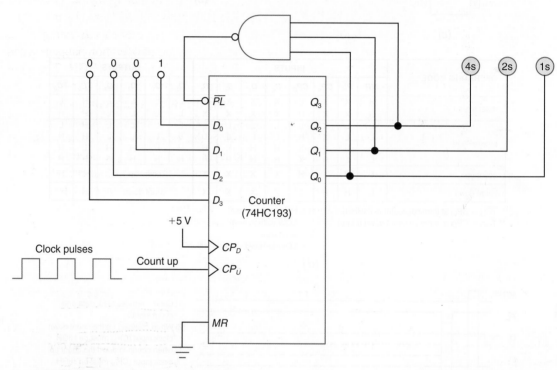

**Mod-6 counter**

Fig. 8-20 A 74HC193 IC wired as a mod-6 counter.

### ✔ Self-Test

*Answer the following questions.*

29. Refer to Fig. 8-17. The 74HC393 IC contains two _____ (4-bit binary, decade) counters.
30. Refer to Fig. 8-17. The reset pin ($MR$) on the 74HC393 counter is an active _____ (HIGH, LOW) input.
31. Refer to Fig. 8-17. The 74HC393 counter's clock inputs are triggered by the _____ (H-to-L, L-to-H) transition of the clock pulse.

32. The circuit drawn in Fig. 8-19 is a mod-_____ (number) _____ (ripple, synchronous) counter.
33. Refer to Fig. 8-18. The 74HC193 is a presettable _____ (ripple, synchronous) 4-bit up/down counter IC.
34. Refer to Fig. 8-18. The reset pin ($MR$) is _____ (asynchronous, synchronous) and overrides all other inputs on the 74HC193 IC.
35. Refer to Fig. 8-18. The outputs of the 74HC193 are labeled _____ ($D_0 - D_3$, $Q_0 - Q_3$).

36. Refer to Fig. 8-20. List the binary counting sequence for this counter circuit.

37. Refer to Fig. 8-20. What is the purpose of the three-input NAND gate in this counter circuit?

38. Refer to Fig. 8-17(a). How do you explain the lack of the > symbol near the clock inputs since the 74HC393 counters are edge-triggered?

## 8-9 A 3-Digit BCD Counter

Historically more and more electronic functions are being embedded in a single IC. The 3-digit BCD counter IC will demonstrate this trend and also feature some devices that you have studied.

The *4553 (MC14553) CMOS 3-Digit BCD Counter* will be featured in this section. A simplified functional block diagram of the 4553 IC is sketched in Fig. 8-21(a). You will notice that the 4553 IC contains three cascaded decade counters. Cascading counters means that the 1s BCD counter triggers the 10s counter as it recirculates from $1001_{BCD}$ to $0000_{BCD}$. In like manner, the 10s counter triggers the 100s counter as it recirculates from $1001_{BCD}$ to $0000_{BCD}$. The BCD output from the three counters are fed through the three 4-bit transparent latches. The BCD data are then transferred to a *display multiplexer* circuit. The display multiplexing circuit will drive three seven-segment displays.

The 4553 BCD counter IC detailed in Fig. 8-21(a) also features a pulse shaper circuit to square up the incoming clock pulses. The *CLK* (clock) input to the 4553 is negative-edge triggered. The display multiplexer circuitry turns on just one of the three decimal displays at a time feeding the correct BCD output to the display. The multiplexing should occur at a rate of about 40- to 80-Hz. An external capacitor ($C_1$) can be attached between the $C_{1A}$ and $C_{1B}$ pins of the IC to set the scan oscillator frequency. Capacitor $C_1$ would typically have a value of about 0.001uF.

The DISABLE clock, MASTER RESET, and LATCH ENABLE inputs to the 4553 counter IC are all active-HIGH inputs and the 4-bit BCD outputs ($Q_0 - Q_3$) are active-HIGH. The digit select ($DS_1$, $DS_2$, and $DS_3$) pins are active-LOW outputs.

A truth table drawn in Fig. 8-21(b) for the 4553 3-digit BCD counter IC shows a few of the modes of operation. These modes of operation are the most useful, but several other combinations of inputs are possible. When the MR input pin goes HIGH, the outputs are reset to $0000\ 0000\ 0000_{BCD}$. The Master Reset mode of operation is shown in line 1 of the truth table in

Fig. 8-21(b). The Count Up mode of operation is detailed in line 2 of the truth table. On the HIGH-to-LOW transition of the clock pulse, the BCD count will advance by 1. Notice from Fig. 8-21(a) that only the 1s counter is triggered by the input clock pulses. The 10s counter is triggered by an output from the 1s counter while the 100s counter is triggered by the output from the 10s counter (called *cascading*). The Disable Clock mode of operation happens when the DISABLE input pin goes HIGH. The input clock pulses are not permitted to reach the 1s counter and the BCD output remains the same.

The three BCD counters are constructed of 12 T-flop-flops that have a memory characteristic. A second layer of memory is provided in the 4553 IC in the form of three 4-bit transparent latches. When the LE (latch enable) input to the 4553 IC is LOW, the three latches pass data directly through to the multiplexer as suggested in the functional block diagram in Fig. 8-21(a). When the latches pass data through, they are said to be transparent. When the LE (latch enable) input to the 4553 IC is activated by HIGH, the last count from the three BCD counter is latched at the inputs to the display multiplexer. It is important to understand that the BCD counters can continue to count upwards even when the LE (latch enable) input is active. However, the BCD output will display the former count frozen in the latches.

One simple application of the 4553 3-digit BCD counter IC is sketched in Fig. 8-22. After activating the MR (master reset) input, the 4553 IC counts the number of input pulses and accumulates the count. The display multiplexer activates one seven-segment LED display at a time in rapid succession. First, as the 1s LED display is turned on by a LOW output from the $\overline{DS1}$ output from the 4553 IC, the correct BCD data from the 1s counter are sent to the 4543 decoder and are translated into seven-segment code lighting the appropriate segments. Second, as the 10s LED display is turned on by a LOW output from the $\overline{DS2}$ output from the 4553 IC, the new BCD data from the 10s counter are decoded by the 4543 IC and the 10s LED display lights. Third,

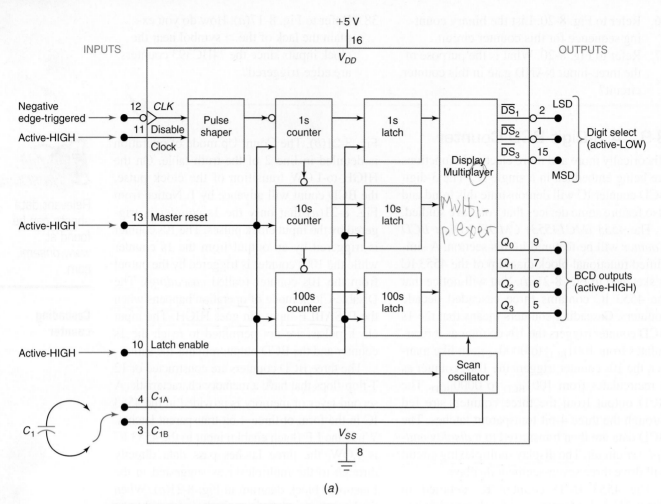

+5 V

16

$V_{DD}$

INPUTS

OUTPUTS

Negative edge-triggered →  12  *CLK*

Active-HIGH →  11  Disable

Clock

Pulse shaper

1s counter

1s latch

$\overline{DS}_1$  2  LSD

$\overline{DS}_2$  1

$\overline{DS}_3$  15  MSD

Display Multiplayer

Multi-plexer

Digit select (active-LOW)

Active-HIGH →  13  Master reset

10s counter

10s latch

Active-HIGH →  10  Latch enable

100s counter

100s latch

$Q_0$  9

$Q_1$  7

$Q_2$  6

$Q_3$  5

BCD outputs (active-HIGH)

Scan oscillator

$C_1$

4  $C_{1A}$

3  $C_{1B}$

$V_{SS}$

8

(a)

Partial Truth Table —4553 3-Digit BCD Counter IC

| Mode of operation | INPUTS | | | | OUTPUTS |
|---|---|---|---|---|---|
| | MR | CLK | DIS | LE | |
| Master reset | 1 | X | X | 0 | 0000 0000 0000$_{BCD}$ |
| Count up | 0 | ↓ | 0 | 0 | Advance count by 1 |
| Disable clock | 0 | X | 1 | 0 | No change |
| Latch outputs | 0 | X | X | 1 | Latches BCD data |
| | Master reset | Clock | Disable the Clock | Latch enable | |

0 = LOW
1 = HIGH
↓ = HIGH-to-LOW transition of clock pulse
X = Irrelevant

(b)

**Fig. 8-21** 4553 3-digit BCD counter IC. (a) Functional block diagram. (b) Partial truth table.

as the 100s LED display is turned on by a LOW output from the $\overline{DS3}$ output of the 4553 IC, the new BCD data from the 100s counter are decoded by the 4543 IC and the 100s LED display lights. The multiplexer section inside the 4553 IC turns only a single display on at a time in rapid succession. To the human eye, the multiplexed seven-segment displays will appear as being lit continuously even if they are being turned ON and OFF many times per second.

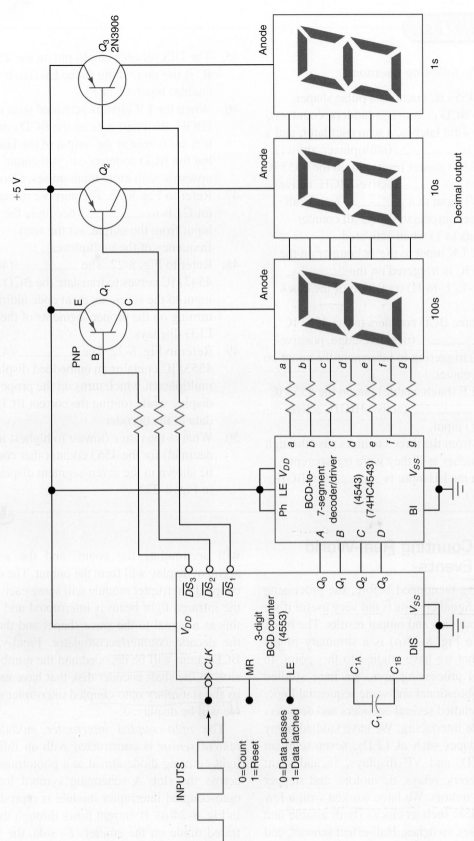

**Fig. 8-22** A 3-digit up counter circuit.

Answer the following questions.

39. The 4553 IC contains a pulse shaper, three BCD _____ (adders, counters), three 4-bit latches, a scan oscillator, and a display _____ (multiplexer, shifter).

40. The MR (master reset) pin on the 4553 IC is a(n) _____ (active-HIGH, active-LOW) input that _____ (resets all counter outputs to 0, sets all counter outputs to 1) when activated.

41. The CLK input to the 1s counter on the 4553 IC is triggered on the _____ (H-to-L, L-to-H) transition of the clock pulse.

42. All three BCD counters in the 4553 IC use _____ (negative-edge, positive-edge) triggering to increment the count on that counter.

43. The LE (latch enable) pin on the 4553 IC is an _____ (active-HIGH, active-LOW) input.

44. Data from the three counters pass through the latches as if they were transparent when the LE input is _____ (HIGH, LOW).

45. The DIS (disable clock) pin on the 4553 IC is the same thing as the LE (latch enable) input. (T or F)

46. When the LE input is activated with a HIGH, the latest data on the BCD counters are frozen at the output of the latches, but the BCD counters can still count upwards with more input pulses. (T or F)

47. Refer to Fig. 8-22. The purpose of capacitor C1 is to _____ (decouple the input from the output, set the scan frequency of the multiplexer).

48. Refer to Fig. 8-22. The _____ (4000, 4543) IC serves to translate the BCD input to the seven-segment code aiding in turning on the proper segments of the LED displays.

49. Refer to Fig. 8-22. The _____ (4543, 4553) IC contains an embedded display multiplexer, which turns on the proper display while routing the correct BCD data to the decoder.

50. What is the range (lowest to highest in decimal) for the 4553 counter that could be shown in the seven-segment displays in Fig. 8-22?

## 8-10 Counting Real-World Events

As we have mentioned before, the processing power of digital circuits is not very useful if we cannot input data and output results. The block diagram in Fig. 8-23(a) is a summary of the systems that we have studied to this point. In the digital processing area, we have studied some combinational and some sequential logic. We have studied several encoders and decoders that handle interfacing. We have studied many output devices such as LEDs, seven-segment LED, LCD, and VF displays, incandescent bulbs, buzzers, relays, dc motors, and stepper and servo motors. We have worked with a few input devices such as clocks (both astable and monostable), switches, Hall-effect sensors, and pulse-width modulators. In this section we will add a new input device.

A block diagram for the system we are studying in this section is drawn in Fig. 8-23(b). We will use *optical encoding* for input, the counter

will accumulate the count, and the seven-segment display will form the output. The opto-coupled interrupter module will sense each time the infrared light beam is interrupted and send this as a signal to the wave shaper and then to the decade counter/accumulator. Finally, the BCD count will be decoded and the number of slots in the shaft encoder disk that have moved by the stationary opto-coupled interrupter module will be displayed.

The *opto-coupled interrupter module* or *optical sensor* is constructed with an infrared light-emitting diode aimed at a phototransistor across the slot. A schematic symbol for the opto-coupled interrupter module is reproduced in Fig. 8-24(a). If current flows through the infrared diode on the emitter (E) side, the NPN phototransistor is activated on the detector (D) side of the module. If the light from the LED is blocked, the phototransistor on the detector side of the module is deactivated (turned off). The H21A1 (ECG3100) opto-coupled interrupter

**Optical sensor**

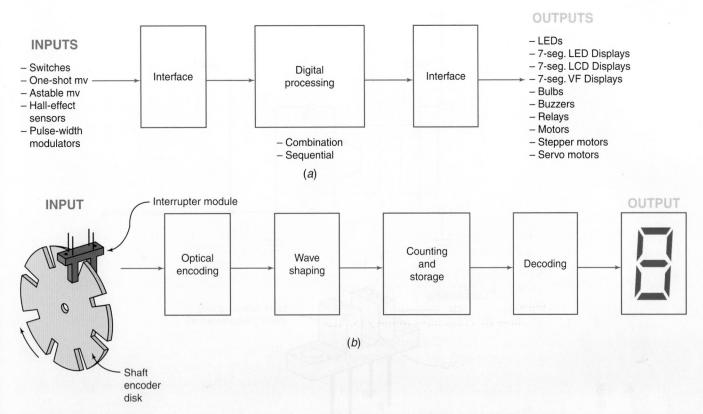

**INPUTS**

- Switches
- One-shot mv
- Astable mv
- Hall-effect sensors
- Pulse-width modulators

Interface → Digital processing → Interface

- Combination
- Sequential

**OUTPUTS**

- LEDs
- 7-seg. LED Displays
- 7-seg. LCD Displays
- 7-seg. VF Displays
- Bulbs
- Buzzers
- Relays
- Motors
- Stepper motors
- Servo motors

(a)

**INPUT**

Interrupter module

Shaft encoder disk

Optical encoding → Wave shaping → Counting and storage → Decoding

**OUTPUT**

(b)

**Fig. 8-23** (a) Typical inputs, processing, and outputs from a digital system. (b) Interrupter module optically encoding disk that drives counter.

module is shown in Fig. 8-24(b). Notice that pins 1 and 2 of the H21A1 interrupter module are for the emitter side or infrared light-emitting diode. Typical wiring of the emitter side of the interrupter module is also detailed in Fig. 8-24(a). Pins 3 and 4 of the H21A1 interrupter module are for the detector side or NPN phototransistor. Typical wiring of the detector side of the interrupter module using a 10-kΩ pull-up resistor is also shown in Fig. 8-24(a). The signal from the detector side of the interrupter module is then sent on to the wave shaping circuit.

A wiring diagram for a simple system that counts the number of pulses coming from the opto-coupled interrupter module is detailed in Fig. 8-25. When an opaque object interrupts the light beam in the module, the phototransistor is deactivated (turned off) and the input to the 7414 Schmitt trigger inverter is pulled HIGH by the 10-k pull-up resistor. The output of the inverter goes from HIGH to LOW. When the opaque object is removed from the slot of the opto-coupled interrupter module, the infrared light crosses the slot striking the base of the phototransistor. The phototransistor is activated (turned on) and the voltage at pin 3 of the inverter goes from HIGH to LOW. Then the

output of the wave shaper goes from LOW to HIGH, which will trigger the 74192 to count upward by 1. The 7447 IC decodes the BCD input to a seven-segment code and lights the appropriate segments on the LED display.

In summary, the optical encoder/counter system in Fig. 8-23(b) *increments* (one count upward) the count each time an opening in the encoder disk passes through the slot in the interrupter module. The opto-coupled interrupted module uses infrared light so the ambient light does not cause false triggering. Remember that the infrared diode emits the correct wavelength of light for the phototransistor to detect.

Two common types of optical sensors are the slot-type module used in the last optical encoder/counter system and the *reflective-type sensor.* A sketch of a common reflective-type optical sensor is shown in Fig. 8-26(a). Notice that it has two holes in the front. One is an infrared emitting diode while the other is the receiver part of the optical sensor, which is a phototransistor. This reflective-type optical sensor is carefully aimed at a target such as the disk shown in Fig. 8-26(b). The white areas reflect light and turn on the output phototransistor while the dark stripes absorb light and turn off the phototransistor.

**Types of optical sensors**

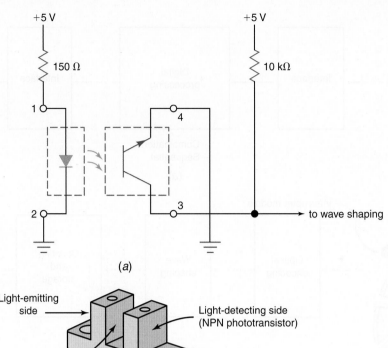

(a)

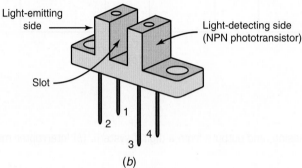

(b)

**Opto-coupled interrupter module**

Fig. 8-24 (a) Schematic diagram of opto-coupled interrupter module's emitter and light detecting sides wired. (b) Drawing of H21A1 opto-coupled interrupter module (slot type).

## Self-Test

*Supply the missing word or words in each statement.*

51. Refer to Fig. 8-23(b). The _____ (decoder, interrupter module) is the device that does the job of optical encoding in this circuit.
52. Refer to Fig. 8-24(a). The diode on the emitter side of the interrupter module gives off _____ (infrared, ultraviolet) light.
53. Refer to Fig. 8-24(a). The _____ (phototransistor, germanium transistor) on the detector side of the interrupter module is sensitive to infrared light and does not trigger because of white room light.

54. Refer to Fig. 8-23(b). The optical sensor being used in this system is classified as a _____ (reflective-type, slot-type) module.
55. Refer to Fig. 8-25. The count on the display increments when the output of the Schmitt trigger inverter goes from _____ (H to L, L to H).
56. Refer to Fig. 8-23(b). The count on the display increments when the encoder disk opening just _____ (enters, leaves) the interrupter module.
57. Refer to Fig. 8-25. The 74192 operates as a _____ (decade, mod-16) counter in this circuit and it also performs the task of _____ (decoding the count, temporarily storing the count).

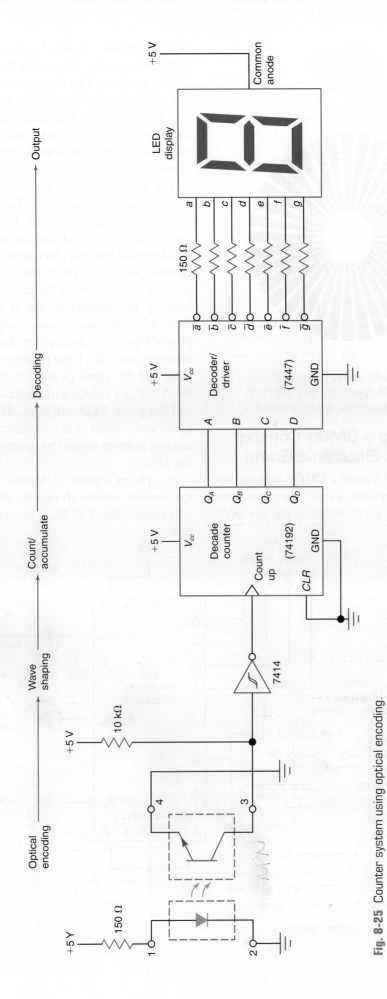

**Fig. 8-25** Counter system using optical encoding.

**Optical encoding**

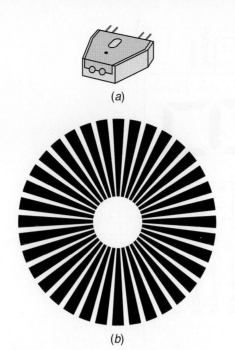

(a)

(b)

74HC85 4-bit magnitude comparator

Fig. 8-26 (a) Reflective-type optical encoder. (b) Shaft encoder disk used with reflective-type optical encoder.

## 8-11 Using a CMOS Counter in an Electronic Game

This section will feature a CMOS counter being used in an electronic game. The game is the classic computer game of "guess the number."

In the computer version, a random number is generated and the player tries to guess the unknown number. The computer responds with one of three responses: correct, too high, or too low. The player can then guess again until he or she zeros in on the unknown number. The player who uses the fewest guesses wins the game.

The schematic for a simple electronic version of this game is drawn in Fig. 8-27. To operate the game, first press the push-button switch ($SW_1$). This allows the approximately 1-kHz signal into the clock input of the binary counter. When the push button is released, a random binary number (from 0000 to 1111) is held by the counter at the $B$ inputs to the *74HC85 4-bit magnitude comparator*. The player's guess is entered at the $A$ inputs to the comparator IC. If the random number ($B$ inputs) and the guess ($A$ inputs) are equal, then the $A = B_{OUT}$ output will be activated (HIGH) and the green LED will light. This means the guess was correct. After a correct guess a new random number should be generated by pressing $SW_1$.

If a player's guess ($A$ inputs) is lower than the random number ($B$ inputs), the comparator will activate the $A < B_{OUT}$ output. The yellow

Electronic "guess the number" game

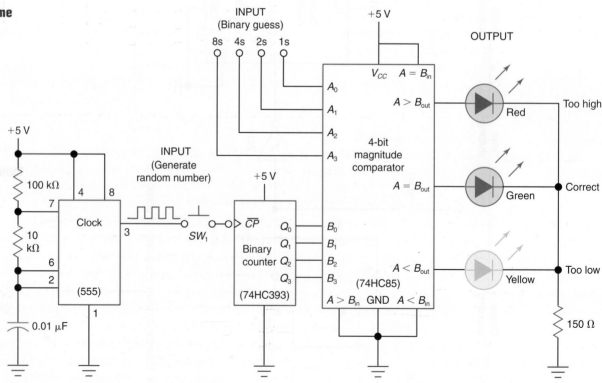

Fig. 8-27 Electronic "guess the number" game.

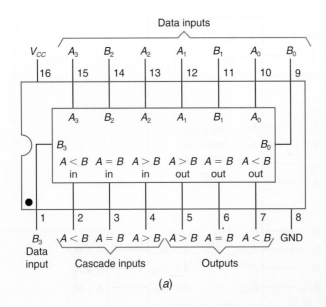

Data inputs

(a)

Truth Table—74HC85 Magnitude Comparator IC

| COMPARING INPUTS | | | | CASCADING INPUTS | | | OUTPUTS | | |
|---|---|---|---|---|---|---|---|---|---|
| $A_3, B_3$ | $A_2, B_2$ | $A_1, B_1$ | $A_0, B_0$ | $A > B$ | $A < B$ | $A = B$ | $A > B$ | $A < B$ | $A = B$ |
| $A_3 > B_3$ | X | X | X | X | X | X | H | L | L |
| $A_3 < B_3$ | X | X | X | X | X | X | L | H | L |
| $A_3 = B_3$ | $A_2 > B_2$ | X | X | X | X | X | H | L | L |
| $A_3 = B_3$ | $A_2 < B_2$ | X | X | X | X | X | L | H | L |
| $A_3 = B_3$ | $A_2 = B_2$ | $A_1 > B_1$ | X | X | X | X | H | L | L |
| $A_3 = B_3$ | $A_2 = B_2$ | $A_1 < B_1$ | X | X | X | X | L | H | L |
| $A_3 = B_3$ | $A_2 = B_2$ | $A_1 = B_1$ | $A_0 > B_0$ | X | X | X | H | L | L |
| $A_3 = B_3$ | $A_2 = B_2$ | $A_1 = B_1$ | $A_0 < B_0$ | X | X | X | L | H | L |
| $A_3 = B_3$ | $A_2 = B_2$ | $A_1 = B_1$ | $A_0 = B_0$ | H | L | L | H | L | L |
| $A_3 = B_3$ | $A_2 = B_2$ | $A_1 = B_1$ | $A_0 = B_0$ | L | H | L | L | H | L |
| $A_3 = B_3$ | $A_2 = B_2$ | $A_1 = B_1$ | $A_0 = B_0$ | X | X | H | L | L | H |
| $A_3 = B_3$ | $A_2 = B_2$ | $A_1 = B_1$ | $A_0 = B_0$ | H | H | L | L | L | L |
| $A_3 = B_3$ | $A_2 = B_2$ | $A_1 = B_1$ | $A_0 = B_0$ | L | L | L | H | H | L |

(b)

**Fig. 8-28** CMOS magnitude comparator IC (74HC85). (a) Pin diagram. (b) Truth table.

LED will light. This means that the guess was too low and the player should try again by entering a somewhat higher number.

Finally, if a player's guess (A inputs) is higher than the random number (B inputs), the comparator will activate the $A > B_{OUT}$ output. The red LED will light. This means that the guess was too high and the player should try again.

Figure 8-28 provides greater detail on the operation of the *74HC85 4-bit magnitude comparator IC*. The pin diagram is shown in Fig. 8-28(a).

This is the top view of the DIP 74HC85 CMOS IC. The truth table for the 74HC85 comparator is reproduced in Fig. 8-28(b).

The 74HC85 comparator has three "extra" inputs used for *cascading comparators*. Typical cascading of 74HC85 magnitude comparators is shown in Fig. 8-29. This circuit compares the magnitude of the two 8-bit binary words $A_7 A_6 A_5 A_4 A_3 A_2 A_1 A_0$ and $B_7 B_6 B_5 B_4 B_3 B_2 B_1 B_0$. The output from $IC_2$ is one of three responses ($A > B$, $A = B$, or $A < B$).

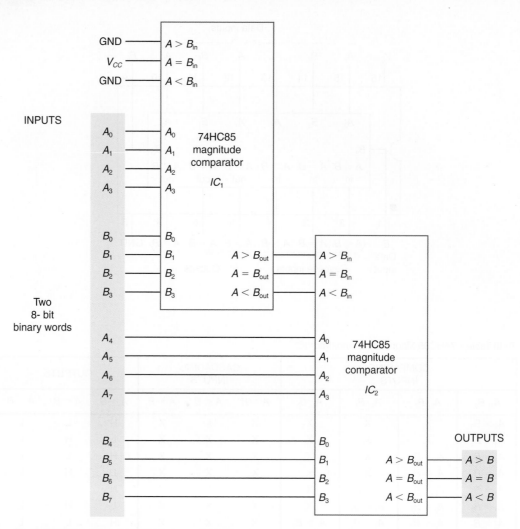

**Fig. 8-29** Cascaded magnitude comparators.

*Answer the following questions.*

58. Refer to Fig. 8-27. If the binary counter holds the number 1001 and your guess is 1011, the _____ (color) LED will light indicating your guess is _____ (correct, too high, too low).

59. Refer to Fig. 8-27. How do you generate a random number before guessing?

60. Refer to Fig. 8-27. The 555 timer is wired as a(n) _____ (astable, monostable) multivibrator.

61. Refer to Fig. 8-30. List the *color* of the output LED that is lit for each time period ($t_1$ to $t_6$).

## 8-12 Using Counters—An Experimental Tachometer

This section shows how you might integrate subsystems to form an *experimental electronic tachometer* system. You will combine known input and output as well as digital devices into a design that will indicate the *angular velocity of a shaft* measured in *rpm (revolutions per minute)*.

The first concept of an experimental tachometer using some of the subsystems studied is sketched in Fig. 8-31(*a*). A 4-digit BCD counter is the heart of the system. The idea is to

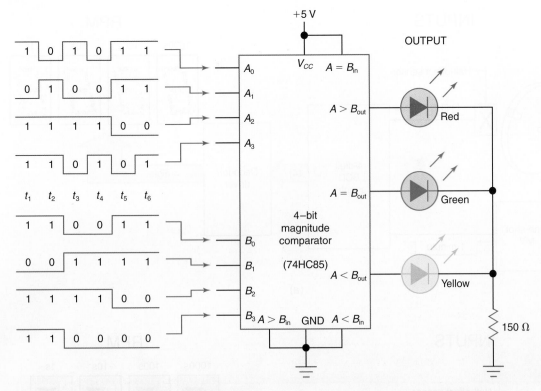

**Fig. 8-30** Comparator problem for test item 61.

count each of the input pulses from a Hall-effect sensor and accumulate that count for a given time such as one minute. The counter would be turned on and count upwards in BCD for one minute and then turned off. The count held in the 4-digit BCD counter would then be decoded and the seven-segment display would present the shaft speed in rpm.

The second concept of the experimental tachometer is sketched in Fig. 8-31(*b*). This time a 3-digit BCD counter becomes the heart of the system. The idea is count the number of input pulses from the Hall-effect sensor in 1/10 of a minute (6 seconds). This would be like counting rpm by 10s. The 1000s, 100s, and 10s seven-segment displays would present the latest accumulated count from the BCD counters and the 0s place always be considered a 0. Therefore, if the input to the tachometer is 1256 rpm, the three active seven-segment displays would show 125 with the 1s place understood to be 0. This would be interpreted as 1250 rpm. The second experimental tachometer sketched in Fig. 8-31(*b*) would have a count-up time of only 6s while the earlier design required a lengthy count time of 1 minute. The sampling time of the second concept experimental tachometer only 6 seconds. The speed that can

be measured by the tachometer in Fig. 8-31(*b*) ranges from 0 to 9990 rpm in increments of 10.

A schematic of an experimental tachometer based on the second concept block diagram [Fig. 8-31(*b*)] is detailed in Fig. 8-32. This unit uses components that you have encountered earlier in your study of electronics.

The input device to the tachometer circuit in Fig. 8-32, which senses the speed of rotation of a shaft, is a Hall-effect switch (3141 IC). A 33k-$\Omega$ pull-up resistor ($R_1$) is required at the output of the 3141 ICs open-collector NPN transistor. The pulses generated by the Hall-effect switch are fed directly into the *CLK* input of the 4553 3-digit BCD counter. Each is pulse from the Hall-effect switch is counted and accumulated in the BCD counter. The 3-digit BCD counter can count from 0000 0000 0000$_{BCD}$ to 1001 1001 1001$_{BCD}$ (from 0 to 999 in decimal).

A second input shown at the left in Fig. 8-32 is called the trigger pulse. This short negative pulse first clears the counters to 0000 0000 0000$_{BCD}$ by activating the MASTER RESET (MR) with a short positive pulse emitted by the 74HC04 inverter. Second, the trigger pulse activates the 555 IC wired as a one-shot multivibrator (MV). When triggered, the one-shot MV emits a 6 second (1/10th of a minute) positive

RPM

1000s   100s   10s   1s

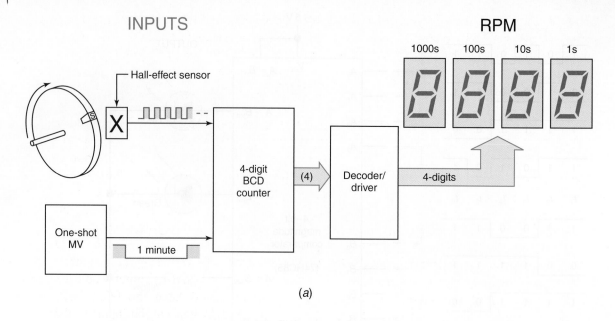

Hall-effect sensor

4-digit
BCD
counter

(4)

Decoder/
driver

4-digits

One-shot
MV

1 minute

(a)

INPUTS

RPM

1000s   100s   10s   1s

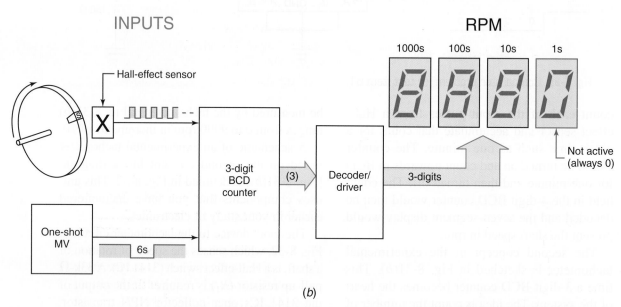

Hall-effect sensor

3-digit
BCD
counter

(3)

Decoder/
driver

3-digits

One-shot
MV

6s

Not active
(always 0)

(b)

**Fig. 8-31** (a) First concept of an experimental tachometer. (b) Second concept of an experimental tachometer.

pulse, which is inverted by the 74HC04 inverter. The precise length of the *count-up pulse* can be adjusted using the 500k-$\Omega$ potentiometer ($R_2$). The 6-second negative count-up pulse enters the LATCH ENABLE (LE) input of the 4553 counter IC. This LOW at the LE input makes the latches transparent and the outputs of the three BCD counters pass through to the multiplexer. When the count-up pulse is LOW, the displays will be observed counting upwards. At the end of the 6-second count-up pulse, the LATCH ENABLE input is driven active by a HIGH and the last data accumulated in the three BCD counters are latched at the inputs to the display multiplexer.

The *display multiplexer* section of the 4553 IC shown in Fig. 8-32 lights the three seven-segment LED displays in a rapidly rotating sequence. Details for the lighting sequence are as follows:

*1000s Display.* The display multiplexer first activates (turns on) PNP transistor $Q_1$. This allows +5V to appear at the anode of the 1000s seven-segment LED display. At this time the display multiplexer sends the segment information for the 1000s display to the 4543 IC for decoding (BCD-to-7-segment code). The 4543 IC drives the segments of all three LED displays at once with the 1000s data but only the 1000s display is activated by display driver transistor $Q_1$.

*100s Display.* Second, the display multiplexer activates (turns on) PNP transistor $Q_2$. This allows +5V to appear at the anode of the 100s seven-segment LED display. At this time, the display multiplexer sends the segment information for the 100s display to the 4543 IC for decoding (BCD-to-seven-segment code). The 4543 IC drives the segments of all three LED displays at once with the 100s data but only the 100s display is activated by display driver transistor $Q_2$.

*10s Display.* Third, the display multiplexer activates (turns on) PNP transistor $Q_3$. This allows +5V to appear at the anode of the 10s seven-segment LED display. At this time, the display multiplexer sends the segment information for the 10s display to the 4543 IC for decoding (BCD-to-seven-segment code). The 4543 IC drives the segments of all three LED displays at once with the 10s data, but only the 10s display is activated by display driver transistor $Q_3$.

The display multiplexing in Fig. 8-32 continues at a frequency high enough so the human eye will not detect the displays turning on and off. The display multiplexing frequency is set by the external capacitor $C_3$. This frequency is set at about 70-Hz in this circuit. In other words, each seven-segment display turns on and off 70 times each second but they all appear to be lit continuously.

As a working example, suppose the speed of rotation of input shaft is 1250 rpm. On an external trigger pulse the experimental tachometer shown in Fig. 8-32 would clear the counters to $000_{10}$ and the one-shot MV would emit the 6-second count-up pulse. During the 6-second count-up time period, 125 pulses would enter the CLK input and be accumulated in the BCD counters as $0001\ 0010\ 0101_{BCD}$ (125 in decimal). The LATCH ENABLE input would go HIGH, which would latch the $0001\ 0010\ 0101_{BCD}$ (125 in decimal) at the inputs to the 4553 ICs display multiplexer. The seven-segment displays would be multiplexed with the reading of 125 shown on the left three displays. This would be interpreted as 1250 rpm with the inactive 1s display assumed to be 0.

# Self-Test

*Answer the following questions.*

62. An instrument that measures the angular velocity of a rotating shaft in revolutions per minute (rpm) is called a(n) _____ (tachometer, Vu-meter).

63. Refer to Fig. 8-32. The interface device used by the tachometer circuit to translate from shaft rotation to digital pulses is a(n) _____ (Hall-effect switch, optical encoder).

64. Refer to Fig. 8-32. The 4553 IC records the number of rotations of the shaft by counting the number of pulses entering the _____ (CLK, MASTER RESET) input of the 3-digit BCD counter.

65. Refer to Fig. 8-32. The negative trigger pulse input first _____ (resets the counter to $000_{10}$, sets the counter to $111_{10}$) and then causes, the _____ (555, 4543) IC wired as a one-shot MV to generate a 6-second count-up pulse.

66. Refer to Fig. 8-32. If the constant rotational speed of the input shaft were 2350 rpm, how many pulses would be counted and accumulated in the BCD counters of the 4553 IC and be displayed after the 6-second count up pulse goes HIGH?

67. Refer to Fig. 8-32. Immediately after the 6-second count-up pulse, the LATCH ENABLE input is _____ (activated, deactivated) by the HIGH and the accumulated count is frozen at the inputs to the _____ (display multiplexer, pulse shaper) of the 4553 IC.

68. Refer to Fig. 8-32. The _____ (counter, display multiplexer) section of the 4553 IC enabled the lighting of the three seven-segment displays in a rapidly rotating sequence.

69. Refer to Fig. 8-32. The three PNP transistors (Q1, Q2, and Q3) could be described as _____ (digit, segment) drivers during display multiplexing.

70. Refer to Fig. 8-32. The external component that sets the frequency of the display multiplexing of the 4553 IC is _____.

71. Refer to Fig. 8-32. The _____ (4543, 4553) IC decodes and drives the segments of the LED displays.

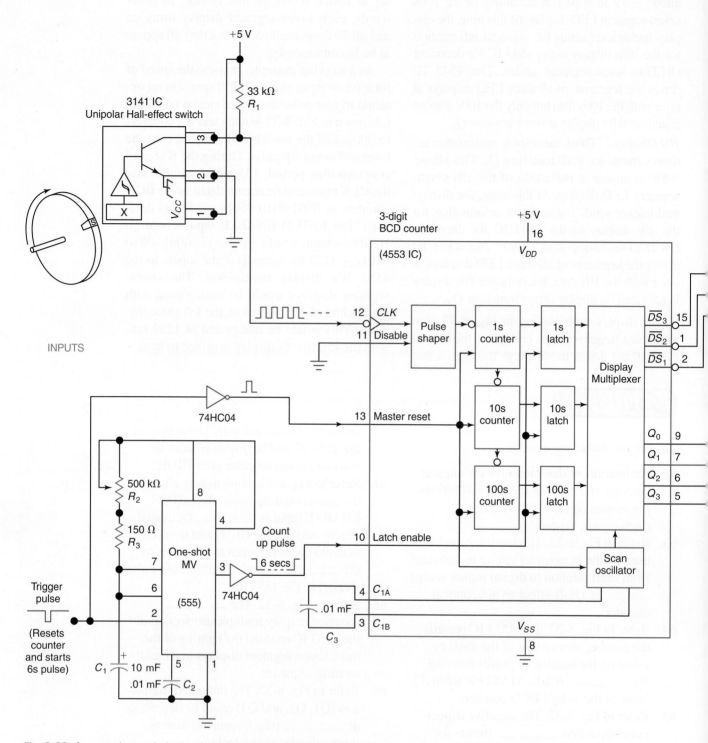

INPUTS

**Fig. 8-32** An experimental electronic tachometer using counters, latched outputs, and multiplexed displays.

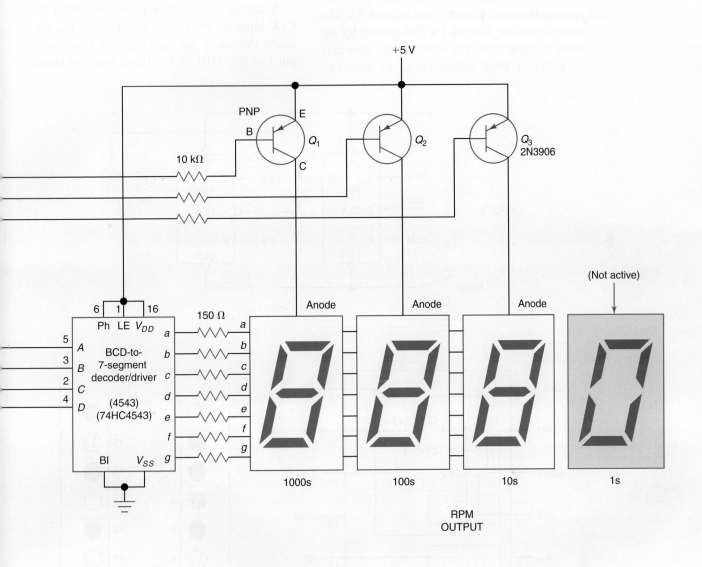

+5 V

PNP

E

B

C

$Q_1$

$Q_2$

$Q_3$
2N3906

10 kΩ

6  1  16

Ph  LE  $V_{DD}$

A

B

C

D

BCD-to-
7-segment
decoder/driver

(4543)
(74HC4543)

BI

$V_{SS}$

5  A

3  B

2  C

4  D

150 Ω

a
b
c
d
e
f
g

Anode

Anode

Anode

(Not active)

1000s

100s

10s

1s

RPM
OUTPUT

## 8-13  Troubleshooting a Counter

Consider the job of troubleshooting the faulty *2-bit ripple counter* shown in Fig. 8-33(*a*). For your convenience, a pin diagram for the 74HC76 IC used in this circuit is shown in Fig. 8-33(*b*). Note that all of the input and output markings in Fig. 8-33(*a*) and (*b*) are *not* the same. For instance, the asynchronous preset inputs on the logic diagram are labeled *PS*. The same inputs are labeled *PR* (for preset) by another manufacturer. The labels on the pins may be different from manufacturer to manufac-

turer. However, the pins on an IC labeled as a 74HC76 serve the same *function* even if the labeling is different.

It is found that the faulty 2-bit counter circuit can be cleared to 00 by the reset switch at the left in Fig. 8-33(*a*). The IC seems to be operating at the correct temperature, and the technician can see no signs of trouble.

A digital logic pulser is used to pulse the *CLK* input on FF 1. According to the pin diagram, the tip of the digital pulser must touch pin 1 of the 74HC76 IC. Upon repeated single

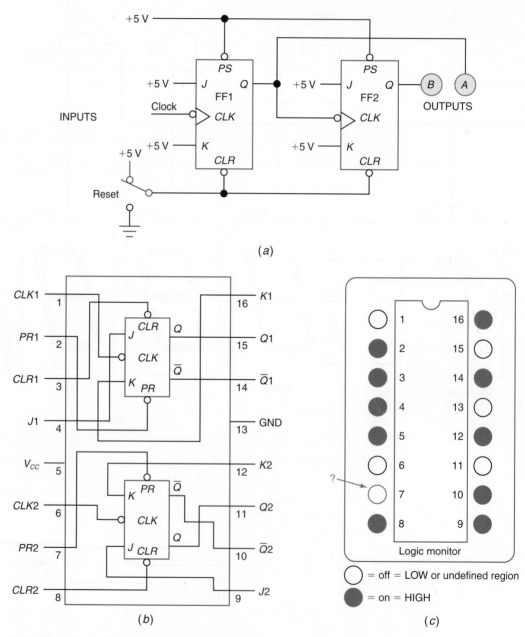

(*a*)

(*b*)

(*c*)

= off = LOW or undefined region

= on = HIGH

**Fig. 8-33**  (*a*) Faulty 2-bit ripple counter circuit used in troubleshooting example. (*b*) Pin diagram for 7476 J-K flip-flop IC. (*c*) Logic clip reading after momentarily resetting the faulty 2-bit counter.

pulses, the counting sequence is 00 (reset), 01, 10, 11, 10, 11, 10, 11, and so on. The *Q* output of FF 2 seems to be "stuck HIGH"; however, the asynchronous clear (*CLR*), or reset, switch can drive it LOW.

Power is then turned off in the circuit in Fig. 8-33(*a*). A logic clip is clipped over the pins on the 74HC76 IC. Power is turned on again. The reset switch is activated. The results displayed on the logic monitor after the reset are shown in Fig. 8-33(*c*). Compare the logic levels shown on the logic monitor with your expectations. You must use the manufacturer's pin diagram furnished in Fig. 8-33(*b*). In looking over the logic levels pin by pin, the LOW or undefined logic level at pin 7 should cause concern. This is the asynchronous preset (*PS* or *PR*) input and should be HIGH according to the logic diagram in Fig. 8-33(*a*). If it is LOW or in the undefined region it can cause the *Q* output of FF 2 to be in the "stuck HIGH" condition.

A logic probe is used to check pin 7 of the 74HC76 IC. Both LEDs on the logic probe remain off. This means neither a LOW nor a HIGH logic level is present. Pin 7 appears to be floating in the undefined region between LOW and HIGH. The IC is interpreting this as LOW at some times and as HIGH at other times.

The IC is removed from the 16-pin DIP IC socket. It is found that pin 7 of the IC is bent under and not making contact with the IC socket. This caused it to float. The fault is illustrated in Fig. 8-34. This common fault is very hard to see when the IC is seated in the IC socket.

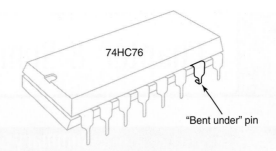

Fig. 8-34 Bent pin caused input to float.

In this example, several tools were used in troubleshooting. First, the logic diagram and your knowledge of how it works are most important. Second, a manufacturer's pin diagram was used. Third, a digital logic pulser was used to inject single pulses. Fourth, a logic clip checked the logic level at all pins of the 74HC76 IC. Fifth, a logic probe was used to check the suspected pin of the IC. Finally, your knowledge of the circuit and visual observation solved the problem. *Your knowledge of the circuit's normal operation and your powers of observation are probably the most important troubleshooting tools.* Logic pulser, logic probes, logic clips, DMMs, logic analyzers, IC testers, and oscilloscopes are only aids to your knowledge and powers of observation.

The example of a floating input caused by a bent-under pin is a very common problem in student-constructed circuits. It is good practice to make sure all inputs go to the proper logic level. This is true for TTL and especially true for CMOS circuits.

**Most important troubleshooting tools**

**Self-Test**

*Supply the missing word or words in each statement.*

72. Refer to Fig. 8-33. Pins 4, 9, 12, and 16 to the *J* and *K* inputs of the flip-flops should all be _____ (HIGH, LOW) in this circuit.
73. Refer to Fig. 8-33. Pins 3 and 8 to the _____ inputs of the flip-flops

should follow the logic state of the reset switch.
74. Refer to Fig. 8-33. Pins 2 and 7 to the _____ inputs of the flip-flops should be _____ (HIGH, LOW) in this circuit.
75. Refer to Fig. 8-33. The fault in this circuit was located at pin _____ (number). It was _____ rather than being at a HIGH logic level.

For related information, visit the website for Texas Instruments.

# Chapter 8 Summary and Review

## Summary

1. Flip-flops are wired together to form binary counters.

2. Counters can operate asynchronously or synchronously. Asynchronous counters are called ripple counters and are simpler to construct than synchronous counters.

3. The modulus of a counter is how many different states it goes through in its counting cycle. A mod-5 counter counts 000, 001, 010, 011, 100 (0, 1, 2, 3, 4 in decimal).

4. A 4-bit binary counter has four binary place values and counts from 0000 to 1111 (decimal 0 to 15).

5. Gates can be added to the basic flip-flops in counters to add features. Counters can be made to stop at a certain number. The modulus of a counter can be changed.

6. Counters are designed to count either up or down. Some counters have both features built into their circuitry.

7. Counters are used as frequency dividers. Counters are also widely used to count or sequence events and temporarily store data.

8. Manufacturers produce a wide variety of self-contained IC counters. They produce detailed data sheets for each counter IC. Several TTL and CMOS counter ICs were studied in this chapter.

9. Many variations in pin labeling and logic symbols occur from manufacturer to manufacturer.

10. A transducer such as an optical sensor can be used to count real-world events such as in shaft encoding.

Optical encoders come in slot types and reflective types, and both are based on an infrared diode shining on an output phototransistor.

11. A magnitude comparator will compare two binary numbers and decide if $A = B$, $A > B$, or $A < B$. Magnitude comparator ICs can be cascaded to compare larger binary numbers.

12. A decade counter counts from 0 through 9 (0000–1001 in binary). Decade counters are also commonly called BCD (binary-coded decimal) counters.

13. One trend in the manufacture of digital IC is to include more functions on one chip. As a simple example, the 4553 3-Digit BCD Counter IC contains three BCD counters, wave shaping, twelve transparent latches, and display multiplexing.

14. A transducer such as a Hall-effect switch may be used to sense rotations of a shaft which can be counted in a given time period yielding an output in revolutions per minute. An instrument that measures speed of rotation of a shaft is called a tachometer.

15. The technician's knowledge of the circuit and powers of observation are the most important tools in troubleshooting. The logic probe, voltmeter, DMM, logic clip, digital pulser, logic analyzer, IC tester, and oscilloscope aid the technician's observations when troubleshooting sequential logic circuits.

## Chapter Review Questions

*Answer the following questions.*

8-1. Draw a logic symbol diagram of a mod-8 ripple up counter. Use three J-K flip-flops. Show input *CLK* pulses and three output indicators labeled *C*, *B*, and *A* (*C* indicator is MSB).

8-2. Draw a table (similar to Fig. 8-1) showing the binary and decimal counting sequence of the mod-8 counter in question 8-1.

8-3. Draw a waveform diagram (similar to Fig. 8-2(*b*)) showing the eight *CLK* pulses and the outputs (*Q*) of FF 1, FF 2, and FF 3 of the mod-8 counter from question 8-1. Assume you are using negative-edge-triggered flip-flops.

.8-4. A(n) _____ (asynchronous, synchronous) counter is the more complex circuit.

8-5. Synchronous counters have the *CLK* inputs connected in _____ (parallel, series).

8-6. Draw a logic symbol diagram for a 4-bit ripple down counter. Use four J-K flip-flops in this mod-16 counter. Show the input *CLK* pulses, *PS* input, and four output indicators labeled *D, C, B,* and *A*.

8-7. If the ripple down counter in question 8-6 is a recirculating type, what are the next three counts after 0011, 0010, and 0001?

8-8. Redesign the 4-bit counter in question 8-6 to count from binary 1111 to 0000 and then *stop*. Add a four-input OR gate to your existing circuit to add this self-stopping feature.

8-9. Draw a block diagram (similar to Fig. 8-12) showing how you would use two counters to get an output of 1 Hz with an input of 100 Hz. Label your diagram.

8-10. Refer to Fig. 8-13 for questions **a** to **f** on the 7493 IC counter:

   a. What is the maximum count length of this counter?

   b. This is a _____ (ripple, synchronous) counter.

   c. What must be the conditions of the reset inputs for the 7493 to count?

   d. This is a(n) _____ (down, up) counter.

e. The 7493 IC contains _____ (number) flip-flops.

f. What is the purpose of the NAND gate in the 7493 counter?

8-11. Refer to Fig. 8-14 for questions **a** to **f** on the 74192 counter:

   a. What is the maximum count length of this counter?

   b. This is a _____ (ripple, synchronous) counter.

   c. A logical _____ (0, 1) is needed to clear the counter to 0000.

   d. This is a(n) _____ (down, up, both up and down) counter.

   e. How could we preset the outputs of the 74192 IC to 1001?

   f. How do we get the counter to count downward?

8-12. Draw a diagram [similar to Fig. 8-15(*a*)] showing how you would wire the 7493 counter as a 4-bit (mod-16) ripple counter. Refer to Fig. 8-13.

8-13. Refer to Fig. 8-35. The 74192 counter is in the _____ (clear, count up, load) mode during pulse $t_1$.

8-14. List the binary output from the 74192 counter IC after each of the eight input pulses shown in Fig. 8-35. Start with $t_1$ and end with $t_8$.

8-15. Refer to Fig. 8-17 for questions **a** to **e** on the 74HC393 IC counter:

   a. This is a _____ (ripple, synchronous) counter.

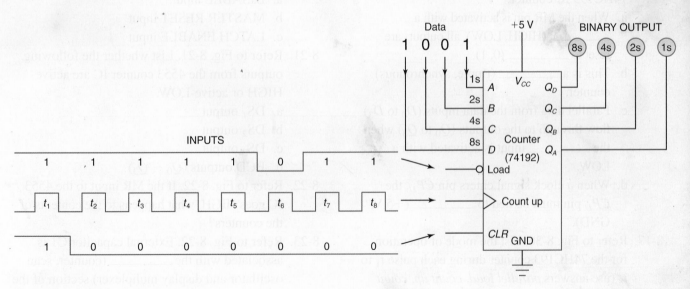

**Fig. 8-35** Counter pulse-train problem.

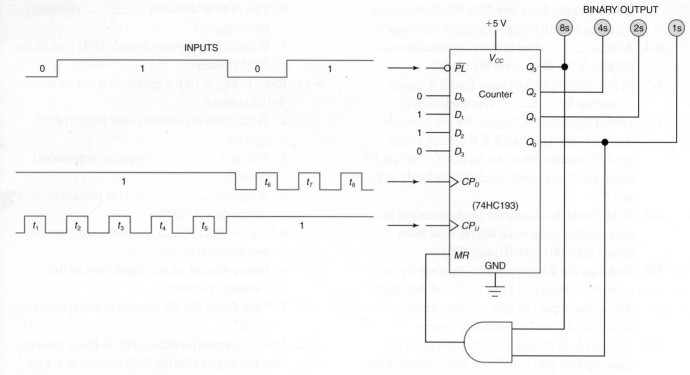

**Fig. 8-36** Counter pulse-train problem.

b. This is a(n) _____ (down, up, either up or down) counter.

c. The MR pins are _____ (asynchronous, synchronous) active _____ (HIGH, LOW) inputs that clear the outputs.

d. Each counter contains four _____ (R-S, T) flip-flops.

e. This is a _____ (CMOS, TTL) counter.

8-16. Refer to Fig. 8-18 for questions **a** to **e** on the 74HC193 IC counter:

a. When the MR pin is activated with a _____ (HIGH, LOW), all outputs are reset to _____ (0, 1).

b. This is a _____ (ripple, synchronous) counter.

c. Parallel data from the data inputs ($D_0$ to $D_3$) flow through to the outputs ($Q_0$ to $Q_3$) when the _____ input is activated with a LOW.

d. When a clock signal enters pin $CP_U$, the $CP_D$ pin must be tied to _____ (+5 V, GND).

8-17. Refer to Fig. 8-36. List the mode of operation for the 74HC193 counter during each pulse $t_1$ to $t_8$ (use answers *parallel load, count up, count down*).

8-18. Refer to Fig. 8-36. List the binary output for the 74HC193 counter IC after each pulse $t_1$ to $t_8$.

8-19. Refer to Fig. 8-21. The CLK input to the 4553 counter IC is triggered on the _____ (H-to-L, L-to-H) transition of the input pulse.

8-20. Refer to Fig. 8-21. List whether the following inputs to the 4553 counter IC are active-HIGH or active-LOW.

a. DISABLE input

b. MASTER RESET input

c. LATCH ENABLE input

8-21. Refer to Fig. 8-21. List whether the following outputs from the 4553 counter IC are active-HIGH or active-LOW.

a. $DS_1$ output

b. $DS_2$ output

c. $DS_3$ output

d. BCD outputs ($Q_0 - Q_3$)

8-22. Refer to Fig. 8-22. If the MR input to the 4553 IC goes HIGH, what happens to the contents of the counters?

8-23. Refer to Fig. 8-22. External capacitor C1 is associated with the _____ (counter, scan oscillator and display muliplexer) section of the 4553 IC.

8-24. Refer to Fig. 8-22. The 12 latches in the 4553 IC are said to be _____ (latched, transparent) when the LE input is LOW.

8-25. Refer to Fig. 8-22. The 4543 decoder IC is closely associated with segment driving while the three PNP transistors are associated with display driving. (T or F)

8-26. Refer to Fig. 8-23(b). The device optically sensing the opening in the shaft encoder disk and sending a signal to the wave-shaping circuit is a(n) _____.

8-27. Refer to Fig. 8-23(b). The optical encoder at the top of the shaft encoder disk is of the _____ (reflective type, slot type).

8-28. Refer to Fig. 8-24(b). The H21A1 interrupter module contains a(n) _____ on the emitter side and a phototransistor on the detector side of the optical sensor.

8-29. Refer to Fig. 8-25. The device that performs wave shaping in this circuit is the _____ (7414, 74192) IC.

8-30. Refer to Fig. 8-25. The device that performs as a decade counter in this circuit is the _____ (7447, 74192) IC.

8-31. Refer to Fig. 8-25. The 7447 IC is a decoder/driver that translates BCD data to _____ code and drives the display.

8-32. Refer to Fig. 8-26(b). The disk shown (black and white strips) would be an encoder disk used by a _____ (reflective-type, slot-type) optical sensor.

8-33. Refer to Fig. 8-29. The two 74HC85 magnitude comparator ICs are said to be _____ (cascaded, subdivided) so they can compare two _____ (number)-bit binary numbers.

8-34. Refer to Fig. 8-37. List the *color* of the output LED that is lit for each time period ($t_1$ to $t_6$).

8-35. A tachometer is an instrument that measures the speed of rotation of a shaft in revolutions per minutes. (T or F)

8-36. Refer to Fig. 8-32. Each rotation of the input shaft will be converted into four pulses that enter the *CLK* input of the 4553 counter IC. (T or F)

8-37. Refer to Fig. 8-32. What two things happen when the negative input trigger pulse occurs?

8-38. Refer to Fig. 8-32. When the output of the one-shot MV goes LOW for 6 seconds the 4553 IC _____ [counts the input pulses, latches (freezes) the outputs of the counters].

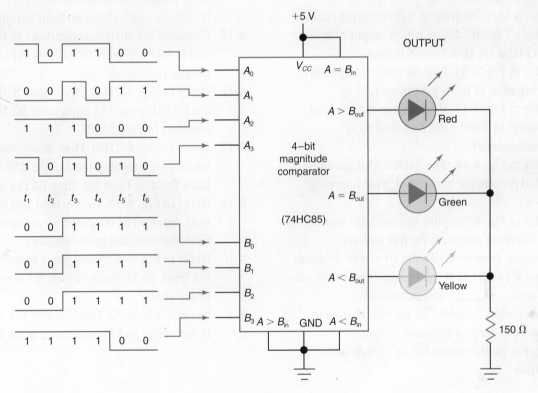

**Fig. 8-37** Magnitude comparator pulse-train problem.

8-39. Refer to Fig. 8-32. The _____ (display multiplexing, pulse shaper) section of the 4553 IC coordinates the lighting of the three seven-segment LED displays in a rapidly rotating sequence so only one display is turned on at a time.

8-40. Refer to Fig. 8-32. If the 4543 decoder/driver IC is called the *segment driver*, then the three PNP transistors ($Q_1$, $Q_2$, and $Q_3$) would be called the _____ (display, scanner) drivers.

8-41. Refer to Fig. 8-32. The purpose of the external capacitor $C_1$ is to decouple the internal circuitry of the 4553 counter IC from ground (GND). (T or F)

8-42. Refer to Fig. 8-32. At a given instant, only one DS output from the 4553 counter IC is _____ (HIGH, LOW) at a given time to turn on its respective PNP transistor and display.

8-43. Refer to Fig. 8-32. When the _____ (CLK, LATCH ENABLE) input goes HIGH, the data accumulated in the BCD counters are frozen at the inputs to the display multiplexer while the counters can continue to count pulses.

8-44. A digital _____ (IC tester, pulser) is an instrument for injecting a signal into a circuit.

## Critical Thinking Questions

8-1. What types of flip-flops are useful in wiring counters because they have a toggle mode?

8-2. Draw a logic symbol diagram of a mod-5 ripple up counter. Use three J-K flip-flops and a two-input NAND gate. Show input *CLK* pulses and three output indicators labeled *C, B,* and *A* (*C* indicator is MSB).

8-3. Draw a logic diagram for a mod-10 counter using a 7493 IC.

8-4. Draw a logic diagram of a divide-by-8 counter using a 7493 IC. Show which output of the 7493 IC is the divide-by-8 output.

8-5. Refer to Fig. 8-35. List the mode of operation during each of the input pulses $t_1$ to $t_8$.

8-6. Refer to Fig. 8-17. Why are the master reset inputs (1MR and 2MR) referred to as asynchronous?

8-7. Refer to Fig. 8-18. The 74HC193 IC counter is called presettable because of what operating mode?

8-8. Refer to Fig. 8-36. Give the modulus and list the counting sequence for this counter.

8-9. Design a decade up counter (0 to 9 in decimal) using a 74HC193 IC and a two-input AND gate.

8-10. On a(n) _____ (asynchronous, synchronous) counter, all outputs change to their new states at the same instant.

8-11. What is another name for an asynchronous counter?

8-12. If we refer to a divide-by-6 counter, the circuit will probably be used for what purpose?

8-13. Refer to Fig. 8-25. The counter and display will increase by one when the output of the inverter goes from _____ (H-to-L, L-to-H).

8-14. Refer to Fig. 8-25. The counter and display will increment when the beam of infrared light across the slot _____.
   a. is broken (from light to no light)
   b. begins again (from no light to light)

8-15. Compare the shaft encoder disks in Fig. 8-23(*b*) and 8-26(*b*). Which disk will provide the greater resolution?

8-16. Refer to Fig. 8-21(*a*). How many T flip-flops are probably used to implement the three BCD counters in the 4553 IC?

8-17. Refer to Fig. 8-21(*a*). How many transparent latches are probably used in the 4553 IC to latch the data from the three BCD counters?

8-18. Refer to Fig. 8-32. How would you adjust the time duration of the *count up pulse* emitted from the one-shot multivibrator?

8-19. Refer to Fig. 8-32. How many pulses are emitted from the Hall-effect switch for each rotation of the shaft?

8-20. Refer to Fig. 8-32. Explain why the 1s display is not active and is considered to be a zero (0).

1. two
2. 4
3. toggle
4. pulse $t_1 = 00$
   pulse $t_2 = 01$
   pulse $t_3 = 10$
   pulse $t_4 = 11$
   pulse $t_5 = 00$
   pulse $t_6 = 01$
5. ripple, decade
6. ripple, 5
7. pulse $t_1 = 111$, then cleared to 000 just before pulse $t_2$
   pulse $t_2 = 001$
   pulse $t_3 = 010$
   pulse $t_4 = 011$
   pulse $t_5 = 100$
   pulse $t_6 = 000$
8. synchronous
9. parallel
10. toggle
11. all the flip-flops toggle
12. FF3
13. toggle
14. HIGH-to-LOW
15. only FF 1 toggles
16. pulse $t_1 = 00$

pulse $t_2 = 11$
pulse $t_3 = 10$
pulse $t_4 = 01$
pulse $t_5 = 00$
pulse $t_6 = 11$
17. down
18. LOW, hold
19. HIGH, toggle
20. 1000
21. 2
22. 0000 (reset)
23. four, up
24. decade, synchronous
25. 5
26. HIGH
27. point $B = 200$ Hz
    point $C = 100$ Hz
    point $D = 50$ Hz
28. 8
29. 4-bit binary
30. HIGH
31. H-to-L
32. 16, ripple
33. synchronous
34. asynchronous
35. $Q_0$–$Q_3$
36. 0001, 0010, 0011, 0100, 0101, 0110 (1 to 6 in decimal)

37. to preset the counter to 0001 after the highest count of 0110
38. manufacturers use different standards when drawing logic symbols and labeling
39. counters, multiplexer
40. active-HIGH, resets all counter outputs to 0
41. H-to-L
42. negative-edge
43. active-HIGH
44. LOW
45. F
46. T
47. set the scan frequency of the multiplexer
48. 4543
49. 4553
50. 000 to 999
51. interrupter module
52. infrared
53. phototransistor
54. slot-type
55. L-to-H (LOW-to-HIGH)
56. enters

57. decade, temporarily storing the count
58. red, too high
59. press and release switch $SW_1$
60. astable
61. $t_1$ = green
    $t_2$ = red
    $t_3$ = yellow
    $t_4$ = red
    $t_5$ = green
    $t_6$ = red
62. tachometer
63. Hall-effect switch
64. *CLK*
65. resets the counter to $000_{10}$, 555
66. 235
67. activated, display multiplexer
68. display multiplexer
69. digit
70. $C_3$ or capacitor $C_3$
71. 4543
72. HIGH
73. clear
74. preset (asynchronous), HIGH
75. 7, floating

# Shift Registers

## Chapter Objectives

*This chapter will help you to:*

1. *Draw* a circuit diagram of a serial-load shift register using D flip-flops.

2. *Define* terms such as *shift right, shift left, parallel load,* and *serial load* and *describe* procedures for performing these operations on various shift registers.

3. *Interpret* data sheets for several commercial TTL and CMOS shift register ICs.

4. *Predict* the operation of several TTL and CMOS shift register ICs based on a series of inputs.

5. *Analyze* the operation of a digital roulette game containing a voltage-controlled oscillator (VCO), a ring counter, a power-up initializing circuit, and an audio amplifier.

6. *Troubleshoot* a faulty shift register circuit.

A *register* is a group of memory cells grouped together and considered a single unit. For example, an 8-bit register could be used to store a byte of data. The register can be used to simply store information for later use or the register can be designed to act on the data as is the case of a *shift register*. A shift register may modify the contents by shifting data right or left.

The term *latch* may be used to describe the register used to store data. You may have used several transparent latches in previous chapters and know that they are commonly constructed using flip-flops (such as D flip-flops). A *buffer register* is a specific use of a storage device that holds data that is waiting to be transferred. For instance, a buffer is used to temporarily store data while they are waiting to be used by a printer.

A typical example of a *shift register* at work is found within a calculator. As you enter each digit on the keyboard, the numbers shift to the left on the display. In other words, to enter the number 268 you must do the following. First, you press and release the 2 on the keyboard; a 2 appears at the extreme right on the display. Next, you press and release the 6 on the keyboard causing the 2 to shift one place to the left allowing 6 to appear on the extreme right; 26 appears on the display. Finally, you press and release the 8 on the keyboard; 268 appears on the display. This example shows two important characteristics of a shift register: (1) It is a *temporary memory* and thus holds the numbers on the display (even if you release the keyboard number) and (2) it shifts the numbers to the left on the display each time you press a new digit

**Memory and shifting characteristics**

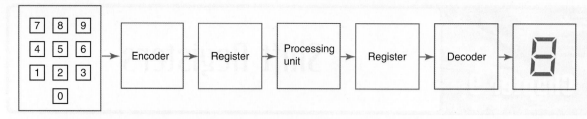

**Fig. 9-1** A digital system using registers.

**Serial load shift registers**

**4-bit shift register**

**Shift registers**

on the keyboard. These *memory* and *shifting characteristics* make the shift register extremely valuable in most digital electronic systems. This chapter introduces you to shift registers and explains their operations.

---

*Shift registers* are constructed by wiring flip-flops together. We mentioned before that flip-flops have a memory characteristic. This memory characteristic is put to good use in a shift register. Instead of wiring shift registers by using individual gates or flip-flops, you can buy shift registers in IC form. In larger-scale digital devices (microcontrollers, microprocessors), the registers are integrated into the chip design.

Registers often are used to store data momentarily. Figure 9-1 shows a typical example of where registers might be used in a digital system. This system could be that of a calculator. Notice the use of registers to hold information from the encoder for the processing unit. A register is also being employed for temporary storage between the processing unit and the decoder. Registers are also used at other locations within a digital system.

One method of describing shift register characteristics is by how data is *loaded into* and *read from* the storage units. Four categories of shift registers are illustrated in Fig. 9-2. Each storage device in Fig. 9-2 is an 8-bit register. The registers are classified as:

**Serial in–serial out**

**Serial in–parallel out**

**Parallel in–serial out**

**Parallel in–parallel out**

1. Serial in–serial out [Fig. 9-2(a)]
2. Serial in–parallel out [Fig. 9-2(b)]
3. Parallel in–serial out [Fig. 9-2(c)]
4. Parallel in–parallel out [Fig. 9-2(d)]

The diagrams in Fig. 9-2 illustrate the fundamental idea of each type of register. These classifications are often used in a manufacturer's literature.

## 9-1 Serial Load Shift Registers

A basic shift register is shown in Fig. 9-3. This shift register is constructed from four D flip-flops. This register is called a *4-bit shift register* because it has four places to store data: $A, B, C, D$.

With the aid of Table 9-1 and Fig. 9-3, let us operate this shift register. First, clear (*CLR* input to 0) all the outputs ($A, B, C, D$) to 0000. (This situation is shown in line 1, Table 9-1.) The outputs remain 0000 while they await a clock pulse. Pulse the *CLK* input once; the output now shows 1000 (line 3, Table 9-1) because the 1 from the *D* input of FF *A* has been transferred to the *Q* output on the clock pulse. Now enter 1s on the data input (clock pulses 2 and 3, Table 9-1); these 1s shift across the display to the right. Next, enter 0s on the data input (clock pulses 4 to 8, Table 9-1); you can see the 0s being shifted across the display (lines 6 to 10, Table 9-1). On clock pulse 9 (Table 9-1) enter a 1 at the data input. On pulse 10 the data input is returned to 0. Pulses 9 to 13 show the single 1 on display being shifted to the right. Line 15 shows the 1 being shifted out the right end of the shift register and being lost.

Remember that the D flip-flop is also called a *delay* flip-flop. Recall that it simply transfers the data from input *D* to output *Q after a delay of one clock pulse.*

The circuit diagrammed in Fig. 9-3 is referred to as a *serial load shift register.* The term "serial load" comes from the fact that only one bit of data at a time can be entered in the register. For instance, to enter 0111 in the register, we had to go through the sequence from lines 3 through 6 in Table 9-1. It took four steps to serially load 0111 into the serial load shift register. To enter 0001 in this serial load shift register we need four steps, as shown in Table 9-1, lines 11 to 14. According to the classifications in Fig. 9-2, this would be a serial in–parallel out

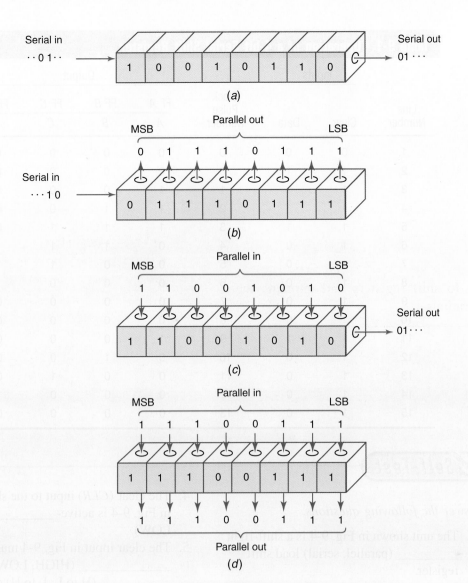

Serial in
···0 1··

Serial out
01···

1 0 0 1 0 1 1 0

(a)

Parallel out

MSB                                              LSB

0   1   1   1   0   1   1   1

Serial in
···1 0

0   1   1   1   0   1   1   1

(b)

Parallel in

MSB                                              LSB

1   1   0   0   1   0   1   0

Serial out
01···

1   1   0   0   1   0   1   0

(c)

Parallel in

MSB                                              LSB

1   1   1   0   0   1   1   1

1   1   1   0   0   1   1   1

1   1   1   0   0   1   1   1

Parallel out

(d)

**Fig. 9-2** Shift register characteristics. (a) Serial in–serial out. (b) Serial in–parallel out. (c) Parallel in–serial out. (d) Parallel in–parallel out.

register. However, if data were taken from only FF *D*, it becomes a serial in–serial out register.

The shift register in Fig. 9-3 could become a 5-bit shift register just by adding one more D flip-flop. Shift registers typically come in 4-, 5-, and 8-bit sizes. Shift registers also can be wired using other flip-flops. J-K flip-flops and clocked R-S flip-flops are also used to wire shift registers.

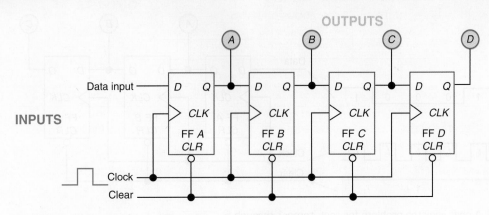

OUTPUTS

INPUTS

Data input

Clock

Clear

**Fig. 9-3** A 4-bit serial load shift register using D flip-flops.

## Table 9-1   Operation of a 4-bit Serial Shift Register

| | Inputs | | | Output | | | |
|---|---|---|---|---|---|---|---|
| Line Number | Clear | Data | Clock Pulse Number | FF A  A | FF B  B | FF C  C | FF D  D |
| 1 | 0 | 0 | 0 | 0 | 0 | 0 | 0 |
| 2 | 1 | 1 | 0 | 0 | 0 | 0 | 0 |
| 3 | 1 | 1 | 1 | 1 | 0 | 0 | 0 |
| 4 | 1 | 1 | 2 | 1 | 1 | 0 | 0 |
| 5 | 1 | 1 | 3 | 1 | 1 | 1 | 0 |
| 6 | 1 | 0 | 4 | 0 | 1 | 1 | 1 |
| 7 | 1 | 0 | 5 | 0 | 0 | 1 | 1 |
| 8 | 1 | 0 | 6 | 0 | 0 | 0 | 1 |
| 9 | 1 | 0 | 7 | 0 | 0 | 0 | 0 |
| 10 | 1 | 0 | 8 | 0 | 0 | 0 | 0 |
| 11 | 1 | 1 | 9 | 1 | 0 | 0 | 0 |
| 12 | 1 | 0 | 10 | 0 | 1 | 0 | 0 |
| 13 | 1 | 0 | 11 | 0 | 0 | 1 | 0 |
| 14 | 1 | 0 | 12 | 0 | 0 | 0 | 1 |
| 15 | 1 | 0 | 13 | 0 | 0 | 0 | 0 |

### ✓ Self-Test

*Answer the following questions.*

1. The unit shown in Fig. 9-4 is a shift-right _____ (parallel, serial) load shift register.

2. List the contents of the register in Fig. 9-4 after each of the six clock pulses starting with $t_1$ ($A$ = left bit, $C$ = right bit).

3. A(n) _____ (entire 3-bit group, single bit) is loaded on each clock pulse in the serial load shift register in Fig. 9-4.

4. The clear ($CLR$) input to the shift register in Fig. 9-4 is active- _____ (HIGH, LOW).

5. The clear input in Fig. 9-4 must be _____ (HIGH, LOW) and a _____ (H to L, L to H) clock pulse at the $CLK$ input will trigger a right shift in this register.

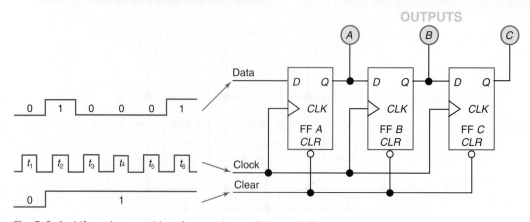

Fig. 9-4 A shift register problem for test items 1 through 5.

## 9-2 Parallel Load Shift Registers

The serial load shift register we studied in the last section has two disadvantages: it permits only one bit of information to be entered at a time, and it loses all its data out the right side when it shifts right. Figure 9-5(a) illustrates a system that permits *parallel loading* of four bits at once. These inputs are the data inputs *A*, *B*, *C*, and *D* in Fig. 9-5. This system could also incorporate a *recirculating feature* that would put the output data back into the input so that they are not lost.

A wiring diagram of the *4-bit parallel load recirculating shift register* is drawn in Fig. 9-5(b). This shift register uses four J-K flip-flops. Notice the recirculating lines leading from the *Q* and *Q̄* outputs of FF *D* back to the *J* and *K* inputs of FF *A*. These feedback lines cause the data that would normally be lost out

of FF *D* to recirculate through the shift register. The *CLR* input clears the outputs to 0000 when enabled by a logical 0. The parallel load data inputs *A*, *B*, *C*, and *D* are connected to the active-LOW preset (*PS*) inputs of the flip-flops to set 1s at any output position (*A*, *B*, *C*, *D*). If the switches attached to the parallel load data inputs are even temporarily switched to a 0, that output will be preset to a logical 1. The clock pulsing the *CLK* inputs of the J-K flip-flops will cause data to be shifted to the right. The data from FF *D* will be recirculated back to FF *A*.

Table 9-2 on page 298 will help you understand the operation of the parallel load shift register. As you turn on the power, the outputs may assume any combination. Line 2 shows the register being cleared with the *CLR* input. Line 3 shows 0100 being loaded into the register using the parallel load data switches. An asynchronous parallel load occurs whenever a parallel

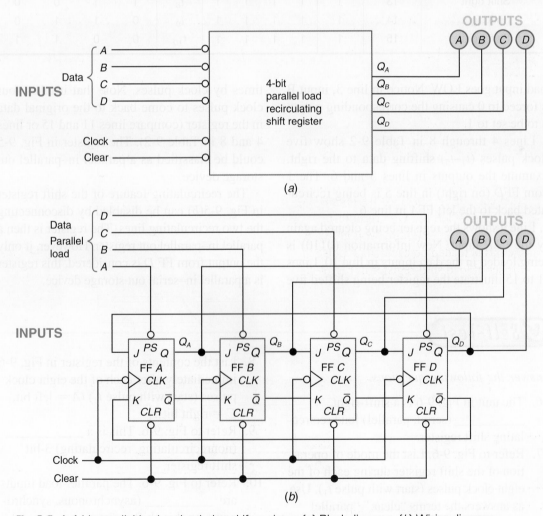

(a)

(b)

**Fig. 9-5** A 4-bit parallel load recirculating shift register. (a) Block diagram. (b) Wiring diagram.

## Table 9-2 Operation of a 4-Bit Parallel Load Recirculating Shift Register

| Mode of operation | Line number | Inputs | | | | | | Output | | | |
|---|---|---|---|---|---|---|---|---|---|---|---|
| | | | Parallel load | | | | Clock pulse | FF$A$ | FF$B$ | FF$C$ | FF$D$ |
| | | Clear | $A$ | $B$ | $C$ | $D$ | | $A$ | $B$ | $C$ | $D$ |
| Power up | 1 | 1 | 1 | 1 | 1 | 1 | | (Random outputs) | | | |
| Clear (asynchronous) | 2 | 0 | 1 | 1 | 1 | 1 | | 0 | 0 | 0 | 0 |
| Parallel load (asynchronous) | 3 | 1 | 1 | 0 | 1 | 1 | | 0 | 1 | 0 | 0 |
| Shift right | 4 | 1 | 1 | 1 | 1 | 1 | $t_1$ | 0 | 0 | 1 | 0 |
| Shift right | 5 | 1 | 1 | 1 | 1 | 1 | $t_2$ | 0 | 0 | 0 | 1 |
| Shift right | 6 | 1 | 1 | 1 | 1 | 1 | $t_3$ | 1 | 0 | 0 | 0 |
| Shift right | 7 | 1 | 1 | 1 | 1 | 1 | $t_4$ | 0 | 1 | 0 | 0 |
| Shift right | 8 | 1 | 1 | 1 | 1 | 1 | $t_5$ | 0 | 0 | 1 | 0 |
| Clear (asynchronous) | 9 | 0 | 1 | 1 | 1 | 1 | | 0 | 0 | 0 | 0 |
| Parallel load (asynchronous) | 10 | 1 | 1 | 0 | 0 | 1 | | 0 | 1 | 1 | 0 |
| Shift right | 11 | 1 | 1 | 1 | 1 | 1 | $t_6$ | 0 | 0 | 1 | 1 |
| Shift right | 12 | 1 | 1 | 1 | 1 | 1 | $t_7$ | 1 | 0 | 0 | 1 |
| Shift right | 13 | 1 | 1 | 1 | 1 | 1 | $t_8$ | 1 | 1 | 0 | 0 |
| Shift right | 14 | 1 | 1 | 1 | 1 | 1 | $t_9$ | 0 | 1 | 1 | 0 |
| Shift right | 15 | 1 | 1 | 1 | 1 | 1 | $t_{10}$ | 0 | 0 | 1 | 1 |

load input goes LOW. Notice in line 3, input B is forced to 0 causing the corresponding output B to be set to 1.

Lines 4 through 8 in Table 9-2 show five clock pulses ($t_1$–$t_5$) shifting data to the right. Examine the outputs in lines 5 and 6. The 1 from FF$D$ (on right) in line 5 is being recirculated back to the left FF$A$ in line 6.

Line 9 shows the register being cleared again by the *CLR* input. New information (0110) is being loaded in the data inputs in line 10. Lines 11 to 15 illustrate the register being shifted five

times by clock pulses. Note that it takes four clock pulses to come back to the original data in the register (compare lines 11 and 15 or lines 4 and 8 in Table 9-2). The register in Fig. 9-5 could be classified as a parallel in–parallel out storage device.

The recirculating feature of the shift register in Fig. 9-5(*b*) can be disabled by disconnecting the two recirculating lines. The register is then a parallel in–parallel out register. However, if only the output from FF *D* is considered, this register is a parallel in–serial out storage device.

*Answer the following questions.*

6. The unit in Fig. 9-6 is a shift-right _____ (serial, parallel) load recirculating shift register.

7. Refer to Fig. 9-6. List the mode of operation of the shift register during each of the eight clock pulses (start with pulse $t_1$). Use as answers the terms "clear," "parallel load," and "shift-right."

8. List the contents of the register in Fig. 9-6 immediately after each of the eight clock pulses (start with pulse $t_1$) ($A$ = left bit, $C$ = right bit).

9. Refer to Fig. 9-6. This is a _____ (nonrecirculating, recirculating) 3-bit shift register.

10. Refer to Fig. 9-6. The parallel-load inputs are _____ (asynchronous, synchronous) on this shift register.

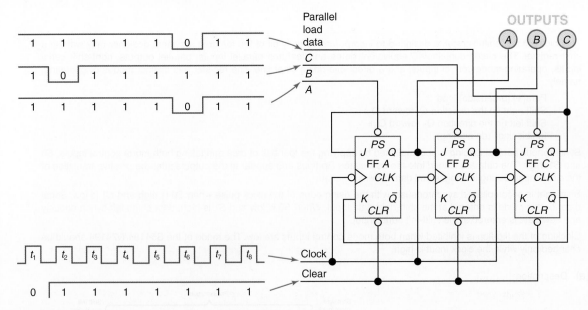

**Fig. 9-6** A shift register problem for test items 6 through 10.

11. Refer to Table 9-2. The shift register is in the clear mode during what two lines on the table?

12. Refer to Table 9-2. The shift register is in the parallel-load mode during what two lines on the table? ✔

## 9-3 A Universal Shift Register

When reviewing data manuals, you will see that manufacturers produce many shift registers in IC form. In this section, one such IC shift register will be studied: the *74194 4-bit bidirectional universal shift register.*

The 74194 IC is a very adaptable shift register and has most of the features we have seen so far in one IC package. A 74194 IC register can shift right or left. It can be loaded serially or in parallel. Several 4-bit 74194 IC registers can be cascaded to make an 8-bit or longer shift register. And this register can be made to recirculate data.

Read the description of the 74194 shift register in Fig. 9-7(*a*) for a good overview of what this shift register can do.

A logic diagram of the 74194 register is reproduced in Fig. 9-7(*b*). Because it is a 4-bit register, the circuit contains four flip-flops. Extra gating circuitry is needed for the many features of this universal shift register. The pin configuration in Fig. 9-7(*c*) will help you determine the labeling of each input and output. Of course, the pin diagram is also a must when actually wiring a 74194 IC.

The truth table and waveform diagrams in Fig. 9-7(*d*) and (*e*) are very helpful in determining exactly how the 74194 IC register works because they illustrate the clear, load, shift-right, shift-left, and inhibit modes of operation. If you use the 74194 universal shift register, you will have occasion to look quite carefully at the truth table and waveform diagrams.

**74194 4-bit bidirectional universal shift register**

## Self-Test

*Answer the following questions.*

13. List the five modes of operation for the 74194 universal shift register IC.
14. Refer to Fig. 9-7. If both mode control inputs (*S0, S1*) to the 74194 are HIGH, the unit is in the _____ mode.

15. Refer to Fig. 9-7. If both mode control inputs (*S0, S1*) to the 74194 IC are LOW, the unit is in the _____ mode.
16. Refer to Fig. 9-7. Shift right on the 74194 IC is accomplished when *S0* is _____ (HIGH, LOW) and *S1* is _____ (HIGH, LOW) and when the clock pulse goes from _____ to _____. ✔

This bidirectional shift register is designed to incorporate virtually all of the features a system designer may want in a shift register. The circuit contains 45 equivalent gates and features parallel inputs, parallel outputs, right-shift serial inputs, operating-mode-control inputs, and a direct overriding clear line. The register has distinct modes of operating, namely:

Parallel (broadside) load
Shift right (in the direction $Q_A$ toward $Q_D$)
Shift left (in the direction $Q_D$ toward $Q_A$)
Inhibit clock (do nothing)

Synchronous parallel loading is accomplished by applying the four bits of data and taking both mode control inputs, $S0$ and $S1$, high. The data are loaded into the associated flip-flops and appear at the outputs after the positive transition of the clock input. During loading, serial data flow is inhibited.

Shift right is accomplished synchronously with the rising edge of the clock pulse when $S0$ is high and $S1$ is low. Serial data for this mode is entered at the shift-right data input. When $S0$ is low and $S1$ is high, data shifts left synchronously and new data is entered at the shift-left serial input.

Clocking of the flip-flop is inhibited when both mode control inputs are low. The mode of the S54194/N74194 should be changed only while the clock input is high.

(a)  Description

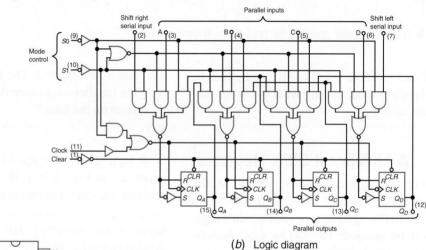

(b)  Logic diagram

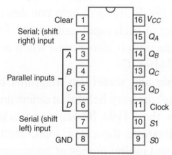

(c)  Pin configuration

| | INPUTS | | | | | | | | | OUTPUTS | | | |
|---|---|---|---|---|---|---|---|---|---|---|---|---|---|
| | MODE | | | SERIAL | | PARALLEL | | | | | | | |
| CLEAR | S1 | S0 | CLOCK | LEFT | RIGHT | A | B | C | D | $Q_A$ | $Q_B$ | $Q_C$ | $Q_D$ |
| L | X | X | X | X | X | X | X | X | X | L | L | L | L |
| H | X | X | L | X | X | X | X | X | X | $Q_{A0}$ | $Q_{B0}$ | $Q_{C0}$ | $Q_{D0}$ |
| H | H | H | ↑ | X | X | a | b | c | d | a | b | c | d |
| H | L | H | ↑ | X | H | X | X | X | X | H | $Q_{An}$ | $Q_{Bn}$ | $Q_{Cn}$ |
| H | L | H | ↑ | X | L | X | X | X | X | L | $Q_{An}$ | $Q_{Bn}$ | $Q_{Cn}$ |
| H | H | L | ↑ | H | X | X | X | X | X | $Q_{Bn}$ | $Q_{Cn}$ | $Q_{Dn}$ | H |
| H | H | L | ↑ | L | X | X | X | X | X | $Q_{Bn}$ | $Q_{Cn}$ | $Q_{Dn}$ | L |
| H | L | L | X | X | X | X | X | X | X | $Q_{A0}$ | $Q_{B0}$ | $Q_{C0}$ | $Q_{D0}$ |

(d)  Function table

H = high level (steady state)
L = low level (steady state)
X = irrelevant (any input, including transitions)
↑ = transition from low to high level
a,b,c,d, = the level of steady state input at inputs A,B,C, or D, respectively
$Q_{A0}$, $Q_{B0}$, $Q_{C0}$, $Q_{D0}$ = the level of $Q_A$, $Q_B$, $Q_C$, $Q_D$, respectively, before the indicated steady state input conditions were established
$Q_{An}$, $Q_{Bn}$, $Q_{Cn}$, $Q_{Dn}$ = the level of $Q_A$, $Q_B$, $Q_C$, $Q_D$, respectively before the most recent ↑ transition of the clock

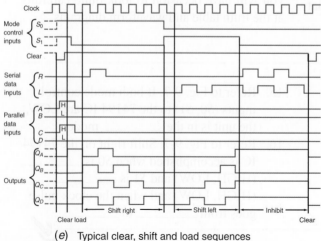

(e)  Typical clear, shift and load sequences

**Fig. 9-7** A 4-bit TTL universal shift register (74194). (a) Description. (b) Logic diagram. (c) Pin configuration. (d) Function (truth) table. (e) Waveforms.

**74194 shift register**

## 9-4 Using the 74194 IC Shift Register

In this section we shall use the 74194 universal shift register in several ways. Figure 9-8(a) and (b) shows the 74194 IC being used as serial load registers. A *serial load shift-right register* is shown in Fig. 9-8(a). This register operates exactly like the serial shift register in Fig. 9-3. Table 9-1 could also be used to chart the performance of this new shift register. Notice that the *mode control inputs* (S0, S1) must be in the positions shown for the 74194 IC to operate in its shift-right mode. Shifting to the right is defined by the manufacturer as shifting from $Q_A$

to $Q_D$. The register in Fig. 9-8(a) shifts data to the right, and as it leaves $Q_D$ the data are lost.

The 74194 IC has been revised slightly in Fig. 9-8(b). The shift-left serial input is used, and the mode control inputs *have been changed*. This register enters data at D ($Q_D$) and shifts them toward A ($Q_A$) with each pulse of the clock. This register is a *serial load shift-left register.*

In Fig. 9-9 on page 302 the 74194 IC is wired as a *parallel load shift-right/left register*. With a single clock pulse, the data from the parallel load inputs A, B, C, and D appear on the display. The loading happens only when the mode controls (S0, S1) are set at 1, as shown. The mode control

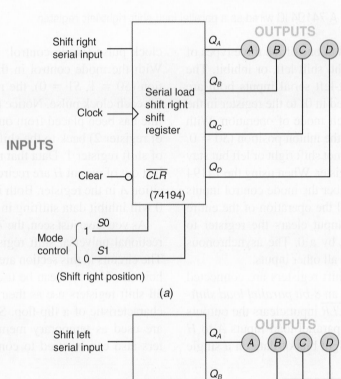

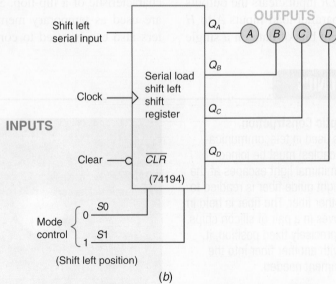

**Fig. 9-8** (*a*) A 74194 IC wired as a 4-bit serial load shift-right register. (*b*) A 74194 IC wired as a 4-bit serial load shift-left register.

*Serial load shift-right register*

*Mode control inputs*

*Serial load shift-left register*

*Parallel load shift-right/left register*

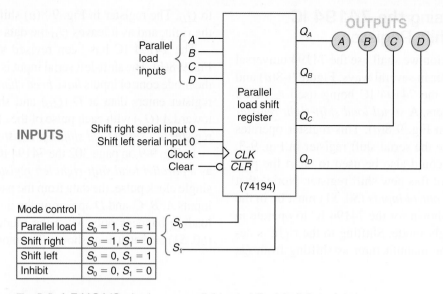

**INPUTS**

Parallel load inputs {
A
B
C
D
}

Shift right serial input 0
Shift left serial input 0
Clock
Clear

Parallel load shift register

CLK
CLR

(74194)

$Q_A$
$Q_B$
$Q_C$
$Q_D$

**OUTPUTS**

Ⓐ Ⓑ Ⓒ Ⓓ

Mode control

| Parallel load | $S_0 = 1, S_1 = 1$ |
|---|---|
| Shift right | $S_0 = 1, S_1 = 0$ |
| Shift left | $S_0 = 0, S_1 = 1$ |
| Inhibit | $S_0 = 0, S_1 = 0$ |

} $S_0$
} $S_1$

**Fig. 9-9** A 74194 IC wired as a parallel load shift-right/left register.

can then be changed to one of the three types of operations: shift right, shift left, or inhibit. The shift-right and shift-left serial inputs both are connected to 0 to feed in 0s to the register in the shift-right or shift-left mode of operation. With the mode control in the inhibit position ($S0 = 0$, $S1 = 0$), the data do not shift right or left but stay in position in the register. When using the 74194 IC you must remember the mode control inputs because they control the operation of the entire register. The $\overline{CLR}$ input clears the register to 0000 when enabled by a 0. The asynchronous $\overline{CLR}$ input overrides all other inputs.

Two 74194 IC shift registers are connected in Fig. 9-10 to form an *8-bit parallel load shift-right register*. The $\overline{CLR}$ input clears the outputs to 0000 0000. The parallel load inputs *A* to *H* allow entry of all eight bits of data on a single

clock pulse (mode control: $S0 = 1$, $S1 = 1$). With the mode control in the shift-right position ($S0 = 1$, $S1 = 0$), the register shifts right for each clock pulse. Notice that a recirculating line has been placed from output *H* (output $Q_D$ of register 2) back to the shift-right serial input of shift register 1. Data that normally would be lost out of output *H* are recirculated back to position *A* in the register. Both inputs $S0$ and $S1$ at 0 will inhibit data shifting in the shift register.

As you have just seen, the 74194 IC 4-bit bidirectional universal shift register is very useful. The circuits in this section are some examples of how the 74194 IC can be used. Remember that all shift registers use as their basis the memory characteristic of a flip-flop. Shift registers often are used as temporary memories. Shift registers also can be used to convert serial data to

**8-bit parallel load shift-right register**

## ABOUT ELECTRONICS

**Guiding Fiber-Optic Construction.**
Light guide fibers used in telecommunications (fiber-optic cables) must be joined carefully so that minimal light escapes at the junction. Here a light guide fiber is readied to be spliced to another fiber. The fiber is held in place by the grooves in a pair of silicon chips. Once it is in this precisely fixed position, it can be merged with another fiber into the near-perfect alignment needed.

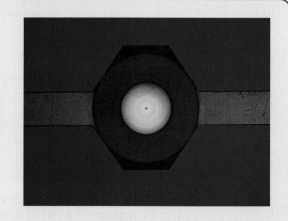

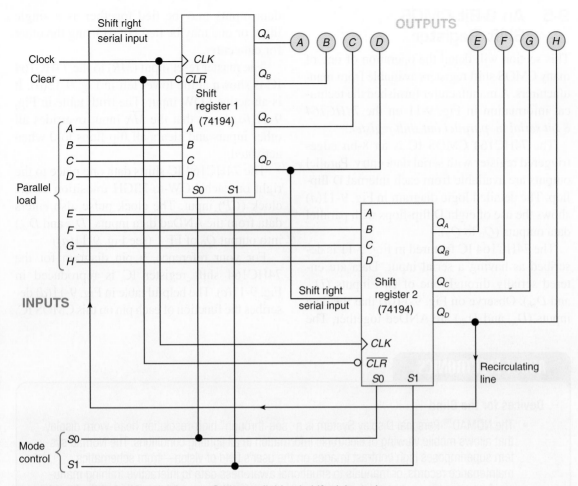

Fig. 9-10 Two 74194 ICs wired as an 8-bit parallel load shift-right register.

parallel data or parallel data to serial data. And shift registers can be used to delay information (delay lines). Shift registers are also used in some arithmetic circuits. Microprocessors and microprocessor-based systems make extensive use of registers similar to the ones used in this chapter. Counterparts to the 74194 are the 74S194, 74LS194A, 74F194, and 74HC194 ICs.

## Self-Test

*Supply the missing word or words in each statement.*

17. The 74194 IC is in the parallel load mode when both mode control inputs (*S0, S1*) are _____ (HIGH, LOW). The four bits of data at the parallel load inputs are loaded into the registers by applying _____ (number) clock pulse(s) to the *CLK* input.

18. If the mode control inputs (*S0, S1*) to the 74194 IC are both LOW, the shift register is in the _____ mode.

19. For the 74194 IC to shift right, the mode controls are *S0* = _____ and *S1* =

_____ and the serial data enter the _____ input.

20. Refer to Fig. 9-9. If *S0* = 1, *S1* = 1, shift left serial input = 1, and clear input = 0, then the outputs are _____.

21. Refer to Fig. 9-7. The 74194 IC is triggered on the _____ (H to L, L to H) transition of the clock pulse.

22. Refer to Fig. 9-7. An active _____ (clear, shift-left serial) input overrides all other inputs and resets the register outputs to 0000 on the 74194 IC.

23. Refer to Fig. 9-7. To _____ (shift left, shift right) means to shift data from $Q_D$ toward the $Q_A$ output on the 74194 IC.

## 9-5 An 8-Bit CMOS Shift Register

This section will detail the operation of one of many CMOS shift registers available from manufacturers. A manufacturer furnished the technical information in Fig. 9-11 on the *74HC164 8-bit serial in–parallel out shift register.*

The 74HC164 CMOS IC is an 8-bit edge-triggered register with serial data entry. Parallel outputs are available from each internal D flip-flop. The detailed logic diagram in Fig. 9-11(*a*) shows the use of eight D flip-flops with parallel data outputs ($Q_0$ to $Q_7$).

The 74HC164 IC featured in Fig. 9-11 is described as having a serial input. Data are entered serially through one of two inputs ($D_{sa}$ and $D_{sb}$). Observe on Fig. 9-11(*a*) that the data inputs ($D_{sa}$ and $D_{sb}$) are ANDed together. The

**74HC164 8-bit serial in–parallel out shift register**

data inputs may be tied together as a single input or one may be tied HIGH using the other for data entry.

The master reset input ($\overline{MR}$) to the 74HC164 IC is shown at the lower left in Fig. 9-11(*a*). It is an active LOW input. The truth table in Fig. 9-11(*b*) shows that the $\overline{MR}$ input overrides all other inputs and clears all flip-flops to 0 when activated.

The 74HC164 IC shifts data one place to the right on each LOW-to-HIGH transition of the clock (*CP*) input. The clock pulse also enters data from the ANDed data inputs ($D_{sa}$ and $D_{sb}$) into output $Q_0$ of FF 1 (see Fig. 9-11(*a*)).

For your reference, a pin diagram for the 74HC164 shift register IC is reproduced in Fig. 9-11(*c*). The helpful table in Fig. 9-11(*d*) describes the function of each pin on this CMOS IC.

## ABOUT ELECTRONICS

### Devices for the Blind

- The NOMAD™ Personal Display System is a "see-through" high-resolution head-worn display that allows mobile viewing of electronic information in all lighting conditions. The Nomad system superimposes high contrast images on the user's field of vision—from schematics, maintenance records, or manuals to situational awareness data to interactive training manuals. A crew can now access information while on scaffolding or working in other difficult positions, *all the time,* working with their hands while viewing critical information.

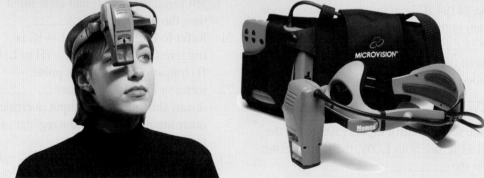

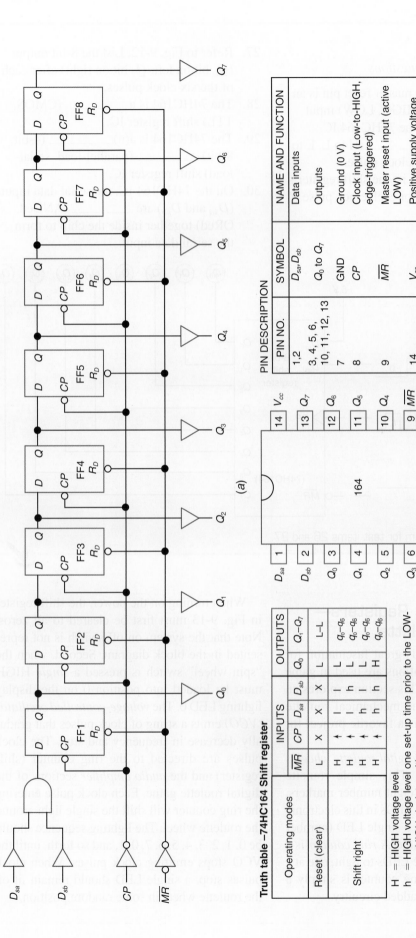

$D_{sa}$ $D_{sb}$ $CP$ $\overline{MR}$

FF1 ... FF8 with $D$, $CP$, $Q$, $R_D$

$Q_0$ $Q_1$ $Q_2$ $Q_3$ $Q_4$ $Q_5$ $Q_6$ $Q_7$

**Truth table –74HC164 Shift register**

| Operating modes | INPUTS | | | | OUTPUTS | | |
|---|---|---|---|---|---|---|---|
| | $\overline{MR}$ | $CP$ | $D_{sa}$ | $D_{sb}$ | $Q_0$ | $Q_1$–$Q_7$ | |
| Reset (clear) | L | X | X | X | L | L–L | |
| Shift right | H | ↑ | l | l | L | $q_0$–$q_6$ | |
| | H | ↑ | l | h | L | $q_0$–$q_6$ | |
| | H | ↑ | h | l | L | $q_0$–$q_6$ | |
| | H | ↑ | h | h | H | $q_0$–$q_6$ | |

H = HIGH voltage level
h = HIGH voltage level one set-up time prior to the LOW-
     to-HIGH clock transition
L = LOW voltage level
l = LOW voltage level one set-up time prior to the LOW-
     to-HIGH clock transition
q = lowercase letters indicate the state of the referenced
     input one set-up time prior to the LOW-to-HIGH
     clock transition
↑ = LOW-to-HIGH clock transition

*(b)*

**PIN DESCRIPTION**

| PIN NO. | SYMBOL | NAME AND FUNCTION |
|---|---|---|
| 1,2 | $D_{sa}, D_{sb}$ | Data inputs |
| 3, 4, 5, 6, 10, 11, 12, 13 | $Q_0$ to $Q_7$ | Outputs |
| 7 | GND | Ground (0 V) |
| 8 | $CP$ | Clock input (Low-to-HIGH, edge-triggered) |
| 9 | $\overline{MR}$ | Master reset input (active LOW) |
| 14 | $V_{cc}$ | Positive supply voltage |

*(d)*

Pin diagram (c):

| 14 $V_{cc}$ | 1 $D_{sa}$ |
| 13 $Q_7$ | 2 $D_{sb}$ |
| 12 $Q_6$ | 3 $Q_0$ |
| 11 $Q_5$ | 4 $Q_1$ |
| 10 $Q_4$ | 5 $Q_2$ |
| 9 $\overline{MR}$ | 6 $Q_3$ |
| 8 $CP$ | 7 GND |

164

*(a)*   *(c)*

**Fig. 9-11** An 8-bit CMOS serial in–parallel out shift register (74HC164). (*a*) Detailed logic diagram. (*b*) Truth table. (*c*) Pin diagram. (*d*) Pin descriptions.

**74HC164 8-bit shift register IC**

*Answer the following questions.*

24. The 74HC164 IC's master reset pin is an active _____ (HIGH, LOW) input.
25. The clock input to the 74HC164 IC responds to a(n) _____ (H to L, L to H) transition of the clock pulse.
26. Refer to Fig. 9-12. List the shift register's mode of operation for each clock pulse ($t_1$ through $t_6$).

27. Refer to Fig. 9-12. List the 8-bit output ($Q_0$ bit on left, $Q_7$ bit on right) after each of the six clock pulses.
28. The 74HC164 is a _____ (CMOS, TTL) shift register IC.
29. The 74HC164 is a(n) _____ (4-bit, 8-bit) _____ (parallel-load, serial-load) shift register IC.
30. On the 74HC164 IC, the serial data inputs ($D_{sa}$ and $D_{sb}$) are _____ (ANDed, ORed) together inside the chip to form the serial data input.

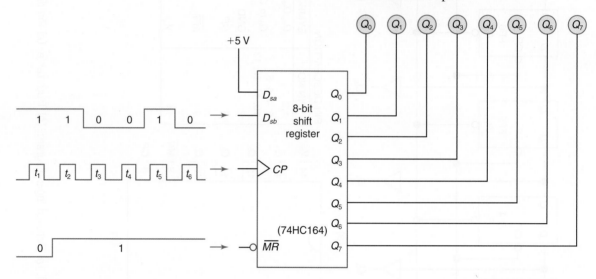

**Fig. 9-12** Shift-register problem for test items 26 and 27.

## 9-6 Using Shift Registers— Digital Roulette

The roulette wheel holds great fascination for people of all ages. Variations are used in game shows and in gaming. This section explores an electronic version of the mechanical roulette wheel. Digital roulette is a favorite project for many students.

A block diagram of a *digital roulette wheel* is sketched in Fig. 9-13. This simple roulette wheel design uses only eight number markers. The number markers are LEDs in this electronic version of roulette. Only a single LED (number marker) must light at a time. A *ring counter* is a circuit that will cause the LEDs to light, one at a time, in sequence. A ring counter is simply a shift register with some added circuitry.

When turning on the power, the shift register in Fig. 9-13 must first be cleared to all zeros. Note that the system on-off switch is not represented in the block diagram. Second, when the "spin wheel" switch is pressed a *single* HIGH must be loaded into position 0 on the display lighting LED 0. The *voltage-controlled oscillator (VCO)* emits a string of clock pulses that gradually decrease in frequency and stop. The clock pulses are directed to the ring counter (shift register) and the *audio amplifier* sections of the digital roulette game. Each clock pulse entering the ring counter will shift the single light around the roulette wheel. The lighting sequence should be 0, 1, 2, 3, 4, 5, 6, 7, 0, 1, and so forth, until the VCO stops emitting clock pulses. When clock pulses stop, a single LED should remain lit on the roulette wheel in some random position.

**Voltage-controlled oscillator (VCO)**

**Digital roulette wheel**

**Audio amplifier**

**Ring counter**

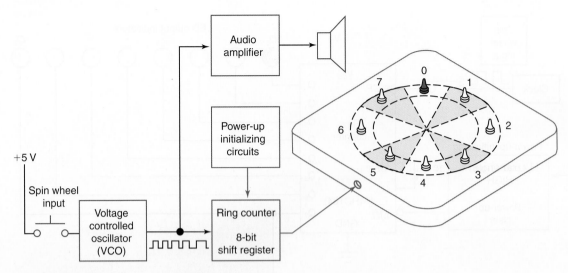

**Fig. 9-13** Block diagram for simplified electronic digital roulette wheel.

The VCO in Fig. 9-13 also sends clock pulses to the audio amplifier section. Each clock pulse is amplified to sound like the click of a roulette wheel. The frequency gradually decreases and stops, simulating a mechanical wheel coasting to a stop.

The ring counter block of the digital roulette game is detailed in Fig. 9-14(a) on page 308. Notice that the ring counter makes use of the 74HC164 8-bit serial in–parallel out shift register IC studied earlier. When power is turned on, the circuits in the power-up initializing block clear all outputs to zero (all LEDs are off). Upon pressing the "spin wheel" input switch, the first pulse loads a single HIGH into the shift register. This situation is illustrated in Fig. 9-14(a). The clock pulses that follow move the single light across the display. This is illustrated in Fig. 9-14(b). Notice that on each L-to-H transition of the clock the single HIGH in the 74HC164 8-bit register shifts one position to the right. When the HIGH reaches output $Q_7$ (after clock pulse eight in Fig. 9-14(b)), a *recirculating line (feedback)* is run back to the data inputs to transfer the HIGH back to the left LED (output $Q_0$). In the example in Fig. 9-14(b), the switch is opened after the twelfth pulse. This stops the light at $Q_3$. This is the "winning number" on the roulette wheel for this spin.

The *74HC164 8-bit shift register IC* is wired as a ring counter in Fig. 9-14(a). The circuit has the two characteristics that make it a ring counter. First, it has feedback from the last flip-flop ($Q_7$) to the first FF ($Q_0$). Second, it is loaded with a given pattern of 1s and 0s and these recirculate as long as clock pulses reach the *CP* input of the shift register. In this case, a single 1 is loaded into the shift register and is recirculated.

In summary, the circuit in Fig. 9-14(a) is a very simple *electronic roulette wheel.* Pressing the spin wheel input causes the single light to be circulated through the LEDs. When the switch opens the shifting stops.

For added appeal, the simple digital roulette circuit in Fig. 9-14 can be changed by adding a clock that will continue to run for a time after the push button is released. Sound could also be added for a more realistic simulation. Figure 9-15 on page 309 adds both features to the digital roulette wheel.

The versatile 555 timer IC is wired as a VCO in Fig. 9-15. Pressing the spin wheel input switch turns on transistor $Q_1$. The 555 timer operates as a free-running MV. This square-wave output from the *VCO* drives both the clock input (*CP*) of the *ring counter* and the *audio amplifier.* Pulses from the VCO alternately turn transistor $Q_2$ on and off, clicking the speaker.

When the spin wheel input switch is opened the 47-μF capacitor holds a positive charge for a time which is applied to the base (*B*) of transistor $Q_1$. This keeps the transistor turned on for several seconds before the capacitor becomes discharged. As the 47-μF capacitor discharges, the voltage at the base of $Q_1$ becomes less and the resistance of the transistor (from emitter to collector) increases. This decreases the frequency

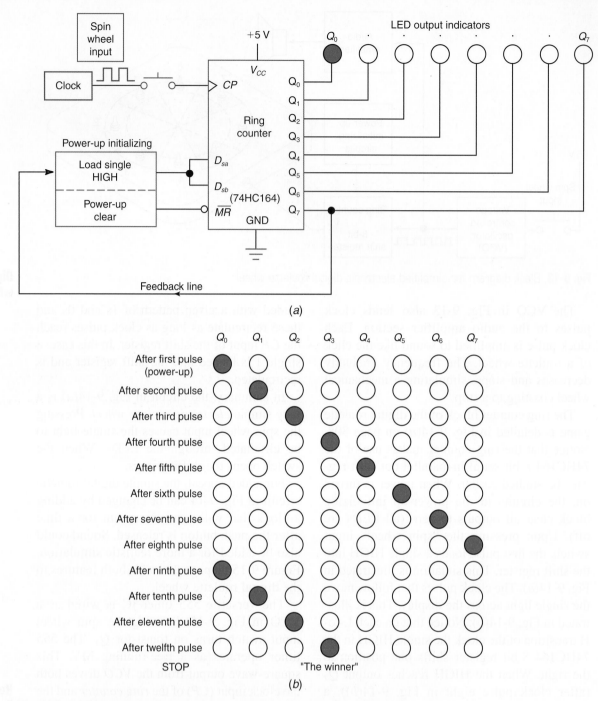

**Fig. 9-14** (a) Ring counter circuit detail in digital roulette wheel. (b) Output from ring counter for first 12 clock pulses.

**Automatic clear circuit**

**Power-up initializing circuitry**

of the oscillator. This causes the shifting light to slow down. The clicking from the speaker also decreases in frequency. This simulates the slowing of a mechanical roulette wheel.

To review, the *power-up initializing circuitry* block in Fig. 9-15 must first clear the shift register and then set only the first output HIGH. These two circuits have been added to the digital roulette wheel in Fig. 9-16 on page 310.

An *automatic clear circuit* has been added to the roulette wheel in Fig. 9-16. It consists of the resistor-capacitor combination ($R_7$ and $C_4$). When power is turned on, the voltage at the top of the 0.01-$\mu$F capacitor starts LOW and increases quickly to a HIGH as it charges through resistor $R_7$. The master reset ($\overline{MR}$) input to the 74HC164 register is held LOW just long enough for the output of the shift register to be

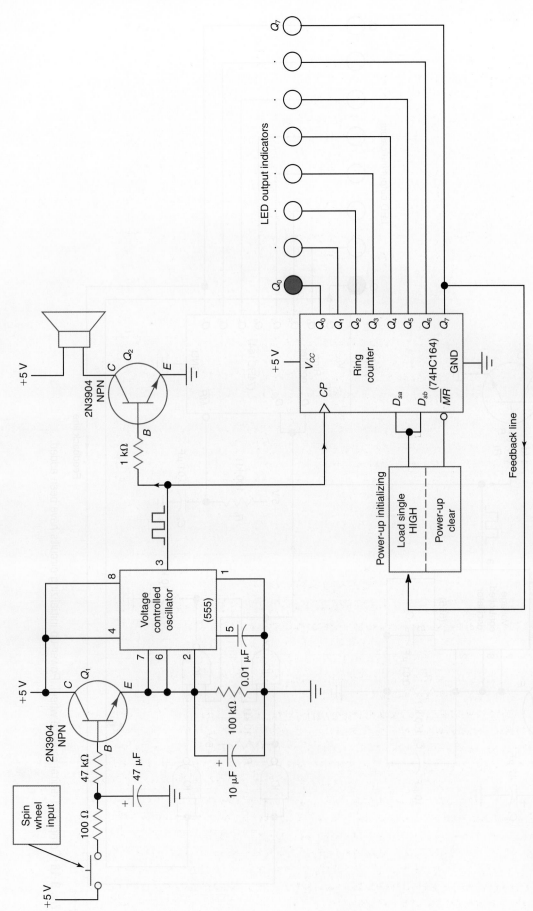

**Fig. 9-15** Voltage-controlled oscillator circuit detail in digital roulette wheel.

**Digital roulette wheel circuit**

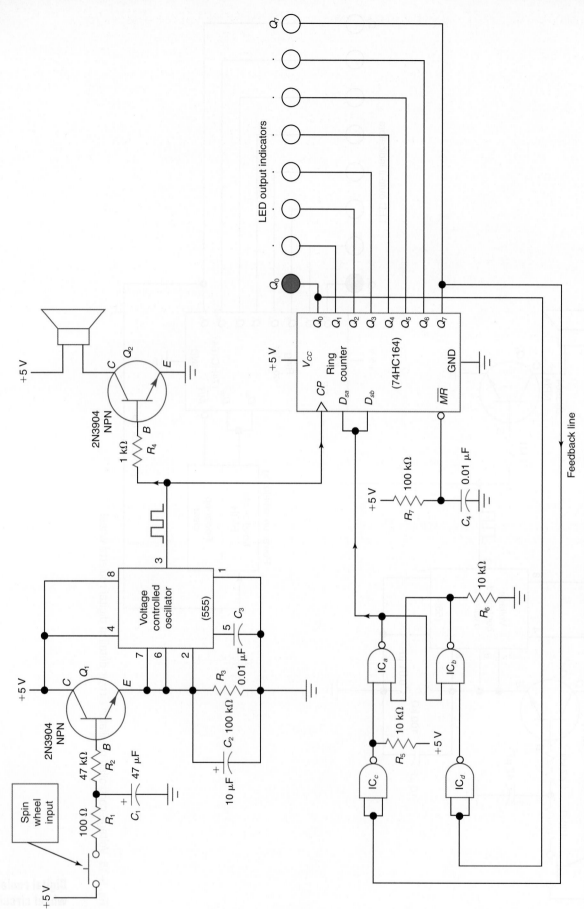

**Fig. 9-16** Completed digital roulette wheel. Power-up initializing circuits have been added.

cleared to 00000000. At this point all the LEDs are off.

The circuit that loads a single 1 into the ring counter consists of the four NAND gates and two resistors ($R_5$ and $R_6$). The NAND gates are wired as an R-S latch. The two resistors ($R_5$ and $R_6$) force the output of the NAND gate (IC*a*) HIGH when the power is first turned on. This HIGH is applied to the data inputs ($D_{sa}$ and $D_{sb}$) of the *ring counter*. On the very first L-to-H transition of the clock the HIGH at the data inputs is transferred to output $Q_0$ of the

74HC164 IC. Immediately this HIGH is fed back to the input of IC*d* and resets the latch so that a LOW now appears at the data inputs ($D_{sa}$ and $D_{sb}$). Only a single HIGH was loaded into the ring counter. Repeated clock pulses move the HIGH (light) across the display until $Q_7$ of the ring counter goes HIGH. This HIGH is fed back to the input of IC*c* setting the latch so that a 1 appears at the data inputs of the ring counter. The single HIGH has been recirculated back to $Q_0$.

**Ring counter**

---

## ✓ Self-Test

*Answer the following questions.*

31. Refer to Fig. 9-16. Components $R_4$, speaker and $Q_2$ form the _____ block of the digital roulette wheel circuit.
32. Refer to Fig. 9-16. The 74HC164 8-bit shift register is wired as a(n) _____ in this circuit.

33. Refer to Fig. 9-16. What components cause the 74HC164 IC to reset all outputs to 0 when the power is first turned on?
34. Refer to Fig. 9-16. Clock pulses are fed to the ring counter by the output of the 555 timer IC wired as a(n) _____.
35. Refer to Fig. 9-16. The four NAND gates used for loading a single 1 in the ring counter are wired as a(n) _____ circuit.

✔

---

## 9-7 Troubleshooting a Simple Shift Register

Consider the faulty serial load shift-right register drawn in Fig. 9-17 on page 312. Four D flip-flops (two 7474 ICs) have been wired together to form this 4-bit register.

After checking for obvious mechanical and temperature problems, the student or technician runs the following sequence of tests to observe the problem:

1. *Action:* Clear input to 0 and back to 1.
   *Result:* Output indicators = 0000 (not lit).
   *Conclusion:* Clear function operating correctly.
2. *Action:* Data input = 1.
   Single pulse to *CLK* of flip-flops from logic pulser.
   *Result:* Output indicators = 1000.
   *Conclusion:* FF A loading 1s properly.
3. *Action:* Data input = 1.
   Single pulse to *CLK* of flip-flops from logic pulser.

*Result:* Output indicators = 1100.
*Conclusion:* FF A and FF B loading 1s correctly.
4. *Action:* Data input = 1.
   Single pulse to *CLK* of flip-flops from logic pulser.
   *Result:* Output indicators = 1110.
   *Conclusion:* FF A, FF B, and FF C loading 1s correctly.
5. *Action:* Data input = 1.
   Single pulse to *CLK* of flip-flops from logic pulser.
   *Result:* Output indicators = 1110.
   *Conclusion:* Suspect problem near or in FF D since it did not load a HIGH properly.
6. *Action:* Logic probe at D input to FF D to see if D = 1.
   *Result:* D = 1 on FF D.
   *Conclusion:* HIGH data at D of FF D is correct.
7. *Action:* One pulse to *CLK* (pin 11) of FF D from logic pulser.

**Troubleshooting**

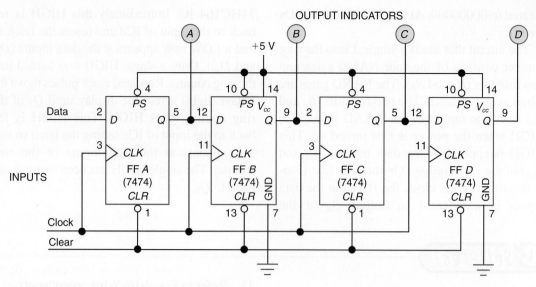

OUTPUT INDICATORS

Fig. 9-17 Faulty 4-bit serial load shift-right register used as a troubleshooting example.

**Faulty 4-bit serial load shift-right register**

**Redundant circuitry**

*Result:* Output indicator remains at 1110.
*Conclusion:* Data not being transferred from input D of FF D to output Q on a clock pulse.

8. *Action:* Logic probe to output Q of FF D (pin 9).
*Result:* Neither HIGH nor LOW indicator lights on logic probe.
*Conclusion:* Output Q (pin 9) of FF D floating between HIGH and LOW. Probably a faulty FF D in second 7474 IC.

9. *Action:* Remove and replace second 7474 IC (FF C and FF D) with exact replacement.

10. *Action:* Retest circuit, starting at step 1.

*Result:* All flip-flops load 1s and 0s.
*Conclusion:* Shift register circuit is now operating properly.

According to the sequence of tests, the Q output of FF D seemed to be stuck LOW while it was actually floating between LOW and HIGH. This fact made our conclusion in step 1 incorrect. This fault was caused by an open circuit within the second 7474 IC itself. Again, the technician's knowledge of how the circuit operates along with observations helped locate the fault. The logic probe and digital logic pulser aided the technician in making observations.

Sometimes the technician is not exactly sure of the appropriate logic level. In a circuit with *redundant circuitry* (circuits repeated over and over), the technician could go back to FF A and FF B and compare these readings with those on FF C and FF D. Digital circuits have much redundant circuitry, and at times this technique is helpful in troubleshooting.

Self-Test

*Answer the following questions.*

36. Refer to Fig. 9-17. Describe the observed problem in this circuit.

37. Refer to Fig. 9-17. What is wrong with this circuit?

38. Refer to Fig. 9-17. How can the fault in this circuit be repaired?

39. What test equipment can be used to troubleshoot this shift register circuit?

# Chapter 9 Summary and Review

## Summary

1. Register is a generic name for a group of memory cells (such as flip-flops) that are considered as a single unit. A few other names used for registers are buffer registers, shift registers and latches.
2. Flip-flops are wired together to form shift registers.
3. A shift register has both a memory and a shift characteristic.
4. A serial load shift register is one that permits only one bit of data to be entered per clock pulse.
5. A parallel load shift register is one that permits all data bits to be entered at one time (one clock pulse).
6. A recirculating register feeds output data back into the input.

7. Shift registers can be designed to shift either left or right.
8. Manufacturers produce many adaptable universal shift registers.
9. Shift registers are widely used as temporary memories and for shifting data. They also have other uses in digital electronic systems.
10. A ring counter is a shift register that (1) has a recirculating line, and (2) is loaded with a pattern of 0s and 1s, which is repeated over and over as the unit is clocked.

## Chapter Review Questions

*Answer the following questions.*

9-1. Draw a logic symbol diagram of a 5-bit serial load shift-right register. Use five D flip-flops. Label inputs data, *CLK*, and *CLR*. Label outputs *A*, *B*, *C*, *D*, and *E*. The circuit will be similar to the one in Fig. 9-3.

9-2. Explain how you would clear to 00000 the 5-bit register you drew in question 9-1.

9-3. After clearing the 5-bit register, explain how you would enter (load) 10000 into the register you drew in question 9-1.

9-4. After clearing the 5-bit register, explain how you would enter (load) 00111 into the register you drew in question 9-1.

9-5. Refer to the register you drew in question 9-1. List the contents of the register after each clock pulse shown in **b** to **e** (assume data input = 0).
   a. Original output = 01001 ($A = 0$, $B = 1$, $C = 0$, $D = 0$, $E = 1$)
   b. After one clock pulse =
   c. After two clock pulses =
   d. After three clock pulses =
   e. After four clock pulses =

9-6. Refer to Fig. 9-9. The parallel load register using the 74194 IC needs _____ (no, one, three, four) clock pulse(s) to load data from the parallel load inputs.

9-7. A _____ (serial, parallel) load shift register is the simplest circuit to wire.

9-8. A _____ (serial, parallel) load shift register is the easiest to load.

9-9. Refer to Fig. 9-7 for questions **a** to **i** on the 74194 IC shift register:
   a. How many bits of information can this register hold?
   b. List the four modes of operation for this register.
   c. What is the purpose of the mode control inputs (*S0*, *S1*)?
   d. The _____ input overrides all other inputs on this register.
   e. What type are the flip-flops and how many are used in this shift register?
   f. The register shifts on the _____ (negative-, positive-) going edge of the clock pulse.

g. What does the inhibit mode of operation mean?

h. By definition, to shift left means to shift data from _____ to _____ (use letters).

i. This register can be loaded _____ (serially, in parallel, either serially or in parallel).

9-10. Refer to Fig. 9-18. List the 74194 shift register's mode of operation during each of the eight clock pulses. Use as answers "clear," "inhibit," "shift right," "shift left," and "parallel load."

9-11. Refer to Fig. 9-11 for questions **a** to **f** on the 74HC164 shift register.

a. How many bits of information can this register store?

b. This shift register is a _____ (CMOS, TTL) IC.

c. This is a _____ (parallel, serial)-load shift register.

d. The master reset is an active _____ (HIGH, LOW) input.

e. The register shifts data on the _____ (H to L, L to H) transition of the clock pulse.

f. The register has two data inputs, which are _____ (ANDed, ORed) together for loading data into FF 1.

9-12. Refer to Fig. 9-19. List the contents of the register during each of the eight clock pulses ($Q_0$ = left bit, $Q_7$ = right bit).

9-13. Refer to Fig. 9-13. The device that generates clock pulses in the digital roulette circuit is called a(n) _____.

9-14. Refer to Fig. 9-14($a$). The 74HC164 shift register is wired as a(n) _____ in this circuit.

9-15. Refer to Fig. 9-16. The frequency of the VCO decreases as the voltage at the top of capacitor _____ ($C_1$, $C_2$, $C_4$) decreases.

9-16. Refer to Fig. 9-16. What is the purpose of resistor $R_7$ and capacitor $C_4$?

9-17. Refer to Fig. 9-16. Resistors $R_5$ and $R_6$ force the output of IC$a$ _____ (HIGH, LOW) when the power is first turned on.

9-18. Refer to Fig. 9-16. If only $Q_0$ of the ring counter is HIGH (as shown), the R-S latch forces the output of IC$a$ _____ (HIGH, LOW).

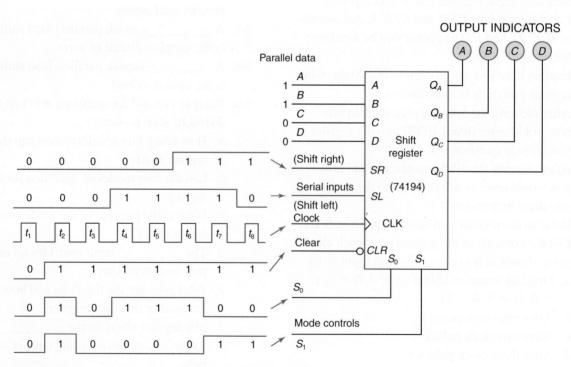

**Fig. 9-18** Shift register problem for critical thinking question 4.

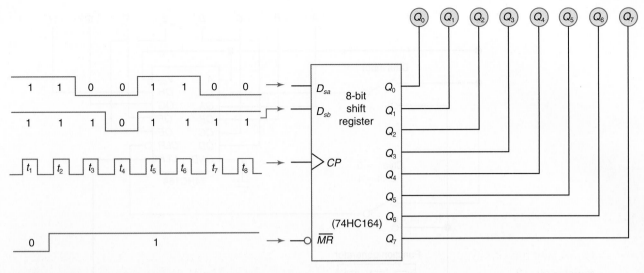

**Fig. 9-19** Shift register problem for review question 12.

# Critical Thinking Questions

9-1. The shift register in Fig. 9-5(*b*) needs _____ (no, one, four) clock pulse(s) to load data from the parallel data inputs.

9-2. The shift register in Fig. 9-5(*b*) can only load _____ (0, 1s) using the parallel load inputs.

9-3. List several uses of shift registers in digital systems.

9-4. List the contents of the register in Fig. 9-18 after each of the eight clock pulses (*A* = left bit, *D* = right bit).

9-5. Describe in general terms the nature of the output from the VCO in Fig. 9-13.

9-6. Refer to Fig. 9-5. Describe the procedure you would follow when loading the data 1101 into this 4-bit parallel-load shift register. Hint: Remember to clear the register to 0000 before activating the asynchronous parallel inputs.

9-7. Refer to Fig. 9-9. Parallel loading of data is a(n) _____ (asynchronous, synchronous) operation when using the 74194 shift register IC.

9-8. A ring counter is classified as a type of _____ (shift register, VCO).

9-9. Draw a block diagram of a 16-bit electronic roulette wheel with VCO, audio amplifier, power-up initializing circuits, and ring counter blocks. It should look something like the 8-bit electronic roulette wheel in Fig. 9-13.

9-10. At the option of your instructor, use Electronics Workbench® or Multisim® circuit simulation software to (1) draw the 8-bit serial-load shift register drawn in Fig. 9-20, (2) test the operation of the shift register, and (3) save the circuit and show your instructor your design.

9-11. At the option of your instructor, use Electronics Workbench® or Multisim® circuit simulation software to (1) add a recirculating line to your 8-bit shift register you designed in question 9-10 (Hint: OR recirculating line and data input), (2) test the operation of the shift register with recirculating feature, and (3) save the circuit and show your instructor your design.

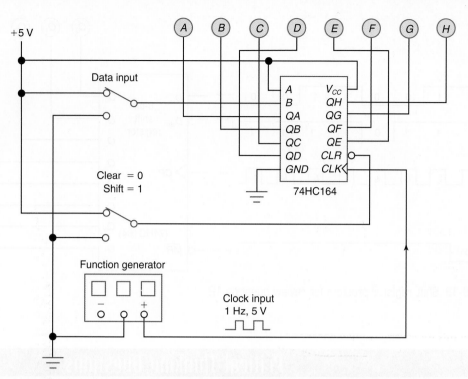

**Fig. 9-20** EWB circuit problem. (Prepared using Electronics Workbench®, version 5)

## Answers to Self-Tests

1. serial
2. after pulse $t_1 = 000$
   after pulse $t_2 = 100$
   after pulse $t_3 = 010$
   after pulse $t_4 = 001$
   after pulse $t_5 = 000$
   after pulse $t_6 = 100$
3. single bit
4. LOW
5. HIGH, L-to-H
6. parallel
7. pulse $t_1$ = clear
   pulse $t_2$ = parallel load
   pulse $t_3$ = shift-right
   pulse $t_4$ = shift-right
   pulse $t_5$ = shift-right
   pulse $t_6$ = parallel load
   pulse $t_7$ = shift-right
   pulse $t_8$ = shift-right
8. after pulse $t_1 = 000$
   after pulse $t_2 = 010$
   after pulse $t_3 = 001$

after pulse $t_4 = 100$
after pulse $t_5 = 10$
after pulse $t_6 = 101$
after pulse $t_7 = 110$
after pulse $t_8 = 011$
9. recirculating
10. asynchronous
11. lines 2 and 9
12. lines 3 and 10
13. 1. clear
    2. parallel load
    3. shift-right
    4. shift-left
    5. inhibit (do nothing)
14. parallel load
15. inhibit
16. HIGH, LOW, LOW, HIGH
17. HIGH, one
18. inhibit
19. 1, 0, shift-right serial
20. 0000 (cleared)
21. L-to-H
22. clear

23. shift left
24. LOW
25. L-to-H or LOW-to-HIGH
26. during pulse $t_1$ = reset
    during pulse $t_2$ = shift-right
    during pulse $t_3$ = shift-right
    during pulse $t_4$ = shift-right
    during pulse $t_5$ = shift-right
    during pulse $t_6$ = shift-right
27. during pulse $t_1$ = 00000000
    during pulse $t_2$ = 10000000
    during pulse $t_3$ = 01000000
    during pulse $t_4$ = 00100000

during pulse $t_5$ = 10010000
during pulse $t_6$ = 01001000
28. CMOS
29. 8-bit, serial-load
30. ANDed
31. audio amplifier
32. ring counter
33. $R_7$ and $C_4$
34. voltage-controlled oscillator or VCO
35. R-S latch or latch
36. will not shift a HIGH into the D position
37. Output Q (pin 9) of FF D floating; 7474 IC that contains FF C and FF D faulty
38. A new 7474 IC should be inserted, replacing FF C and FF D.
39. logic pulser, logic probe

# Arithmetic Circuits

## Chapter Objectives

*This chapter will help you to:*

1. *Memorize* and *draw* the block diagrams for a half adder, a full adder, a half subtractor, and a full subtractor.

2. *Solve* binary addition and subtraction problems by hand and from a truth table.

3. *Design* and *draw* block-style logic diagrams for several parallel adder and subtractor circuits using half adders, full adders, and gates.

4. *Use* the 7483 IC as a 4-bit adder and *cascade* two 7483 ICs to form an 8-bit binary adder circuit.

5. *Solve* several binary multiplication problems.

6. *Convert* decimal numbers to 2s complement notation and 2s complement to decimal numbers.

7. *Add* and *subtract* signed numbers using 2s complement addition and subtraction.

8. *Troubleshoot* a faulty full-adder circuit.

The public's imagination has been captured by computers and modern-day calculators, probably because these machines perform arithmetic tasks with such fantastic speed and accuracy. This chapter deals with some logic circuits that can add and subtract. (Of course, the adding and subtracting is done in binary.) Regular logic gates will be wired together to form *adders* and *subtractors*. Basic adder and subtractor circuits are combinational logic circuits, but they are commonly used with various latches and registers to hold data.

In the *central processing unit (CPU)* of a computer, arithmetic is handled in a section commonly called the *arithmetic-logic unit (ALU)*. This section within the CPU can usually add and subtract, multiply and divide, complement, compare, shift and rotate, increment and decrement, and perform logic operations such as AND, OR, and XOR. Many older *microprocessors* and several modern *microcontrollers* (a miniature microprocessor used mainly for control purposes) do not have multiply and divide commands in their instruction set.

**Central processing unit (CPU)**

**Arithmetic-logic unit (ALU)**

**Microcontroller**

## 10-1 Binary Addition

**Binary addition**

Remember that in a binary number, such as 101011, the leftmost digit is the *MSB* and the rightmost digit is the *LSB*. Also remember the place values given to the binary number is: 1s, 2s, 4s, 8s, 16s, and 32s.

**MSB**

**LSB**

You probably still recall learning your addition and subtraction tables when you were in elementary school. This is a difficult task in the decimal number system because there are so many combinations. This section deals with the simple task of adding numbers in binary.

$$\begin{array}{cccc} 0 & 1 & 0 & 1 \\ +0 & +0 & +1 & +1 \\ \hline 0 & 1 & 1 & 0_{\text{carry 1}} \end{array}$$

(a)

$$
\begin{array}{ccc}
\phantom{00}1\phantom{00} & & \\
101 & 5 & \\
+\phantom{0}10 & +2 & \\
\hline
111 & 7 &
\end{array}
\qquad
\begin{array}{ccc}
\overset{\text{carry}}{\phantom{0}1\phantom{000}} & & \\
10|10 & 10 & \\
\phantom{0}|11 & +\phantom{0}3 & \\
\hline
11|01 & 13 &
\end{array}
\qquad
\begin{array}{ccc}
\overset{\text{carry}\quad\text{carry}}{1\,1\phantom{00}} & & \\
1|1010 & 26 & \\
|1100 & +12 & \\
\hline
1|00110 & 38 &
\end{array}
$$

(b)

**Binary addition tables**

**Fig. 10-1** (a) Binary addition tables. (b) Sample binary addition problems.

Because they have only two digits (0 and 1), the binary addition tables are simple. Figure 10-1(a) shows the binary addition tables. Just as in the case of adding with decimals, the first three problems are easy. The next problem is 1 + 1. In decimal that would be 2. In binary a 2 is written as 10. Therefore, in binary 1 + 1 = 0, with a carry of 1 to the next most significant place value.

Figure 10-1(b) shows some examples of adding numbers in binary. The problems are also shown in decimal so that you can check your understanding of binary addition. The first problem is adding binary 101 to 10, which equals 111 (decimal 7). This problem is simple using the addition tables in Fig. 10-1(a). The second problem in Fig. 10-1(b) is adding binary 1010 to 11. Here you must notice that a 1 + 1 = 0 plus a carry from the 2s place to the 4s place, as shown in the diagram. The answer to this problem is 1101 (decimal 13). In the third problem in Fig. 10-1(b), the binary

number 11010 is added to 1100. In the figure, note two carries with the solution as 100110 (decimal 38).

Another sample addition problem is shown in Fig. 10-2(a). The solution looks simple until we get to the 2s column and find 1 + 1 + 1 in binary. This equals 3 in decimal, which is 11 in binary. This one situation we left out of the first group of binary addition tables. Looking carefully at Fig. 10-2, you see that the 1 + 1 + 1 situation can arise in any column except the 1s column. So the binary addition table in Fig. 10-1(a) is complete for the *1s column only*. The new short-form addition table in Fig. 10-2(b) adds the other possible combination of 1 + 1 + 1. The addition table in Fig. 10-2(b), then, is for all the place values (2s, 4s, 8s, 16s, and so on) except the 1s column.

To be an intelligent worker on digital equipment, you must master binary addition. Several practice problems are provided in the first test.

*Answer the following questions.*

1. What is the sum of binary 1010 + 0100? (Check your answer using decimal addition.)

2. What is the sum of binary 1010 + 0111?
3. What is the sum of binary 1111 + 1001?
4. What is the sum of binary 10011 + 0111?

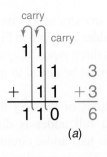

$$\begin{array}{r} \text{carry} \\ \overbrace{\phantom{1}1}^{\curvearrowleft}\text{carry} \\ 1\ 1 \qquad\quad \\ 1\ 1 \qquad 3 \\ +\ 1\ 1 \qquad +3 \\ \hline 1\ 1\ 0 \qquad 6 \end{array}$$

(a)

$$\begin{array}{cccc}
 & & & 1 \\
0 & 1 & 1 & 1 \\
+0 & +0 & +1 & +1 \\
\hline
0 & 1 & 0\ \text{carry 1} & 1\ \text{carry 1}
\end{array}$$

(b)

**Fig. 10-2** (a) Sample binary addition problem. (b) Short-form addition table.

## 10-2 Half Adders

The addition table in Fig. 10-1(a) can be thought of as a truth table. The numbers being added are on the input side of the table. In Fig. 10-3(a), these are the A and B input columns. The truth table needs *two* output columns, one column for the sum and one column for the carry. The sum column is labeled with the summation symbol $\Sigma$. The carry column is labeled with a $C_O$. The $C_O$ stands for carry output or *carry out*. A convenient block symbol for the adder that performs the job of the truth table is shown in Fig. 10-3(b). This circuit is called a *half-adder circuit*. The half-adder circuit has two inputs (A, B) and two outputs, ($\Sigma$, $C_O$).

Take a careful look at the half-adder truth table in Fig. 10-3(a). What is the Boolean expression needed for the $C_O$ output? The Boolean expression is $A \cdot B = C_O$. You need a two-input AND gate to take care of output $C_O$.

Now what is the Boolean expression for the sum ($\Sigma$) output of the half adder in Fig. 10-3(a)? The Boolean expression is $\overline{A} \cdot B + A \cdot \overline{B} = \Sigma$. Two AND gates, two inverters, and

one OR gate will do the job. If you look closely, you will notice that this pattern is also that of an XOR gate. The simplified Boolean expression is then $A \oplus B = \Sigma$. In other words, we find that only one 2-input XOR gate is needed to produce the sum output.

Using a two-input AND gate and a two-input XOR gate, a logic symbol diagram for a half adder is drawn in Fig. 10-3(c). The half-adder circuit adds only the LSB column (1s column) in a binary addition problem. A circuit called a *full adder* must be used for the 2s, 4s, 8s, and 16s, and higher places in binary addition.

| INPUTS | | OUTPUTS | |
|---|---|---|---|
| **B** | **A** | $\Sigma$ | $C_O$ |
| 0 | 0 | 0 | 0 |
| 0 | 1 | 1 | 0 |
| 1 | 0 | 1 | 0 |
| 1 | 1 | 0 | 1 |
| Binary digits to be added | | Sum | Carry out |
| | | XOR | AND |

(a)

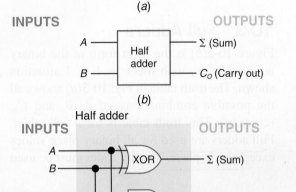

(b)

(c)

**Fig. 10-3** Half adder. (a) Truth table. (b) Block symbol. (c) Logic diagram.

**Half adder**

**Half-adder circuit**

**Full adder**

*Answer the following questions.*

5. Draw a block diagram of a half adder. Label inputs A and B; label outputs $\Sigma$ and $C_O$.
6. Draw a truth table for a half adder.

7. A half-adder circuit is used for adding only the _____ (1s, 2s, 4s, 8s) column of a binary addition problem.
8. Refer to Fig. 10-4. List the outputs from both the sum ($\Sigma$) and carry out ($C_O$) terminals of the half-adder circuit for each input pulse ($t_1$ to $t_4$).

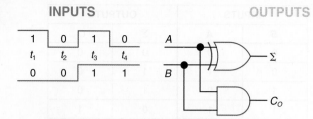

INPUTS | OUTPUTS

| 1 | 0 | 1 | 0 |
|---|---|---|---|
| $t_1$ | $t_2$ | $t_3$ | $t_4$ |
| 0 | 0 | 1 | 1 |

**Fig. 10-4** Half adder pulse-train problem for test question 8.

## 10-3 Full Adders

Figure 10-2(b) is the short form of the binary addition table, with the $1 + 1 + 1$ situation shown. The truth table in Fig. 10-5(a) shows all the possible combinations, of $A$, $B$, and $C_{in}$ (carry in). This truth table is for a full adder. Full adders are used for all binary place values except the 1s place. The full adder must be used

when it is possible to have an extra *carry input*. A block diagram of a full adder is shown in Fig. 10-5(b). The full adder has three inputs: $C_{in}$, $A$, and $B$. These three inputs must be added to get the $\Sigma$ and $C_O$ outputs.

One of the easiest methods of forming the combinational logic for a full adder is diagramed in Fig. 10-5(c); two half-adder circuits and an OR gate are used. The expression for this arrangement is $A \oplus B \oplus C = \Sigma$. The expression for the carry out is $A \cdot B + C_{in} \cdot (A \oplus B) = C_O$. The logic circuit in Fig. 10-6(a) is a full adder. This circuit is based upon the block diagram using two half adders shown in Fig. 10-5(c). Directly below this logic diagram is a logic circuit that is somewhat easier to wire. Figure 10-6(b) contains two XOR gates and three NAND gates, which makes the circuit fairly easy to wire. Notice that the circuit in Fig 10-6(b) is exactly the same as the one in

| INPUTS | | | OUTPUTS | |
|---|---|---|---|---|
| $C_{in}$ | $B$ | $A$ | $\Sigma$ | $C_O$ |
| 0 | 0 | 0 | 0 | 0 |
| 0 | 0 | 1 | 1 | 0 |
| 0 | 1 | 0 | 1 | 0 |
| 0 | 1 | 1 | 0 | 1 |
| 1 | 0 | 0 | 1 | 0 |
| 1 | 0 | 1 | 0 | 1 |
| 1 | 1 | 0 | 0 | 1 |
| 1 | 1 | 1 | 1 | 1 |
| Carry + B + A | | | Sum | Carry out |

(a)

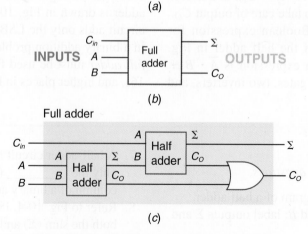

(b)

Full adder

(c)

**Fig. 10-5** Full adder. (a) Truth table. (b) Block symbol. (c) Constructed from half adders and an OR gate.

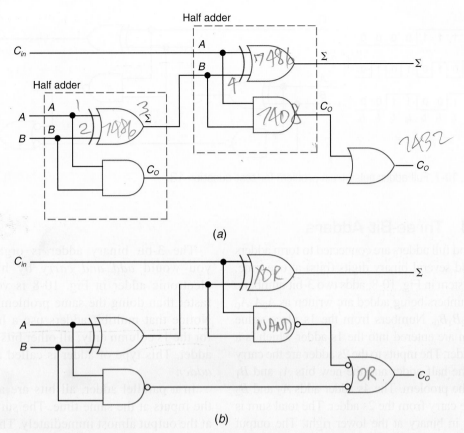

Half adder

$C_{in}$

A

B

Σ

Σ

Half adder

A

A

B

B

Σ

$C_O$

$C_O$

$C_O$

(a)

$C_{in}$

XOR

Σ

A

B

NAND

OR

$C_O$

(b)

**Fig. 10-6** Full adder. (a) Logic diagram. (b) Logic diagram using XOR and NAND gates.

Fig. 10-6(a), except that NAND gates have been substituted for AND and OR gates.

Half and full adders are used together. For the problem in Fig. 10-2(a) we need one half adder for the 1s place and two full adders for the 2s place and the 4s place value. Half and full adders are rather simple circuits. However, many of these circuits are needed to add longer problems (more binary digits).

Many circuits similar to half and full adders are part of a microprocessor's *arithmetic-logic unit (ALU)*. These circuits are then used for adding 8-bit or 16- or 32-bit binary numbers in a microcomputer system. The microprocessor's ALU can also subtract using the same half- and full-adder circuits. Later in this chapter, you will use adders to perform binary subtraction.

*Answer the following questions.*

9. Draw a block diagram of a full adder. Label inputs *A*, *B*, and $C_{in}$; label outputs Σ and $C_O$.
10. Draw a truth table for a full adder.
11. Adder circuits are widely used in the _____ section of a microprocessor.
12. A _____ (half-adder, full-adder) circuit must be used for the 2s, 4s, 8s, and more significant bits in a binary addition problem.
13. Refer to Fig. 10-7 on page 322. List the outputs from both the sum Σ) and carry out ($C_O$) terminals of the full-adder circuit for each of the input pulses ($t_1$ to $t_8$).

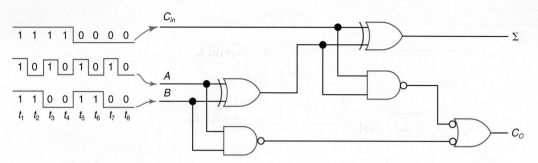

$t_1$ $t_2$ $t_3$ $t_4$ $t_5$ $t_6$ $t_7$ $t_8$

**Fig. 10-7** Full adder pulse-train problem for test question 13.

**3-bit adders**

## 10-4 Three-Bit Adders

Half and full adders are connected to form adders that add several binary digits (bits) at one time. The system in Fig. 10-8, adds two 3-bit numbers. The numbers being added are written as $A_2A_1A_0$ and $B_2B_1B_0$. Numbers from the 1s place value column are entered into the 1s adder which is a

**Parallel adder**

half adder. The inputs to the 2s adder are the carry from the half adder and the new bits $A_1$ and $B_1$ from the problem. The 4s adder adds $A_2$ and $B_2$ and the carry from the 2s adder. The total sum is shown in binary at the lower right. The output

**Combinational logic circuit**

also has an 8s place value to take care of any binary number over 111 in the sum. Notice that the 4s adder's output ($C_O$) is connected to the 8s sum indicator.

The 3-bit binary adder is organized as you would *add and carry* by hand. The electronic adder in Fig. 10-8 is very much faster than doing the same problem by hand. Notice that multibit adders use a half adder for the 1s column only; all other bits use a full adder. This type of adder is called a *parallel adder*.

In a parallel adder, all bits are applied to the inputs at the same time. The sum appears at the output almost immediately. The parallel adder shown in Fig. 10-8 is a *combinational logic circuit* and typically needs various registers to latch data at the inputs and outputs.

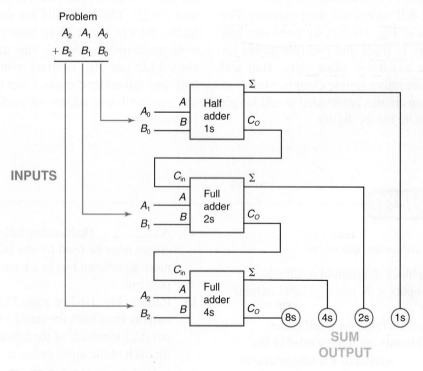

**Fig. 10-8** A 3-bit parallel adder.

*Supply the missing word (or words) in each statement.*

14. The unit in Fig. 10-8 uses a(n) _____ for adding the 1s column and _____ for the more significant columns.

15. Parallel adders are _____ (combinational, sequential) logic circuits.

16. If the inputs to the 3-bit binary adder in Fig. 10-8 are $110_2$ and $111_2$, the output indicators will show a sum of _____ in binary.

17. If the inputs to the 3-bit binary adder in Fig. 10-8 are $010_2$ and $110_2$, the output indicators will show a sum of _____ in binary.

18. If the inputs to the 3-bit binary adder in Fig. 10-8 are $111_2$ and $111_2$, the output indicators will show a sum of _____ in binary.

## 10-5 Binary Subtraction

You will find that *adders* and *subtractors* are very similar. You use *half subtractors* and *full subtractors* just as you use half and full adders. Binary subtraction tables are shown in Fig. 10-9(*a*). Converting these rules to truth-table form gives the table in Fig. 10-9(*b*). On the input side, *B* is subtracted from *A* to give output *Di* (difference). If *B* is larger than *A*, such as in line 2, we need a *borrow*, which is shown in the column labeled $B_O$ (borrow out).

A block diagram of a half subtractor is shown in Fig. 10-9(*c*). Inputs *A* and *B* are on the left. Outputs *Di* and $B_O$ are on the right side of the diagram. Looking at the truth table in Fig. 10-9(*b*), we can determine the Boolean expressions for the half subtractor. The expression for the *Di* column is $A \oplus B = Di$. This is the same as for the half adder (see Fig. 10-3(*a*)). The Boolean expression for the $B_O$ column is $\overline{A} \cdot B = B_O$. Combining these two expressions in a logic diagram gives the logic circuit in Fig. 10-9(*d*). This is the logic circuit for a half subtractor; notice how much it looks like the half-adder circuit in Fig. 10-4.

When you subtract several columns of binary digits, you must take into account the borrowing. Suppose you are subtracting the numbers in Fig. 10-10(*a*) on page 324. You might keep track of the differences and borrows as shown in the figure. Look over the subtraction problem carefully, and check if you can do binary subtraction by this longhand method. (You can check yourself on the next test.)

**Binary subtraction**
**Half subtractors**
**Full subtractors**

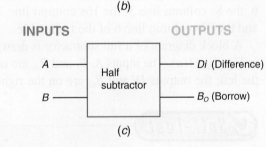

(a)

| INPUTS | | OUTPUTS | |
|---|---|---|---|
| **A** | **B** | *Di* | $B_O$ |
| 0 | 0 | 0 | 0 |
| 0 | 1 | 1 | 1 |
| 1 | 0 | 1 | 0 |
| 1 | 1 | 0 | 0 |
| A − B | | Difference | Borrow out |

(b)

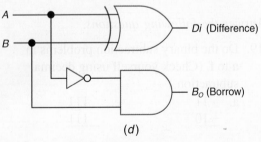

(c)

(d)

**Fig. 10-9** (*a*) Binary subtraction tables. (*b*) Truth table for the half subtractor. (*c*) Block symbol of half subtractor. (*d*) Logic diagram for half subtractor.

**Binary subtraction tables**

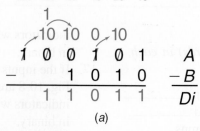

```
      32s 16s 8s  4s  2s  1s
        1
      ↗10  10  0 ↗10
    1   0   0   1   0   1      A
 –          1   0   1   0    – B
    1   1   0   1   1         Di
```

(a)

| INPUTS | | | OUTPUTS | |
|---|---|---|---|---|
| **A** | **B** | **B_in** | **Di** | **B_O** |
| 0 | 0 | 0 | 0 | 0 |
| 0 | 0 | 1 | 1 | 1 |
| 0 | 1 | 0 | 1 | 1 |
| 0 | 1 | 1 | 0 | 1 |
| 1 | 0 | 0 | 1 | 0 |
| 1 | 0 | 1 | 0 | 0 |
| 1 | 1 | 0 | 0 | 0 |
| 1 | 1 | 1 | 1 | 1 |
| $A - B - B_{in}$ | | | Difference | Borrow out |

(b)

**Fig. 10-10** (a) Sample binary subtraction problem.
(b) Truth table for a full subtractor.

**Truth table for a full subtractor**

A truth table that considers all the possible combinations in binary subtraction is shown in Fig. 10-10(b). For instance, line 5 of the table is the situation in the 1s column in Fig. 10-10(a). The 2s column equals line 3, the 4s column line 6, the 8s column line 3, the 16s column line 2, and the 32s column line 6 of the truth table.

A block diagram of a full subtractor is drawn in Fig. 10-11(a). The inputs $A$, $B$, and $B_{in}$ are on the left; the outputs $Di$ and $B_O$ are on the right.

Like the full adder, the full subtractor can be wired using two half subtractors and an OR gate. Figure 10-11(b) is a full subtractor showing how half subtractors are used. A logic diagram for a full subtractor is shown in Fig. 10-11(c). This circuit performs as a full subtractor as specified in the truth table in Fig. 10-10(b). The AND-OR circuit on the $B_O$ output can be converted to three NAND gates if you want. The circuit would then be similar to the full-adder circuit in Fig. 10-6(b).

**Self-Test**

*Answer the following questions.*

19. Do the binary subtraction problems in **a** to **f**. (Check yourself using decimal subtraction.)

**Half subtractors**

a.    11
    – 10

b.   100
    – 10

c.   111
   – 111

d. 1010
  – 101

e. 10010
  –   11

f. 1000
 –  01

20. Draw a block diagram of a half subtractor. Label inputs $A$ and $B$; outputs $Di$ and $B_O$.
21. Draw a truth table for a *half subtractor*.
22. Draw a block diagram for a full subtractor. Label inputs $A$, $B$, and $B_{in}$; label outputs $Di$ and $B_O$.
23. Draw a truth table for a full subtractor.

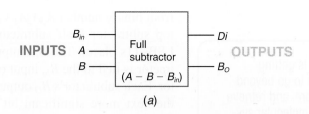

INPUTS    OUTPUTS

(a)

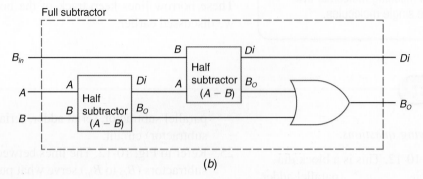

(b)

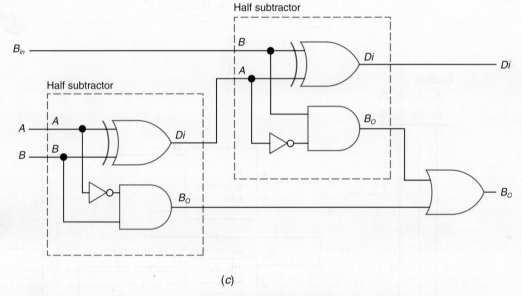

(c)

**Fig. 10-11** Full subtractor. (a) Block symbol. (b) Constructed with half subtractors and an OR gate. (c) Logic diagram.

## 10-6 Parallel Subtractors

Half and full subtractors are wired together to perform as a *parallel subtractor*. You have already seen adders connected as parallel adders. An example of a parallel adder is the 3-bit adder in Fig. 10-8. A parallel subtractor is wired in a similar manner. The adder in Fig. 10-8 is considered a parallel adder because all the digits from the problem flow into the adder at the same time.

Figure 10-12 on page 326 diagrams the wiring of a single half subtractor and three full subtractors. This forms a 4-bit parallel subtractor that can subtract binary number $B_3B_2B_1B_0$

### ABOUT ELECTRONICS

**Wilderness Locator** One of the most dangerous aspects of a hike in the wilderness has always been the problem of finding help in the event of an accident—at least until recently. New handheld devices, such as this one from Magellan, utilize satellite technology to send messages such as distress signals and to identify the exact coordinates of the device to rescuers.

from binary number $A_3A_2A_1A_0$. Notice that the top subtractor (half subtractor) subtracts the LSBs (1s place). The $B_O$ output of the 1s subtractor is tied to the $B_{in}$ input of the 2s subtractor. Each subtractor's $B_O$ output is connected to the next more significant bit's borrow input. These borrow lines keep track of the borrows we discussed earlier.

**Self-Test**

*Answer the following questions.*

24. Refer to Fig. 10-12. This is a block diagram of a 4-bit _____ (parallel adder, parallel subtractor, serial adder, serial subtractor) circuit.

25. Refer to Fig. 10-12. The lines between subtractors ($B_O$ to $B_{in}$) serve what purpose in this circuit?

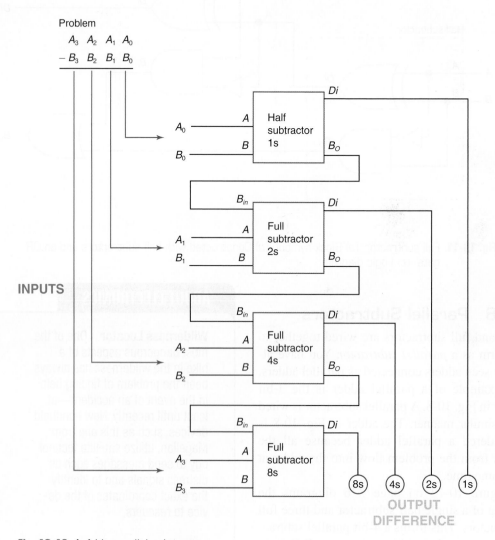

Problem

$A_3$  $A_2$  $A_1$  $A_0$
$- B_3$  $B_2$  $B_1$  $B_0$

INPUTS

**Fig. 10-12** A 4-bit parallel subtractor.

## 10-7  IC Adders

IC manufacturers produce several adders. One elementary arithmetic IC is the *TTL 7483 4-bit binary full adder.* A block symbol for the 7483 IC adder is drawn in Fig. 10-13. The problem of addition of the two 4-bit binary numbers $(A_3A_2A_1A_0$ and $B_3B_2B_1B_0)$ is shown being entered into the eight inputs of the 7483 IC. Notice a difference in numbering systems on the problem and the IC (subscripts don't match). For adding just two 4-bit numbers, the $C_O$ input is held at 0. The $C_O$ input is marked as the $C_{in}$ input by some manufacturers. The sum outputs are shown attached to output indicators. The $C_4$ output is attached to the 16s output indicator. The $C_4$ output is marked as the $C_O$ output by some manufacturers. This binary adder can indicate a sum as high as 11110 (decimal 30) when adding binary 1111 to 1111.

The internal organization of the 7483 adder IC is detailed in Fig. 10-14. The 7483 IC is a combinational logic circuit with no memory capabilities. The pin numbers used on DIP 7483 ICs are shown on the logic diagram in Fig. 10-14 with numbers in parenthesis. For instance, data input A1 is pin 10 on the DIP version of the 7483 adder IC. You will observe from the logic diagram in Fig. 10-14 that the circuitry is fairly complex.

The 7483 adder can be *cascaded* by connecting $C_4$ (carry out) output of *IC1* to the $C_o$ (carry input) of the next 7483 IC (*IC2*). The details for cascading of two 7483 adders is shown in Fig. 10-15. This circuit is an *8-bit binary adder.* This circuit will add the 8-bit binary inputs $A_7A_6A_5A_4A_3A_2A_1A_0$ to $B_7B_6B_5B_4B_3B_2B_1B_0$ yielding a 9-bit binary sum. The 8-bit binary adder can handle a maximum 9-bit sum of $111111110_2$ ($1FE_{16}$ or $510_{10}$). For instance, if the inputs are $00011100_2$ and $11100011_2$ then the output will be $11111111_2$ (in hexadecimal that would be $1C + E3 = FF$).

Counterparts to the 7483 4-bit adder are the 74LS83, 74C83, and 4008 ICs. Other 4-bit adders that function the same as the 7483 IC but have a different pin configuration are the 74283, 74LS283, 74S283, 74F283, and 74HC283.

A more complex arithmetic chip is the 74LS181 IC. The 74LS181 and its relatives, the 74LS381, are described as *arithmetic-logic units/function generators.* These units perform

**TTL 7483 4-bit binary full adder**

**Cascading adders**

**8-bit binary adder**

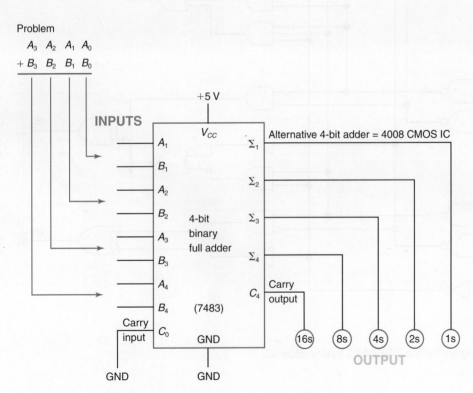

**Fig. 10-13** The 7483 4-bit binary adder IC.

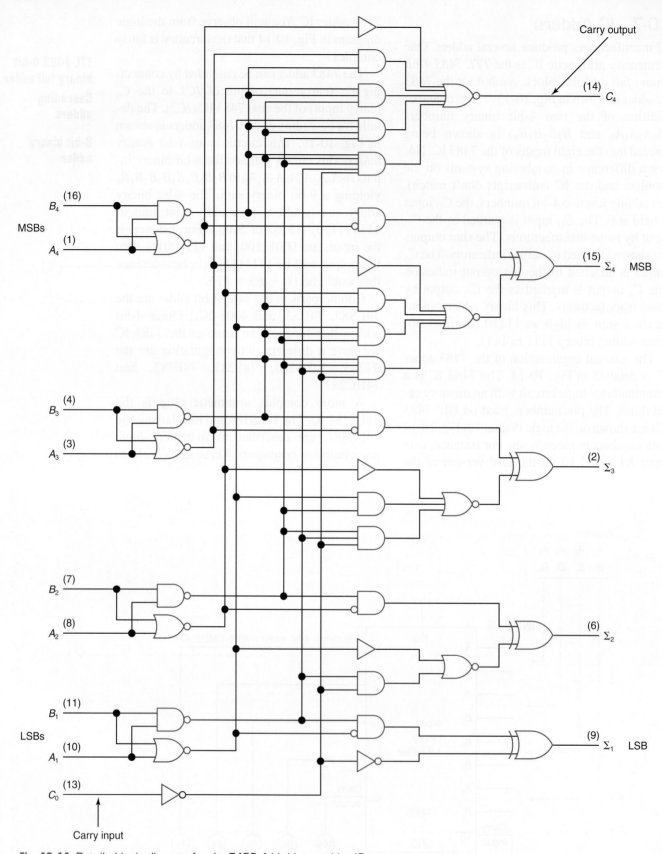

Carry output

$C_4$ (14)

$B_4$ (16)

MSBs

$A_4$ (1)

$\Sigma_4$ (15) MSB

$B_3$ (4)

$A_3$ (3)

$\Sigma_3$ (2)

$B_2$ (7)

$A_2$ (8)

$\Sigma_2$ (6)

$B_1$ (11)

LSBs

$A_1$ (10)

$\Sigma_1$ (9) LSB

$C_0$ (13)

Carry input

**Fig. 10-14** Detailed logic diagram for the 7483 4-bit binary adder IC.

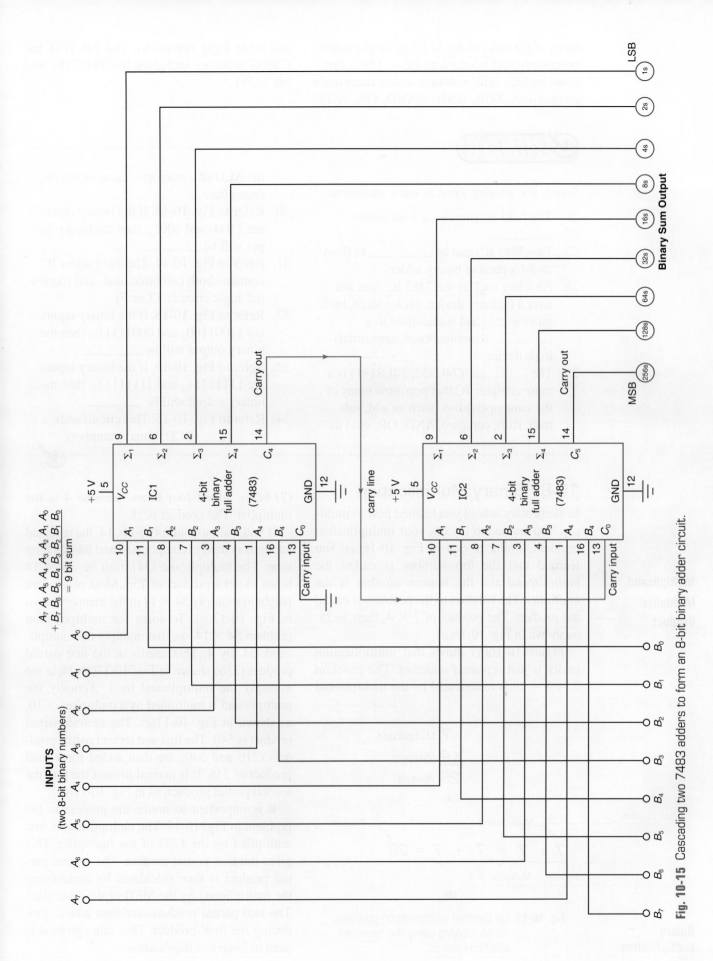

**Fig. 10-15** Cascading two 7483 adders to form an 8-bit binary adder circuit.

many of the tasks of the ALUs in simple microprocessors and microcontrollers. These functions include add, subtract, shift, magnitude comparison, XOR, AND, NAND, OR, NOR, and other logic operations. The 74LS181 has CMOS relatives including the 74HC181 and MC14581.

## ✓ Self-Test

*Supply the missing word in each statement.*

26. The 7483 IC contains a 4-bit binary _____ .

27. Two 7483 ICs can be _____ to form an 8-bit parallel binary adder.

28. An adder such as the 7483 IC does not have a memory device, such a latch, built into the chip and is classified as a _____ (combinational, sequential) logic device.

29. The _____ (74LS32, 74LS181) is a more complex IC that performs many of the same operations (such as add, subtract, shift, compare, AND, OR, etc.) as the ALU of a microprocessor or microcontroller.

30. Refer to Fig. 10-13. If the binary inputs are $1100_2$ and $1001_2$, then the binary output will be _____ .

31. Refer to Fig. 10-14. The 7483 adder IC contains both combinational- and sequential-logic circuits. (T or F).

32. Refer to Fig. 10-15. If the binary inputs are $11001100_2$ and $00011111_2$, then the binary output will be _____ .

33. Refer to Fig. 10-15. If the binary inputs are $11111111_2$ and $11111111_2$, then the binary output will be _____ .

34. Refer to Fig. 10-15. This circuit adds _____ (BCD, binary) numbers.

## 10-8 Binary Multiplication

In elementary school you learned how to multiply. You learned to lay out your multiplication problem similar to that in Fig. 10-16(*a*). You learned that the top number is called the *multiplicand* and the bottom number is the *multiplier.* The solution to the problem is called the *product.* The product of $7 \times 4$, then, is 28, as shown in Fig. 10-16(*a*).

Figure 10-16(*b*) shows that multiplication really is just *repeated addition.* The problem $7 \times 4 = 28$ is represented by the multiplicand (7) being added four times, because 4 is the multiplier. The product is 28.

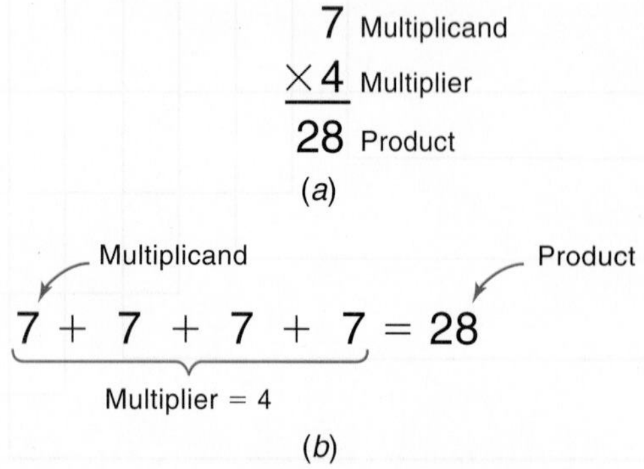

If you want to multiply $54 \times 14$, the repeated addition system is complicated and takes a long time. The multiplicand (54) must be added 14 times to get a product of 756. Most of us were taught to multiply $54 \times 14$ in the manner shown in Fig. 10-17(*a*). To solve the multiplication problem $54 \times 14$, we first multiply the multiplicand, 54, by 4. This results in the first partial product (216) shown in Fig. 10-17(*b*). Next we multiply the multiplicand by 1. Actually the multiplicand is multiplied by a multiplier of 10, as shown in Fig. 10-17(*c*). The second partial product is 540. The first and second partial products (216 and 540) are then added for a final product of 756. It is normal to omit the 0 in the second partial product, as in Fig. 10-17(*a*).

It is important to notice the *process* in the problem in Fig. 10-17. The multiplicand is first multiplied by the LSD of the multiplier. This gives the first partial product. The second partial product is then calculated by multiplying the multiplicand by the MSD of the multiplier. The two partial products are then added, producing the final product. This same process is used in *binary multiplication.*

**Multiplicand**
**Multiplier**
**Product**

$$7 \quad \text{Multiplicand}$$
$$\times 4 \quad \text{Multiplier}$$
$$28 \quad \text{Product}$$

(*a*)

Multiplicand ⎞     ⎛ Product

$$\underbrace{7 + 7 + 7 + 7}_{\text{Multiplier} = 4} = 28$$

(*b*)

**Fig. 10-16** (*a*) Decimal multiplication problem. (*b*) Multiplying using the repeated addition method.

**Binary multiplication**

```
       54   Multiplicand
    × 14   Multiplier
     216
      54
     756   Product
        (a)

       54
    × 14
     216   First partial product
        (b)

       54
    × 10
     216   First partial product
     540   Second partial product
        (c)
```

**Fig. 10-17** (a) Decimal multiplication problem.
(b) Calculating the first partial product.
(c) Calculating the second partial product.

Binary multiplication is much simpler than decimal multiplication. The binary system has only two digits (0 and 1), which makes the rules for multiplying simple. Figure 10-18(a) shows the rules for binary multiplication.

Multiplication with binary numbers is done just as with decimal numbers. Figure 10-18(b) details a problem where binary 111 is multiplied by binary 101. First, the multiplicand (111) is multiplied by the 1s bit of the multiplier. The result is the first *partial product,* shown as 111 in Fig. 10-18(b). Next, the multiplicand is multiplied by the 2s bit of the multiplier. The result is the second partial product (0000). Notice that the LSB of the second partial product, 0000, is left off in Fig. 10-18(b). Third, the multiplicand is multiplied by the 4s bit of the multiplier. The result is the third partial product of 11100, shown in Fig. 10-18(b) as 111, with the two blank spaces in the 1s and 2s places. Finally, the first, second, and third partial products are added, resulting in a product of binary 100011.

Rules for binary multiplication

```
    0      0      1      1
  × 0    × 1    × 0    × 1
  ───    ───    ───    ───
    0      0      0      1
            (a)
```

```
Decimal          Binary
   7             1 1 1    Multiplicand
 × 5           × 1 0 1    Multiplier
 ───           ─────────
  35             1 1 1    First partial product
               0 0 0      Second partial product
             1 1 1        Third partial product
             ─────────────
             1 0 0 0 1 1  Product
                 (b)
```

**Fig. 10-18** (a) Rules for binary multiplication.
(b) Sample multiplication problem.

```
Decimal              Binary
  2 7            1 1 0 1 1    Multiplicand
× 1 2          × 1 1 0 0      Multiplier
─────          ─────────────
  5 4          1 1 0 1 1 0 0  Third partial product
2 7            1 1 0 1 1      Fourth partial product
─────          ───────────────
3 2 4          1 0 1 0 0 0 1 0 0  Product
```

**Fig. 10-19** Sample multiplication problem.

Notice that the same multiplication problem in decimal is shown at the left of Fig. 10-18(b) for your convenience. The binary product 100011 equals the decimal product 35.

Another binary multiplication problem is shown in Fig. 10-19. At the left the problem is in the familiar decimal form; the same problem is repeated in binary form at the right, where binary 11011 is multiplied by 1100. As in decimal multiplication, the 0s in the multiplier can simply be brought down to hold the 1s and 2s places in the binary number. The binary product is shown as 101000100, which equals decimal 324.

You will gain experience in solving binary multiplication problems by answering the questions that follow.

**Rules for binary multiplication**

**Partial product**

*Answer the following questions.*

35. Find the product for binary 111 × 10.

36. Find the product for binary 1101 × 101.
37. Find the product for binary 1100 × 1110.

## 10-9 Binary Multipliers

We can multiply numbers by repeated addition, as illustrated in Fig. 10-16(b). The multiplicand (7) could be added four times to obtain the product of 28. A block diagram of a circuit that performs repeated addition is shown in Fig. 10-20. The multiplicand is held in the top register. In our example the multiplicand is a decimal 7, or a binary 111. The multiplier is held in the down counter shown on the left in Fig. 10-20. The multiplier in our example is a decimal 4, or a binary 100. The lower product register holds the product.

The repeated addition technique is shown in operation in Fig. 10-21. This chart shows how the multiplicand (binary 111) is multiplied by the multiplier (binary 100). The product register is cleared to 00000. After one count downward, a partial product of 00111 (decimal 7) appears in the products register. After the second count downward, a partial product of 01110 (decimal 14) appears in the product register. After the third count downward, a partial product of 10101 (decimal 21) appears in the product register. After the fourth downward count, the *final product* of 11100 (decimal 28) appears in the product register. The multiplication problem ($7 \times 4 = 28$) is complete. The circuit of Fig. 10-20 has added 7 four times for a total of 28.

This type of circuit is not widely used because of the long time it takes to do the repeated addition when large numbers are multiplied. A more practical method of multiplying in digital electronic circuits is the *add and shift method* (also called the shift and add method). Figure 10-22 shows a binary multiplication problem. In this problem binary 111 is multiplied by 101 ($7 \times 5$

**Add and shift method**

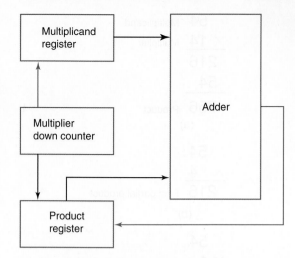

**Fig. 10-20** Block diagram of a repeated addition-type multiplier system.

in decimal). This hand-done procedure is standard except for the temporary product in line 5. Line 5 has been added to help you understand how multiplication might be done by digital circuits. Close observation of binary multiplication shows the following three important facts:

1. Partial products are always 000 if the multiplier is 0 and equal to the multiplicand if the multiplier is 1.
2. The product register needs twice as many bits as the multiplicand register assuming the multiplier has the same or a fewer number of bits.
3. The first partial product is shifted one place to the *right* (*relative to* the second partial product) when adding.

You can observe each characteristic by looking at the sample problem in Fig. 10-22.

| | Load with binary | After 1 down count | After 2 down counts | After 3 down counts | After 4 down counts |
|---|---|---|---|---|---|
| Multiplicand register | 111 | 111 | 111 | 111 | 111 |
| Multiplier counter | 100 | 011 | 010 | 001 | 000 |
| Product register | 00000 | 00111 | 01110 | 10101 | 11100 |
| | Load | | | | Stop |

**Fig. 10-21** Multiplying binary 111 and 100 using the repeated addition circuit.

| | | |
|---|---|---|
| Line 1 | **1 1 1** | Multiplicand |
| Line 2 | × **1 0 1** | Multiplier |
| Line 3 | 1 1 1 | First partial product |
| Line 4 | 0 0 0 | Second partial product |
| Line 5 | 0 1 1 1 | Temporary product (line 3 + line 4) |
| Line 6 | 1 1 1 | Third partial product |
| Line 7 | 1 0 0 0 1 1 | Product |

**Fig. 10-22** Binary multiplication problem.

The important characteristics of longhand multiplication have been given. A binary multiplication circuit can be designed by using these characteristics. Figure 10-23(*a*) on page 334 shows a circuit that does binary multiplication. Notice that the multiplicand (111) is loaded into the register at the upper left. The accumulator register is cleared to 0000. The multiplier (101) is loaded into the register at the lower right. Notice, too, that the accumulator and the multiplier are considered together. This is shown by the shading connecting the two registers.

Let us use the circuit in Fig. 10-23(*a*) to demonstrate the detailed procedure for multiplying. The diagram in Fig. 10-23(*b*) is a step-by-step review of how binary 111 is multiplied by 101 using the add and shift method. The binary 111 is loaded into the multiplicand register. The accumulator and multiplier registers are loaded in step *A* in Fig. 10-23(*b*). Step *B* shows the 0000 and the 111 from the accumulator and multiplicand registers being added when the 1 is applied to the control line. This is comparable to line 3 of the multiplication problem in Fig. 10-22. Step *C* shifts both the accumulator and the multiplier register one place to the right. The LSB of the multiplier (1) is shifted out the right end and lost. Step *D* represents another *add* step. This time a 0 is applied to the control line. A 0 on the control line means *no* addition. The register contents remain the same. Step *D* is comparable to lines 4 and 5, Fig. 10-22. Step *E* shows the registers being shifted one place to the right. This time the 2s bit of the multiplier is lost as it is shifted out the right end of the register. Step *F* shows the 4s bit of the multiplier (1) commanding the adder to add. The accumulator contents (0001) and the multiplicand (111) are added. The result of that addition is deposited in the accumu-

lator register (1000). This step is comparable to the left section of lines 5 to 7, Fig. 10-22. Step *G* is the final step in the *add and shift multiplication*; it shows a single shift to the right for both registers. The 4s bit of the multiplier is lost out of the right end of the register. The final product appears across both registers as 100011. Binary 111 multiplied by 101 resulted in a product of 100011 (7 × 5 = 35 in decimal). The final product calculated by the multiplier circuit is the same result we got in line 7, Fig. 10-22, when we multiplied by hand.

Two types of multiplier circuits have just been illustrated. The first uses repeated addition to arrive at the product. That system is shown in Fig. 10-20. The second circuit uses the add and shift method of multiplying. The add and shift system is shown in Fig. 10-23.

In many computers *the procedure,* such as the add and shift method, can be programmed into the machine. Instead of permanently wiring the circuit, we simply *program,* or instruct, the computer to follow the procedure shown in Fig. 10-23(*b*). We are thus using *software* (a program) to do multiplication. This use of software cuts down on the amount of electronic circuits needed in the CPU of a computer.

Simpler 8-bit microprocessors, such as the obsolete Intel 8080/8085, Motorola 6800, and the 6502/65C02 do not have circuitry in their ALUs to do multiplication. To perform binary multiplication on these processors, the programmer must write a program (a list of instructions) that multiplies numbers. Either the add and shift or the repeated addition method can be used for programming these microprocessor-based machines to do multiplication. Most advanced microprocessors do have multiply instructions. Some more expensive microcontrollers also have a multiply instruction.

**Add and shift method of multiplying**

**Microprocessors**

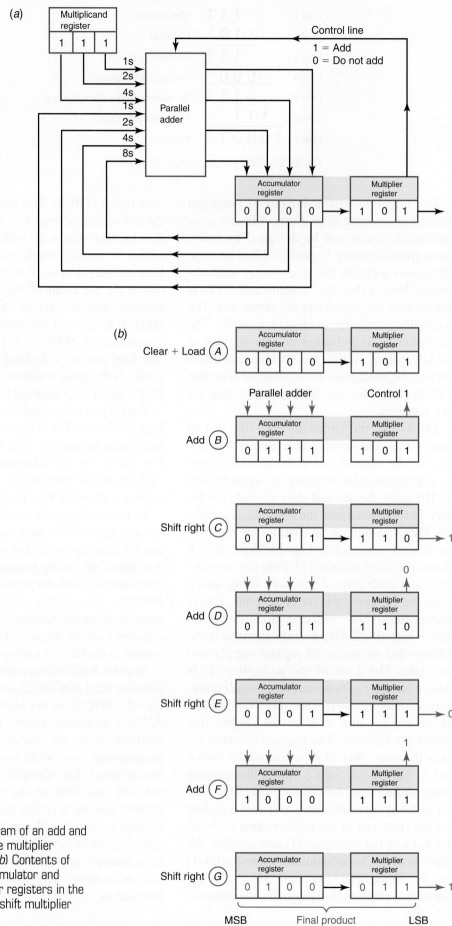

(a)

Multiplicand register

| 1 | 1 | 1 |

Control line
1 = Add
0 = Do not add

1s
2s
4s
1s
2s
4s
8s

Parallel adder

Accumulator register

| 0 | 0 | 0 | 0 |

Multiplier register

| 1 | 0 | 1 |

(b)

Clear + Load (A)

| Accumulator register | | | | Multiplier register | | |
| 0 | 0 | 0 | 0 | 1 | 0 | 1 |

Parallel adder          Control 1

Add (B)

| Accumulator register | | | | Multiplier register | | |
| 0 | 1 | 1 | 1 | 1 | 0 | 1 |

Shift right (C)

| Accumulator register | | | | Multiplier register | | |
| 0 | 0 | 1 | 1 | 1 | 1 | 0 | → 1

0

Add (D)

| Accumulator register | | | | Multiplier register | | |
| 0 | 0 | 1 | 1 | 1 | 1 | 0 |

Shift right (E)

| Accumulator register | | | | Multiplier register | | |
| 0 | 0 | 0 | 1 | 1 | 1 | 1 | → 0

1

Add (F)

| Accumulator register | | | | Multiplier register | | |
| 1 | 0 | 0 | 0 | 1 | 1 | 1 |

Shift right (G)

| Accumulator register | | | | Multiplier register | | |
| 0 | 1 | 0 | 0 | 0 | 1 | 1 | → 1

MSB          Final product          LSB

**Fig. 10-23** (a) Diagram of an add and shift-type multiplier circuit. (b) Contents of the accumulator and multiplier registers in the add and shift multiplier circuit.

**Add and shift-type multiplier circuit**

*Answer the following questions.*

38. Refer to Fig. 10-20. This circuit uses what method of binary multiplication?
39. A widely used technique for multiplying using digital circuits is the _____ method.
40. Refer to Fig. 10-23. This circuit uses what method of binary multiplication?
41. All microcontrollers have a multiply instruction. (T or F)

## 10-10 2s Complement Notation, Addition, and Subtraction

The 2s complement method of representing numbers is widely used in microprocessors. To now, we have assumed that all numbers are positive. However, microprocessors must process both positive and negative numbers. Using *2s complement representations*, the *sign as well as the magnitude* of a number can be determined.

### 2s Complement 4-bit

For simplicity, assume we are using a 4-bit processor. This means that all data are transferred and processed in groups of four. The MSB is the *sign bit* of the number. This is shown in Fig. 10-24(*a*). A 0 sign bit means a positive number, while a 1 sign bit means a negative number.

The table in Fig. 10-24(*b*) shows the *2s complement representation* for all the 4-bit positive and negative numbers from +7 to −8. The MSBs in Fig. 10-24(*b*) of the positive 2s complement numbers are 0s. All negative numbers (−1 to −8) start with a 1. Note that 2s complement representations of positive numbers are the same as binary. Therefore, +7 (decimal) = 0111 (2s complement) = 0111 (binary).

The 2s complement representation of a negative number is found by first taking the 1s complement of the number and then adding 1. An example of this process is shown in Fig. 10-25(*a*). The negative decimal number −4 is to be converted to its 2s complement form:

1. Convert the decimal number to its binary equivalent. In this example, convert $-4_{10}$ to $0100_2$.
2. Convert the binary number to its 1s complement by changing all 1s to 0s and all 0s to 1s. In this example, convert $0100_2$ to 1011 (1s complement).

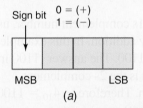

*(a)*

**2s complement representations**

**Sign bit**

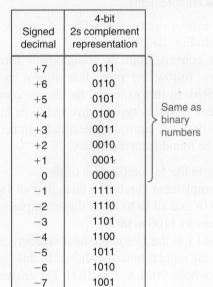

| Signed decimal | 4-bit 2s complement representation |
|:---:|:---:|
| +7 | 0111 |
| +6 | 0110 |
| +5 | 0101 |
| +4 | 0100 |
| +3 | 0011 |
| +2 | 0010 |
| +1 | 0001 |
| 0 | 0000 |
| −1 | 1111 |
| −2 | 1110 |
| −3 | 1101 |
| −4 | 1100 |
| −5 | 1011 |
| −6 | 1010 |
| −7 | 1001 |
| −8 | 1000 |

Same as binary numbers

*(b)*

**Fig. 10-24** (*a*) MSB of 4-bit register is a sign bit. (*b*) 2s complement representation of positive and negative numbers.

−4 (Decimal)

Step (1) Convert decimal to binary

0100 (Binary)

Step (2) 1s complement

1011 (1s complement)

Step (3) Add + 1 (1011 + 1 = 1100)

−4₁₀ = 1100 (2s complement)

$-4_{10} = 1100$ (2s complement)

(a)

1100 (2s complement)

Step (1) 1s complement

0011 (1s complement)

Step (2) Add + 1 (0011 + 1 = 0100)

$4_{10} = 0100$ (Binary)

(b)

**Converting signed decimal numbers to 2s complement form**

**Fig. 10-25** (a) Converting signed decimal numbers to 2s complement form. (b) Converting from 2s complement form to binary numbers.

3. Add 1 to the 1s complement number, using regular binary addition. In this example, $1011 + 1 = 1100$. The answer (1100 in this example) is the 2s complement representation. Therefore, $-4_{10} = 1100$ (2s complement).

This answer can be verified by referring to the table in Fig. 10-24(b).

To convert from 2s complement form to binary, follow the procedure shown in Fig. 10-25(b). In this example, the 2s complement number (1100) is being converted to its binary equivalent. Its equivalent decimal number can then be found from the binary.

1. Form the 1s complement of the 2s complement number by changing all 1s to 0s and all 0s to 1s. In this example, convert 1100 to 0011.
2. Add 1 to the 1s complement number, using regular binary addition. In this example, $0011 + 1 = 0100$. The answer (0100 in this example) is in binary. Therefore, $0100_2 = 4_{10}$.

Because the MSB of the 2s complement number (1100) is a 1, the number is negative. Therefore, 2s complement 1100 equals $-4_{10}$.

For information on new product developments, visit the website for the Consumer Electronics Association (CEA).

## 2s Complement Addition

2s complement notation is widely employed because it makes it easy to add and subtract signed numbers. Four examples of adding 2s complement numbers are shown in Fig. 10-26. Two positive numbers are added in Fig. 10-26(a). 2s complement addition looks just like adding in binary in this example. Two negative numbers ($-1_{10}$ and $-2_{10}$) are added in Fig. 10-26(b). The 2s complement numbers representing $-1$ and $-2$ are given as 1111 and 1110. The MSB (overflow from 4-bit register) is discarded, leaving the 2s complement sum of 1101, or $-3$ in decimal. Look over examples (c) and (d) in Fig. 10-26 to see if you understand the procedure for adding signed numbers using 2s complement notation.

## 2s Complement Subtraction

2s complement notation is also useful in subtracting signed numbers. Four subtraction problems are shown in Fig. 10-27. The first problem is $(+7) - (+3) = +4_{10}$. The subtrahend

$$
\begin{array}{r}
(+4) \\
+ (+3) \\
\hline
+7_{10}
\end{array}
\qquad
\begin{array}{r}
0100 \\
+ \ 0011 \\
\hline
0111
\end{array} \text{ (2s complement SUM)}
$$

(a)

$$
\begin{array}{r}
(-1) \\
+ (-2) \\
\hline
-3_{10}
\end{array}
\qquad
\begin{array}{r}
1111 \\
+ \ 1110 \\
\hline
1\ 1101
\end{array} \text{ (2s complement SUM)}
$$

Discard

(b)

$$
\begin{array}{r}
(+1) \\
+ (-3) \\
\hline
-2_{10}
\end{array}
\qquad
\begin{array}{r}
0001 \\
+ \ 1101 \\
\hline
1110
\end{array} \text{ (2s complement SUM)}
$$

(c)

$$
\begin{array}{r}
(+5) \\
+ (-4) \\
\hline
+1_{10}
\end{array}
\qquad
\begin{array}{r}
0101 \\
+ \ 1100 \\
\hline
1\ 0001
\end{array} \text{ (2s complement SUM)}
$$

Discard

(d)

**Fig. 10-26** Four sample signed addition problems using 4-bit 2s complement numbers.

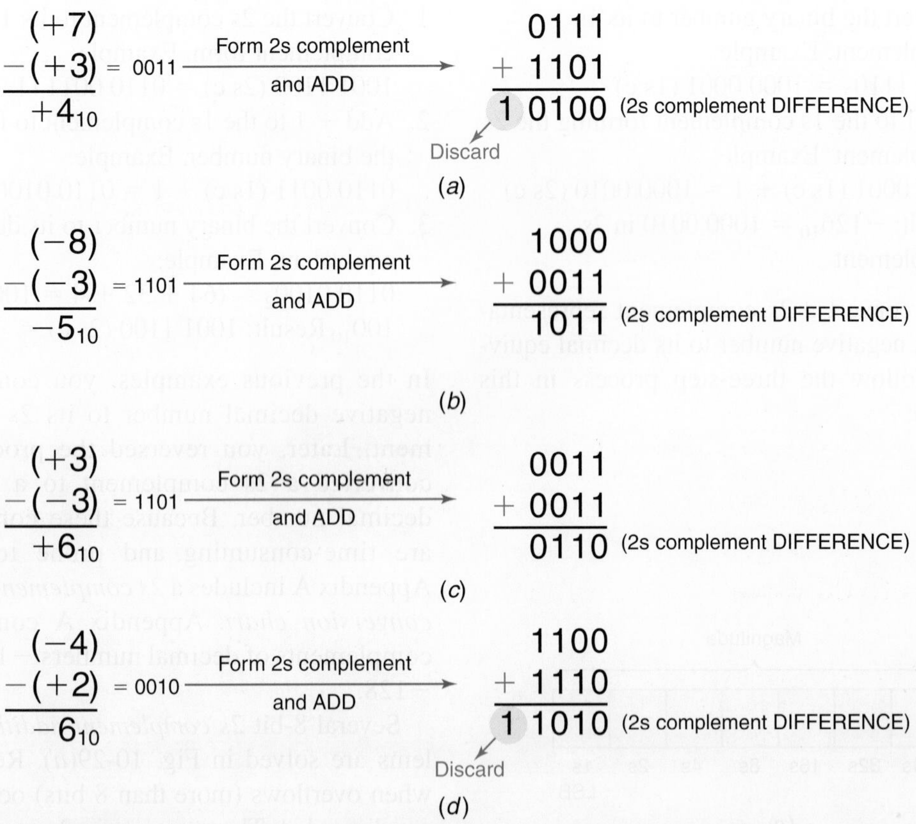

$$\begin{array}{r}(+7)\\-(+3)\\\hline +4_{10}\end{array} = 0011 \xrightarrow[\text{and ADD}]{\text{Form 2s complement}} \begin{array}{r}0111\\+\ 1101\\\hline 1\ 0100\end{array} \text{(2s complement DIFFERENCE)}$$

Discard

(a)

$$\begin{array}{r}(-8)\\-(-3)\\\hline -5_{10}\end{array} = 1101 \xrightarrow[\text{and ADD}]{\text{Form 2s complement}} \begin{array}{r}1000\\+\ 0011\\\hline 1011\end{array} \text{(2s complement DIFFERENCE)}$$

(b)

$$\begin{array}{r}(+3)\\-(-3)\\\hline +6_{10}\end{array} = 1101 \xrightarrow[\text{and ADD}]{\text{Form 2s complement}} \begin{array}{r}0011\\+\ 0011\\\hline 0110\end{array} \text{(2s complement DIFFERENCE)}$$

(c)

$$\begin{array}{r}(-4)\\-(+2)\\\hline -6_{10}\end{array} = 0010 \xrightarrow[\text{and ADD}]{\text{Form 2s complement}} \begin{array}{r}1100\\+\ 1110\\\hline 1\ 1010\end{array} \text{(2s complement DIFFERENCE)}$$

Discard

(d)

**Fig. 10-27** Four sample signed subtraction problems using 4-bit 2s complement numbers.

($+3$ in this case) is converted to its binary form. Next, the 2s complement of this is formed, yielding 1101. Then 0111 is *added* to 1101, yielding 1 0100. The MSB (overflow from 4-bit register) is discarded, leaving the *difference* of 0100, or $+4_{10}$. Note that an adder is used for subtraction. This is done by converting the subtrahend to its 2s complement and adding. Any carry or overflow into the fifth binary place is discarded.

Look over the sample 2s complement subtraction problems using an adder in Fig. 10-27(b), (c), and (d). See if you can follow the procedure in these remaining subtraction problems.

## 2s Complement 8-bit

Only 4-bit 2s complement representations have been used in previous examples. Most microprocessors and microcontrollers use 8-, 16-, or 32-bit groupings. The procedures used with 4-bit 2s complement descriptions of binary numbers also apply to 8-bit, 16-bit, or 32-bit representations.

In an 8-bit 2s complement of a number, the MSB is the sign bit as illustrated in Fig. 10-28(a) on page 338. This allows both the sign and magnitude of the number to be represented. A sampling of some 8-bit 2s complement representations of positive and negative numbers are shown in Fig. 10-28(b). Notice that the range of numbers for an 8-bit 2s complement is from $-128$ to $+127$. Notice from the top half of the chart in Fig. 10-28(b) that *decimal numbers from 0 through $+127$ (positive numbers) have 2s complements that are the same as binary numbers.* As an example, $+125$ is represented by 0111 1101 in either binary or 2s complement.

Converting a negative decimal number (from $-1$ to $-128$) to its 8-bit 2s complement is accomplished by the same process shown earlier in Fig. 10-25(a). Follow the three-step process in the example below:

1. Convert the decimal number $-126$ to its binary equivalent.
   Example: $126_{10} = 0111\ 1110_2$

**Add or subtract signed numbers**

2. Convert the binary number to its 1s complement. Example:
   $0111\ 1110_2 = 1000\ 0001$ (1s c)
3. Add 1 to the 1s complement forming the 2s complement. Example:
   $1000\ 0001$ (1s c) $+ 1 = 1000\ 0010$ (2s c)
   Result: $-126_{10} = 1000\ 0010$ in 2s complement

Next convert a 2s complement representation of a negative number to its decimal equivalent. Follow the three-step process in this example:

1. Convert the 2s complement to its 1s complement form. Example:
   $1001\ 1100$ (2s c) $= 0110\ 0011$ (1s c)
2. Add $+\ 1$ to the 1s complement to form the binary number. Example:
   $0110\ 0011$ (1s c) $+ 1 = 0110\ 0100_2$
3. Convert the binary number to its decimal equivalent. Example:
   $0110\ 0100_2 = (64 + 32 + 4 = 100) = 100_{10}$ Result: $1001\ 1100$ (2s c) $= -100_{10}$

In the previous examples, you converted a negative decimal number to its 2s complement. Later, you reversed the process and converted a 2s complement to a negative decimal number. Because these conversions are time-consuming and prone to errors, Appendix A includes a *2s complement number conversion chart*. Appendix A contains 2s complements of decimal numbers $-1$ through $-128$.

Several 8-bit *2s complement addition* problems are solved in Fig. 10-29(*a*). Remember when overflows (more than 8 bits) occur, they are discarded. The sums are in 2s complement notation, but remember that for positive numbers the 2s complement and binary number are the same. Review these addition problems to see if you understand the procedure. You will have practice problems later.

Several 8-bit *2s complement subtraction* problems are solved in Fig. 10-29(*b*). Remember when overflows (more than 8 bits) occur, they are discarded. Notice that only the subtrahends are 2s complemented before they are added to the minuend. The differences are in 2s complement notation but remember that for positive numbers, the 2s complement and binary number are the same. Review these subtraction problems to see if you understand the procedure. You will have practice problems later.

In summary, 2s complement notation is used because it shows both the sign and magnitude of a number. Remember that 2s complement and binary numbers are identical for positive numbers. Two's complement numbers can be used with adders to either *add* or *subtract signed numbers*. The next section in the textbook will diagram an adder/subtractor system that makes use of 2s complement notation.

**2s complement addition**

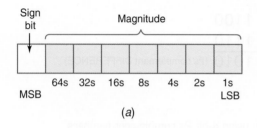

(a)

**2s complement subtraction**

| Signed decimal | 8-bit 2s complement representation |
|---|---|
| +127 | 0111 1111 |
| +126 | 0111 1110 |
| +125 | 0111 1101 |
| • | • |
| • | • |
| • | • |
| +5 | 0000 0101 |
| +4 | 0000 0100 |
| +3 | 0000 0011 |
| +2 | 0000 0010 |
| +1 | 0000 0001 |
| 0 | 0000 0000 |
| −1 | 1111 1111 |
| −2 | 1111 1110 |
| −3 | 1111 1101 |
| −4 | 1111 1100 |
| −5 | 1111 1011 |
| −6 | 1111 1010 |
| • | • |
| • | • |
| • | • |
| −125 | 1000 0011 |
| −126 | 1000 0010 |
| −127 | 1000 0001 |
| −128 | 1000 0000 |

Same as binary numbers

(b)

**Fig. 10-28** (*a*) MSB of 8-bit register is a sign bit. (*b*) 2s complement representation of selected positive and negative numbers.

$$\begin{array}{r}(+60) \\ +\,(+20) \\ \hline +80_{10}\end{array} \qquad \begin{array}{r}0011\ 1100 \\ +\ \ 0001\ 0100 \\ \hline 0101\ 0000 \end{array} \text{ (2s complement SUM)}$$

$$\begin{array}{r}(-50) \\ +\,(-30) \\ \hline -80_{10}\end{array} \qquad \begin{array}{r}1100\ 1110 \\ +\ \ 1110\ 0010 \\ \hline 1\ 1011\ 0000 \end{array} \text{ (2s complement SUM)}$$

↙ Discard

$$\begin{array}{r}(+30) \\ +\,(-90) \\ \hline -60_{10}\end{array} \qquad \begin{array}{r}0001\ 1110 \\ +\ \ 1010\ 0110 \\ \hline 1100\ 0100 \end{array} \text{ (2s complement SUM)}$$

$$\begin{array}{r}(+90) \\ +\,(-80) \\ \hline +10_{10}\end{array} \qquad \begin{array}{r}0101\ 1010 \\ +\ \ 1011\ 0000 \\ \hline 1\ 0000\ 1010 \end{array} \text{ (2s complement SUM)}$$

↙ Discard

(a)

$$\begin{array}{r}(+65) \\ -\,(+35) \\ \hline +30_{10}\end{array} = 0010\ 0011 \xrightarrow[\text{and ADD}]{\text{Form 2s complement}} \begin{array}{r}0100\ 0001 \\ +\ \ 1101\ 1101 \\ \hline 1\ 0001\ 1110 \end{array} \text{ (2s complement DIFFERENCE)}$$

↙ Discard

$$\begin{array}{r}(-78) \\ -\,(-35) \\ \hline -43_{10}\end{array} = 1101\ 1101 \xrightarrow[\text{and ADD}]{\text{Form 2s complement}} \begin{array}{r}1011\ 0010 \\ +\ \ 0010\ 0011 \\ \hline 1101\ 0101 \end{array} \text{ (2s complement DIFFERENCE)}$$

$$\begin{array}{r}(+40) \\ -\,(-21) \\ \hline +61_{10}\end{array} = 1110\ 1011 \xrightarrow[\text{and ADD}]{\text{Form 2s complement}} \begin{array}{r}0010\ 1000 \\ +\ \ 0001\ 0101 \\ \hline 0011\ 1101 \end{array} \text{ (2s complement DIFFERENCE)}$$

$$\begin{array}{r}(-45) \\ -\,(+22) \\ \hline -67_{10}\end{array} = 0001\ 0110 \xrightarrow[\text{and ADD}]{\text{Form 2s complement}} \begin{array}{r}1101\ 0011 \\ +\ \ 1110\ 1010 \\ \hline 1\ 1011\ 1101 \end{array} \text{ (2s complement DIFFERENCE)}$$

↙ Discard

(b)

**Fig. 10-29** (a) Four sample signed addition problems using 8-bit 2s complement numbers. (b) Four sample signed subtraction problems using 8-bit 2s complement numbers.

*Answer the following questions.*

42. When microprocessors process both positive and negative numbers, _____ representations are used.

43. The 2s complement number 0111 represents _____ in binary and _____ in decimal.

44. The 2s complement number 1111 represents _____ in decimal.

45. In 2s complement representation, the MSB is the _____ bit. If the MSB is 0, the number is _____ (negative, positive), whereas if the MSB is 1, the number is _____ (negative, positive).

46. The decimal number −6 equals _____ in 2s complement 4-bit representation.

47. The decimal number +5 equals _____ in 2s complement 4-bit representation.

48. Calculate the sum of the 2s complement numbers 1110 and 1101. Give the answer in 2s complement and in decimal.

49. Calculate the sum of the 2s complement numbers 0110 and 1100. Give the answer in 2s complement and in decimal.

50. Decimal 90 equals _____ in binary and _____ in 2s complement.

51. Decimal −90 equals _____ in 2s complement.

52. Adding 0111 1111 (2s c) and 1111 0000 (2s c) yields _____ in 2s complement or _____ in decimal.

53. Adding 1000 0000 (2s c) and 0000 1111 (2s c) yields _____ in 2s complement or _____ in decimal.

54. Subtracting 0001 0000 (2s c) from 1110 0000 (2s c) yields _____ in 2s complement or _____ in decimal.

55. Subtracting 1111 1111 (2s c) from 0011 0000 (2s c) yields _____ in 2s complement or _____ in decimal.

## 10-11 2s Complement Adders/Subtractors

**2s complement 4-bit adder/subtractor system**

A *2s complement 4-bit adder/subtractor system* is drawn in Fig. 10-30. Note the use of four full adders to handle the two 4-bit numbers. XOR gates have been added to the $B$ inputs of each full adder to control the mode of operation of the unit. With the mode control at 0, the system *adds* the 2s complement numbers $A_3A_2A_1A_0$ and $B_3B_2B_1B_0$. The sum appears in 2s complement notation at the output indicators at the lower right. The LOW at the $A$ inputs of the XOR gates permit the $B$ data to flow through the gate with *no inversion*. If a HIGH enters $B_0$ input to the XOR gate, then a HIGH exits the gate at $Y$. The $C_{in}$ input to the top 1s full adder is held at a 0 during the time the mode control is in the add position. In the add mode, the 2s complement adder operates just like a binary adder except that the carry out ($C_O$) from the 8s full adder is discarded. In Fig. 10-30, the $C_O$ output from the 8s full adder is left disconnected.

The mode control input is placed at logical 1 for the unit to subtract 2s complement numbers. This causes the XOR gates *to invert the data at the B inputs*. The $C_{in}$ input to the 1s full adder also receives a HIGH. The combination of the XOR gate's inversion plus adding the 1 at the $C_{in}$ input of the 1s full adder is the same as complementing and adding 1. This is comparable to forming the 2s complement of the subtrahend ($B$ number in Fig. 10-30).

Remember that the system in Fig. 10-30 only uses 2s complement numbers. The 4-bit adder/subtractor system in Fig. 10-30 could be extended to 8 bits, 16 bits or 32 bits to handle larger 2s complement numbers.

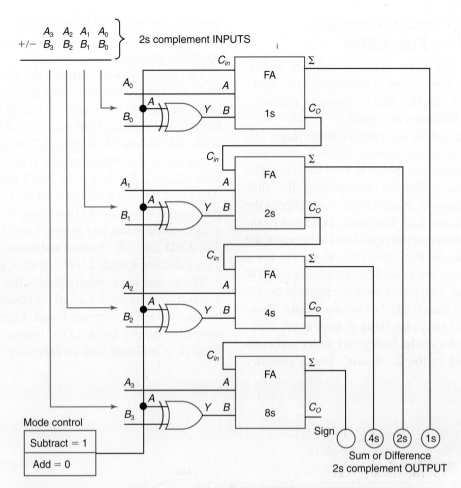

$A_3$ $A_2$ $A_1$ $A_0$
$+/-$ $B_3$ $B_2$ $B_1$ $B_0$ } 2s complement INPUTS

$C_{in}$

FA
1s

$A_0$   A
$B_0$   Y   B
$C_O$
Σ

$C_{in}$

FA
2s

$A_1$   A
$B_1$   Y   B
$C_O$
Σ

$C_{in}$

FA
4s

$A_2$   A
$B_2$   Y   B
$C_O$
Σ

$C_{in}$

FA
8s

$A_3$   A
$B_3$   Y   B
$C_O$
Σ

Mode control

Subtract = 1

Add = 0

Sign   (4s) (2s) (1s)

Sum or Difference
2s complement OUTPUT

**Fig. 10-30** Adder/subtractor system using 2s complement numbers.

**Self-Test**

*Supply the missing word or words in each statement.*

56. Refer to Fig. 10-30. The numbers to be added or subtracted in this system must be in _____ (binary, BCD, 1s complement, 2s complement) form.
57. Refer to Fig. 10-30. The sum or difference output from this system will be in _____ (binary, BCD, 1s complement, 2s complement) form.
58. Refer to Fig. 10-30. This system can add or subtract _____ (signed, only unsigned) numbers.
59. Refer to Fig. 10-30. If the system is adding 0011 2sC to 1100 2sC, the output will read _____. This is the 2s

complement representation for decimal _____.
60. Refer to Fig. 10-30. If the system is subtracting 0010 2sC from 0101 2sC, the output will read _____. This is the 2s complement representation for decimal _____.
61. Refer to Fig. 10-30. If the system is adding 1010 2sC to 0100 2sC, the output will read _____. This is the 2s complement representation for decimal _____.
62. Refer to Fig. 10-30. If the system is subtracting 1110 2sC from 1001 2sC, the output will read _____. This is the 2s complement representation for decimal _____.

## 10-12 Troubleshooting a Full Adder

A faulty full adder circuit is sketched in Fig. 10-31(*a*). The student or technician first checks the circuit visually and for signs of excessive heat. No problems are found.

The full-adder is a *combinational logic circuit*. For your convenience, its truth table with normal outputs is shown in Fig. 10-31(*b*). The student or technician manipulates the full-adder inputs and using a logic probe checks the outputs ($\Sigma$ and $C_O$). The actual logic probe outputs are shown in the right-hand columns of the truth table in Fig. 10-31(*b*). H stands for a HIGH logic level, while L stands for a LOW logic level. Two errors seem to appear in the $C_O$ column in lines 6 and 7 of the truth table. These are noted in Fig. 10-31(*b*). A look at the truth-table results of the faulty full adder indicates no trouble in the $\Sigma$ column. The $\Sigma$ circuitry involves the two XOR gates labeled 1 and 2 in Fig. 10-31(*a*). It appears that these gates are operating properly.

The troubleshooter expects the problem to be in the OR gate or two AND gates. The bottom line of the truth table suggests that the bottom AND gate and OR gate work. The upper AND gate (labeled 4) is suspect. The technician manipulates the inputs to line 6 on the truth table ($C_{in} = 1, B = 0, A = 1$). Pins 1 and 2 of the AND gate labeled 4 should both be 1. Both inputs to gate 4 indicate a HIGH logic level when a logic probe is touched to pins 1 and 2. Output 3 of AND gate 4 is checked and remains LOW. This indicates a stuck LOW output at gate 4.

The technician carefully checks the 7408 IC and surrounding circuit board for possible short circuits to GND. None are found. Gate 4 is assumed to have a stuck LOW output, and the 7408 IC is replaced with an exact duplicate.

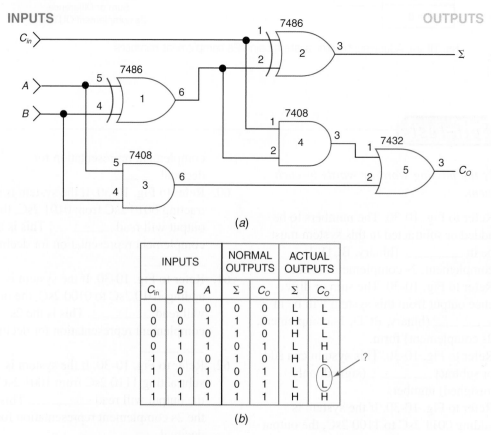

(*a*)

| INPUTS | | | NORMAL OUTPUTS | | ACTUAL OUTPUTS | |
|---|---|---|---|---|---|---|
| $C_{in}$ | $B$ | $A$ | $\Sigma$ | $C_O$ | $\Sigma$ | $C_O$ |
| 0 | 0 | 0 | 0 | 0 | L | L |
| 0 | 0 | 1 | 1 | 0 | H | L |
| 0 | 1 | 0 | 1 | 0 | H | L |
| 0 | 1 | 1 | 0 | 1 | L | H |
| 1 | 0 | 0 | 1 | 0 | H | L |
| 1 | 0 | 1 | 0 | 1 | L | L |
| 1 | 1 | 0 | 0 | 1 | L | L |
| 1 | 1 | 1 | 1 | 1 | H | H |

(*b*)

**Fig. 10-31** (*a*) Faulty full-adder circuit used for troubleshooting problem. (*b*) Full-adder truth table with normal and actual outputs.

After replacement of the 7408 IC, the troubleshooter checks the full-adder circuit for proper operation. The circuit works according to its normal truth table. Truth tables help both technicians and students with troubleshooting. Such tables define how a *normal circuit should respond*. The truth table becomes part of the technician's knowledge of the circuit. Knowledge of normal circuit operation is critical to good troubleshooting.

To review, six hints for successful troubleshooting are:

1. *Know* the normal operation of the circuit.
2. *Feel* to top of the IC to determine if it is hot.
3. *Look* for broken connections or signs of excessive heat.
4. *Smell* for overheating.
5. *Check* the power source and power to ICs.
6. *Trace* the path of logic through the circuit and isolate the faulty section.

*Supply the missing word or words in each statement.*

63. Refer to Fig. 10-31. The fault in the _____ (combinational, sequential) logic circuit seems to be in the _____ (carry out, sum) part of the circuit.

64. Refer to Fig. 10-31. The fault in the circuit is in gate _____ (number); the output is stuck _____ (HIGH, LOW).

65. Knowledge of normal circuit operation is critical to good troubleshooting. (T or F)

66. List several hints for successful troubleshooting.

# Chapter 10 Summary and Review

## Summary

1. Arithmetic circuits, such as adders and subtractors, are combinational logic circuits constructed with logic gates.
2. The basic addition circuit is called a half adder. Two half adders and an OR gate can be wired to form a full adder.
3. The basic subtraction circuit is called a half subtractor. Two half subtractors and an OR gate can be wired to form a full subtractor.
4. Adders (or subtractors) can be wired together to form parallel adders.
5. A 4-bit parallel adder adds two 4-bit binary numbers at one time. This adder contains a single half adder (1s place) and three full adders.

6. Manufacturers produce several arithmetic ICs.
7. Adder/subtractor units are often part of the CPU of calculating machines.
8. Binary multiplication performed by digital circuits may use repeated additions or the add and shift method.
9. Microprocessors use 2s complement notation when dealing with signed numbers. Adders can be used to perform addition and subtraction using 2s complement numbers.
10. Truth tables are a great aid in troubleshooting combinational logic circuits since they define the normal operation of the circuits.

## Chapter Review Questions

*Answer the following questions.*

10-1. Do binary addition problems **a** to **h** (show your work):
   a. 101 + 011 =
   b. 110 + 101 =
   c. 111 + 111 =
   d. 1000 + 0011 =
   e. 1000 + 1000 =
   f. 1001 + 0111 =
   g. 1010 + 0101 =
   h. 1100 + 0101 =

10-2. Draw a block diagram for a half adder (label two inputs and two outputs).

10-3. Draw a block diagram for a full adder (label three inputs and two outputs).

10-4. Do binary subtraction problems **a** to **h** (show your work):
   a. 1100 − 0010 =
   b. 1101 − 1010 =
   c. 1110 − 0011 =
   d. 1111 − 0110 =
   e. 10000 − 0011 =
   f. 1000 − 0101 =
   g. 10010 − 1011 =
   h. 1001 − 0010 =

10-5. Draw a block diagram of a half subtractor (label two inputs and two outputs).

10-6. Draw a block diagram of a full subtractor (label three inputs and two outputs).

10-7. Draw a block diagram of a 2-bit parallel adder (use a half and a full adder).

10-8. Use circuit simulation software to (1) construct a full-adder circuit like the one in Fig. 10-6(*a*) (2) test the circuit, and (3) show your instructor your circuit and results.

10-9. Do binary multiplication problems **a** to **h** (show your work). Check your answers using decimal multiplication.
   a. 101 × 011 =
   b. 111 × 011 =
   c. 1000 × 101 =
   d. 1001 × 010 =
   e. 1010 × 011 =
   f. 110 × 111 =
   g. 1100 × 1000 =
   h. 1010 × 1001 =

10-10. List two methods of doing binary multiplication with digital electronic circuits.

10-11. Convert the following signed decimal numbers to their 4-bit 2s complement form:
   a. +1 =
   b. +7 =
   c. −1 =
   d. −7 =

10-12. Convert the following 4-bit 2s complement numbers to their signed decimal form:
    a. 0101 =    c. 1110 =
    b. 0011 =    d. 1000 =

10-13. Convert the following 8-bit 2s complement numbers to their signed decimal form:
    a. 0111 0000    c. 1000 0001
    b. 1111 1111    d. 1100 0001

10-14. Convert the following signed decimal numbers to their 8-bit 2s complement form:
    a. +50    c. −50
    b. −32    d. −96

10-15. Add the following 4-bit 2s complement numbers. Give each sum as a *4-bit 2s complement number.* Also give each sum as a signed decimal number.
    a. 0110 + 0001 =
    b. 1101 + 1011 =
    c. 0001 + 1100 =
    d. 0100 + 1110 =

10-16. Subtract the following 4-bit 2s complement numbers. Give each difference as a *4-bit 2s complement number.* Also give each difference as a signed decimal number.
    a. 0110 − 0010 =
    b. 1001 − 1110 =
    c. 0010 − 1101 =
    d. 1101 − 0001 =

10-17. Add the following 8-bit 2s complement numbers. Give the sum in 8-bit 2s complement notation. Also give each sum as a signed decimal number.
    a. 0001 0101 + 0000 1111 =
    b. 1111 0000 + 1111 1000 =

    c. 0000 1111 + 1111 1100 =
    d. 1101 1111 + 0000 0011 =

10-18. Subtract the following 8-bit 2s complement numbers. Give the difference in 2s complement notation. Also give each difference as a signed decimal number.
    a. 0111 0000 − 0001 1111 =
    b. 1100 1111 − 1111 0000 =
    c. 0001 1100 − 1110 1111 =
    d. 1111 1100 − 0000 0010 =

10-19. See Table 10-1. The problem with the faulty half-adder circuit appears to be in the _____ ($C_O$, sum) output, which seems to be _____ (stuck HIGH, stuck LOW).

10-20. See Table 10-1. In an attempt to repair the faulty half-adder circuit, you might start by substituting a good _____ (AND gate IC, XOR gate IC) and then testing the circuit for correct operation.

### Table 10-1 Logic Probe Results on Faulty Half-Adder Circuit

| Inputs | | Outputs | |
| --- | --- | --- | --- |
| B | A | Sum | CO |
| L | L | L | H |
| L | H | H | H |
| H | L | H | H |
| H | H | L | H |

## Critical Thinking Questions

10-1. Draw a logic symbol diagram of a 2-bit parallel adder using XOR, AND, and OR gates.

10-2. Draw a logic symbol diagram of a full subtractor circuit using XOR, NOT, and NAND gates. Use Fig. 10-11 as a guide.

10-3. Draw a logic diagram of an 8-bit binary adder using two 7483 4-bit adder ICs.

10-4. Convert the signed number +127 to its 8-bit 2s complement form. Remember that the left-most bit will be 0, which means the number is positive.

10-5. Convert the signed number −25 to its 8-bit 2s complement form. Remember that the left-most bit will be 1, which means the number is negative.

10-6. Twos-complement numbers are widely used in digital systems (such as microprocessors) because they can be used to represent _____ numbers.

10-7. Describe how you would form a 2s complement from a binary number.

10-8. The negative of a binary number is its _____ (2s complement, 9s complement).

10-9. Why might we say that decimal 0 would be represented as a positive number in 2s complement notation?

10-10. At the option of your instructor, use circuit simulation software to (1) construct a 4-bit binary adder using an adder IC (see Fig. 10-13), (2) test the circuit by adding several 4-bit binary numbers, and (3) show your instructor your circuit and results. You may substitute a 4008 4-bit adder IC in place of the 7483 TTL IC.

10-11. At the option of your instructor, use circuit simulation software to (1) construct an adder/subtractor system using 2s complement numbers (see Fig. 10-30), (2) test the circuit by adding and subtracting 2s complement numbers (see Figs. 10-26 and 10-27 for samples), and (3) show your instructor your circuit and results.

## Answers to Self-Tests

1. 1110
2. 10001
3. 11000
4. 11010
5.

| A — HA | Σ |
| B — | $C_O$ |

6.

| B | A | Σ | $C_O$ |
|---|---|---|---|
| 0 | 0 | 0 | 0 |
| 0 | 1 | 1 | 0 |
| 1 | 0 | 1 | 0 |
| 1 | 1 | 0 | 1 |

7. 1s
8. $t_1$: sum = 1, $C_O$ = 0
   $t_2$: sum = 0, $C_O$ = 0
   $t_3$: sum = 0, $C_O$ = 1
   $t_4$: sum = 1, $C_O$ = 0
9.

| $C_{in}$ — | | Σ |
| A — FA | | |
| B — | | $C_O$ |

10.

| $C_{in}$ | B | A | Σ | $C_O$ |
|---|---|---|---|---|
| 0 | 0 | 0 | 0 | 0 |
| 0 | 0 | 1 | 1 | 0 |
| 0 | 1 | 0 | 1 | 0 |
| 0 | 1 | 1 | 0 | 1 |
| 1 | 0 | 0 | 1 | 0 |
| 1 | 0 | 1 | 0 | 1 |
| 1 | 1 | 0 | 0 | 1 |
| 1 | 1 | 1 | 1 | 1 |

11. arithmetic-logic unit (ALU)
12. full-adder

13. $t_1$: sum = 1, $C_O$ = 1
    $t_2$: sum = 0, $C_O$ = 1
    $t_3$: sum = 0, $C_O$ = 1
    $t_4$: sum = 1, $C_O$ = 0
    $t_5$: sum = 0, $C_O$ = 1
    $t_6$: sum = 1, $C_O$ = 0
    $t_7$: sum = 1, $C_O$ = 0
    $t_8$: sum = 0, $C_O$ = 0
14. half adder, full adders
15. combinational
16. 1101
17. 1000
18. 1110
19. a. 01
    b. 10
    c. 000
    d. 101
    e. 1111
    f. 111
20.

| A — HS | Di |
| B — | $B_O$ |

21.

| A | B | Di | $B_O$ |
|---|---|---|---|
| 0 | 0 | 0 | 0 |
| 0 | 1 | 1 | 1 |
| 1 | 0 | 1 | 0 |
| 1 | 1 | 0 | 0 |

22.

| $B_{in}$ — | Di |
| A — FS | |
| B — | $B_O$ |

23.

| A | B | $B_{in}$ | Di | $B_O$ |
|---|---|---|---|---|
| 0 | 0 | 0 | 0 | 0 |
| 0 | 0 | 1 | 1 | 1 |
| 0 | 1 | 0 | 1 | 1 |
| 0 | 1 | 1 | 0 | 1 |
| 1 | 0 | 0 | 1 | 0 |
| 1 | 0 | 1 | 0 | 0 |
| 1 | 1 | 0 | 0 | 0 |
| 1 | 1 | 1 | 1 | 1 |

24. parallel subtractor
25. borrow lines
26. adder
27. cascaded
28. combinational
29. 74LS181
30. 10101
31. F
32. 11101011
33. 1 1111 1110
34. binary
35. 1110
36. 1000001
37. 10101000
38. repeated addition
39. add and shift
40. add and shift
41. F
42. 2s complement
43. 0111, +7
44. −1
45. sign, positive, negative

46. 1010
47. 0101
48. 1011, −5
49. 0010, +2
50. 0101 1010, 0101 1010
51. 1010 0110
52. 0110 1111, +111
53. 1000 1111, −113
54. 1101 0000, −48
55. 0011 0001, +49
56. 2s complement
57. 2s complement
58. signed
59. 1111, −1
60. 0011, +3
61. 1110, −2
62. 1011, −5
63. combinational, carry out
64. 4, LOW
65. T
66. Know the normal operation of circuit, feel top of IC, look for broken connections or signs of excessive heat, smell for overheating, check the power sources, trace path of logic, and isolate the faulty section.

# Memories

## Chapter Objectives

*This chapter will help you to:*

1. *List* and *characterize* common memory and storage devices used in a microcomputer system.

2. *Sketch* the general organization of a computer, including the CPU, control bus, address bus, data bus, internal RAM, ROM, NVRAM, and bulk storage memory devices.

3. *Match* certain semiconductor memory cell types with specific characteristics and common uses.

4. *Associate* specific storage devices with their fundamental technology, such as magnetic, mechanical, optical, or semiconductor.

5. Given small semiconductor memory organization, *draw* the memory in table form, *sketch* a logic symbol for the memory, and *explain* the programming of the memory.

6. *Define* the read and write processes in a memory and *describe* how these processes might be carried out using a specific device.

7. *Identify* several specifications often associated with semiconductor memories.

8. *Identify* several common memory packages.

9. *List* and *describe* several bulk-storage methods.

10. *List* several emerging memory technologies that hold great promise.

It has been said that the most important characteristic that a digital system has over an analog system is its *ability to store data* for short or long periods. The availability and use of memory and digital storage devices has fueled what writers have called the *information revolution*. The entire Internet system is dependent on the transfer of data from one storage/memory device to another. Of course, computers and telecommunication systems are dependent on large amounts of digital storage.

The *compact disk read-only memory* (CD-ROM) is one example of an optical storage device that features high storage capacity at low cost. A single CD-ROM can store the equivalent of more than 200,000 typed pages of information, which is more data storage than is available on more than 400 standard floppy disks. The CD-ROM has a storage capacity of 650 Mbytes per single 4.75-in. diameter disk. The manufacturing cost of a high-quality CD-ROM is inexpensive. A higher capacity CD is the DVD (*digital versatile disk*).

The flip-flop, which we have already studied, forms a basic "memory cell" in some *semiconductor memories*. You have already used a simple shift register, latches, and counters, which use the flip-flop as a temporary memory. Several other types of semiconductor memory cells will be investigated in this chapter. Several types of bulk storage devices will also be surveyed. Bulk storage devices are commonly classified as either magnetic, mechanical, optical, or semiconductor in nature.

**Semiconductor memory**

## 11-1 Overview of Memory

### Memory Devices in Computer

The sketch in Fig. 11-1 is an overview of a typical microcomputer system featuring the many types of memory and storage devices used in an everyday machine. The *CPU* is the *central processing unit,* which is the section of a computer or microprocessor that contains the arithmetic, logic, and control sections. The CPU is the focus of most data transfers. Flowing from the CPU in Fig. 11-1 are the *address bus* and *control bus* lines. A *bus* is a group of parallel conductors whose job it is to transfer information to other parts of the computer or microprocessor. The address bus and control bus are one-way communication lines that tell memory, storage, and other peripheral devices *who does what and when.* The data bus is a two-way communication channel for sending information to and receiving information from memory, storage, and other peripheral devices. The simplified block diagram in Fig. 11-1 shows some of the common internal semiconductor memory devices used in computers such as the *RAM, ROM,* and *NVRAM.* Notice that data from the data bus can flow into *(to write in memory)* or out of

*(to read from memory)* both *random-access memory (RAM)* and *nonvolatile RAM (NVRAM).* The *read-only memory (ROM)* is different because it is permanently programmed and data can flow out of this semiconductor device only as shown by the arrow in Fig. 11-1. A variety of semiconductor read-only memory devices such as PROMs or EPROMs could be substituted for the ROM in this computer system.

Other memory components commonly associated with a modern microcomputer are listed under bulk storage devices in Fig. 11-1. They are divided according to the type of storage medium such as magnetic, optical, or semiconductor.

### Magnetic Storage

*Floppy disk drives* have been standard on most personal computers in the past. They are still common on desktop computers. The 3.5-in. type is the most common type of floppy disk. They store data in a thin coating of metal oxide on a flexible plastic disk. Digital 0s and 1s are represented by the alignment of magnetic domains in one direction or the other in the metal oxide surface. A write head places 0s and 1s as the magnetic material, and a read head detects or reads a 0 or 1 as the

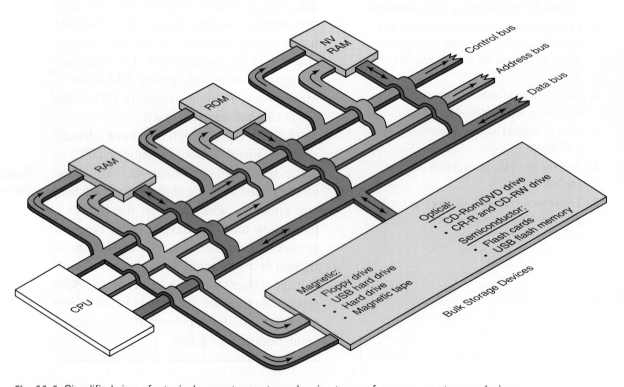

**Fig. 11-1** Simplified view of a typical computer system showing types of memory or storage devices.

magnetic medium moves past it. A typical floppy disk might have a storage capacity of 1 to 2 MB.

The *hard drive* is an almost universal magnetic bulk storage device in personal computers. It operates on the same principles as the floppy disk except its storage capacity is much greater. A typical *internal hard drive* on a PC may have a huge storage capacity of 60 to 500 GB or more.

On the hard drive, fine-grained magnetic material is coated on rigid disks that rotate at higher speeds than floppy disk drives. Floppy disks have the advantage that they can be removed and transported. However, small *portable hard drives* (usually USB hard drives) are becoming quite common as a high-capacity portable storage device.

On large corporate and government computer networks, *magnetic tape* may be used. Magnetic tape is useful in periodically backing up important files.

## Optical Storage

Many new computer systems contain a CD reader drive capable of reading information from several types of CDs (compact disks). The music industry started using the CD (compact disk) in the early 1980s.

Common optical media include CD-ROM (CD read-only memory), the CD-DA (CD digital audio), CD-R (CD recordable), CD-RW (CD rewritable), and DVD (digital versatile disc). DVDs come in a wide variety including DVD-video (digital video disc), DVD-audio, DVD-ROM, and DVD-RAM.

Manufactured CD-ROM and DVD disks are produced using expensive industrial plastics injection equipment. During manufacture, tiny pits and lands (no pit) are molded into the shiny side of the CD. The CD reader drive aims a laser beam at a track on the spinning CD. The reflected light bounced off the pits and lands are interpreted as logical 0s and 1s.

A high capacity version of the CD-ROM is the *digital versatile disk (DVD)*. DVDs are most commonly associated with video productions (movies). DVD-video standards are used when the disk holds only audio/video (such as movies). DVD-ROM standards are used when the digital versatile disk is used for

data storage as with a computer. The pits and lands are smaller with the DVD, which yields a greater storage capacity than older CD-ROMs. A simple one-sided, single-layer, 4.75-inch DVD has a capacity of about 4.7 GB.

## Semiconductor Storage

A single semiconductor-type of bulk storage device is listed in the microcomputer system sketched in Fig. 11-1. Flash memories can appear in regular IC packages or in memory card form. A memory card looks something like a thick credit card. Digital cameras commonly use flash memory cards to store photos. A decade ago, flash memories were available only in small sizes, but recently large capacity chips have become available. Ultimately, semiconductor flash memories may become *solid-state drives* as they replace the hard drive in some portable computers and other devices such as personal organizers. Flash memories commonly take the form of small *USB flash memory* devices. USB flash memory modules are removable and are commonly used like floppy disks.

## Semiconductor Storage Cells

Semiconductor storage devices are commonly classified in about six categories: SRAM, DRAM, ROM, EPROM, EEPROM, and flash memory (flash EEPROM). Some of these technologies are better than others for certain jobs in a digital system. Following is a brief description of these technologies:

- *SRAM (static random-access memory)*— high access speed, read or write, requires continuous power (volatile memory), low density, high cost, associated with high-speed *cache* memory in microprocessors.

- *DRAM (dynamic random-access memory)*—good access speed, read or write, volatile memory plus a need for refresh circuitry, high density, lower cost, RAM type used in most modern PCs

- *ROM (read-only memory)*—high density, nonvolatile (cannot be altered), reliable, low cost especially at high volumes

- *EPROM (electrically programmable read-only memory)*—high density, nonvolatile (can be updated although not easily), ultraviolet light erasable before reprogramming

- *EEPROM (electrically erasable programmable read-only memory)*—nonvolatile but electrically erasable by bytes for reprogramming, lower density, high cost

- *Flash Memory*—very high density, low power, nonvolatile but rewritable (bit-by-bit) within the digital system, fairly new and developing technology holding great promise as a solid-state hard drive, can be portable (like floppy disk) in memory card form or USB flash memory

- *FRAM (ferroelectric RAM)*—nonvolatile RAM, in-circuit programmable, good access speed (reading and writing), low density, high cost, FRAM memory cells based on ferroelectric capacitor and MOS transistor.

- *MRAM (magnetoresistive RAM or magnetic RAM)*—nonvolatile RAM, in-circuit programmable, excellent access speed, high density, nanotechnology used in fabrication; cost has not been determined because it is a new technology.

The diagram in Fig. 11-2 suggests three important characteristics of a semiconductor memory represented by the three large circles: nonvolatility, high density, and the capacity of

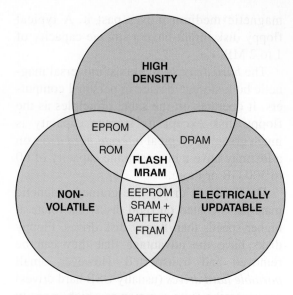

**Fig. 11-2** Important semiconductor memory characteristics.

being electrically updated. Notice in Fig. 11-2 that the newer flash memory has the best combination of nonvolatility, high density, and read/write capability (electrically updatable). Flash memory is a developing technology, and it can be expected that densities will go up and the price will fall, making the technology widely applied.

Consider the advantage of using flash memory in the system sketched in Fig. 11-1. In a common microcomputer system, the control unit of the computer would direct the floppy disk drive to transfer a file or files to the RAM (probably DRAM in most systems). This takes quite a bit of time. If the disk drive were replaced with flash memory this seek time (disk-to-DRAM loading) is eliminated making users experience higher speed operation.

*Answer the following questions.*

1. The section of a computer system that contains the arithmetic, logic, and control section and is the center of many data transfers is called the _____.
2. List two one-way buses in a microcomputer system that direct memory, storage, and peripheral devices.
3. List three general categories of bulk storage devices based on the technology each uses.
4. List at least two bulk storage devices commonly found in microcomputer systems.
5. Spell out the full term for each of the following abbreviations.
   a. RAM       d. EEPROM
   b. ROM       e. SRAM
   c. EPROM     f. DRAM

6. A DVD is an optical disk that has more storage capacity than a CD-ROM. (T or F)
7. A CD writer drive is used on many personal computers to write data on _____ (CD-ROMs, CD-R or CD-RWs).

8. Based on the information in Fig. 11-2, which semiconductor memory type would be the best choice if you wanted a non-volatile memory with read/write capabilities, and high density (memory cells are very small)?

## 11-2 Random-Access Memory (RAM)

One type of semiconductor memory device used in digital electronics is the *random-access memory*. The *RAM* is a memory that you can "teach." After the "teaching-learning" process (called *writing*), the RAM remembers the information for a while and the RAM's stored information can be recalled, or "remembered," at any time. We say that we can *write* information (0s and 1s) into the memory and *read out,* or recall, information. The RAM is also called a *read/write memory* or a *scratch-pad memory.*

A semiconductor memory with 64 cells in which to place 0s and 1s is illustrated in Fig. 11-3. The 64 squares (mostly blank) represent the 64 cells that can be filled with data. Notice that the 64 bits are organized into 16

groups called *words*. Each of the 16 words contains 4 *bits* of information. This memory is said to be organized as a $16 \times 4$ memory. That is, it contains 16 words, and each word is 4 bits long. A 64-bit memory could be organized as a $32 \times 2$ memory (32 words of 2 bits each), a $64 \times 1$ memory (64 words of 1 bit each), or an $8 \times 8$ memory (8 words of 8 bits each).

The memory in Fig. 11-3 looks very much like a truth table on a scratch pad. On the table after word 3 we have written the contents of word 3 (0110). We say we have stored, or *written,* a word into the memory; this is the *write operation*. To see what is in the memory at word location 3, just read from the table in Fig. 11-3, this is the *read operation*. The write operation is the process of putting new information into the memory. The read operation is the process of copying information from memory. The read operation is also referred to as the *sense* operation because it senses, or reads, the contents of the memory.

You could write any combination of 0s and 1s in the table in Fig. 11-3 rather like writing on a scratch pad. You could then read any word(s) from the memory, as from a scratch pad. Notice that the information in the memory remains even after it is read. Now it should be obvious why this memory is sometimes called a 64-bit scratch-pad memory. The memory has a place for 64 bits of information, and the memory can be written into or read from very much like a scratch pad.

The memory in Fig. 11-3 is called a random-access memory because you can go directly to word 3 or word 15 and read its contents. In other words, you have access to any bit (or word) at any instant. You merely skip down to its word location and read that word. A location in the memory, such as word 3, is referred to as the storage location or *address*. In the case of

**Memory organization**

**Random-access memory**

**Read/write memory**

**Write operation**

**Read operation**

**Address**

| Address | Bit D | Bit C | Bit B | Bit A |
|---------|-------|-------|-------|-------|
| Word 0  |       |       |       |       |
| Word 1  |       |       |       |       |
| Word 2  |       |       |       |       |
| Word 3  | 0     | 1     | 1     | 0     |
| Word 4  |       |       |       |       |
| Word 5  |       |       |       |       |
| Word 6  |       |       |       |       |
| Word 7  |       |       |       |       |
| Word 8  |       |       |       |       |
| Word 9  |       |       |       |       |
| Word 10 |       |       |       |       |
| Word 11 |       |       |       |       |
| Word 12 |       |       |       |       |
| Word 13 |       |       |       |       |
| Word 14 |       |       |       |       |
| Word 15 |       |       |       |       |

**Fig. 11-3** Organization of a 64-bit memory.

**Nonvolatile storage devices**

**Volatile memory**

Fig. 11-3, the address of word 3 is $0011_2$ ($3_{10}$). However, the data stored at this address is 0110.

The RAM cannot be used for permanent memory because it loses its data when the power to the IC is turned off. The RAM is considered a *volatile memory* because of this loss of data. Volatile memories are thus used for the *temporary* storage of data. However, some memories are permanent; they do not "forget" or lose their data when the power goes off. Such permanent memories are called *nonvolatile storage devices*.

RAMs are used where only a temporary memory is needed. RAMs are used for calculator memories, buffer memories, and cache memories.

Modern personal computers implement random-access memory using both SRAM (static RAM) and DRAM (dynamic RAM).

---

**✓ Self-Test**

*Supply the missing word or words in each statement.*

9. The letters "RAM" stand for _____.
10. Copying information into a storage location is called _____ into memory.
11. Copying information from a storage location is called _____ from memory.
12. A RAM is also called a(n) _____ or scratch-pad memory.

13. Refer to Fig. 11-3. This 64-bit unit is organized as a(n) _____ memory.
14. A disadvantage of the RAM is that it is _____; it loses its data when the power is turned _____ (off, on).
15. The RAM section of a PC usually consists of both SRAM and _____ (DRAM, PXRAM).

---

**ABOUT ELECTRONICS**

**CD-Rewritable Drives**   CD-Rewritable drives are a versatile piece of computer hardware. They are like having three drives in one. (1) A CD-Rewritable disc can be used like a 700-megabyte floppy disk. (2) With special software, the CD-Rewritable disc drive can be used to record audio CDs. (3) Finally, a CD-Rewritable disc drive can double as a CD-ROM drive.

## 11-3  Static RAM ICs

The *7489 read/write TTL RAM* is a 64-bit data storage unit in IC form. Figure 11-4(*a*) is a logic symbol for the 7489 RAM. The memory cells are arranged like the layout of the table in Fig. 11-3. The memory can hold 16 words; each word in the 7489 IC is 4 bits wide. The 7489 RAM is said to be organized as a 16 × 4-bit memory. A pin diagram for the 7489 IC is given in Fig. 11-4(*b*).

A simplified truth table for the 7489 RAM is shown in Fig. 11-4(*c*). The *memory enable* ($\overline{ME}$) input is used to "turn on" or "select" the RAM for either reading or writing. The top line in the truth table shows both the $\overline{ME}$ and the *write enable* ($\overline{WE}$) inputs LOW. The 4 bits at the data inputs ($D_1$ to $D_4$) are stored in the memory location selected by the address inputs ($A_3$ to $A_0$). The RAM is in the *write mode.*

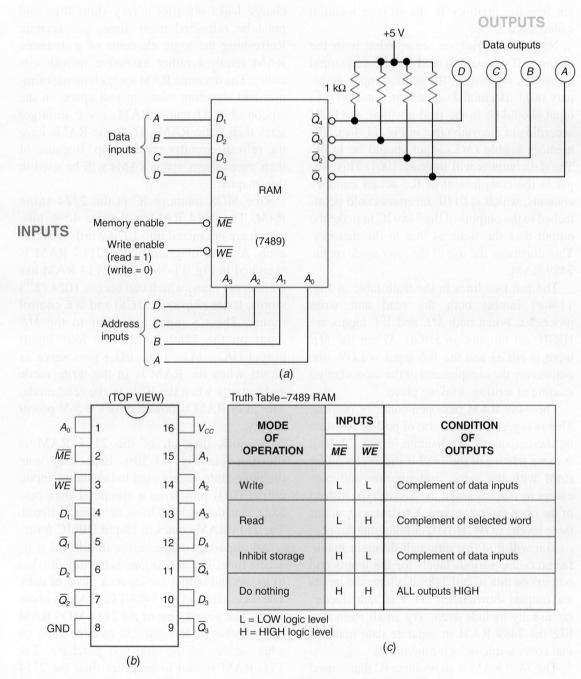

Fig. 11-4  7489 64-bit RAM TTL IC. (*a*) Logic diagram. (*b*) Pin diagram. (*c*) Truth table.

Let us write data into the 7489 memory chip. Suppose we want to write 0110 into word 3 location, as shown in Fig. 11-3. The address for word 3 is $A_3 = 0$, $A_2 = 0$, $A_1 = 1$, and $A_0 = 1$. Word 3 is located in the memory by placing a binary 0011 on the *address inputs* of the 7489 RAM (see Fig. 11-4($a$)). Next, place the correct input data at the *data inputs*. To enter 0110, place a 0 at input *A,* a 1 at input *B,* a 1 at input *C,* and a 0 at input *D.* Next, place a LOW at the write enable ($\overline{WE}$) input. Last, place a LOW at the memory enable ($\overline{ME}$) input. Data are written into the memory in the storage location called word 3.

Now let us *read,* or *sense,* what is in the memory. If we want to read out the data stored at word 3, we first set the address inputs to binary 0011 (decimal 3). The write enable ($\overline{WE}$) input should be in the read position, or HIGH according to the truth table in Fig. 11-4($c$). The memory enable ($\overline{ME}$) input should be LOW. The data outputs will indicate 1001. This output is the *complement* of the actual memory contents, which is 0110. Inverters could be attached to the outputs of the 7489 IC to make the output data the same as that in the memory. This illustrates the use of the *read mode* on the 7489 RAM.

The last two lines in the truth table in Fig. 11-4($c$) inhibit both the read and write processes. When both $\overline{ME}$ and $\overline{WE}$ inputs are HIGH, all outputs go HIGH. When the $\overline{ME}$ input is HIGH and the $\overline{WE}$ input is LOW, the outputs are the complement of the inputs but no reading or writing is taking place.

The 7489 RAM has open-collector outputs. This is suggested by the use of pull-up resistors on the outputs in the diagram in Fig. 11-4($a$). A close relative of the 7489 is the *74189 64-bit RAM* with the same configuration and pins except its outputs are of the tristate type instead of the open-collector type. A *tristate output* has three levels: LOW, HIGH, or high impedance.

You will find that although different manufacturers use various labels for the inputs and outputs on this IC, all 7489 ICs have the inputs and outputs shown in Fig. 11-4. IC manufacturers usually include even very small memories like the 7489 RAM in separate data manuals that cover semiconductor memories.

The 7489 RAM is an obsolete IC that is used for experimental purposes in lab experiments to help show how many semiconductor memory chips are addressed, read from, and written to. Microprocessor-based equipment makes extensive use of semiconductor read/write RAMs in IC form.

Semiconductor RAM ICs are subdivided by manufacturers into static and dynamic types. The *static RAM* stores data in a flip-flop-like element. It is called a static RAM because it holds its 0 or 1 as long as the IC has power. The *dynamic RAM* IC stores its logic state as an electric charge in an MOS device. The stored charge leaks off after a very short time and must be refreshed many times per second. Refreshing the logic elements of a dynamic RAM requires rather extensive refresh circuitry. The dynamic RAM logic element is simpler and therefore takes up less space on the silicon chip. Dynamic RAMs come in larger sizes than static RAMs. Dynamic RAMs have the refresh circuitry on the chip. Because of their ease of use, static RAMs will be used in this chapter.

One MOS memory IC is the *2114 static RAM.* The 2114 RAM will store 4096 bits, which are organized into 1024 words of 4 bits each. A logic diagram of the 2114 RAM is sketched in Fig. 11-5($a$). The 2114 RAM has 10 address lines, which can access 1024 ($2^{10}$) words. It has *chip select* ($\overline{CS}$) and $\overline{WE}$ control inputs. The $\overline{CS}$ input is similar to the $\overline{ME}$ input on the 7489 RAM. The four input/output ($I/O_1$, $I/O_2$, $I/O_3$, $I/O_4$) pins serve as inputs when the RAM is in the write mode and outputs when the IC is in the read mode. The 2114 RAM is powered by a +5-V power supply.

A block diagram of the 2114 RAM is illustrated in Fig. 11-5($b$). Especially note the *three-state buffers* used to isolate the input/output (I/O) pins from a computer data bus. Note that the address lines are also buffered. The 2114 RAM comes in 18-pin DIP IC form.

An important characteristic of a RAM is its access time. The *access time* is the time it takes to locate and output (or input) a piece of data. The access time of the 7489 TTL RAM is about 33 ns. The access time of the 2114 MOS RAM ranges between 100 and 250 ns depending on what version of the chip you purchase. The TTL RAM is said to be faster than the 2114 memory chip because of its shorter access time.

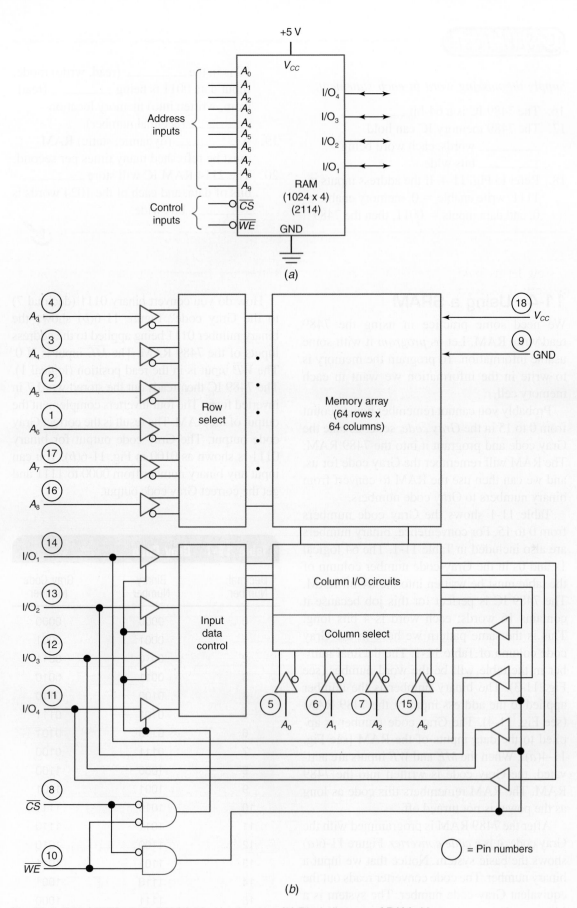

Fig. 11-5 2114 MOS static RAM (a) Logic diagram. (b) Block diagram of RAM chip.

2114 MOS
static RAM

Memories    Chapter 11    355

## 11-4 Using a SRAM

**Gray code**

We need some practice in using the 7489 read/write RAM. Let us *program* it with some usable information. To program the memory is to write in the information we want in each memory cell.

Probably you cannot remember how to count from 0 to 15 in the *Gray code,* so let us take the Gray code and program it into the 7489 RAM. The RAM will remember the Gray code for us, and we can then use the RAM to convert from binary numbers to Gray code numbers.

Table 11-1 shows the Gray code numbers from 0 to 15. For convenience, binary numbers are also included in Table 11-1. The 64 logical 1s and 0s in the Gray code number column of the table must be written into the 64-bit RAM. The 7489 IC is perfect for this job because it contains 16 words; each word is 4 bits long. This is the same pattern we have in the Gray code column of Table 11-1. The decimal number in the table will be the word number (see Fig. 11-3). The binary number is the number applied to the address input of the 7489 RAM (see Fig. 11-4). The Gray code number is applied to the data inputs of the RAM [see Fig. 11-4(*a*)]. When the $\overline{ME}$ and $\overline{WE}$ inputs are activated, the Gray code is written into the 7489 RAM. The RAM remembers this code as long as the power is not turned off.

**Programming RAM**

After the 7489 RAM is programmed with the Gray code, it is a *code converter.* Figure 11-6(*a*) shows the basic system. Notice that we input a binary number. The code converter reads out the equivalent Gray code number. The system is a *binary-to-Gray-code converter.*

**Binary-to-Gray-code converter**

How do you convert binary 0111 (decimal 7) to the Gray code? Figure 11-6(*b*) shows the binary number 0111 being applied to the address inputs of the 7489 RAM. The $\overline{ME}$ input is at 0. The $\overline{WE}$ input is in the read position (logical 1). The 7489 IC then reads out the stored word 7 in inverted form. The four inverters complement the output of the RAM. The result is the correct Gray code output. The Gray code output for binary 0111 is shown as 0100 in Fig. 11-6(*b*). You can input any binary number from 0000 to 1111 and get the correct Gray code output.

| Table 11-1 | Gray Code | |
|---|---|---|
| Decimal Number | Binary Number | Gray Code Number |
| 0 | 0000 | 0000 |
| 1 | 0001 | 0001 |
| 2 | 0010 | 0011 |
| 3 | 0011 | 0010 |
| 4 | 0100 | 0110 |
| 5 | 0101 | 0111 |
| 6 | 0110 | 0101 |
| 7 | 0111 | 0100 |
| 8 | 1000 | 1100 |
| 9 | 1001 | 1101 |
| 10 | 1010 | 1111 |
| 11 | 1011 | 1110 |
| 12 | 1100 | 1010 |
| 13 | 1101 | 1011 |
| 14 | 1110 | 1001 |
| 15 | 1111 | 1000 |

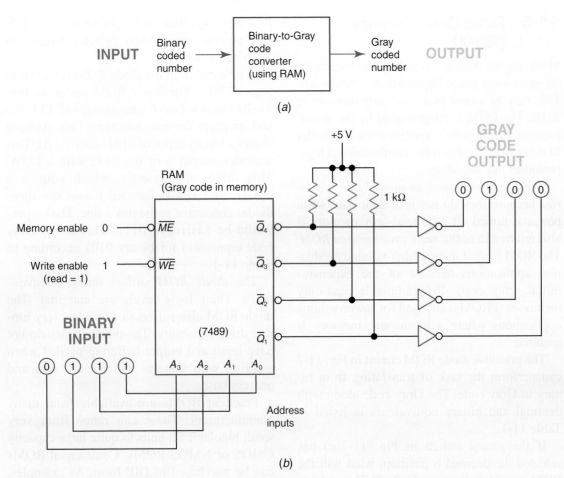

**Fig. 11-6** Binary-to-Gray-code converter. (*a*) System diagram. (*b*) Wiring diagram using RAM.

The binary-to-Gray-code converter in Fig. 11-6 works fine. It demonstrates how you can program and use the 7489 RAM. It is not practical, however, because the RAM is a volatile memory. If the power is turned off for even an instant, the storage unit loses all its memory and "forgets" the Gray code. We say the memory has

been *erased*. You then have to again program, or teach, the Gray code to the 7489 RAM.

Each time your home or school computer boots up when it is first started, it loads codes/programs into its RAM section of memory. This is much like loading the Gray code into the tiny 7489 RAM.

*Supply the missing word or words in each statement.*

21. Refer to Fig. 11-6. The RAM is programmed as a(n) _____-code converter in this example.
22. Refer to Fig. 11-6. If the address inputs = 1000, $\overline{WE} = 1$, and $\overline{ME} = 0$, then

the output at the displays on the right will be _____. This is the _____ code equivalent of binary _____.

23. If power to the 7489 IC in Fig. 11-6 is turned off for an instant, the RAM will _____ (lose its program and have to be reprogrammed, still hold the Gray code in its memory cells).

Read-only
memory (ROM)
Address

1-of-10 decoder

Nonvolatile
memories

Mask-
programmed
ROM

Diode ROM
disadvantages

## 11-5 Read-Only Memory (ROM)

Many digital devices including microcomputers must store some information permanently. This may be stored in a *read-only memory* or *ROM*. The ROM is programmed by the manufacturer to the user's specifications. Smaller ROMs can be used to solve combinational logic problems like decoding.

ROMs are classified as *nonvolatile memories* because they do not lose their data when power is turned off. The read-only memory is also referred to as the *mask-programmed ROM*. The ROM is used in only high-volume production applications because of the expensive initial setup costs. Programmable read-only memories (PROMs) are used for lower-volume applications where a permanent memory is required.

The primitive diode ROM circuit in Fig. 11-7 can perform the task of translating from binary to Gray code. The Gray code along with decimal and binary equivalents is listed in Table 11-1.

If the rotary switch in Fig. 11-7(*a*) has selected the decimal 6 position, what will the ROM output indicators display? The outputs (*D, C, B, A*) will indicate LHLH or 0101. The *D* and *B* outputs are connected directly to ground through the resistors and read LOW. The *C* and *A* outputs are connected to +5 V through two forward-biased diodes and the output voltage will read about +2 to +3 V, which is a logical HIGH. Notice that the pattern of diodes in the diode ROM matrix in Fig. 11-7(*a*) is similar to the pattern of 1s in the Gray code column in Table 11-1. Each new position of the rotary switch will give the correct Gray code output.

In a memory, such as the ROM in Fig. 11-7, each position of the rotary switch is referred to as an *address*.

A refinement in the diode ROM is shown in Fig. 11-7(*b*). The diode ROM circuit in Fig. 11-7(*b*) uses a *1-of-10 decoder* (7442 TTL IC) and inverters for row selection. This example shows a binary input of 0101 (decimal 5). This activates output 5 of the 7442 with a LOW. This drives the inverter, which outputs a HIGH. The HIGH forward biases the three diodes connected to the row 5 line. The outputs would be LHHH or 0111. This is the Gray code equivalent for binary 0101 according to Table 11-1.

The *diode ROM* suffers many *disadvantages*. Their logic levels are marginal. The diode ROM also suffers in that it has very limited drive capability. The diode ROMs do not have input and output buffering needed when working with systems that contain data and address buses.

Practical ROMs are available from many manufacturers. These can range from very small bipolar TTL units to quite large capacity CMOS or NMOS ROMs. Commercial ROMs can be purchased in DIP form. As examples, a very small capacity unit might be the TTL 74S370 2048-bit ROM organized as a 512-word by 4-bit memory. A larger-capacity unit might be CMOS TMS47C512 524,288-bit ROM organized as a 65,536-word by 8-bit memory. The $65,536 \times 8$ unit has an access time of from 200 to 350 ns depending on the version you purchase. Personal computers have ROMs of larger capacity.

As an example of a commercial product, the TMS4764 ROM will be featured. The *TMS4764* is an *8192-word by 8-bit ROM*. Its $8192 \times 8$ organization makes it useful in systems that might store data in 8-bit groups, or bytes.

A pin diagram for the TMS4764 ROM is reproduced in Fig. 11-8(*a*). The ROM is housed in a 24-pin DIP. The names and functions of the pins are given in the chart in Fig. 11-8(*b*). Notice that a total of 13 addresses lines ($A_0$ to $A_{12}$) are needed to address the 8192 ($2^{13}$) memory locations. $A_0$ is the LSB and $A_{12}$ is the MSB of the word address. The access time of the TMS4764 ROM varies from 150 to 250 ns depending on the version of the chip you

### ABOUT ELECTRONICS

**Protein-Based Memory** Are protein-based 3D RAM memories in the future? A small cube of optically sensitive protein (such as Rhodopsin), suspended in a transparent plastic, might be the basis for a 20 gigabit RAM memory. Two laser beams might intersect at a point in the cube of protein to switch that "organic memory cell" from one logic state to another.

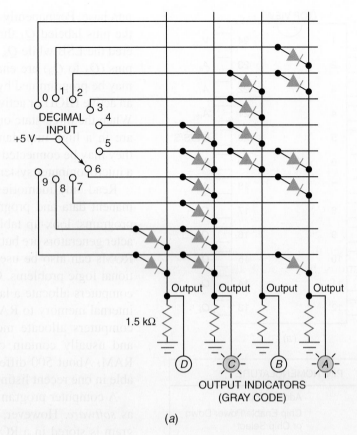

(a)

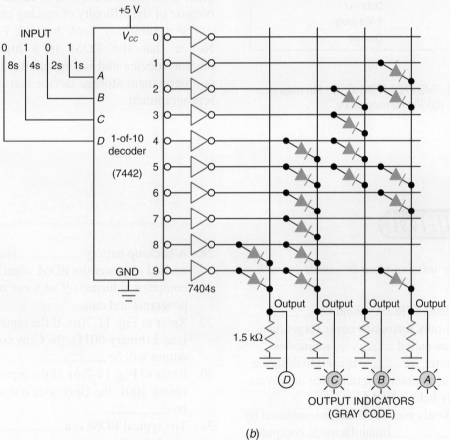

(b)

**Fig. 11-7** Diode ROMs. (a) Primitive diode ROM programmed with Gray code. (b) Diode ROM with input decoding (programmed with Gray code).

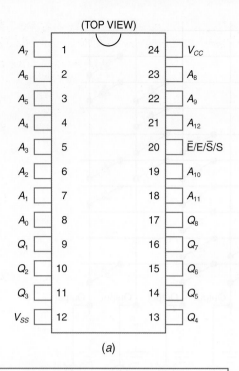

(TOP VIEW)

| | | |
|---|---|---|
| $A_7$ | 1 | 24 | $V_{CC}$ |
| $A_6$ | 2 | 23 | $A_8$ |
| $A_5$ | 3 | 22 | $A_9$ |
| $A_4$ | 4 | 21 | $A_{12}$ |
| $A_3$ | 5 | 20 | $\bar{E}/E/\bar{S}/S$ |
| $A_2$ | 6 | 19 | $A_{10}$ |
| $A_1$ | 7 | 18 | $A_{11}$ |
| $A_0$ | 8 | 17 | $Q_8$ |
| $Q_1$ | 9 | 16 | $Q_7$ |
| $Q_2$ | 10 | 15 | $Q_6$ |
| $Q_3$ | 11 | 14 | $Q_5$ |
| $V_{SS}$ | 12 | 13 | $Q_4$ |

(a)

**PIN NOMENCLATURE**

| | |
|---|---|
| $A_0$–$A_{12}$ | Address inputs |
| $\bar{E}/E/\bar{S}/S$ | Chip Enable/Power Down or Chip Select |
| $Q_1$–$Q_8$ | Data out |
| $V_{CC}$ | 5-V supply |
| $V_{SS}$ | Ground |

(b)

**Fig. 11-8** TMS4764 ROM IC. (a) Pin diagram. (b) Pin nomenclature.

**Software**

**Firmware**

purchase. Permanently stored data is output via the pins labeled $Q_1$ through $Q_8$. $Q_1$ is considered the LSB while $Q_8$ is the MSB. The output pins ($Q_1$ to $Q_8$) are enabled by pin 20. Pin 20 may be programmed by the manufacturer to be an active HIGH or active LOW $\overline{CS}$ or $\overline{CE}$ input. When the three-state outputs are disabled, they are in a high-impedance state, which means they may be connected directly to a data bus in a microcomputer system.

Read-only memories are used to store permanent data and programs. Computer system programs, look-up tables, decoders, and character generators are but a few uses of the ROM. ROMs can also be used for solving combinational logic problems. General-purpose microcomputers allocate a larger proportion of their internal memory to RAM. However, dedicated computers allocate more addresses to ROM and usually contain only small amounts of RAM. About 500 different ROMs were available in one recent listing.

A computer program is typically referred to as *software*. However, when a computer program is stored in a ROM it is called *firmware* because of the difficulty of making changes.

For a summary, look back at Fig. 11-2. Notice that the ROM is a high-density memory device and is nonvolatile. The ROM is a permanent storage device that cannot be reprogrammed.

---

**Self-Test**

*Supply the missing word or words in each statement.*

24. The letters "ROM" stand for _____.
25. Read-only memories never forget data and are called _____ memories.
26. The term _____ is used to describe microcomputer programs that are permanently held in ROM.
27. Read-only memories are programmed by the _____ (manufacturer, computer operator) to your specifications.

28. A back-up battery _____ (is, is not) needed to power the ROM when the computer is turned off so it can retain its programs and data.
29. Refer to Fig. 11-7(a). If the input switch is at 3 (binary 0011), the Gray code output will be _____.
30. Refer to Fig. 11-7(b). If the input is binary 1001, the Gray code output will be _____.
31. The typical ROM is a _____ (high-, low-) density memory device.

## 11-6 Using a ROM

Suppose you have to design a device that will give the decimal counting sequence shown in Table 11-2: 1, 117, 22, 6, 114, 44, 140, 17, 0, 14, 162, 146, 134, 64, 160, 177, and then back to 1. These numbers are to read out on seven-segment displays and must appear in the order shown.

Knowing you will use digital circuits, you convert the decimal numbers to BCD numbers. This is shown in Table 11-2. You find you have 16 rows and 7 columns of logical 0s and 1s. This section forms a truth table. As you look at the truth table, the problem seems quite complicated to solve with logic gates or data selectors. You decide to try a ROM. You think of the inside of a memory as a truth table. The BCD section of Table 11-2 reminds you that a memory organized as a 16 × 7 storage unit will do the job. This 16 × 7 ROM will have 16 words for the 16 rows on the truth table. Each word will contain seven bits of data for seven columns on the truth table. This will take a 112-bit ROM.

A 112-bit ROM is shown in Fig. 11-9. Notice that it has four address inputs to select one of the 16 possible words stored in the

### Table 11-2 Counting Sequence Problem

| Decimal Readout | | | Binary-Coded Decimal Number | | |
| --- | --- | --- | --- | --- | --- |
| 100s | 10s | 1s | 100s | 10s | 1s |
| | | 1 | 0 | 000 | 001 |
| 1 | 1 | 7 | 1 | 001 | 111 |
| | 2 | 2 | 0 | 010 | 010 |
| | | 6 | 0 | 000 | 110 |
| 1 | 1 | 4 | 1 | 001 | 100 |
| | 4 | 4 | 0 | 100 | 100 |
| 1 | 4 | 0 | 1 | 100 | 000 |
| | 1 | 7 | 0 | 001 | 111 |
| | | 0 | 0 | 000 | 000 |
| | 1 | 4 | 0 | 001 | 100 |
| 1 | 6 | 2 | 1 | 110 | 010 |
| 1 | 4 | 6 | 1 | 100 | 110 |
| 1 | 3 | 4 | 1 | 011 | 100 |
| | 6 | 4 | 0 | 110 | 100 |
| 1 | 6 | 0 | 1 | 110 | 000 |
| 1 | 7 | 7 | 1 | 111 | 111 |

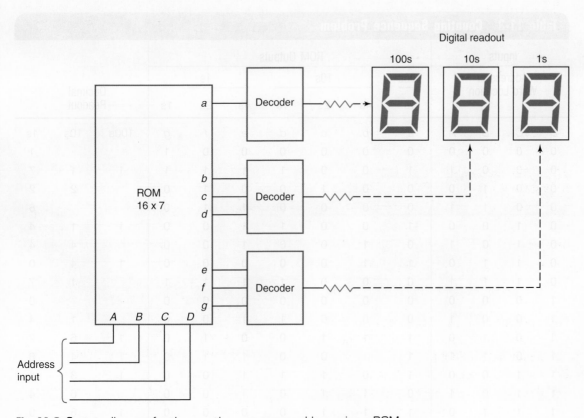

**Fig. 11-9** System diagram for the counting sequence problem using a ROM.

ROM. The 16 different addresses are shown in the left columns of Table 11-3. Suppose the address inputs are binary 0000. Then the first line in Table 11-3 shows that the stored word is 0 000 001 (a to g). After decoding in Fig. 11-9, this stored word reads out on the digital displays as a decimal 1 (100s = 0, 10s = 0, 1s = 1).

Let us consider another example. Apply binary 0001 to the address inputs of the ROM in Fig. 11-9. The second row on Table 11-3 shows us that the stored word is 1 001 111 (a to g). When decoded, this word reads out on the digital display as decimal 117 (100s = 1, 10s = 1, 1s = 7). Remember that the 0s and 1s in the center section of Table 11-3 are *permanently* stored in the ROM. When the address at the left appears at the address input of the ROM, a row of 0s and 1s (7-bit word) appears at the outputs.

You have solved the difficult counting sequence problem. Figure 11-9 diagrams the basic system to be used. The information in Table 11-3 shows the addressing and programming of the 112-bit ROM and the decoded BCD as a decimal readout. You would give the information in Table 11-3 to a manufacturer, who would custom-make as many ROMs as you need with the correct pattern of 0s and 1s.

It is quite expensive to have just a few ROMs custom-programmed by a manufacturer. You probably would not use the ROM if you did not have need for many of these memory units. Remember that this problem also could have been solved by a combinational logic circuit using logic gates.

Semiconductor memories usually come in $2^n$ sizes or 64-, 256-, 1024-, 4096-, 8192-bit and larger units. A 112-bit memory is an unusual size. The 112-bit memory was used in the example because its truth table in Table 11-3 is exactly the truth table of the 7447 IC. You used the 7447 IC as BCD-to-seven-segment decoder earlier. You will want to use the 7447 IC as a ROM in the laboratory.

Read-only memories are used for encoders, code converters, look-up tables, microprograms, character generators, function generators, microcomputer system firmware, and microcontroller firmware.

## Table 11-3 Counting Sequence Problem

| Inputs: Address or Word Location | | | | ROM Outputs 100s 1s | ROM Outputs 10s 4s | 10s 2s | 10s 1s | 1s 4s | 1s 2s | 1s 1s | Decimal Readout 100s | Decimal Readout 10s | Decimal Readout 1s |
|---|---|---|---|---|---|---|---|---|---|---|---|---|---|
| D | C | B | A | a | b | c | d | e | f | g | 100s | 10s | 1s |
| 0 | 0 | 0 | 0 | 0 | 0 | 0 | 0 | 0 | 0 | 1 | | | 1 |
| 0 | 0 | 0 | 1 | 1 | 0 | 0 | 1 | 1 | 1 | 1 | 1 | 1 | 7 |
| 0 | 0 | 1 | 0 | 0 | 0 | 1 | 0 | 0 | 1 | 0 | | 2 | 2 |
| 0 | 0 | 1 | 1 | 0 | 0 | 0 | 0 | 1 | 1 | 0 | | | 6 |
| 0 | 1 | 0 | 0 | 1 | 0 | 0 | 1 | 1 | 0 | 0 | 1 | 1 | 4 |
| 0 | 1 | 0 | 1 | 0 | 1 | 0 | 0 | 1 | 0 | 0 | | 4 | 4 |
| 0 | 1 | 1 | 0 | 1 | 1 | 0 | 0 | 0 | 0 | 0 | 1 | 4 | 0 |
| 0 | 1 | 1 | 1 | 0 | 0 | 0 | 1 | 1 | 1 | 1 | | 1 | 7 |
| 1 | 0 | 0 | 0 | 0 | 0 | 0 | 0 | 0 | 0 | 0 | | | 0 |
| 1 | 0 | 0 | 1 | 0 | 0 | 0 | 1 | 1 | 0 | 0 | | 1 | 4 |
| 1 | 0 | 1 | 0 | 1 | 1 | 1 | 0 | 0 | 1 | 0 | 1 | 6 | 2 |
| 1 | 0 | 1 | 1 | 1 | 1 | 0 | 0 | 1 | 1 | 0 | 1 | 4 | 6 |
| 1 | 1 | 0 | 0 | 1 | 0 | 1 | 1 | 1 | 0 | 0 | 1 | 3 | 4 |
| 1 | 1 | 0 | 1 | 0 | 1 | 1 | 0 | 1 | 0 | 0 | | 6 | 4 |
| 1 | 1 | 1 | 0 | 1 | 1 | 1 | 0 | 0 | 0 | 0 | 1 | 6 | 0 |
| 1 | 1 | 1 | 1 | 1 | 1 | 1 | 1 | 1 | 1 | 1 | 1 | 7 | 7 |

*Supply the missing word or words in each statement.*

32. Refer to Fig. 11-9. If power is turned off and then back on, the counting sequence programmed into the ROM will _____ (be lost from, remain in) memory.

33. Refer to Table 11-3 and Fig. 11-9. If the ROM address input = 1111, the digital readout will be _____.

34. Refer to Table 11-3 and Fig. 11-9. If the ROM address input = 1001, the digital readout will be _____.

35. A mask-programmable ROM is programmed by the _____ (manufacturer, user).

36. A group of programs and data held permanently in a microcomputer's _____ (RAM, ROM) would be called firmware.

## 11-7 Programmable Read-Only Memory [PROM]

Mask-programmable ROMs are programmed by the manufacturer using photographic masks to expose the silicon die. *Mask-programmable ROMs* have long development times and the initial costs are high. Mask-programmable ROMs are usually simply called ROMs.

*Field-programmable ROMs (PROMs)* are also available. They shorten development time and many times lower costs. It is also much easier to correct program errors and update products when PROMs can be programmed (burned) by the local developer. The regular PROM can only be programmed once, but its advantage is that it can be made in limited quantities and can be programmed in the local lab or shop. The PROM is also called a *fusible-link PROM.*

The *EPROM (erasable programmable read-only memory)* is a variation of the PROM. The EPROM is programmed or burned in the local lab using a *PROM burner.* If an EPROM needs to be reprogrammed, a special window on the top of the IC is used. Ultraviolet (UV) light is directed at the chip under the window of the EPROM. The UV light erases the EPROM by setting all the memory cells to a logical 1. The EPROM can then be reprogrammed. A 24-pin EPROM DIP IC is shown in Fig. 11-10. The actual EPROM chip is visible through the window on top of the IC. These units are sometimes called *UV erasable PROMs* or *UV EPROMs.*

The *EEPROM* is a third variation of a programmable read-only memory. The EEPROM is an *electrically erasable PROM* also referred to as an $E^2PROM$. Because EEPROMs can be erased electrically, it is possible to erase and reprogram

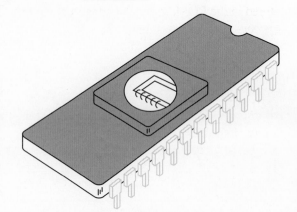

**Fig. 11-10** EPROM. Note window in top used to erase EPROM with ultraviolet light.

them without removing them from the circuit board. The EEPROM can be reprogrammed one byte at a time.

The *flash EEPROM* is a fourth variation of a programmable read-only memory. The newer flash EEPROM is like an EEPROM in that it can be erased and reprogrammed while on the circuit board. Flash EEPROMs are gaining favor because they use a simpler storage cell, thereby allowing more memory cells on a single chip. We say they have greater density. Flash EEPROMs can be erased sector-by-sector and reprogrammed faster than EEPROMs. While parts of the code can be erased and reprogrammed on an EEPROM, the entire flash EEPROM must be erased and reprogrammed.

The basic idea of a PROM is illustrated in Fig. 11-11. This simplified 16-bit (4 × 4) PROM is similar to the diode ROM studied in the previous section. In Fig. 11-11(a), each memory cell contains a diode and a good fuse. This indicates that all of the memory cells are

**Mask-programmable ROMs**

**Field-programmable ROMs (PROMs)**

**EPROM (erasable programmable read-only memory)**

**PROM burner**

**Flash EEPROM**

**EEPROM**

**Electrically erasable PROM**

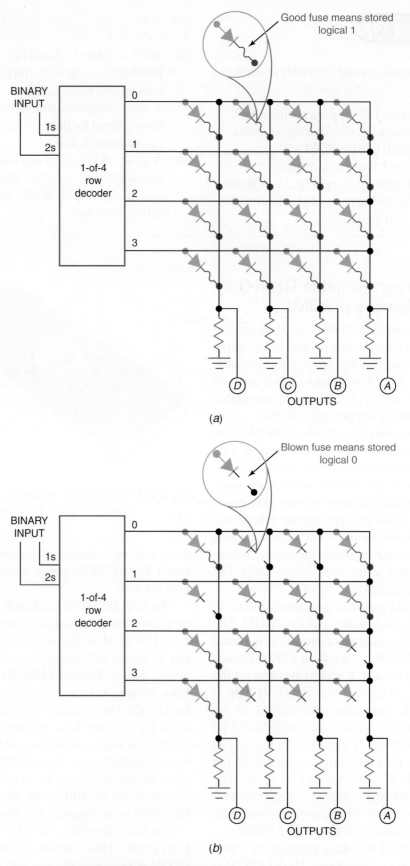

Good fuse means stored logical 1

BINARY INPUT

1s

2s

1-of-4 row decoder

0

1

2

3

D    C    B    A

OUTPUTS

(a)

Blown fuse means stored logical 0

BINARY INPUT

1s

2s

1-of-4 row decoder

0

1

2

3

D    C    B    A

OUTPUTS

(b)

**Simplified PROM**

Fig. 11-11 Simplified PROM. (a) PROM before programming. All fuses good (all 1s). (b) PROM after programming. Seven fuses blown (seven 0s programmed).

storing a logical 1. This is how the PROM might look before programming.

The PROM in Fig. 11-11(b) has been programmed with seven 0s. To program or *burn the PROM*, tiny fuses must be blown as shown in Fig. 11-11(b). A blown fuse in this case disconnects the diode and means a logical 0 is permanently stored in this memory cell. Because of the permanent nature of burning a PROM, the unit cannot be reprogrammed. A PROM of the type illustrated in Fig. 11-11 can only be programmed once.

A popular *EPROM family* is the *27XXX series*. These are available from many manufacturers. A short summary of some models in the 27XXX series is shown in Table 11-4. Notice that they are all organized with byte-wide (8-bit-wide) outputs, making them compatible with many digital systems. Many versions of each of these basic numbers are available such as low-power CMOS units, EPROMs with different access times, and even pin-compatible PROMs, EEPROMs, and ROMs.

A sample IC from the 27XXX series EPROM family is illustrated in Fig. 11-12. The pin diagram in Fig. 11-12(a) represents the *2732A 32K (4K × 8) ultraviolet-erasable PROM*. The 2732 EPROM has 12 address pins ($A_0$ to $A_{11}$) which can access 4096 ($2^{12}$) byte-wide words in the memory. The 2732 EPROM uses a 5-V power supply and can be erased using UV light. The $\overline{CE}$ input is like the chip select *CS* inputs on some other memory chips. The $\overline{CE}$ input is activated with a LOW. The $\overline{OE}/V_{PP}$ pin serves a dual purpose. It has one purpose during reading and another during writing. Under normal use the EPROM is being read. A LOW at the output enable ($\overline{OE}$) pin during a memory read activates the outputs driving the data bus of the computer system. The eight output pins are labeled $O_0$ to $O_7$ on the 2732 EPROM. A block diagram is

drawn in Fig. 11-12(b) to show the organization of the 2732 EPROM chip.

When the 2732 EPROM is erased, all memory cells are returned to logical 1. Data is introduced by changing selected memory cells to 0s. The 2732 is in the *programming mode* (writing into the EPROM) when the dual-purpose $\overline{OE}/V_{PP}$ input is at 21 V. During programming (writing), the input data is applied to the data output pins ($O_0$ to $O_7$). The word to be programmed into the EPROM is addressed using the 12 address lines. A very short (less than 55 ms) TTL level LOW pulse is then applied to the $\overline{CE}$ input to complete the write process.

Programming an EPROM is handled by special equipment called PROM burners. After erasing and reprogramming it is common to protect the EPROM window (see Fig. 11-10) with an opaque sticker. The sticker over the EPROM window protects the chip from UV light from fluorescent lights and sunlight. The EPROM can be erased by direct sunlight in about one week or room-level fluorescent lighting in about three years.

## Table 11-4    27XXX Series EPROMs

| EPROM 27XXX | Organization | Number of Bits |
|---|---|---|
| 2708 | 1024 × 8 | 8192 |
| 2716 | 2048 × 8 | 16384 |
| 2732 | 4096 × 8 | 32768 |
| 2764 | 8192 × 8 | 65536 |
| 27128 | 16384 × 8 | 131072 |
| 27256 | 32768 × 8 | 262144 |
| 27512 | 65536 × 8 | 524288 |

"Burning" a PROM

27XXX-series EPROM family

2732A 32K (4K × 8) ultraviolet-erasable PROM

Erasing the UV EPROM

---

![Self-Test]

*Supply the missing word or words in each statement.*

37. The letters "PROM" stand for _____.
38. The letters "EPROM" stand for _____.
39. The letters "EEPROM" stand for _____.

40. Erasing EPROMs can be done by shining _____ light through a special window in the top of the IC.
41. See Table 11-4. The 27512 EPROM can store a total of _____ bits of data organized as _____ words, each 8 bits wide.

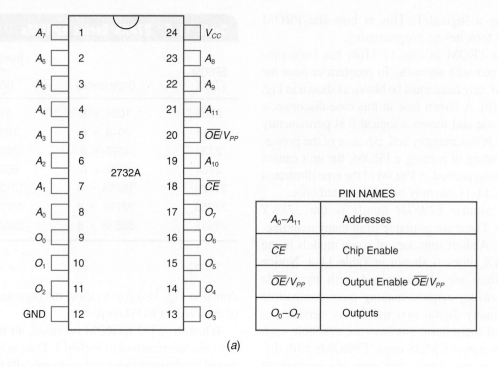

| PIN NAMES | |
|---|---|
| $A_0$–$A_{11}$ | Addresses |
| $\overline{CE}$ | Chip Enable |
| $\overline{OE}/V_{PP}$ | Output Enable $\overline{OE}/V_{PP}$ |
| $O_0$–$O_7$ | Outputs |

(a)

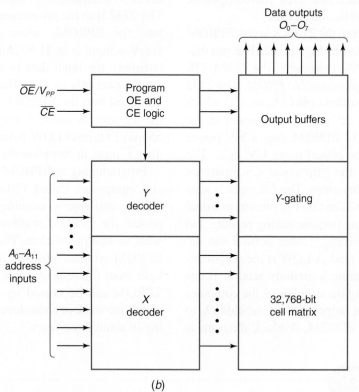

(b)

**Fig. 11-12** 2732 EPROM IC. (*a*) Pin diagram. (*b*) Pin names. (*c*) Block diagram.

## 11-8 Nonvolatile Read/ Write Memory

Both static and dynamic RAMs have the disadvantage of being volatile. When power is turned off, the data is lost. To solve this problem, *nonvolatile read/write memories* were developed. These are currently implemented by (1) using *battery backup* for a CMOS SRAM, (2) using a *nonvolatile static RAM (NVSRAM)*, (3) using a *flash EEPROM* or flash memory or

(4) newer FRAM (ferroelectric random-access memory).

The new MRAM features high speed, high density, nonvolatility, low power, and endurance (unlimited reads and writes).

## SRAM-Battery Backup

Battery backup is a common method of solving the volatility problem of a SRAM. CMOS RAMs are used with battery backup systems because they consume little power. A long-life lithium battery is typically used to back up data on the CMOS SRAM. Backup batteries have a life expectancy of about 10 years and may be imbedded in the memory package. Under normal operating conditions, the SRAM is powered by the equipment's power supply. When the power supply voltage drops to some predetermined lower level, voltage-sensing circuitry switches to backup battery power to maintain the contents of the SRAM until power is restored. Battery backup SRAMs are common in microcomputer systems.

## NVSRAM

Nonvolatile RAMs can solve the volatility problem. The nonvolatile RAM may be referred to as a *NVRAM (nonvolatile RAM), NOVRAM (nonvolatile RAM), NVSRAM (nonvolatile static RAM).* The NVRAM combines the read/write capabilities of a SRAM with the nonvolatility of an EEPROM. A block diagram of a small NVSRAM is detailed in Fig. 11-13. Note that the NVSRAM has two parallel memory arrays. The front array is a SRAM, while the back is a shadow EEPROM. During normal operation, the read/write SRAM is used. When the power supply voltage drops, a duplicate of all data in the SRAM is automatically stored in the nonvolatile EEPROM array. The *store* operation is represented in Fig. 11-13 with an arrow pointing toward the EEPROM array. On power up, the NVSRAM automatically executes the recall operation, which copies all data from the EEPROM to the SRAM. The *recall* operation is symbolized in Fig. 11-13 by the arrow pointing toward the front static RAM array.

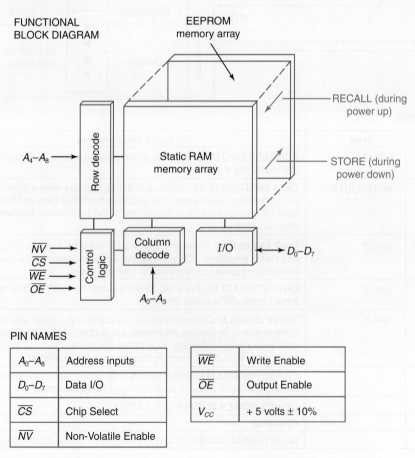

**FUNCTIONAL BLOCK DIAGRAM**

**PIN NAMES**

| $A_0$–$A_8$ | Address inputs | $\overline{WE}$ | Write Enable |
|---|---|---|---|
| $D_0$–$D_7$ | Data I/O | $\overline{OE}$ | Output Enable |
| $\overline{CS}$ | Chip Select | $V_{CC}$ | + 5 volts ± 10% |
| $\overline{NV}$ | Non-Volatile Enable | | |

**Fig. 11-13** Block diagram and pin names on a typical NVSRAM.

NVSRAMs seem to have a slight advantage over battery backup SRAMs. NVSRAMs have better access speed and generally better overall life. NVSRAM ICs are smaller than the more bulky battery backup SRAM packages and therefore save PC board space. Currently, NVSRAMs are more expensive and are manufactured in limited sizes.

## Flash Memory

*Flash EEPROMS* may become a *low-cost* alternative to battery backup SRAMs and NVSRAMs. Flash memories are expected to be widely used in laptop computers.

The commercial flash memory by Intel is featured in Fig. 11-14. Intel's *28F512 512K*

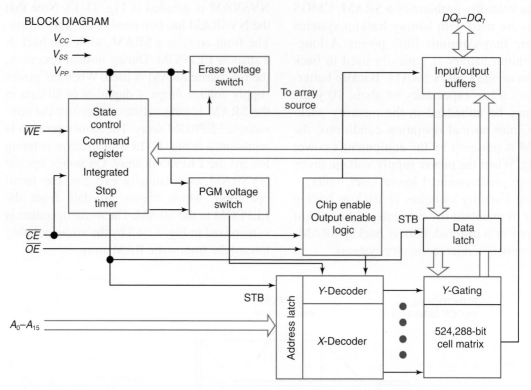

BLOCK DIAGRAM

PIN DIRETIONS

| Symbol | Type | Name and Function |
|---|---|---|
| $A_0$–$A_{15}$ | INPUT | **ADDRESS INPUTS** for memory addresses. Addresses are internally latched during a write cycle. |
| $DQ_0$–$DQ_7$ | INPUT/OUTPUT | **DATA INPUT/OUTPUT:** Inputs data during memory write cycles; outputs data during memory read cycles. The data pins are active high and float to tri-state OFF when the chip is deselected or the outputs are disabled. Data is internally latched during a write cycle. |
| $\overline{CE}$ | INPUT | **CHIP ENABLE:** Activates the device's control logic, input buffers, decoders and sense amplifiers. $\overline{CE}$ is active low; $\overline{CE}$ high deselects the memory device and reduces power consumption to standby levels. |
| $\overline{OE}$ | INPUT | **OUTPUT ENABLE:** Gates the device's output through the data buffers during a read cycle. $\overline{OE}$ is active low. |
| $\overline{WE}$ | INPUT | **WRITE ENABLE:** Controls writes to the command register and the array. Write enable is active low. Addresses are latched on the falling edge and data is latched on the rising edge of the $\overline{WE}$ pulse. **Note:** With $V_{PP} \leq 6.5$ V, memory contents cannot be altered. |
| $V_{PP}$ | | **ERASE/PROGRAM POWER SUPPLY** for writing the command register, erasing the entire array, or programming bytes in the array. |
| $V_{CC}$ | | **DEVICE POWER SUPPLY** (5 V ± 10%) |
| $V_{SS}$ | | **GROUND** |
| NC | | **NO INTERNAL CONNECTION** to device. Pin may be driven or left floating. |

**28F512 flash memory IC**

**Fig. 11-14** Block diagram and pin descriptions for 28F512 512K CMOS Flash Memory. *(Courtesy of Intel Corporation.)*

*(64K × 8) CMOS Flash Memory* will store 524,288 ($2^{19}$) bits organized into 65,536 ($2^{16}$) words, each 8 bits wide. The block diagram and pin descriptions in Fig. 11-14 give an overview of the flash memory. The 28F512 flash memory reacts like a read-only memory when the $V_{pp}$ erase/program power supply pin is LOW. When the $V_{PP}$ pin goes HIGH (about +12 V), the memory can be quickly erased or programmed based on commands sent to the command register by the attached microprocessor or microcontroller. The 28F512 flash memory uses a 5-V supply to power the chip, but +12 V is required at the $V_{PP}$ pin during erasing and programming.

In summary, flash EEPROMs or flash memories are an emerging memory technology which will become even more popular in the future. Flash memories have many desirable characteristics including being nonvolatile, in-system rewritable (read/write), highly reliable and having low power consumption. Flash memories currently boast high densities (the single transistor storage cells are very tiny). Recent developments by Intel suggest that even higher densities will be available in flash memories. Intel has announced the Strata-Flash™ memory which will store multiple bits of information in each cell. Intel is now producing to 128-Mbit StrataFlash™ memory chips.

### Ferroelectric RAM

*Ferroelectric RAM* (FeRAM or FRAM) is a high-speed memory similar to SRAM or DRAM but it is nonvolatile. The FRAM is faster than flash EEPROM memory. The FRAM does not require constant battery power like SRAM with battery backup. Low power consumption makes the ferroelectric RAM an excellent choice for portable digital devices. A semiconductor memory like the FRAM may be integrated into microcontrollers and other chips with a few added manufacturing steps. The FRAM memory cell consists of a *ferroelectric capacitor* and MOS transistor. The ferroelectric capacitors (memory cells) do not all need to be periodically refreshed as in the popular DRAM. A FRAM memory cell *needs refreshing only after a read* of a specific cell.

Currently, ferroelectric RAM densities are somewhat low and prices are high. However, FRAM technology is fairly new and it is expected that densities will increase and prices will fall. A major developer and manufacturer of FRAMs is Ramtron International.

### Magnetoresistive RAM

*Magnetoresistive RAM* (MRAM) is an emerging semiconductor memory technology. Magnetoresistive RAM has wonderful characteristics combining the access speed of the SRAM, the density of the DRAM, and the nonvolatility of flash EEPROM memory. The MRAM's small memory cell is based on a single transistor and a *magnetic tunnel junction (MTJ)* structure. The cell changes resistance representing different logic states (0 or 1). The MRAM has fast read and write speeds. The MRAM has almost unlimited read and write cycles and has low power requirements. MRAM is compatible with CMOS processes allowing processors (such as a microcontroller) and memory to be fabricated on the same chip. Some sources suggest that MRAM has the potential to be a "universal semiconductor memory." MRAM is sometimes also referred to as *magnetic random access memory*.

Ramtron International's site www.ramtron.com.

MRAM info www.freescale.com.

Operation, applications, and benefits of Nanotube RAM (NRAM™) www.nantero.com.

*Supply the missing word or words in each statement.*

42. The abbreviation "NVRAM" stands for _____.

43. Battery backup SRAMs commonly use a _____ (carbon-zinc, lithium) battery, which has a long life and maintains the data in the memory when power is lost.

44. A NVSRAM contains a static RAM array and a shadow _____ (EEPROM, ROM) memory array.

45. During power up using a NVSRAM, the _____ (recall, store) operation automatically occurs, duplicating all the data from the EEPROM into the SRAM memory array.

46. Refer to Fig. 11-14. The 28F512 flash memory can be erased/reprogrammed when +12 V is applied to the _____ pin of the IC.

47. The _____ (ROM, flash memory) has high density, is reliable, and is rewritable.

48. The _____ (SRAM, flash memory) is a read/write memory that is quite expensive and very fast.

49. Flash EEPROMs are a good substitute for a ROM but cannot replace DRAMs because they are not rewritable. (T or F)

50. The new _____ (MRAM, ZDRAM) has the potential of replacing many types of semiconductor memory because it is fast, low power, high density, has good endurance, and is nonvolatile.

51. In semiconductor memory jargon, the acronym MRAM stands for _____.

52. In semiconductor memory jargon, the acronym FeRAM stands for _____.

53. The flash EEPROM is like a SRAM because it is high density, low speed, and a volatile semiconductor memory. (T or F)

## 11-9 Memory Packaging

A general evolution of memory packaging is depicted in Fig. 11-15. The dual in-line package (DIP) is represented in Fig. 11-15(a). The dual in-line package (DIP) is the traditional IC package. DIPs occupy a fair amount of surface area on a printed circuit board. The DIP sketched in Fig. 11-15(a) may be either a surface-mount or a through-the-hole type. Small-outline ICs (SOIC) are smaller and reduce the board area used by the DIP package. In less-complex systems (such as

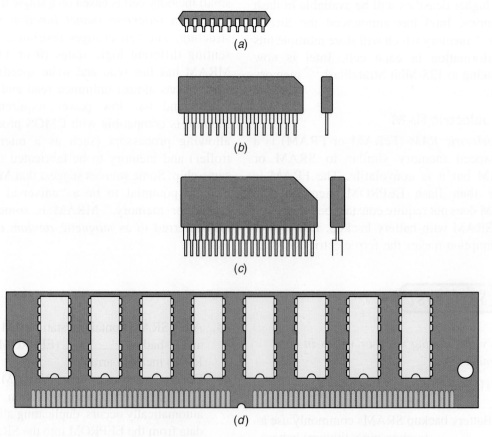

(a)

(b)

(c)

(d)

Fig. 11-15 Evolution of memory packages. (a) Dual in-line package (DIP). (b) Single in-line package (SIP). (c) Zig-zag in-line package (ZIP). (d) Single-in-line memory module (SIMM).

microcontroller-based pc boards), DIPs in SOIC form are mounted directly on the main pc board. In larger systems (such as microcomputers), DIP memory ICs are not mounted directly on the main motherboard. Memory modules (boards holding many DIP SOIC memory ICs) are inserted into sockets on the computer's motherboard.

Two types of memory modules used on some older computers are represented in Fig. 11-15(b) and (c). These packages are the SIP (single in-line package) and the ZIP (zig-zag in-line package). These may be found in older equipment, but not used in new designs.

Some older microcomputers may have SIMM memory modules. The older memory module sketched in Fig. 11-15(d) is a 72-pin SIMM (*single-in-line memory module*). Notice the 72 contacts-pads on the bottom edge of the SIMM. These contacts are located on *only one side* of the SIMM. An earlier version of this type of memory module was the 30-pin SIMM. Notice the notches on the left side and bottom of the SIMM in Fig. 11-15(d). These notches help the technician install the SIMM in the socket properly.

Newer microcomputers commonly use memory modules that look something like the DIMM (*dual-in-line memory module*) depicted in Fig. 11-16(a). This is a 168-pin DIMM memory module. The DIMM sketched in Fig. 11-16(a) has 84 contacts on each side of the bottom edge of the pc board for a total of 168 pins. Installation of a typical DIMM is depicted in Fig. 11-16(b). Notice the tabs at each end of the socket. These units help lock the memory module in place when it is pressed firmly downward in the socket or the levers act as ejectors when removing a seated DIMM. Notice also the notches along the bottom of the memory board in Fig. 11-16(b). These notches slide over solid raised areas in the center of the DIMM socket. This allows the DIMM to fit into the socket in only one direction and assures the correct memory module is installed.

The DIMM shown in Fig. 11-16(c) is a 184-pin DDR SDRAM (*double data rate synchronous DRAM*). Currently, the DDR SDRAM is a popular memory module used in PCs. Also widely used in modern PCs is the 184-pin RDRAM (Rambus DRAM). RDRAM is sometimes referred to as RIMM by its manufacturer Rambus, Inc. The 184-pin RDRAM physically looks somewhat different than the DDR SDRAM, and its lower edge notches have different locations. The 184-pin RDRAM and DDR SDRAM cannot be interchanged because the motherboards are designed for one style or the other DIMM. A larger capacity DIMM used in some higher-end PCs might be the 240-pin DDR2 SDRAM.

Portable computers use memory modules that are smaller than those pictured in Fig. 11-16. These look different physically. Some DDR SDRAM modules used in laptops might be the small outline 200-pin SO-DIMM or 172-pin micro-DIMM.

DIMMs have many variations including different physical sizes, voltages, speeds, and memory capacities. Replacement or added memory modules *must be ordered for your specific computer.*

Another packaging method is the memory card. The *Personal Computer Memory Card International Association (PCMCIA)* defines standard physical and electrical characteristics of the PCMCIA card. This memory card is about the width and length of a standard credit card; its thickness varies (in four thicknesses) from about 3 to 19 mm. The memory card can house arrays of memory chips and other electronics using almost any type of memory device (PROM, DRAM, battery-backed SRAM, flash EEPROM, and so on). The flash memory card is very popular because of its very high density, low power consumption, read/write capabilities, nonvolatility, and modest cost. A PCMCIA device containing flash memory would probably be referred to as a *flash memory card*. The memory card enables a method of adding memory to laptop and palmtop computing devices or even a copier. Large-capacity flash memory cards can be used as *solid-state disk drives*. The standard PCMCIA memory card has an edge connector with 68 pins, which are assigned tasks (address lines, data lines, power supply, ground, and so forth). The 68-pin PCMCIA memory card allocates 26 pins for address lines, which allows addressing of a large memory ($2^{26}$ = 64 Mbytes). You should show some caution when plugging in a memory card because other standards are used, such as the PCMCIA 88-pin interface,

See www. pctechguide. com for good information on computer memory.

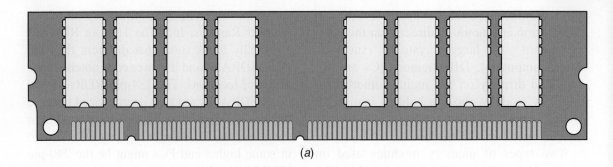

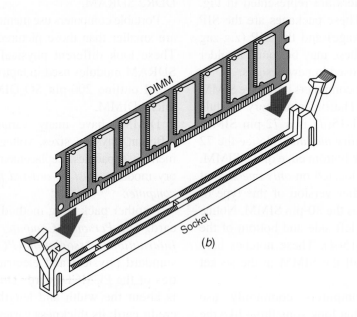

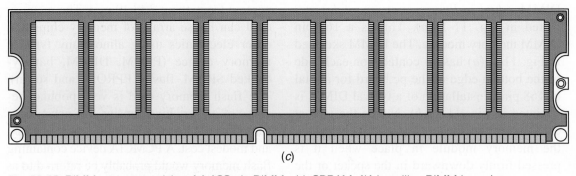

**Fig. 11-16** DIMM memory modules. (*a*) 168-pin DIMM with SDRAM. (*b*) Installing DIMM in socket. (*c*) 184-pin DIMM with DDR SDRAM.

the Panasonic 34-pin interface, the Maxwell 36- or 38-pin interface, the Epson 40- or 50-pin interface, and others.

The disk drive (rigid or floppy disk) is an electromagnetic device that consumes much power and can mechanically wear, thereby causing problems. Disk drives are particularly vulnerable to shock, vibration, dust, and dirt. The *solid-state disk* using flash memory cards or similar devices that appear

in portable computers and other equipment must be very small. They use very little power, and withstand shock and vibration. In a *solid-state computer,* the traditional DRAM and magnetic drive (floppy and/or rigid) combination would be replaced by some fast SRAM and flash memory. The SRAM/flash memory combination is very fast when transferring data from the solid-state disk to RAM.

*Answer the following questions.*

54. The memory package in Fig. 11-15(*a*), which is the most traditional, is called the _____ (DIP, SIP, ZIP).
55. The acronym DIMM stands for what when referring to a computer memory module package?
56. The acronym DDR SDRAM stands for what when referring to computer memory modules?
57. The acronym SIMM stands for what when referring to computer memory modules?

58. The acronym RDRAM stands for what when referring to computer memory modules?
59. DIMMs have many variations including different physical sizes, voltages, speeds, and memory capacities. (T or F)
60. The letters "PCMCIA" stand for what when referring to a memory card?
61. A PCMCIA flash memory card is about the size of a thick _____ (credit card, 5.25-in. floppy disk).
62. All memory cards follow the PCMCIA 68-pin standard for electrical connections and physical dimensions. (T or F)

## 11-10 Computer Bulk Storage Devices

Generally, semiconductor memories are used for internal storage in most modern computers. The computer's internal storage is also called *primary storage.* It is not possible to store all data inside the computer itself. For instance, it is neither necessary nor desirable to store last month's payroll information inside the computer after the checks are printed and cashed. Thus most data is stored outside the computer. External storage is also called *secondary storage.* Several methods are used to store information for immediate and future use by a computer. External storage devices are usually classified as either *mechanical, magnetic, optical,* or *semiconductor.*

### Mechanical Devices

*Mechanical bulk storage* devices include the punched paper card and punched or perforated paper tape. The punched card was developed before 1900 by Herman Hollerith, who adapted them for use in the 1890 United States census. These cards commonly have holes punched in them to represent alphanumeric data. The code used is called the *Hollerith card code.* A typical Hollerith punched card is made of heavy paper and measures about 3.25 × 7.5 in. A common punched card can hold 80 characters. Punched paper cards are now obsolete.

Perforated paper tape was another method of mechanically storing data. The paper tape is a narrow strip of paper with holes punched across the tape at places selected according to a code. The paper tape can be stored on reels. This method is also obsolete.

### Magnetic Devices

Common *magnetic bulk storage* devices are the magnetic tape, the magnetic disk, and the magnetic drum. Each device operates much like a common tape recorder. Information is recorded (stored) on the magnetic material. Information can also be read from the magnetic material. Magnetic tape has been widely used for many years as a secondary storage medium. It is still very popular for backing up data because it is quite inexpensive. The main disadvantage of

**Computer bulk storage devices**

**Primary storage**

**Magnetic bulk storage**

**Secondary storage**

**Mechanical bulk storage**

**Hollerith card code**

### ABOUT ELECTRONICS

Handheld computing products have taken portability to a new level. Smaller seems to always be the goal in the ever-developing computer industry. With the soaring popularity of palm-size organizers, manufacturers are developing devices that have a phone, organizer, e-mail, instant text messaging, and wireless web all in one product. The Handspring™ Treo, shown on the right, weighs 5.2 ounces and is 4.3 × 2.7 × .07 inches, and runs on a rechargeable lithium ion battery.

**Magnetic disks**

**Rigid or floppy disk**

Up-to-date information about hard drives www. seagate. com.

magnetic tapes is that they are sequential-access devices. That is, to find information on the tape you must search through the tape sequentially, which makes the access time long.

*Magnetic disks* have become particularly popular in recent years. Magnetic disks are random-access devices, which means that any data can be accessed easily and in a short time. Magnetic disks are manufactured in both *rigid* and *floppy* (flexible) disk form. The floppy disk is a popular form of secondary storage used by many microcomputers.

### Hard Disks

The hard disk or rigid disk drive is currently the most important bulk storage memory device used on modern computer systems. One of the original sealed dust-free hard-disk drives was developed by IBM and was referred to as a *Winchester drive* (after famous 30-30 Winchester rifle—30 Mbytes with 30 msec access time). The hard disk has proved to be reliable, fast, and today has a very large storage capacity. A picture of a modern hard-disk drive by Seagate Technology is reproduced in Fig. 11-17. The cover has been removed from this normally sealed unit to expose four 3.5-in. rigid disks (called *platters*) made of aluminum, glass, or ceramic. The platters are probably coated with a *thin-film medium,* which is a microscopic layer of metal bonded to the disk. The hard drive featured in Fig. 11-17 has eight read/write heads (only one is visible), one on each side of the four platters. When the platters spin, the read/write heads float just above the surface of the disk. The read/write arm pivots to locate a specific circular track on the surface of the platters. The spindle speed on many hard drives is 3600 rpm or faster. Seagate Cheetah hard drives similar to the one pictured in Fig. 11-17 have spindle speeds of 10,000 or 15,000 rpm. The higher spindle speed allows the read/write heads to locate data more quickly.

The specification sheet for a hard-disk drive gives information such as the total storage capacity, number and size of platters, number of read/write heads, average seek time (read/write), average latency, spindle speed, physical dimensions, power requirements, and operating temperatures. The organization of the data on the disks might be given as the number of sectors (a sector commonly holds 512 bytes of data plus other information such as an address), number of tracks (concentric circles of data), or number of cylinders (like the number of tracks but three-dimensional, including both sides of all of the platters).

The hard drive pictured in Fig. 11-17 is designed to function as internal storage in a computer system. Currently, internal hard drives in modern PCs have storage capacities ranging from about 50 to 500 Gbytes. A portable hard drive, such as the pocket hard drive sketched in Fig. 11-18, is a popular alternative to floppy disk storage. The pocket hard drive unit has a storage capacity of 2.5 to 5 Gbytes. It is shirt pocket size weighing only a few ounces. It has a built-in retractable connector, which can be hot-plugged into a USB port on your PC with data transfers up to 480 Mbps. The pocket hard drive (Fig. 11-18) is powered by the USB port. Other portable hard drives

**Fig. 11-17** A Cheetah hard-disk drive by Seagate.

3600 rpm-1 inch
Pocket hard drive

USB 2.0
connector

**Fig. 11-18** Pocket hard drive.

are available with larger capacities, but they are larger in size and may require separate power supplies.

## Floppy Disks

Currently the 3.5-in. floppy disk is a common bulk storage devices used for long-term storage and transporting of data. Floppy disks or diskettes come in either 5.25- or 3.5-in. versions. The 3.5-in. floppy disk has become the standard. A diagram of a common 3.5-in. floppy disk is shown in Fig. 11-19(*a*). The drawing shows the bottom view of the disk. It labels the rigid plastic case as well as the sliding metal shutter, both of which help protect the delicate floppy disk housed inside. The sliding metal shutter is shown open, exposing the floppy disk. When released, the cover snaps back to cover the floppy disk inside. The read/write heads of the disk drive can store or retrieve data from both sides of the floppy disk. The center has a metal hub attached to the bottom of the floppy for gripping the disk. The rectangular hole in the hub is an index hole used by some disk drives for timing purposes. The write-protect notch is located at the lower right in Fig. 11-19(*a*). If the write-protect hole is closed (as shown in the drawing), you can both write to or read from the disk. If the hole is open (move plastic slider down), the drive can only read the disk; we say that the disk is "write protected."

The most common 3.5-in. high-density (HD) floppy disk can store up to 2 Mbytes of data. A common format allows 1.44 Mbytes of data to be stored on a high-density, double-sided (HD DS) 3.5-in. diskette. Older 3.5-in. floppy disks can store much less data. Your floppy disk drive can tell which 3.5-in. floppy is inserted by the hole or no hole at the lower left of the disk. This is illustrated in Fig. 11-19(*a*). Many older 3.5-in. floppy disks are called double density (DD). As a practical matter, you will have trouble using a new HD disk in an older disk drive that was designed for older 3.5-in. DD floppy disks.

Data on a floppy disk is organized during the formatting process. The organization can be visualized by looking at the 3.5-in. floppy disk in Fig. 11-19(*b*). Notice that the disk is organized by *tracks* and *sectors*. The 3.5-in. HD disk is commonly organized in concentric circles or 80 tracks; each track then divides into 18 sectors. Remember that both sides of the disk are organized in tracks and sectors. Each sector can hold 512 bytes of data plus other information, such as an address, as illustrated in Fig. 11-18(*b*). An older DD 3.5-in. floppy disk is formatted with 80 tracks, both sides, but with only 9 sectors per track. The track and sector organization is also used by hard disks.

In summary, the 3.5-in floppy disk may be the most recognized storage device in the world. It is a cheap read/write bulk storage device that can be transported and read from or written to by almost any compatible computer. Its storage capacity is approximately 1.4 Mbytes. Many laptop and other small computers do not build in a 3.5-in. floppy disk drive. They depend on the use of the USB and FireWire (IEEE 1394) ports for connecting separate removable media drives. Some of these include floppy disk drives, portable hard drives, or flash memory devices.

## Optical Discs

The *optical disc* has become one of the most familiar bulk-storage devices used with modern personal computers. Optical disc technology is popular because it is (1) reliable, (2) high capacity, (3) transportable, and (4) inexpensive. Optical discs are available in many forms and formats including the popular *CD-ROM* (compact-disc read-only memory), CD-R (compact-disc recordable), and CD-RW (compact-disc rewritable). Newer high capacity optical discs include the DVD-ROM (digital versatile disc read-only memory), DVD-R (digital versatile disc recordable), DVD-RW (digital versatile disc rewritable), and DVD + RW (another version of digital versatile disc rewritable). Optical discs are commonly available in the usual 120 mm (about 4.72 in.)—and the smaller 80 mm (about 3.15 in.)—size. Besides being used for storing computer data, optical discs are also commonly used for storing audio and video.

The CD was first developed for audio and then adapted for computer use in the CD-ROM form in the mid 1980s. The CD-ROM is manufactured using a carefully prepared glass master. The master is pressed into injection molded polycarbonate plastic forming the CD-ROM.

**Optical disc**

**CD-ROM**

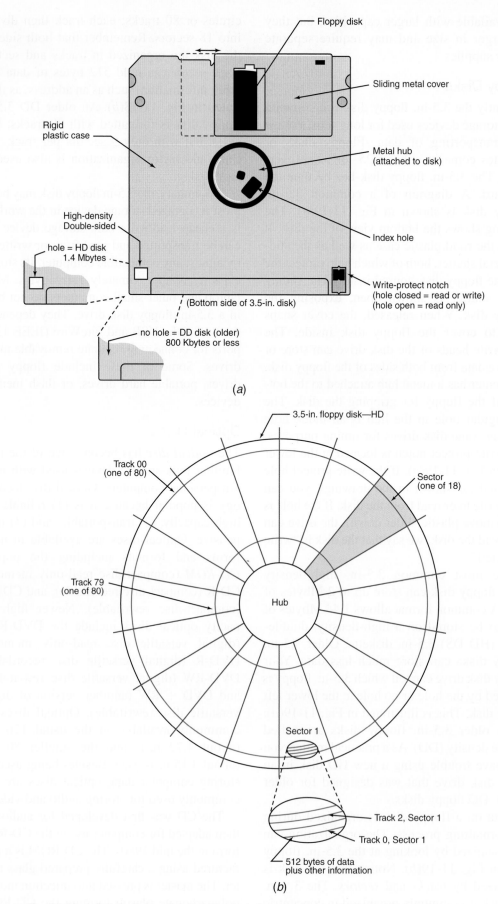

Fig. 11-19 A 3.5-in. floppy disk. (a) Physical characteristics. (b) Typical formatting into 80 tracks and 18 sectors.

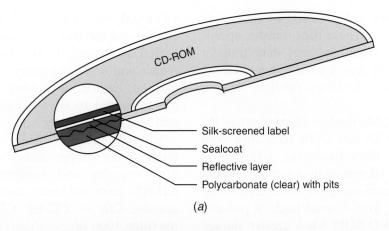

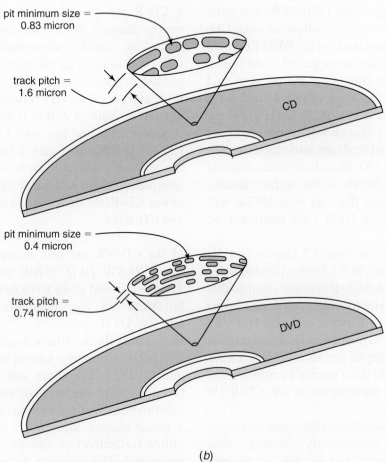

Silk-screened label
Sealcoat
Reflective layer
Polycarbonate (clear) with pits

(a)

pit minimum size = 0.83 micron

track pitch = 1.6 micron

CD

pit minimum size = 0.4 micron

track pitch = 0.74 micron

DVD

(b)

**Fig. 11-20** (a) CD-ROM construction. (b) Comparing track pitch (width) and pit size for CDs and DVDs.

Search www. pctechguide.com for new info on storage devices.

The resulting CD-ROM contains small pits and lands (no pits). A sketch of a CD-ROM is shown in Fig. 11-20(a). When read by the computer system's CD-ROM drive, a laser is aimed at the disc from the bottom and the reflection from the pits and lands on the platter are interpreted by a photodetector and digital circuitry as logical 0s and 1s.

The *data transfer rate* of a CD-ROM drive is indicated by the manufacturer with a designation such as 1x, 2x, 16x, or 32x. A CD-ROM drive with a 1x designation would have a maximum data transfer rate of 150 Kbytes per second. Therefore, a 16x drive would have a maximum data transfer rate of 2400 Kbytes per second

**Data transfer rate**

Memories **Chapter 11** 377

(150 Kbytes/sec $\times$ 16 = 2400 Kbytes/sec or 2.4 Mbytes/sec). These data transfer speeds are maximums and the real data transfer rates are usually less. Currently new PCs are equipped with CD-drives rated at 48x or higher.

DVDs

CD-RW

The CD-ROMs modern counterpart is the DVD-ROM. CDs and *DVDs* look alike both being plastic discs measuring 120 mm in diameter and 1.2 mm in thickness. They both are manufactured using the same technology and read data from a spiral track of pits and lands. The DVD-ROM has a greater storage capacity. A single-layer DVD-ROM can store about seven times more data than the older CD-ROM. The pits and lands on the DVD-ROM are more closely packed as suggested in the sketch in Fig. 11-20(*b*) comparing the pit sizes and track pitch (width) for a CD-ROM and for a DVD-ROM. The tracks on the DVD-ROM are spaced closer together allowing more tracks per disk. The pits and lands are also much smaller. Currently many CD drives have lasers that can read either CD-ROMs or the higher capacity DVDs. CD drives that can read DVDs will commonly have a DVD logo imprinted on the front.

DVD-ROMs can store 4.7 Gbytes (single-sided single-layer), or 9.4 Gbytes (double-sided single layer), or 8.5 Gbytes (single-sided double layer), or 17 Gbytes (double-sided double layer) compared to 0.65 Gbytes for a standard CD-ROM. DVD-ROM drives provide a data transfer rate of 1.385 Mbytes per second. This means that a DVD-ROM drive rated at 1x would transfer data at about the same rate as a 9x CD-ROM drive.

Phase-change technology

CD-R

CD-R (CD-recordable) discs are becoming popular for permanently storing data (archival storage). CD-Rs are *write-once read-many (WORM)* storage devices where the PC operator can "burn" a CD-R using a CD-writer drive. Before burning a CD-R disc, the reflective surface appears to a CD reader as a continuous land (no pits). During the burning process a laser heats a gold reflective layer and a dye layer in the CD-R causing it to have a duller appearance. The reflective layer may be gold or silver while the dye layer may be gold, green or blue depending on the manufacturer. When read by the CD-R drive the dark burned areas (like the pits on a

CD-ROM) reflect less light. The shiny areas (lands) and the dull areas (burned) are interpreted as logical 0s and 1s by the CD-R reader and digital circuitry. The CD-R disc can be burned only once. Most CD-Rs are formatted to have a capacity of about 650 Mbytes.

CD-RW (CD-rewritable) discs are one alternative to floppy disks because of their high capacity and read/write capability. CD-RW discs are sometimes referred to as erasable-CDs or CD-Es. CD-RWs can be rewritten 1000 times or more. When burning a CD-R the photosensitive dye is permanently changed. When burning a CD-RW the recording layer (*silver-indium-antimony-tellurium alloy*) can be recorded and rerecorded making it rewritable. The recording layer alloy is either very reflective in its polycrystalline state or dull in it amorphous state (like the lands and pits on a CD-ROM). The CD-R/CD-RW drive uses a laser to identify the reflective and dull areas on the CD-RW interpreting them as logical 0s and 1s. Many newer CD-ROM drives will also read CD-Rs and CD-RWs.

Three newer high capacity DVD versions of the CD-RW are just emerging. They are the *DVD-RW, DVD + RW,* and *DVD-RAM*. The DVD-RAM disks were developed earlier but were not very compatible with other CD-RW/DVD products. The DVD-RAM with caddy looks like a large floppy disk. DVD-RW (formerly known as DVD-R/W) and DVD + RW discs use *phase-change technology* for reading, writing, and erasing information. During writing the laser heats a *phase-change alloy* so it is either crystalline (reflective) or amorphous (dark, non-reflective). The resulting difference between the reflective and dark areas on the disc can be read by a photodetector and interpreted as logical 0s or 1s. DVD + RW disks are more compatible with both the consumer electronics and personal computer environments which is important in multimedia applications.

It is expected that various forms of optical discs will be used for computer, audio, and video storage for decades due to their reliability, high capacity, low cost and transportability.

## Access Time

Various bulk-storage devices are compared on an access time/storage capacity basis in Fig. 11-21. Whereas access time is given in seconds, storage capacity is graphed in Mbytes. *Access time* is the time in seconds it takes to retrieve a piece of data from memory. The highest performance (shortest access time) device on the chart is the flash memory card. The disadvantage of the flash memory card, however, is the high cost. Mechanical methods of storing data (paper tape and punched cards) have the lowest performance and, as such, have been phased out for most applications. Magnetic tape and digital audiotape (DAT) have poor access time but have very large storage capacity at low cost. Hard disks are extremely popular because of their ease of use, large storage capacities, good access times, reasonable cost, and universal usage. Floppy disks continue to be popular because they are

easy to use, are available for a very low cost, are portable, have medium access times, and are used universally. Optical discs, such as CD-ROMs, CD-Rs, and CD-RWs, have become standard bulkstorage media on personal computers. The higher capacity versions of optical discs, such as DVDs, will probably gain popularity. Optical storage is popular because of their high storage capacities, low cost, transportability, and reliability. Flash cards are an emerging technology, providing excellent access times and good storage capacities, but they are still a bit expensive.

Read/write optical discs have very large storage capacities. The read/write *magneto-optical* disk drive uses a laser and a coil of wire to write to, read from, and erase the metalcoated optical disc. These *rewritable magneto-optical discs* look much like the 3.5-in. floppy disk except they are thicker and house an optical disc.

**Rewritable magneto-optical disc**

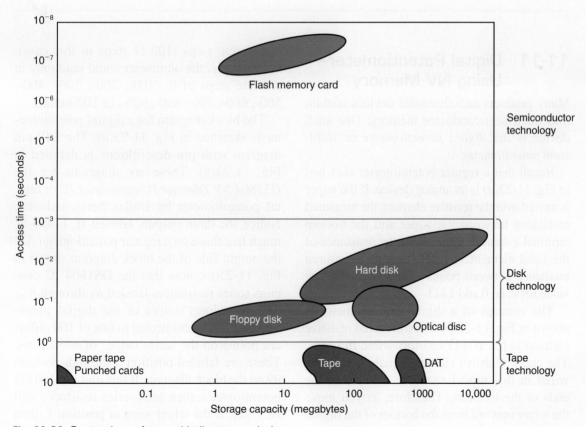

**Fig. 11-21** Comparison of several bulk-storage devices.

*Answer the following questions.*

63. External computer bulk storage devices can be classified as mechanical, _____, _____, or _____.

64. List several computer bulk storage devices.

65. Magnetic disks are manufactured in both floppy (flexible) and _____ (amorphous, rigid) form.

66. The most important bulk storage device used in almost all computer systems is the _____ (magneto-optical disk drive, hard-disk drive).

67. The _____ (floppy disk, rigid disk) is faster and has much greater storage capacity.

68. Refer to Fig. 11-17. The rigid disks are commonly called _____ (platters, spindles) on this disk drive.

69. A gigabyte equals _____ bytes.

70. A Winchester drive was an early name for a(n) _____ (optical-disc drive, hard-disk drive) developed by IBM.

71. Modern personal computers are being shipped with hard disk drives have storage capacities of about _____ (30 Mbytes, 50 to 500 Gbytes).

72. The _____ (flash memory card, optical disc) is a bulk storage device that is low cost, reliable, high capacity, and transportable.

73. The popular Digital Video Disk is referred to as a DVD. The acronym DVD also stands for _____ _____ _____.

74. Both CDs and DVDs are classified as _____ (mechanical, optical) bulk storage devices.

75. The acronym WORM stands for what in reference to an optical disc?

76. The acronym CD-RW stands for what in reference to an optical storage disc?

77. The _____ (CD-ROM, CD-RW) disc is manufactured with small pits and lands which are interpreted by the CD drive as logical 0s and 1s.

## 11-11 Digital Potentiometer— Using NV Memory

Many products and electronic devices contain imbedded semiconductor memory. One such device is the *digital potentiometer* or *solid-state potentiometer.*

Recall that a regular potentiometer sketched in Fig. 11-22(*a*) is an analog device. If the wiper is moved over the resistive element, the measured resistance between the wiper and the bottom terminal *gradually changes*. If the resistance of the fixed element is 1 k$\Omega$, then the measured resistance between points A and B can be any value between 0 and 1 k$\Omega$.

The concept of a digital potentiometer is shown in Fig. 11-22(*b*). Here the fixed resistive element is ten 100-$\Omega$ resistors wired in series. The entire resistive element equals 1 k$\Omega$. The wiper in this model can only connect to the ends of the resistors. Therefore, as you move the wiper upward from the bottom of the digital potentiometer, the measured resistance jumps

in discrete steps (100 $\Omega$ steps in this case). For instance, the ohmmeter could read only in discrete steps of 0-, 100-, 200-, 300-, 400-, 500-, 600-, 700-, 800-, 900-, or 1000-ohms.

The block diagram for a digital potentiometer is sketched in Fig. 11-23(*a*). The DIP pin diagram with pin descriptions is detailed in Fig. 11-23(*b*). These are diagrams for the *DS1804 NV Trimmer Potentiometer IC* or digital potentiometer by Dallas Semiconductor. Notice the three outputs labeled H, L and W much like those on a regular potentiometer. On the output side of the block diagram shown in Fig. 11-23(*a*), note that the DS1804 IC contains series resistances labeled $R_1$ through $R_{99}$. The W (wiper) output of the digital potentiometer can be connected to one of 100 different points on the series ladder of resistances. These are labeled position 0 through position 99 on the block diagram. If this unit is a 100 k$\Omega$ potentiometer, then each series resistance will be 1 k$\Omega$. If the wiper were at position 1, then the resistance between outputs L and W will

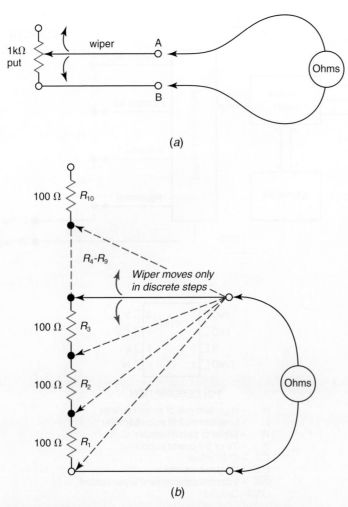

**Fig. 11-22** (a) Analog output from a potentiometer. (b) Digital output from solid-state potentiometer (10 steps, 100 Ω each).

be 1 kΩ. If the wiper were at position 98, then the resistance between outputs L and W will be 98 kΩ.

When the DS1804 IC, shown in Fig. 11-23, is first powered up an *initial wiper position,* stored in nonvolatile EEPROM, is automatically loaded into the control logic section of the chip. This is passed to the multiplexer to locate the starting position of the wiper. The wiper position can be altered by applying signals to the inputs ($\overline{CS}$, $\overline{INC}$, and $U/\overline{D}$).

Consider changing the wiper position of the DS1804-100 IC (100 kΩ digital pot) using the logic diagram in Fig. 11-24. The first example [Fig. 11-24(a)] shows a LOW at the chip select ($\overline{CS}$) input and a HIGH at the up/down ($U/\overline{D}$) input. This will allow the wiper position to move upward one position for each negative pulse entering the increment ($\overline{INC}$) input. In

this example, three negative pulses enter the $\overline{INC}$ input therefore the wiper will move upward three positions from its initial setting. In this example each position is 1 kΩ, therefore the three pulses will increase the resistance between outputs L and W by 3 kΩ.

Disabling the chip select input ($\overline{CS}$ = HIGH) will disable all inputs to the DS1804 IC. This allows for one-time programming (OTP) of the chip. With the $\overline{CS}$ input HIGH, the EEPROM is not written to during power down. The EEPROM still holds the last programmed wiper position and this position is read from the nonvolatile memory during power up.

A second example is shown in Fig. 11-24(b). This example shows the chip select ($\overline{CS}$) input activated with a LOW, the up/down ($U/\overline{D}$) input is LOW selecting the down mode. Two negative pulses are entering the $\overline{INC}$ pin of the IC. This

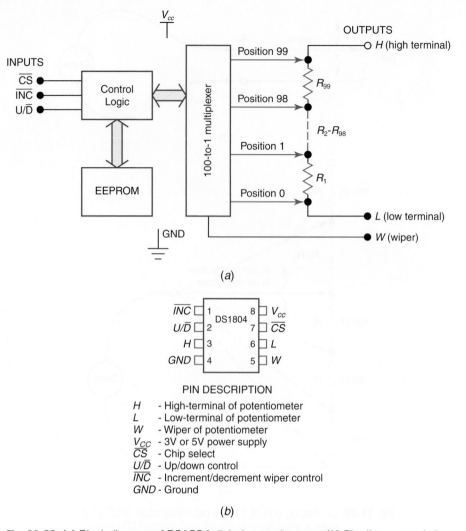

**Fig. 11-23** (*a*) Block diagram of DS1804 digital potentiometer. (*b*) Pin diagram and pin descriptions for DSI804 digital potentiometer (DIP IC). *(Courtesy of Dallas Semiconductor.)*

would cause the wiper position to decrease by two which decreases the resistance from outputs *W* to *L* by 2 kΩ.

The last wiper position is stored in EEPROM using the $\overline{INC}$ and $\overline{CS}$ inputs. Storage of the wiper position occurs whenever the $\overline{CS}$ input changes from LOW-to-HIGH

while the $\overline{INC}$ input is HIGH. The DS1804 IC can accept at least 50,000 writes to EEPROM before a *wear-out condition* occurs. After wear-out the DS1804 will still function and the wiper position can be changed when the IC is powered. However, after wear-out the wiper position may be random on power-up.

*Answer the following questions.*

78. The DS1804 IC is described by the manufacturer as a(n) _____ also sometimes known as a solid-state potentiometer.

79. Refer to Fig. 11-23(*a*). During power-up (power first turned on to IC), the initial position of the wiper is retrieved from _____ (RAM, EEPROM) and the _____ (multiplexer, XOR gate) adjusts the wiper to stored wiper position.

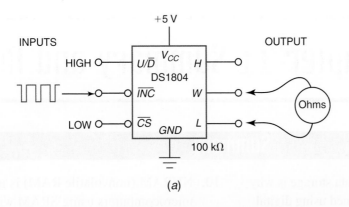

(a)

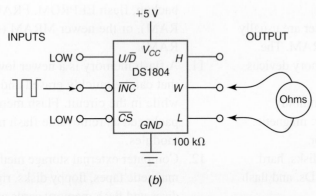

(b)

**Fig. 11-24** Changing the wiper position on the DSI804-100 digital potentiometer.

80. Refer to Fig. 11-25. Assume the initial output resistance is 50 k$\Omega$. The measured output resistance after input pulse $t_1$ is _____ ohms.

81. Refer to Fig. 11-25. The measured output resistance from the DS1804 IC after input pulse $t_2$ is _____ ohms.

82. Refer to Fig. 11-25. The measured output resistance from the DS1804 IC after input pulse $t_3$ is _____ ohms.

83. Refer to Fig. 11-25. The measured output resistance from the DS1804 IC after input pulse $t_4$ is _____ ohms.

84. Refer to Fig. 11-25. The measured output resistance from the DS1804 IC after input pulse $t_5$ is _____ ohms.

85. Refer to Fig. 11-25. The measured output resistance from the DS1804 IC after input pulse $t_6$ is _____ ohms.

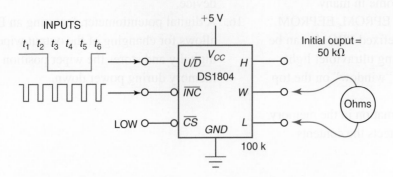

**Fig. 11-25** Digital potentiometer pulse-train problem.

# Chapter 11 Summary and Review

1. The availability of memory and data storage is why many electronic devices are designed using digital instead of analog circuitry.

2. Internal memory devices in a computer are usually in the form of RAM, ROM, and NVRAM. The CPU also contains other smaller memory devices like registers, counters, and latches.

3. External bulk storage devices are commonly classified as to their basic technology: magnetic, mechanical, optical, or semiconductor.

4. Bulk storage devices include floppy disks, hard disks, magnetic tape, CD-ROMs, DVDs, and flash memory modules.

5. Semiconductor storage cells are classified as SRAM, DRAM, SDRAM, ROM, EPROM, EEPROM, flash EEPROM, MRAM, and FRAM. Some important characteristics of semiconductor memory devices are density, reliability, cost, power consumption, read-only or read/write, nonvolatile/volatile, and electrically updateable.

6. A RAM is a semiconductor read/write random-access memory device. RAM comes in two primary forms including SRAM (static RAM) and DRAM (dynamic RAM). Faster SRAM and slower DRAM are both classified as volatile memory.

7. A ROM is considered a permanent storage unit that has a read-only characteristic.

8. A PROM operates like a ROM. PROMs are one-time write devices. PROMs come in many varieties, generally known as EPROM, EEPROM, and NVSRAM. These "E" prefixed PROMs can be erased electrically or by shining ultraviolet light through a special transparent "window" on the top of the IC.

9. The write process stores information in the memory. The read, or sense, process detects the contents of the memory cell.

10. NVRAM (nonvolatile RAM) is implemented in microcomputers using SRAM with battery backup, flash EEPROM, FRAM (ferroelectric RAM), or the newer MRAM (magnetoresistive RAM).

11. A flash memory is a newer low-cost EEPROM that can be quickly erased and reprogrammed while in the circuit. Flash memory chips can be packaged as removable flash memory cards or modules.

12. Computer external storage methods include magnetic tapes, floppy disks, rigid disks, optical discs and flash memory cards or modules.

13. Microcomputers typically use various types of RAM, ROM and NVRAM for internal main memory. Floppy disks, hard disks, CD/DVDs, and flash memory cards and modules are the most common popular bulk storage devices used on smaller computer systems.

14. A byte is an 8-bit word. One Gbyte of memory means one billion bytes (actually $2^{30}$). One Mbyte of memory means one million bytes (actually $2^{20}$). One Kbyte of memory means one thousand (actually $2^{10}$ or 1024) bytes of memory.

15. DIP, SIP, ZIP, SIMM, DIMM, and RIMM are common memory packages. Memory cards are commonly packaged as a PCMCIA (Personal Computer Memory Card International Association) device.

16. A digital potentiometer, featuring an EEPROM, allows for changing of the output wiper's position digitally and stores the wiper position in NV memory during power down.

*Answer the following questions.*

11-1. The most important characteristic of a digital system compared to an analog system is its _____ (ability to store data, ease of interfacing with real-world events).

11-2. The CD-RW is an example of a bulk storage device using _____ (mechanical, optical) technology.

11-3. The _____ (CPU, RAM) is the section of the computer system that contains the arithmetic, logic, and control sections and is the focus of many data transfers.

11-4. Three common internal semiconductor memory devices used in most computer systems are the _____ (floppy, rigid-disk and CD-ROM; RAM, ROM, and NVRAM).

11-5. Semiconductor RAM is a _____ (read-only, read/write) type memory device.

11-6. Semiconductor ROM is a _____ (read-only, read/write) type memory device.

11-7. Semiconductor RAM is a _____ (nonvolatile, volatile) memory device.

11-8. Semiconductor ROM is a _____ (nonvolatile, volatile) memory device.

11-9. Semiconductor NVRAM is a _____ (read-only, read/write) memory device.

11-10. Name the three buses used in a typical personal computer system.

11-11. The _____ (address, data) bus in a typical PC system is a one-way bus used for selecting a specific memory location or peripheral.

11-12. The common floppy used in today's PC systems is described as _____ (HD 3.5-in. disks, DD 5.25-in. disks).

11-13. Both floppy and rigid disks use _____ (magnetic, optical) technology for storing data.

11-14. Compared to the typical floppy disk, the hard-disk drive can store much _____ (less, more) data.

11-15. The CD-ROM is an optical memory device that can store about _____ (30 Mbytes, 650 Mbytes) of data.

11-16. List at least five semiconductor memory devices.

11-17. Press the store key on a calculator. This activates the _____ (read, write) process in the memory section.

11-18. Press the recall key on a calculator. This activates the _____ (read, write) process in the memory section.

11-19. The following abbreviations stand for what?
a. RAM        e. EEPROM
b. ROM        f. NVRAM
c. PROM       g. FRAM
d. EPROM      h. MRAM

11-20. A _____ (RAM, ROM) has both the read and write capability.

11-21. A _____ (RAM, ROM) is a permanent memory.

11-22. A _____ (RAM, PROM) is a nonvolatile memory.

11-23. A _____ (RAM, ROM) has a read/write input control.

11-24. A _____ (RAM, ROM) has data inputs.

11-25. A 32 × 8 memory can hold _____ words. Each word is _____ bits long.

11-26. List at least three advantages of semiconductor memories.

11-27. A _____ (RAM, ROM) can be erased easily.

11-28. A _____ (flash memory, ROM) can be quickly erased and reprogrammed.

11-29. A _____ (flash memory, UV EPROM) can be erased electrically in a very short time.

11-30. A _____ (flash memory, ROM) is a read/write nonvolatile memory device.

11-31. A(n) _____ (EEPROM, UV EPROM) can be erased and reprogrammed byte-by-byte without being removed from the equipment.

11-32. A _____ (DRAM, SRAM with battery backup) is a nonvolatile read/write memory device.

11-33. A _____ (MRAM, SRAM) semiconductor memory is a newer type nonvolatile RAM.

11-34. A _____ (FRAM, SRAM) semiconductor memory is classified as NVRAM.

11-35. Memory or data storage is much easier to implement using _____ (analog, digital) electronic circuitry.

11-36. The 2114 IC is a _____ (dynamic, static) RAM.

11-37. The access time of the TTL 7489 RAM is _____ (faster, slower) than that of the MOS 2114 RAM.

11-38. Refer to Fig. 11-7(b). If the input to the decoder is binary 0010, the output from the ROM will be _____ in Gray code.

11-39. Computer programs that are permanently held in ROM are called _____.

11-40. Refer to Table 11-4. Which 27XXX series EPROM could be used to implement a 16K ROM in a microcomputer?

11-41. Refer to Fig. 11-14. The 16-address line inputs to the 28F512 flash memory IC can address _____ (number) words, each 8 bits wide.

11-42. Refer to Fig. 11-14. To erase and/or reprogram the 28F512 flash memory IC, the _____ ($CE$, $V_{PP}$) input must be pulled to a high of about _____ (+5, +12) volts.

11-43. List at least five common types of computer bulk (external) storage.

11-44. Magnetic disks have much _____ (faster, slower) access time than magnetic tapes.

11-45. Short access time in a memory device is a measure of _____ (good, poor) performance.

11-46. Refer to Fig 11-16(b). The 168-pin DIMM being installed in the socket on the PCs motherboard contains _____ (ROM, SDRAM) memory chips.

11-47. A RIMM refers to a memory module used in modern PCs that contains semiconductor memory chips of the _____ (RDRAM, ROM BIOS) type.

11-48. Replacement or added memory modules (such as DIMMs) are universal and will fit any PC. (T or F)

11-49. A laptop (portable) computer uses the exact same DIMMs and RIMMs as larger desktop PCs. (T or F)

11-50. CD-R and CD-RW discs store data using _____ (magnetic, optical) media.

11-51. The acronym DVD may stand for Digital Video Disc or _____ _____ _____.

11-52. A _____ (CD-R, CD-RW) is an example of a WORM optical disc.

11-53. A digital potentiometer, such as the DS1804 IC, features a _____ (nonvolatile, volatile) memory for storing the wiper position for later retrieval during power-down conditions.

11-54. Generally, an analog input (gradually changing voltage) controls the digital output of a solid-state potentiometer such as the DS1804 IC. (T or F)

## Critical Thinking Questions

11-1. Draw a diagram of how a $32 \times 8$ memory might look in table form. The table will be similar to the one in Fig. 11-4.

11-2. List at least three uses of read-only memories.

11-3. Why are many microcomputer systems equipped with both floppy- and hard-disk drives?

11-4. List several precautions you should take when handling floppy disks.

11-5. If a computer has 4 Mbytes of RAM, how many bytes of read/write memory does it contain?

11-6. Explain the difference between software and firmware.

11-7. Explain the difference between a mask-programmable ROM and a fusible-link PROM.

11-8. Explain the difference between a UV EPROM and an EEPROM.

11-9. Why have hard disks become almost standard bulk storage devices on most microcomputers?

11-10. List several types of nonvolatile read/write memory.

1. CPU or central processing unit
2. address bus, control bus
3. magnetic, optical, semiconductor, or mechanical
4. floppy-disk drive, hard-(rigid) disk drive, CD-ROM, DVD, flash card.
5. a. random-access memory
   b. read-only memory
   c. electrically programmable read-only memory
   d. electrically erasable programmable read-only memory
   e. static RAM
   f. dynamic RAM
6. T
7. CD-R or CD-RWs
8. flash memory or MRAM
9. random-access memory
10. writing
11. reading
12. read/write
13. 16 × 4 bit
14. volatile, off
15. DRAM

16. RAM
17. 16, 4
18. write, written into, 15
19. dynamic
20. 4096, 4
21. binary-to-Gray
22. 1100, Gray, 1000
23. lose its program and have to be reprogrammed
24. read-only memory
25. nonvolatile
26. firmware
27. manufacturer
28. is not
29. 0010
30. 1101
31. high-
32. remain in
33. 177
34. 14
35. manufacturer
36. ROM
37. programmable read-only memory
38. erasable programmable read-only memory
39. electrically erasable programmable read-only memory

40. ultraviolet
41. 524,288, 65,536
42. nonvolatile RAM
43. lithium
44. EEPROM
45. recall
46. $V_{PP}$
47. flash memory
48. SRAM
49. F
50. MRAM
51. magnetoresistive RAM
52. ferroelectric RAM
53. F
54. DIP
55. dual-in-line memory module
56. double data rate synchronous DRAM
57. single in-line memory module
58. Rambus DRAM
59. T
60. Personal Computer Memory Card International Association
61. credit card
62. F
63. magnetic, optical, semiconductor

64. magnetic tapes, magnetic disks (floppy and rigid), magnetic drum, optical discs, flash memory or cards, paper tapes, paper punched cards
65. rigid
66. hard-disk drive
67. rigid disk
68. platters
69. one billion
70. hard-disk drive
71. 50-to-500 Gbytes
72. optical disc
73. digital versatile disc
74. optical
75. write-once read-many
76. compact-disc rewritable
77. CD-ROM
78. NV trimmer potentiometer
79. EEPROM, multiplexer
80. 51 kΩ
81. 52 kΩ
82. 51 kΩ
83. 50 kΩ
84. 49 kΩ
85. 50 kΩ

# Chapter 12

# Digital Systems

## Chapter Objectives

*This chapter will help you to:*

1. *Identify* six elements found in most systems.
2. *Describe* each of the six elements found in most systems.
3. *Describe* each of the five scales of integration of digital ICs.
4. *Analyze* the operation of digital dice game circuits.
5. *Diagram* the organization of a digital clock system.
6. *Analyze* the operation of a digital clock system including frequency divider circuits and display multiplexing.
7. *Analyze* the operation of a digital frequency counter system.
8. *Analyze* the operation of an LCD timer system.
9. *Answer* select questions about terms used in boundary scan technology (JTAG).

**M**any devices we use every day, such as calculators, alarm clocks, digital wristwatches, cellular telephones, MP3 players, and computers, are *digital systems*. Calculators, digital clocks, and computers are assemblies of *subsystems*. Typical subsystems include counters, RAMs, ROMs, encoders, decoders, clocks, and display decoders/drivers. You have already used most of these subsystems. This chapter discusses several digital systems, how they transmit data, and how they can be tested for proper system operation. Digital systems are formed by the proper assembly of digital subsystems.

## 12-1 Elements of a System

Most mechanical, chemical, fluid, and electrical systems have certain features in common. Systems have an *input* and an *output* for their product, power, or information. Systems also act on the product, power, or information; this is called *processing*. The entire system is organized and its operation directed by a *control* function. The *transmission* function transmits products, power, or information. More complicated systems also contain a *storage* function. Figure 12-1 illustrates the overall organization of a system. Look carefully and you can see that this diagram is general enough to apply to nearly any system, whether it is transportation, fluid, school, or electronic. The transmission from device to device is shown by the colored lines and arrows. Notice that the data or whatever is being transferred always moves in one direction. It is common to use double arrows on the control lines to show that the control unit is directing the operation of the system as well as receiving feedback from the system.

**Elements of a system**

**Input**
**Output**

**Processing**
**Control**
**Transmission**

**Storage**

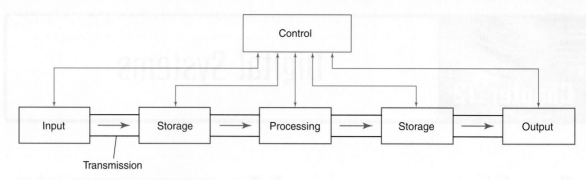

Fig. 12-1 The elements of a system.

Digital systems deal with transmitting digital data, usually as numbers or codes. The general system shown in Fig. 12-1 will help explain several digital systems discussed in this chapter and Chapter 13, Computer Systems.

*Answer the following questions.*

1. There is a two-way path from the _____ section of a system to all other parts.

2. The keyboard of a microcomputer is classified as what part of the system?

3. Digital systems deal with transmitting _____.

## 12-2 A Digital System on an IC

We have learned that all digital systems can be wired from individual AND and OR gates and inverters. We have also learned that manufacturers produce subsystems on a single IC (counters, registers, and so on). Now manufacturers have gone a step further: some ICs contain entire digital systems.

The least complex digital integrated circuits are classified as *small-scale integrations (SSI)*. An SSI contains circuit complexity of less than 12 gates or circuitry of similar complexity. Small-scale integrations include the logic gate and flip-flop ICs you have used.

A *medium-scale integration (MSI)* has a complexity ranging from 12 to 99 gates. ICs that are classified as MSIs belong to the small subsystem group. Typical examples include adders, registers, comparators, code converters, counters, data selectors/multiplexers, and small RAMs. Most of the ICs you have studied and used so far have been either SSIs or MSIs.

A *large-scale integration (LSI)* has the complexity of 100 to 9999 gates. A major subsystem or a simple digital system is fabricated on a single chip. Examples of LSI chips are digital clocks, calculators, microcontrollers, ROMs, RAMs, PROMs, EPROMs, and flash memories.

A *very large-scale integration (VLSI)* has the complexity of 10,000 to 99,999 gates. VLSI ICs are usually digital systems on a chip. The term "chip" refers to the single silicon die (perhaps ¼-in. square) that contains all the electronic circuitry in an IC. Large memory chips and advanced microprocessors are examples of VLSI ICs.

An *ultra-large-scale integration (ULSI)* is the next higher level of circuit complexity and contains more than 100,000 gates on a single chip. Various manufacturers tend to define SSI, MSI, LSI, VLSI, and ULSI differently.

In the 1960s, families of digital ICs were being developed using SSI and MSI technology. Late in the 1960s, large-scale integration technology developed many specialized ICs. Higher production LSIs included single-chip clocks, calculators, and memories. After the development of *calculator chips,* the architecture of a computer was designed into a single chip

**Small-scale integration (SSI)**

**Medium-scale integration (MSI)**

**Large-scale integration (LSI)**

**Very large-scale integration (VLSI)**

**Ultra large-scale integration (ULSI)**

**Calculator chip**

called the microprocessor. The *microprocessor* forms the CPU of a computer system. Improvements in CPU design and chip manufacturing have produced the latest generation of microprocessors that contain the equivalent of tens of millions of transistors. In the 1980s, manufacturers combined some of the separate sections of a computer system (CPU, RAM, ROM, and input/output) into a single inexpensive IC. These "tiny computers on a chip" were used mainly for control purposes and were not used in general-purpose computers. These inexpensive computers on a chip are generally referred to as *microcontrollers*. You will learn more about microcomputers and microcontrollers in Chapter 13.

**Microprocessor**

**Microcontroller**

## ✔ Self-Test

*Supply the missing word in each statement.*

4. A medium-scale integration is an IC that contains the equivalent of _____ gates.
5. A VLSI is an IC that contains the equivalent of more than _____ gates.
6. Using SSI and MSI technology, digital families of ICs (like TTL)

were developed in the _____ (1940s, 1960s).
7. A(n) _____ (adder IC, microcontroller IC) could be described as a digital system on a chip.
8. Identify an ULSI device that forms the CPU of modern general-purpose computers.

## 12-3 Digital Games

Electronics has been a popular hobby for more than a half century. A favorite task for many electronic hobbyists, young and old, is to construct electronic games. Electronic games and toys are also very popular with students studying electronics in high schools, technical schools, and colleges.

Electronic games may be classified as simple self-contained, computer, arcade, or TV games. The simple self-contained type includes the games and toys most often constructed by students and hobbyists. Several simple digital electronic games using SSI and MSI digital ICs will be surveyed in this section.

### Simple Dice Game

A block diagram of a simple *digital dice game* is sketched in Fig. 12-2. When the push button is pressed, a signal from the clock is sent to the counter. The counter is wired to have a counting sequence of 1, 2, 3, 4, 5, 6, 1, 2, 3, and so on. The binary output from the counter is translated to seven-segment code by the decoder block. The decoder block also contains a seven-segment LED display driver. The output device in this circuit is an LED display. When the push-button switch is opened, the counter *stops at a random number* from 1 through 6. This simulates the roll of a single die. The binary number stored in the counter is decoded and shown as a decimal number on the display. This circuit could be doubled to simulate the rolling of a pair of dice.

A wiring diagram for the digital dice game is detailed in Fig. 12-3. Pressing the input switch causes the counter to sequence through the binary numbers 001, 010, 011, 100, 101, 110,

**Digital dice game**

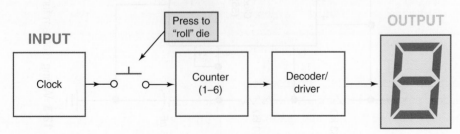

Fig. 12-2 Simple block diagram of a digital dice game.

**Electronic dice simulation game**

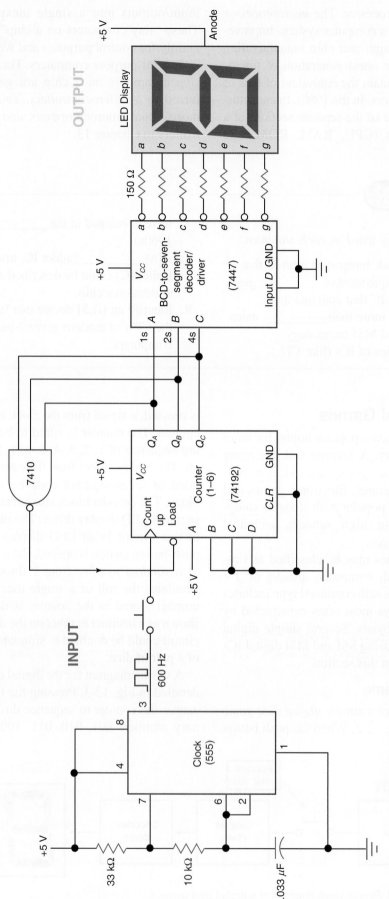

**Fig. 12-3** Wiring diagram for an experimental digital dice game using TTL ICs.

Digital dice
game circuit

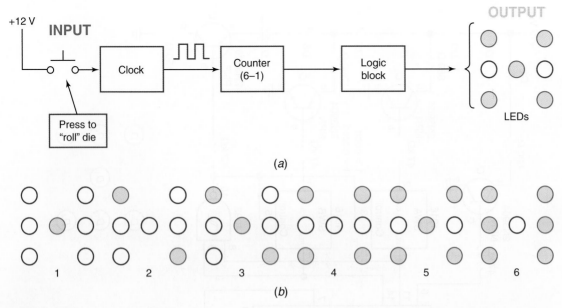

*(a)*

*(b)*

**Fig. 12-4** Electronic dice simulation game (*a*) Simple block diagram. (*b*) LED patterns used for representing dice rolls 1 through 6.

001, 010, 011, etc. When the switch is opened, the last binary count is stored in the flip-flops of the 74192 counter. It is decoded by the 7447 IC and lights the seven-segment LED display.

The 555 timer IC is wired as an astable MV in Fig. 12-3. It generates a 600-Hz rectangular wave signal.

The 74192 IC is wired as a mod-6 (1 to 6) up counter. The three-input NAND gate is activated when the count reaches binary 111. The LOW signal from the NAND gate loads the next number in the count sequence, which is binary 001. It should be noted that the three outputs of the counter ($Q_A$, $Q_B$, $Q_C$) all go HIGH for only an extremely short time (less than a microsecond) while the counter is being loaded with 0001. Therefore, the temporary count of binary 111 never appears as a 7 on the LED display.

The 7447 BCD-to-seven-segment decoder chip translates the binary inputs (*A, B, C*) to seven-segment code. The 7447 IC drives the LED segments with active LOW outputs (*a* to *g*). The seven 150-Ω resistors serve to limit the current flow through the LEDs to a safe level. Note that the seven-segment LED display used in Fig. 12-3 is a common-anode type.

### Another Dice Game

The digital dice game featured in Fig. 12-3 used TTL ICs with an unrealistic seven-segment display. A more realistic dice simula-

tion is implemented in the circuit shown in Figs. 12-4 and 12-5.

A block diagram for a second digital dice game is sketched in Fig. 12-4(*a*). This unit uses individual LEDs for the output device.

Depressing the push button in Fig. 12-4(*a*) causes the clock to generate a square-wave signal. This signal causes the down counter to cycle through the count sequence 6, 5, 4, 3, 2, 1, 6, 5, 4, and so on. The logic block lights the proper LEDs to represent various decimal counts. The LED pattern for each possible decimal output is diagramed in Fig. 12-4(*b*).

The wiring diagram for the second digital dice game is detailed in Fig. 12-5. The circuit features the use of 4000 series CMOS ICs and a 12-V dc power supply. The push-button switch on the left is the input device, while the LEDs (*D*1 to *D*7) on the right form the output. Physically, the LEDs are arranged as shown at the lower right in Fig. 12-5.

When the "roll dice" switch is closed, the two NAND Schmitt trigger gates at the left in Fig. 12-5 produce a 100-Hz square-wave signal. The two NAND gates and associated resistors and capacitors are wired to form an astable MV. The 100-Hz signal is fed into the clock input of the *4029 presettable binary/decade up/down counter*. In this circuit, the 4029 IC is wired as a down counter whose outputs produce binary 110, 101, 100, 011, 010, 001, 110, 101, 100, etc.

**4029 presettable binary/decade up/down counter**

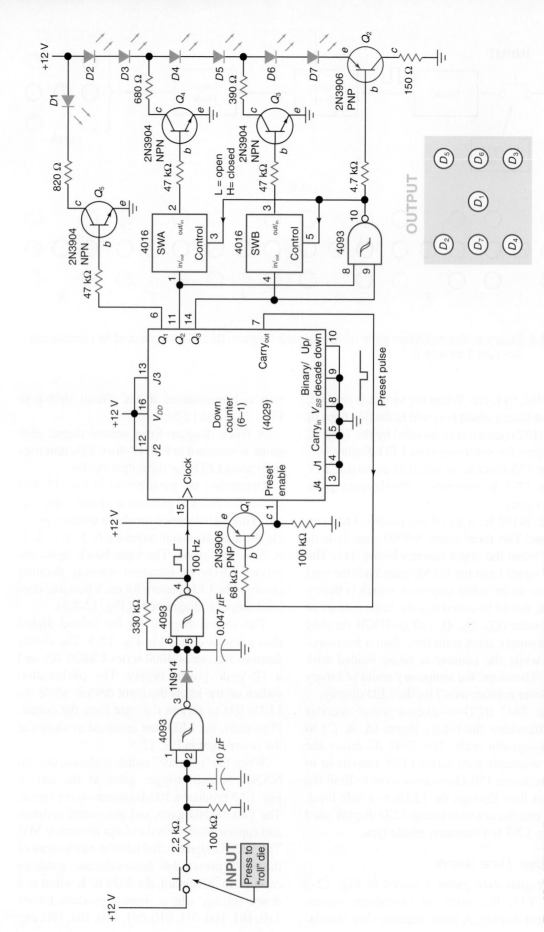

**Fig. 12-5** Wiring diagram for an electronic dice simulation game. *(Courtesy of Graymark, Inc.)*

**Electronic dice game circuit**

| INPUTS | | | ACTIVE COMPONENTS | OUTPUT | |
|---|---|---|---|---|---|
| 4s (pin 14) | 2s (pin 11) | 1s (pin 6) | | LEDs LIT | DECIMAL |
| H | H | L | NAND output LOW<br>transistor $Q_2$ turned on | D2, D3, D4, D5, D6, D7 | 6 |
| H | L | H | transistor $Q_5$ turned on<br>bilateral switch SWB closed<br>transistor $Q_3$ turned on | D1<br>D2, D3, D4, D5 | 5 |
| H | L | L | bilateral switch SWB closed<br>transistor $Q_3$ turned on | D2, D3, D4, D5 | 4 |
| L | H | H | transistor $Q_5$ turned on<br>bilateral switch SWA closed<br>transistor $Q_4$ turned on | D1<br>D2, D3 | 3 |
| L | H | L | bilateral switch SWA closed<br>transistor $Q_4$ turned on | D2, D3 | 2 |
| L | L | H | transistor $Q_5$ turned on | D1 | 1 |

Fig. 12-6 Explaining the logic and output sections of the electronic dice simulation game.

Logic and output sections of dice game

Consider the situation when the down counter in Fig. 12-5 reaches binary 001. On the next LOW-to-HIGH transition of the clock pulse, the *carry out* output (pin 7) of the 4029 IC drops LOW. This signal is fed back and turns on transistor $Q_1$. This causes the *preset enable* input of the 4029 counter to go HIGH. When the preset enable input goes HIGH, the data at inputs J4, J3, J2, and J1 (the "jam inputs") are asynchronously loaded into the counter's flip-flops. In this example, binary 0110 is loaded on the preset pulse. Once the flip-flops have been loaded, the carry out pin goes back HIGH and transistor $Q_1$ turns off.

The final sections of the dice game at the right in Fig. 12-5 contain many components. The table in Fig. 12-6 will help explain the complexities of the *logic* and *output* sections of this dice game.

The input section of the table in Fig. 12-6 indicates the logic levels present at the output of the 4029 counter. The top line of the table shows a binary 110 (HHL) stored in the 4s, 2s, and 1s flip-flops of the counter. The middle column of the table lists only the components that are activated to light the proper LEDs.

Consider the first line of the table in Fig. 12-6. The output of the NAND gate goes LOW, which turns on the PNP transistor $Q_2$. The transistor conducts and all six LEDs (D2 to D7) on the right in Fig. 12-5 light. This would simulate a dice roll of 6.

Consider line 2 of the table in Fig. 12-6. The binary data equals 101 (HLH). The HIGH on the 1s line (pin 6) turns on transistor $Q_5$. Transistor $Q_5$ conducts and lights LED D1. The output of the NAND gate is HIGH, which causes both bilateral switches to be in the closed condition (low impedance from in/out to out/in). The bilateral switches pass the logic level from the 2s and 4s lines to the base of transistors $Q_4$ and $Q_3$. Transistor $Q_3$ is turned on by the HIGH and conducts. Light-emitting diodes D2, D3, D4, and D5 light. A decimal 5 is represented when these five LEDs (D1 to D5) are lit.

You may look over the remaining lines of the table in Fig. 12-6 to determine the operation of the logic and output sections of this CMOS digital dice game.

The 4016 IC used in Fig. 12-5 is described by the manufacturer as a *quad bilateral switch*. It is an electronically operated SPST switch. A HIGH at the control input of the 4016 bilateral switch causes it to be in the "closed" or "ON"

Logic and output sections of dice game

Quad bilateral switch

position. In the "closed" position the internal resistance from in/out to out/in terminals is quite low (400 Ω typical). A LOW at the control input of the bilateral switch causes it to be in the open or OFF position. The 4016 IC acts like an open switch when the control is LOW. Unlike a gate, a bilateral switch can pass data in either direction. It can pass either dc or ac signals. A bilateral switch is also referred to as a *transmission gate*.

**Transmission gate**

✓ Self-Test

*Answer the following questions.*

9. Refer to Fig. 12-3. The 555 timer is wired as a(n) _____ multivibrator in this digital system.
10. Refer to Fig. 12-3. When the 74192 IC increases its count from 110 to 111, the output of the NAND gate goes _____ (HIGH, LOW) immediately loading _____ (binary) into the counter.
11. Refer to Fig. 12-3. List the possible digits that can appear on the seven-segment LED display when the input switch is released.
12. Refer to Fig. 12-5. Two gates packaged in the _____ (4016, 4029, 4093) IC are wired as a free-running MV in this digital dice game circuit.
13. Refer to Fig. 12-5. List the binary counting sequence of the 4029 IC in this circuit.
14. Refer to Fig. 12-5. When the output of the counter is binary 001, only $D1$ lights because only transistor _____ is turned on and conducts.
15. Refer to Fig. 12-5. When the output of the counter is binary 010, LED(s) _____ light because the bilateral switch SWA is _____ (closed, open) and transistor _____ is turned on grounding the cathode of light-emitting diodes $D_2$ and $D_3$.
16. A(n) _____ is an IC device available in CMOS that acts much like a SPST switch that can conduct either ac or dc signals.

## 12-4 The Digital Clock

We introduced a digital electronic clock earlier and noted that various *counters* are the heart of a digital clock system. Figure 12-7($a$) is a simple block diagram of a digital clock system. Some clocks use the power-line frequency of 60 Hz as their input or frequency standard. This frequency is divided into seconds, minutes, and hours by the *frequency divider* section of the clock. The one-per-second, one-per-minute, and one-per-hour pulses are then counted and stored in the *count accumulator* section of the clock. The stored contents of the count accumulators (seconds, minutes, hours) are then *decoded,* and the correct time is shown on the output *time displays.* The digital clock has the typical elements of a system. The input is the 60-Hz alternating current. The processing takes place in the frequency divider, count accumulator, and decoder sections. Storage takes place in the count accumulators. The control section is illustrated by the *time-set* control, as shown in

**Frequency divider**

**Count accumulator**

**Frequency divider circuits**

Fig. 12-7($a$). The output section is the digital time display.

It was mentioned that all digital systems consist of logic gates, flip-flops, and subsystems. The diagram in Fig. 12-7($b$) shows how subsystems are organized to display time in hours, minutes, and seconds. This is a more detailed diagram of a digital clock. The input is still a 60-Hz signal. The 60 Hz may be from the low-voltage secondary coil of a transformer. The 60 Hz is divided by 60 by the first frequency divider. The output of the first divide-by-60 circuit is 1 pulse per second. The 1 pulse per second is fed into an *up counter* that counts upward from 00 through 59 and then resets to 00. The seconds counters are then decoded and displayed on the two 7-segment LED displays at the upper right, Fig. 12-7($b$).

Consider the middle *frequency-divider circuit* in Fig. 12-7($b$). The input to this divide-by-60 circuit is 1 pulse per second; the output is 1 pulse per minute. The 1-pulse-per-minute output is

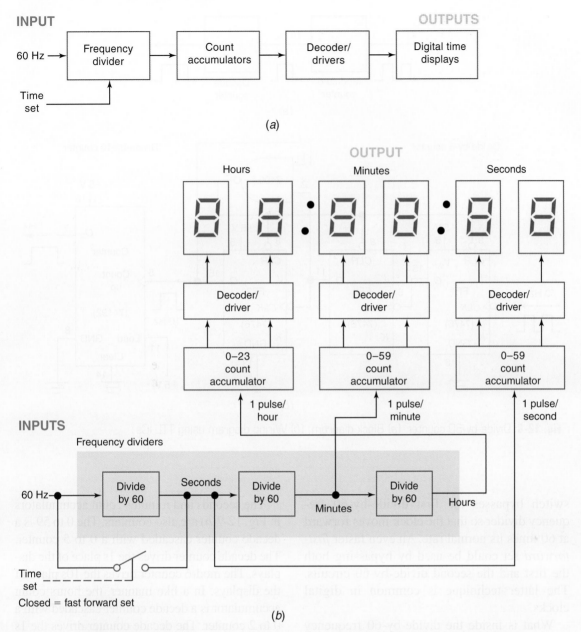

INPUT

60 Hz → **Frequency divider** → **Count accumulators** → **Decoder/ drivers** → **Digital time displays**

Time set

OUTPUTS

*(a)*

OUTPUT

Hours          Minutes          Seconds

**Decoder/ driver**          **Decoder/ driver**          **Decoder/ driver**

**0–23 count accumulator**          **0–59 count accumulator**          **0–59 count accumulator**

1 pulse/ hour          1 pulse/ minute          1 pulse/ second

INPUTS

Frequency dividers

60 Hz → **Divide by 60** — Seconds → **Divide by 60** — Minutes → **Divide by 60** — Hours

Time set

Closed = fast forward set

*(b)*

**Fig. 12-7** *(a)* Simplified block diagram of a digital clock. *(b)* More detailed block diagram of a digital clock.

**Digital clock as a system**

transferred into the 0 to 59 minutes counter. This up counter keeps track of the number of minutes from 00 through 59 and then resets to 00. The output of the minutes count accumulator is decoded and displayed on the two 7-segment LEDs at the top center, Fig. 12-7(*b*).

Now for the divide-by-60 circuit on the right in Fig. 12-7(*b*). The input to this frequency divider is 1 pulse per minute. The output of this circuit is 1 pulse per hour. The 1-pulse-per-hour output is transferred to the hours counter on the left. This hours count accumulator keeps track of

the number of hours from 0 to 23. The output of the hours count accumulator is decoded and transferred to the two 7-segment LED displays at the upper left, Fig. 12-7(*b*). You probably have noticed that this is a 24-h digital clock. It easily could be converted to a 12-h clock by changing the 0 to 23 count accumulator to a 1 to 12 counter.

For setting the time a time-set control has been added to the digital clock in Fig. 12-7(*b*). When the switch is closed (a logic gate may be used), the display counts forward at a fast rate. This enables you to set the time quickly. The

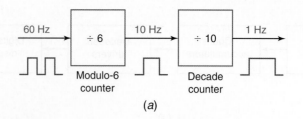

(a)

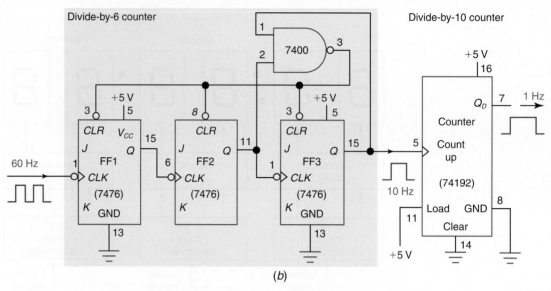

(b)

**Fig. 12-8** Divide-by-60 counter. (a) Block diagram. (b) Wiring diagram using TTL ICs.

switch bypasses the first divide-by-60 frequency divider so that the clock moves forward at 60 times its normal rate. An even faster *fast-forward* set could be used by bypassing both the first and the second divide-by-60 circuits. The latter technique is common in digital clocks.

What is inside the divide-by-60 frequency dividers in Fig. 12-7(b)? In Chapter 8 we spoke of a counter being used to divide frequency. Figure 12-8(a) is a block diagram of how a divide-by-60 frequency divider might be organized. Notice that a divide-by-6 counter is feeding a divide-by-10 counter. The entire unit divides the incoming frequency by 60. In this example, the 60-Hz input is reduced to 1 Hz at the output.

A detailed wiring diagram for a divide-by-60 counter circuit is drawn in Fig. 12-8(b). The three JK flip-flops and NAND gate form the divide-by-6 counter while the 74192 decade counter performs as a divide-by-10 unit. If 60 Hz enters at the left, the frequency will be reduced to 1 Hz at output $Q_D$ of the 74192 counter.

The seconds and minutes count accumulators in Fig. 12-7(b) are also counters. The 0 to 59 is a decade counter cascaded with a 0 to 5 counter. The decade counter drives the 1s place of the displays. The mod-6 counter drives the 10s place of the displays. In a like manner, the hours count accumulator is a decade counter cascaded with a 0 to 2 counter. The decade counter drives the 1s place in the hours display. The mod-3 counter drives the 10s place of the hours display.

In many practical digital clocks the output may be in hours and minutes only. Most digital clocks are based upon one of many inexpensive ICs. Large-scale-integrated *clock chips* have all the frequency dividers, count accumulators, and decoders built into a single IC. For only a few extra dollars, clock chips have other features, such as 12- or 24-h outputs, calendar features, alarm controls, and radio controls.

An added feature you will use when you construct a digital timepiece is shown in Fig. 12-9(a). A *wave-shaping circuit* has been added to the block diagram of our digital clock. The IC counters that make up the frequency-

**Wave-shaping circuit**

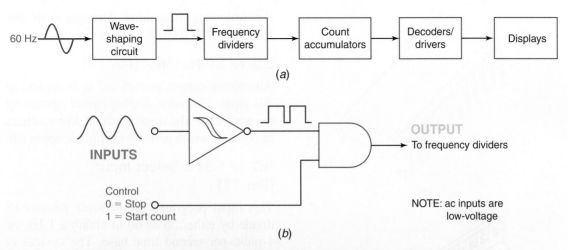

*(a)*

*(b)*

**Fig. 12-9** Wave shaping. *(a)* Adding a wave-shaping circuit to the input of the digital clock system. *(b)* Schmitt trigger inverter used as a wave shaper.

Wave shaping using Schmitt trigger device

Start/stop control

Slow rise time

Schmitt trigger inverter IC

Frequency division

Count accumulation

divider circuit do not work well with a sine-wave input. The sine wave (shown at the left in Fig. 12-9(a)) has a *slow rise time* that does not trigger the counter properly. The sine-wave input must be converted to a square wave. The wave-shaping circuit changes the sine wave to a square wave. The square wave will now properly trigger the frequency-divider circuit.

Commercial LSI clock chips have wave-shaping circuitry built into the IC. In the laboratory you may use a *Schmitt trigger inverter IC* to square up the sine waves as you did in Chapter 7. A simple wave-shaping circuit is shown in Fig. 12-9(b). This circuit uses the TTL 7414 Schmitt trigger inverter IC to convert the sine wave to a square wave. The circuit in Fig. 12-9(b) also contains a *start/stop control*. When the control input is HIGH, the square wave from the Schmitt trigger inverter passes through the AND gate. When the control input goes LOW, the square-wave signal is inhibited and does not pass through the AND gate. The counter is stopped.

You will want to get some practical knowledge of how counters are used in dividing frequency. Remember that the counter subsystem is used for two jobs in the digital timepiece: first to *divide frequency* and second to *count* upward and *accumulate* or store the number of pulses at its input.

Self-Test

*Answer the following questions.*

17. Refer to Fig. 12-7(a). Counters are used in the _____ and _____ sections of a digital clock.
18. Refer to Fig. 12-9. When operating a clock with a sine wave, a(n) _____ circuit is added to the clock.

19. Refer to Fig. 12-9(b). What is the purpose of the Schmitt trigger inverter in this circuit?
20. Refer to Fig. 12-9(b). What is the purpose of the AND gate in this circuit?

## 12-5  The LSI Digital Clock

The LSI clock chip forms the heart of modern digital timepieces. These digital clock chips are made as monolithic MOS ICs. Many times, the MOS LSI chip, or die, is mounted in an 18-, 24-, 28-, or 40-pin DIP IC. Other times, the MOS LSI chip is mounted directly on the PC board of a *clock module*. The tiny silicon die is sealed under an epoxy coating. Examples of the two packaging methods are shown in Fig. 12-10. An

Clock module

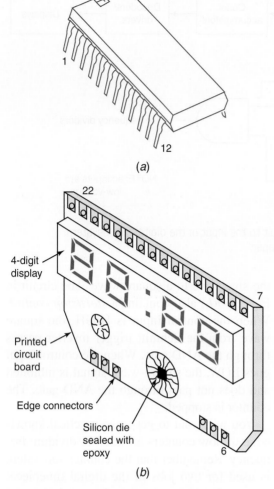

(a)

(b)

**Fig. 12-10** (a) An LSI clock chip in a 24-pin dual in-line package. (b) A typical clock module containing a MOS/LSI die.

MOS LSI clock IC packaged in a 24-pin DIP is illustrated in Fig. 12-10(*a*). Pin 1 of the DIP IC is identified in the normal manner (pin 1 is immediately ccw from the notch). A clock module is sketched in Fig. 12-10(*b*). The back is a PC board with 22 edge connectors. The numbering of the edge connectors is shown. A four-digit LED display is premounted on the board with all connections complete. Some clock modules have some discrete components and a DIP clock IC mounted on the board. The clock module in Fig. 12-10(*b*) has the tiny silicon chip, or die, mounted on the PC board. It is sealed with a protective epoxy coating.

A block diagram of National Semiconductor's *MM5314 MOS LSI clock IC* is shown in Fig. 12-11(*a*). The pin diagram is shown in Fig. 12-11(*b*). Refer to Fig. 12-11(*a*) and (*b*) for

**MM5314 MOS LSI clock IC**

the following functional description of the MM5314 digital clock IC.

### 50- or 60-Hz Input [Pin 16]

Alternating current or rectified ac is applied to this input. The *wave-shaping circuit* squares up the waveform. The shaping circuit drives a chain of counters which perform the time-keeping job.

### 50- or 60-Hz Select Input [Pin 11]

This input programs the *prescale counter* to divide by either 50 or 60 to obtain a 1-Hz, or 1-pulse-per-second time base. The counter is programmed for 60-Hz operation by connecting this input to $V_{DD}$ (GND). If the 50/60-Hz select input pin is left unconnected, the clock is programmed for 50-Hz operation.

### Time-setting Inputs (Pins 13, 14, and 15)

Slow- and fast-setting inputs as well as a hold input are provided on this clock IC. These inputs are enabled when they are connected to $V_{DD}$ (GND). Typically, a normally open push-button switch is connected from these pins to $V_{DD}$. The three gates in the counter chain are used for setting the time. For *slow set,* the prescale counter is bypassed. For *fast set,* the prescale counter and seconds counter are bypassed. The *hold* input inhibits any signal from passing through gate *A* to the prescale counter. This stops the counters, and time does not advance on the output display.

### 12- or 24-H Select Input [Pin 10]

This input is used to program the hours counter to divide by either 12 or 24. The 12-h display format is selected by connecting this input to $V_{DD}$ (GND). Leaving pin 10 unconnected selects the 24-h format.

### Output MUX Operation (Pins 3 to 9 and 17 to 22)

The seconds, minutes, and hours counters continuously reflect the time of day. Outputs from each counter are *multiplexed* to provide digit-by-digit sequential access to the time data. In other words, only one display digit is turned on for a very short time, then the second, then the third, and so forth. By multiplexing the displays instead of using 48 leads to the six displays

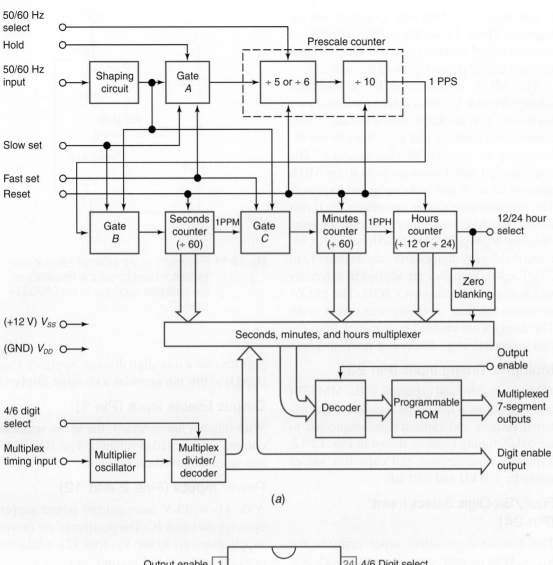

(a)

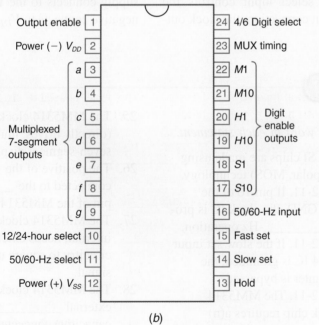

(b)

**Fig. 12-11** (a) Functional block diagram of the MM5314 MOS/LSI clock chip. (b) Pin diagram for the MM5314 digital clock IC.

MM5314
MOS/LSI
clock chip

**Multiplexing and driving displays**

(8 pins each × 6 = 48), only 13 output pins are required. These 13 outputs are the multiplexed seven-segment outputs (pins 3 through 9) and the digit enable outputs (pins 17 through 22).

The MUX is addressed by a *multiplex divider/decoder,* which is driven by a *multiplex oscillator.* The oscillator uses external timing components (resistor and capacitor) to set the frequency of the multiplexing function. The four/six-digit select input controls if the MUX turns on all six or just four displays in sequence. The *zero-blanking* circuit suppresses the 0 that would otherwise sometimes appear in the tens-of-hours display. The MUX addresses also become the display digit enable outputs (pins 17 to 22). The MUX outputs are applied to a decoder which is used to address a *PROM.* The PROM generates the final seven-segment output code. The displays are enabled in sequence from the unit seconds through the tens-of-hours display.

**MM5314 clock IC features PROM**

### Multiplex Timing Input (Pin 23)

**Relaxation oscillator (multiplex oscillator)**

Adding a resistor and capacitor to the MM5314 clock IC forms a *relaxation oscillator.* The external resistor and capacitor are connected to the MUX timing input as shown in Fig. 12-12. Typical timing resistor and capacitor values might be 470 kΩ and 0.01 μF.

### Four/Six-Digit Select Input (Pin 24)

The four/six-digit select input controls the MUX. With no input connection, the clock out-

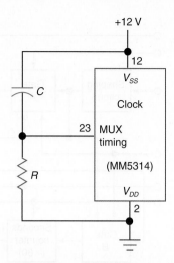

**Fig. 12-12** Placement of the external resistor and capacitor used to set the frequency of the multiplex oscillator in the MM5314 clock IC.

puts data for a four-digit display. Applying $V_{DD}$ (GND) to this pin provides a six-digit display.

### Output Enable Input (Pin 1)

With this pin unconnected, the seven-segment outputs are enabled. Switching $V_{DD}$ (GND) to this input inhibits these outputs.

### Power Inputs (Pins 2 and 12)

A dc 11- to 19-V nonregulated power supply operates the clock IC. The positive of the power supply connects to the $V_{SS}$ (pin 12), while the negative connects to $V_{DD}$ (pin 2).

*Supply the missing word in each statement.*

21. Digital clock LSI chips are made using _____ (bipolar, MOS) technology.
22. Refer to Fig. 12-11. If pin 16 of the MM5314 is at GND, the clock IC is programmed for _____-Hz operation.
23. Refer to Fig. 12-11. If the slow set input to the MM5314 IC is grounded, the _____ counter is bypassed.
24. Refer to Fig. 12-11. The MM5314 MOS/LSI clock chip requires a(n) _____-V nonregulated power supply.

25. The MM5314 clock chip _____ (directly drives, multiplexes) seven-segment displays.
26. The positive of the 12-V power supply is connected to the _____ ($V_{DD}$, $V_{SS}$) pin of the MM5314 clock chip.
27. The MM5314 clock chip _____ (has internal, needs external) wave-shaping circuitry to square up the incoming 60-Hz signal.
28. The MM5314 clock chip needs an external _____ (crystal, resistor and capacitor) connected to the MUX timing pin of the IC.

## 12-6 A Practical LSI Digital Clock System

A *six-digit clock* using the MM5314 IC is sketched in Fig. 12-13(*a*). This student-built unit uses six common-anode seven-segment LED displays. Also notice the many extra parts used in the clock system. A block diagram of this system is shown in Fig. 12-13(*b*). The National Semiconductor MM5314 clock chip is used in this system. The 60 Hz is divided down to seconds, minutes, and hours by counters across the top in Fig. 12-13(*b*). These feed the MUX. The *oscillator* at the lower left produces a frequency of about 1 kHz.

Outside the MM5314 clock chip are 6 seven-segment common-anode LED displays. Because of the higher currents used by the LED displays, *segment drivers* are used to sink the current from the cathodes of the displays. The *digit drivers* furnish an adequate amount of current to the anodes of the selected digit.

To help explain how the *MUX* works, suppose the time is 12:34:56. This information is held in the counters in the clock chip. The multiplex decoder first selects the *S1* display. The MUX takes data from the *S1* counter and places it on the decoder PROM. Segment lines *c, d, e, f,* and *g* are *activated to all the displays*. The multiplex decoder activates *only* the *S1* line of the digit driver. The decimal number 6 flashes on for an instant, as shown in Fig. 12-14(*a*). The segments *c, d, e, f,* and *g* have been activated on all the displays, but only the right *S1* unit has had its common-anode lead activated, or connected to $V_{SS}$. Therefore only the *S1* display lights up.

Second, the clock IC's multiplex decoder selects the *S10* display. The MUX finds that the *S10* counter holds 5. The decoder-PROM-segment driver activates segments *a, c, d, f,* and *g*. Next, the common anode of the *S10* display is activated, or connected to $V_{SS}$. A decimal 5 flashes on the *S10* display. This is shown in Fig. 12-14(*b*).

One at a time, each display is activated by the multiplex decoder and digit driver. At the same time the MUX-decoder-PROM activates the proper segments. This is based on the present contents of the counters. Look over Fig. 12-14. This is one cycle through the six displays. This whole sequence (*a* through *f*)

happens more than 100 times each second. This multiplexing, or scanning, occurs at a fast rate, and so the eye does not notice a flickering in the displays.

A schematic diagram of the digital clock system using the MM5314 IC is shown in Fig. 12-15. The step-down 12-V transformer (*T1*) with the bridge rectifier (*D1–4*) and filter capacitor (*C1*) form the dc power supply section of the clock. Alternating current voltage is taken off the transformer and coupled to the 50/60-Hz input (pin 16) of the clock chip through resistor *R3*. Capacitor *C3* and resistor *R4* determine the frequency of the multiplex oscillator. Placing a much larger value capacitor (perhaps 1 to 5 μF) across *C3* slows down the multiplexing process to a point where you can see each display light in sequence.

The fast-set, slow-set, and hold normally open push-button switches (*S2, S3,* and *S4*) are located at the lower left in Fig. 12-15. The action (fast set, slow set, or hold) is taken when these pins are connected through the switch to $V_{DD}$.

The *segment drivers* are the seven NPN transistors (*Q7* to *Q13*) on the right of the IC in Fig. 12-15. These transistors sink the current from the displays when activated. The *digit drivers* are the six PNP transistors (*Q1* to *Q6*) at the upper left in Fig. 12-15. These transistors connect *only one display anode at a time* to $V_{SS}$. The digit drivers scan the six displays at a frequency of about 500 to 1500 Hz. This activates each display about 100 to 200 times each second.

The two LEDs (*D6* and *D7* in Fig. 12-15) are activated 100 to 200 times each second and appear lit continuously. These LEDs form the colon between the hours and minutes displays on the completed clock. This colon may be seen in Fig. 12-13(*a*). Resistor *R3*, capacitor *C2*, and diode *D5* form an *RC filter network*. This network is used to remove possible line voltage transients that could either cause the clock to gain time or damage the IC.

The 12/24-h select input (pin 10) on the MM5314 clock chip in Fig. 12-15 is connected to $V_{DD}$. This selects the 12-h format. The 50/60-Hz select input (pin 11) is connected to $V_{DD}$. This programs the IC for 60-Hz operation. The four/six-digit select input (pin 24) is connected to $V_{DD}$. This programs the multiplex decoder to provide a six-digit display.

**Six-digit clock**

**Oscillator**

**Segment drivers**

**MUX**

**Digit drivers**

**RC filter network**

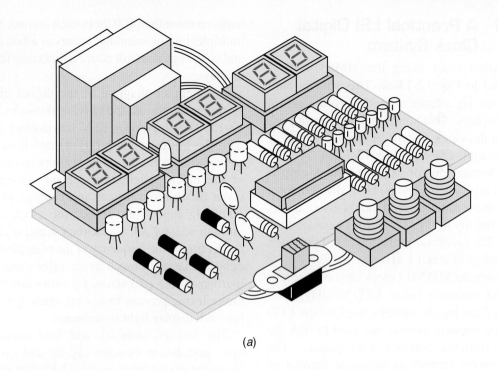

(a)

MM5314 clock chip

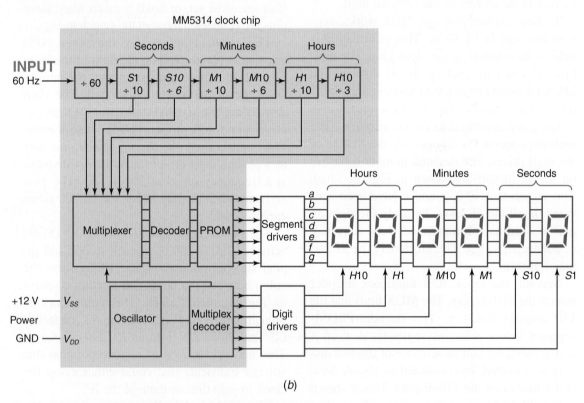

(b)

**Fig. 12-13** (a) Sketch of a practical six-digit clock project. (b) Block diagram of the practical six-digit clock project using the MM5314 clock chip.

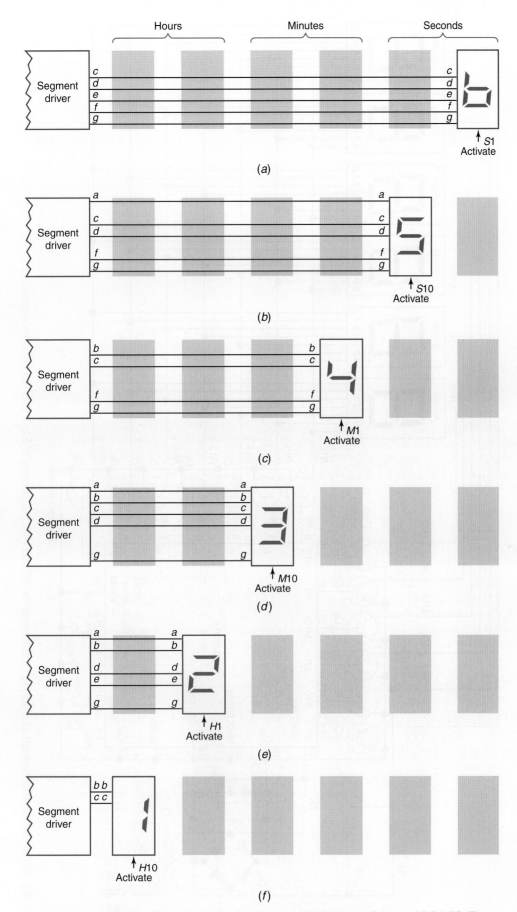

(a)

(b)

(c)

(d)

(e)

(f)

**Fig. 12-14** Example of multiplexing a six-digit display with the time of day at 12:34:56. The entire sequence from (a) through (f) occurs in about 0.01 s.

**Multiplexing a six-digit display**

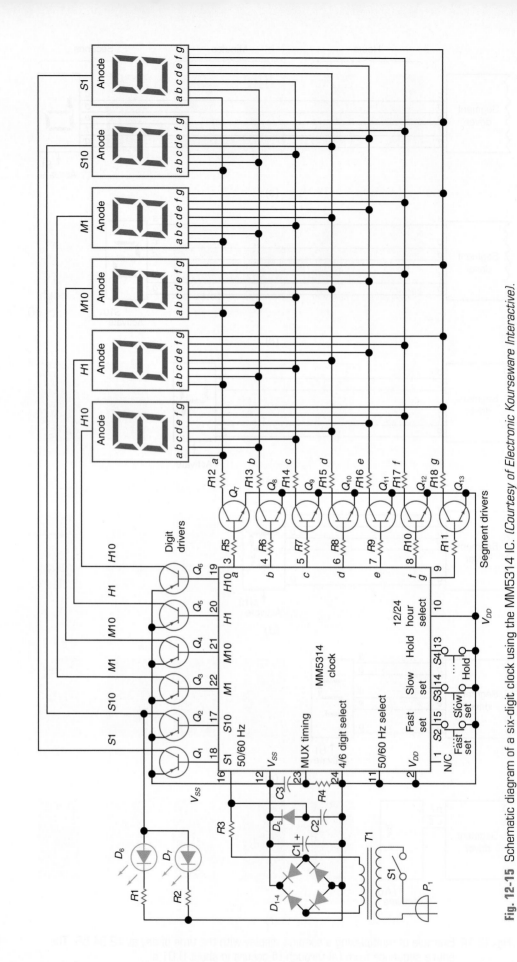

**Fig. 12-15** Schematic diagram of a six-digit clock using the MM5314 IC. *(Courtesy of Electronic Kourseware Interactive).*

**Digital clock circuit**

*Answer the following questions.*

29. Refer to Fig. 12-14. This is an example of the MM5314 clock chip _____ (counting, multiplexing) six LED decimal displays.
30. Refer to Fig. 12-15. The six PNP transistors function as _____ in this digital clock circuit.
31. Refer to Fig. 12-15. The seven NPN transistors function as _____ in this digital clock circuit.
32. Refer to Fig. 12-15. With the 12/24-h select input grounded, the clock is programmed as a(n) _____-h clock.
33. Refer to Fig. 12-15. Which two components (outside the clock chip) are responsible for determining the frequency of the multiplexer?

34. Refer to Fig. 12-15. The 4/6 digital select (pin 24) input to the MM5314 clock chip is connected to _____ ($V_{DD}$, $V_{SS}$), which selects the six-digit mode.
35. Refer to Fig. 12-14. The multiplexing of the seven-segment LED displays occurs at about _____ Hz in this digital clock.
36. Refer to Fig. 12-15. The dc power supply consists of the ac source, step-down transformer, bridge rectifier ($D_1$ to $D_4$), and filter capacitor _____ ($C_1$, $C_3$).
37. Refer to Fig. 12-15. The voltage that is applied to pin 16 of the IC through resistor $R_3$ is _____ (pure dc, 60 Hz pulsating dc or ac).
38. Refer to Fig. 12-13. The _____ (oscillator, PROM) in the clock IC selects which segments of the seven-segment LED display will light.

## 12-7  The Frequency Counter

An instrument used by technicians and engineers is the *frequency counter.* A digital frequency counter shows in decimal numbers the frequency in a circuit. Counters can measure from low frequencies of a few cycles per second (hertz, Hz) up to very high frequencies of thousands of megahertz (MHz). Like a digital clock, the frequency counter uses decade counters.

As a review, the block diagram for a digital clock is shown in Fig. 12-16(*a*). The known frequency is divided properly by the counters in the clock. The counter outputs are decoded and displayed in the time display. Figure 12-16(*b*) shows a block diagram of a frequency counter. Notice that the frequency counter circuit is fed an *unknown* frequency instead of the known frequency in a digital clock. The counter circuit in the frequency counter in Fig. 12-16(*b*) also contains a *start/stop control.*

The frequency counter has been redrawn in Fig. 12-17(*a*). Notice that an AND gate has been added to the circuit. The AND gate controls the input to the decade counters. When the start/stop control is at logical 1, the

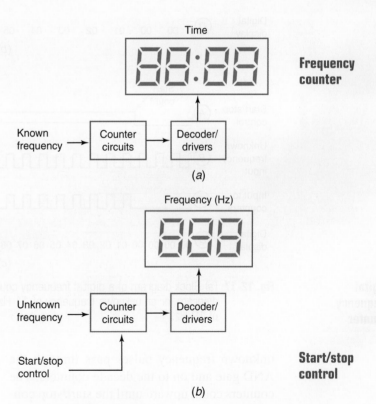

**Frequency counter**

**Start/stop control**

**Fig. 12-16** (*a*) Simplified block diagram of a digital clock. (*b*) Simplified block diagram of a digital frequency counter.

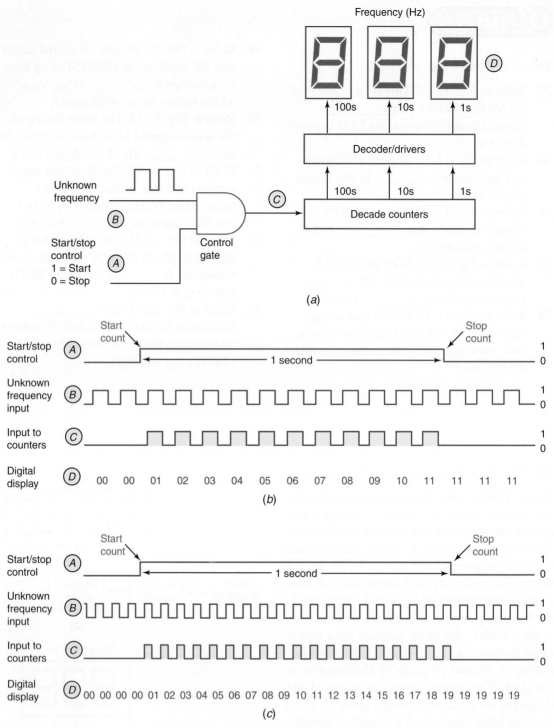

Frequency (Hz)

100s    10s    1s

Decoder/drivers

100s    10s    1s

Decade counters

Unknown
frequency

B

Start/stop
control
1 = Start
0 = Stop

A

Control
gate

C

D

(a)

Start/stop
control

A

Start
count

1 second

Stop
count

1

0

Unknown
frequency
input

B

1

0

Input to
counters

C

1

0

Digital
display

D

00   00   01   02   03   04   05   06   07   08   09   10   11   11   11   11

(b)

Start/stop
control

A

Start
count

1 second

Stop
count

1

0

Unknown
frequency
input

B

1

0

Input to
counters

C

1

0

Digital
display

D

00 00 00 00 01 02 03 04 05 06 07 08 09 10 11 12 13 14 15 16 17 18 19 19 19 19 19

(c)

**Digital
frequency
counter**

**Fig. 12-17** (a) Block diagram of a digital frequency counter showing the start/stop control (b) Waveform diagram for an unknown frequency of 11 Hz. (c) Waveform diagram for an unknown frequency of 19 Hz.

unknown frequency pulses pass through the AND gate and on to the decade counters. The counters count upward until the start/stop control returns to logical 0. The 0 turns off the control gate and stops the pulses from getting to the counters.

Figure 12-17(b) is a more exact timing diagram of what happens in the frequency counter. Line A shows the start/stop control at logical 0 on the left and then going to 1 for *exactly 1 s*. The start/stop control then returns to logical 0. Line B diagrams a continuous string of pulses from the unknown frequency input. The unknown frequency and the start/stop control are ANDed together as we saw in Fig. 12-17(a). Line C in Fig. 12-17(b) shows only the pulses that are allowed through the AND gate. These pulses trigger the up counters. Line D shows the count observed on the displays. Notice that the displays start cleared to 00. The displays then count upward to 11 during the 1 s. The unknown frequency in line B in Fig. 12-17(b) is shown as 11 Hz (11 pulses/s).

A somewhat higher frequency is fed into the frequency counter in Fig. 12-17(c). Again line A shows the start/stop control beginning at 0. It is then switched to logical 1 for *exactly 1 s*. It is then returned to logical 0. Line B in Fig. 12-17(c) shows a string of higher-frequency pulses. This is the unknown frequency being measured by this digital frequency counter. Line C shows the pulses that trigger the decade counters during the 1-s count-up period. The decade counters sequence upward to 19, as shown in line D. The unknown frequency in Fig. 12-17(c) is then 19 Hz.

If the unknown frequency were 870 Hz, the counter would count from 000 to 870 during the 1-s count period. The 870 would be displayed for a time, and then the counters would be reset to 000 and the frequency counted again. This *reset-count-display sequence* is repeated over and over.

Notice that the start/stop control pulse (count pulse) must be *very accurate*. Figure 12-18 shows how a count pulse can be generated by using an accurate known frequency, such as the 60 Hz from the power line. The 60-Hz sine wave is converted to a square wave by the wave-shaping circuit. The 60-Hz square wave triggers a counter that divides the frequency by 60. The output is a pulse*1 s* in length. This *count pulse* turns on the control circuit when it goes high and permits the unknown frequency to trigger the counters. The unknown frequency is applied to the counters for 1 s.

Remember that the frequency counter goes through the reset-count-display sequence. So far we have shown only the count part of this sequence. The *counter reset circuit* is a group of gates that reset, or clear, the decade counters to 000 at the correct time—just before the count starts. Next, the 1-s count pulse permits the counters to count upward. The count pulse ends, and the unknown frequency is *displayed* on the seven-segment displays. In this circuit, the frequency is displayed in hertz. It is convenient to leave this display on the LEDs for a time. To do this, the divide-by-10 counter sends a pulse to the control circuit, which turns off the count sequence for 9 s. Events then happen like this. *Reset* the counters to 000. *Count* upward for 1 s. *Display* the unknown frequency for 9 s with no counts. Repeat the reset-count-display procedure every 10 s.

The frequency counter in Fig. 12-18 measures frequencies from 1 to 999 Hz. Notice the extensive use of counters in the divide-by-60, divide-

**Counter reset circuit**

**Reset-count-display sequence**

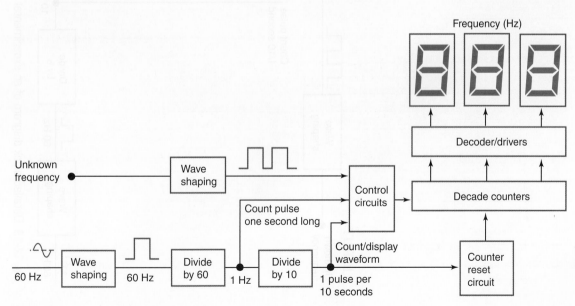

**Fig. 12-18** More detailed block diagram of a digital frequency counter.

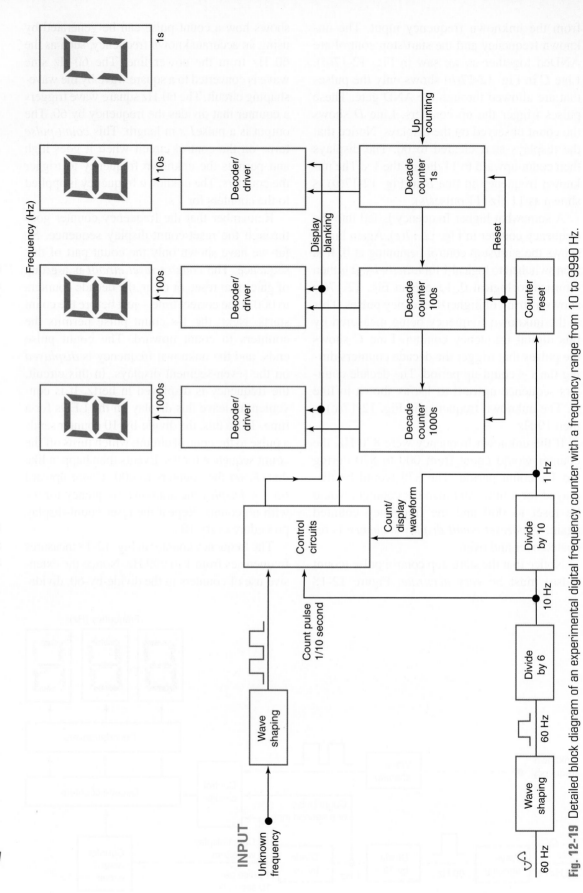

Fig. 12-19 Detailed block diagram of an experimental digital frequency counter with a frequency range from 10 to 9990 Hz.

by-10, and three decade counter circuits—hence the name frequency counter. The digital frequency counter actually counts the pulses in a given amount of time.

One limitation of the counter diagramed in Fig. 12-18 is its top frequency; the top frequency that can be measured is 999 Hz. There are two ways to increase the top frequency of our counter. The first method is to add one or more counter-decoder-display units. We could extend the range of the frequency counter in Fig. 12-18 to a top limit of 9999 Hz by adding a single counter-decoder-display unit.

The second method of increasing the frequency range is to count by 10s instead of 1s. This idea is illustrated in Fig. 12-19. A divide-by-6 counter replaces the divide-by-60 unit in our former circuit. This makes the *count pulse* only 0.1 s long. The count pulse permits only one-tenth as many pulses through the control as with the 1-s pulse. This is the same as counting by 10s. Only three LED displays are used. The 1s display in Fig. 12-19 is only to show that a 0

must be added to the right of the three LED displays. The range of this frequency counter is from 10 to 9990 Hz.

In the circuit in Fig. 12-19, the decade counters count upward for 0.1 s. The display is held on the LEDs for 0.9 s. The counters are then reset to 000. The count-display-reset procedure is then repeated. The circuit in Fig. 12-19 has one other new feature: during the count time the displays are blanked out. They are then turned on again when the unknown frequency is on the display. The sequence for this frequency counter is then reset, count (with displays blank), and, finally, the longer display period. This sequence repeats itself once every second while the instrument is being used.

The frequency counter diagramed in Fig. 12-19 is similar to one you can assemble in the laboratory from gates, flip-flops, and subsystems. It is suggested that you set up this complicated digital system because practical experience will teach you the details of the frequency counter system.

## ✔ Self-Test

*Supply the missing word in each statement.*

39. Refer to Fig. 12-17. This digital frequency counter counts the number of pulses passing through the AND gate in _____ s.

40. Refer to Fig. 12-17. If the start/stop control to the AND gate is LOW, the signal at output *C* is a _____ (HIGH, LOW, square wave).

41. Refer to Fig. 12-18. The wave-shaping blocks might be implemented using TTL _____ (XOR gates, Schmitt trigger inverters).

42. Refer to Fig. 12-18. The divide-by-60 block might be implemented using _____ (counters, shift registers).

43. Refer to Fig. 12-19. The count pulse is _____ s long in this frequency counter.

44. Refer to Fig. 12-19. The unknown frequency input waveform is conditioned by a _____ circuit before entering the control circuitry of the counter.

45. Refer to Fig. 12-19. The decade counters serve the dual purpose of counting upward and _____ the count for display.

## 12-8 An Experimental Frequency Counter

This section is based upon a frequency counter you can construct in the laboratory. Figure 12-20 is a detailed wiring diagram of that frequency counter. This instrument was purposely designed using only components you have already used earlier in the book. This experimental frequency counter is not as accurate or stable as commercial units. Its maximum frequency is also limited to 9990 Hz, and its inputs are somewhat primitive.

**Experimental frequency counter**

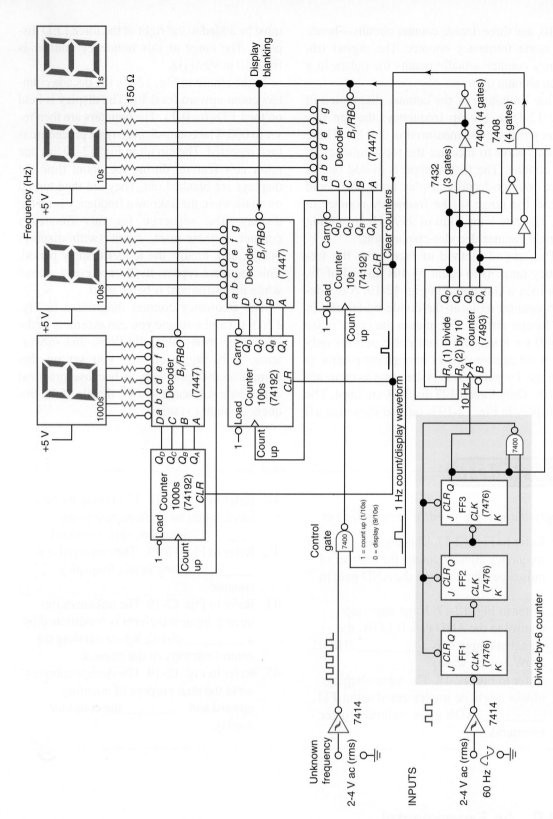

**Fig. 12-20** Wiring diagram for an experimental digital frequency counter.

The purposes for including the experimental frequency counter are as follows:

1. To show how SSI and MSI chips may be used to build digital subsystems and systems.
2. To demonstrate the concepts involved in the design and operation of a frequency counter.

Figure 12-19 is a block diagram of the frequency counter. Most components in the wiring diagram are in the same general position as in the block diagram.

At the lower left in Fig. 12-20, a 60-Hz sine wave is shaped into a square wave. The 60-Hz signal may come from the secondary of a low-voltage power transformer. The *wave shaping* is done by the 7414 Schmitt trigger inverter. This is the same unit we used earlier in Chapter 7 and in the digital clock to square up a sine wave. Remember that the divide-by-6 counter needs a square-wave input to operate properly.

To the right of the lower 7414 inverter is a *divide-by-6 counter*. Three flip-flops (FF1, FF2, and FF3) and a NAND gate are wired to form the mod-6 counter. The frequency going into the divide-by-6 counter is 60 Hz; the frequency coming out of the counter (at $Q$ of FF3) is 10 Hz. The 10 Hz is fed into the 7493 IC wired as a decade or *divide-by-10 counter*.

Figure 12-20 shows that the four outputs from the 7493 counter are NORed together (OR gate and inverter). The four-input NOR gate generates a 1-Hz signal. This 1-Hz signal is called the *count/display waveform*. The count/display waveform is HIGH for exactly 0.1 s and low for 0.9 s. The count/display waveform is fed back into the 7400 control gate. When the counter/display waveform is HIGH for 0.1 s, the unknown frequency passes through the NAND gate on to the clock input of the 10s counter. When the count/display waveform is LOW for 0.9 s, the unknown frequency is blocked from passing through the NAND control gate. It is during the 0.9 s that you may read the frequency off the seven-segment LED displays.

The frequency counter goes through a *reset-count-display sequence*. The *reset* pulse is generated by the five-input AND gate near the lower right in Fig. 12-20. It clears the 10s, 100s, and 1000s counters to zero. The reset (or counter clear pulse) is a very short positive pulse that occurs just before the counting occurs.

Next in the reset-count-display sequence is the *count* or *sampling time*. When the count/display waveform goes HIGH, the control gate is enabled and the unknown frequency passes through the NAND gate to the clock input of the 10s counter. Each pulse during this *sampling time* increments the 10s counter. When the 10s counter goes from 9 to 10, it carries the 1 to the 100s counter. After 0.1 s the count/display waveform goes LOW. This is the end of the sampling time. You will notice that the unknown frequency that was sampled causes the frequency to increase by 10s.

Last in the reset-count-display sequence is the *display time*. When the count/display waveform goes LOW the control gate is disabled. It is during this time that a stable frequency display may be read from the LEDs. Notice that an extra 1s display has been added in Fig. 12-20 to remind you that a 0 must be added to the right of the three active displays for a readout in hertz.

To make the displays look better, *display blanking* occurs during the count time of the reset-count-display sequence. The displays light normally with a stable readout during the display time. The display blanking waveform is a 0.1 s negative pulse generated by a 7404 inverter off the count/display waveform line. It causes the three displays to blank out for 0.1 s during the count time. The blanking does cause the displays to flicker. This problem could be cured by the use of latches to hold data on the inputs of the decoders.

**Wave shaping**

**Sampling time (count time)**

**Divide-by-6 counter**

**Divide-by-10-counter**

**Display blanking**

**Reset-count-display sequence**

For the most part commercial frequency counters operate like the one in Fig. 12-20. Commercial counters usually have more displays and read out in kilohertz and megahertz. The experimental frequency counter needs an input signal of about 3 to 8 V to make it operate. Commercial counters usually have an amplifier circuit before the first waveshaping circuit to amplify weaker signals to the proper level. Overvoltage protection is also provided with a zener diode. To get rid of the blinking of the display, commercial counters

usually use a slightly different method of storing and displaying the contents of the counters. We used the powerline frequency of 60 Hz as our known frequency. Commercial frequency counters usually use an accurate high-frequency crystal oscillator to generate their known frequency.

Some of the important specifications of commercial frequency counters are the *frequency range, input sensitivity, input impedance, input protection, accuracy, gate intervals,* and *display time.*

## ✔ Self-Test

*Supply the missing word(s) or number in each statement.*

46. Refer to Fig. 12-20. FF1, FF2, FF3, and the NAND gate form a(n) _____ counter.
47. Refer to Fig. 12-20. The count time for this frequency counter is _____ s, while the display time is _____ s.
48. Refer to Fig. 12-20. The sampling time of the frequency counter is also called the _____ (count, display) time.
49. Refer to Fig. 12-20. The five-input AND gate generates a counter clear or _____ pulse. This is a _____ (negative, positive) pulse.
50. Refer to Fig. 12-20. The display blanking pulse is generated during the _____ (count, display) time. This is a _____ (negative, positive) pulse.
51. Refer to Fig. 12-20. The 7414 IC is called a _____ _____ inverter. The

7414 inverters are used for _____ shaping in this circuit.
52. Refer to Fig. 12-20. Each pulse from the unknown frequency that reaches the counter increases the frequency reading by _____ (1, 10, 100) Hz.
53. Refer to Fig. 12-20. The frequency range of this experimental counter is from a low of _____ Hz to a high of _____ Hz.
54. Refer to Fig. 12-20. The gate interval for the experimental frequency counter is 0.1 s while the display time is _____ seconds.
55. Refer to Fig. 12-20. The known frequency entering the experimental counter is _____ Hz.
56. Refer to Fig. 12-20. The _____ (7447 decoder ICs, 74192 counter ICs) function as temporary memory devices to hold the highest count during the display time.

## 12-9 LCD Timer with Alarm

Most microwave ovens and kitchen stoves feature at least one timer with an alarm. Older appliances used mechanical timers, but modern microwaves and ranges feature electronic timers using digital circuitry. The concept of a timer system is sketched in Fig. 12-21(*a*). In

this system, the keypad is the input and both the digital display and alarm buzzer are the output devices. The processing and storage of data occur within the digital circuits block in Fig. 12-21(*a*).

A somewhat more detailed block diagram of a digital timer is shown in Fig. 12-21(*b*). The

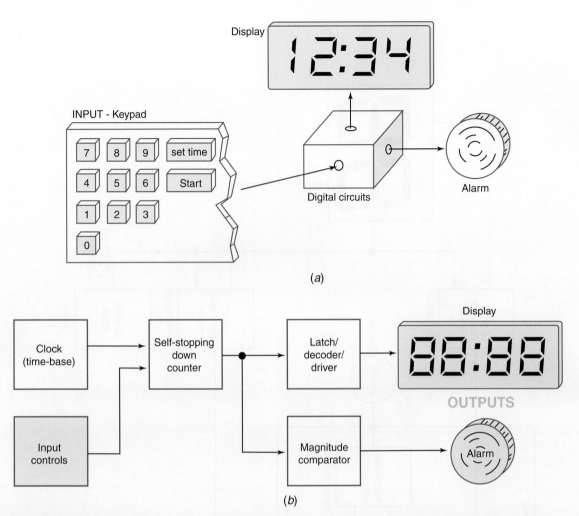

(a)

Display

OUTPUTS

(b)

**Fig. 12-21** Digital timer system. (*a*) Concept sketch of timer with alarm. (*b*) Simple block diagram of timer with alarm.

LCD timer circuit

Time-base clock

digital circuits block has been subdivided into four blocks. They are the *time-base clock,* the *self-stopping down counter,* the *latch/decoder/ driver,* and the *magnitude comparator.* The *input controls* block presets the time held in the down counter. The time base is an astable multivibrator which generates a known frequency. In this case, the signal is a 1-Hz square wave. The accuracy of the entire timer depends on the accuracy of the time-base clock. Activating the *start* input control causes the down counter to decrement. Each lower number is latched and decoded by the latch/decoder/driver. This block also drives the display.

The illustrations in Fig. 12-21 might be preliminary sketches used by a designer in visualizing a timer system. The designer might next decide on what type of input, output, and processing technologies to use to implement the system.

A somewhat more detailed block diagram of a digital electronic timer system is drawn in Fig. 12-22. The designer decided to use a two-digit LCD along with low-power CMOS ICs. This system was designed with your lab trainers in mind, so the inputs are logic switches to simplify the input section. The designer decided on seconds as the time interval. Notice that each block roughly corresponds to an MSI digital IC or input/output device. A wiring diagram could then be developed from the detailed block diagram in Fig. 12-22.

The block diagram in Fig. 12-22 represents an experimental LCD timer with alarm that you might construct in the laboratory. The timer is operated as follows:

1. Set the load/start control to 0 (load mode).
2. Load the 1s counter by setting a BCD number using the top four switches.

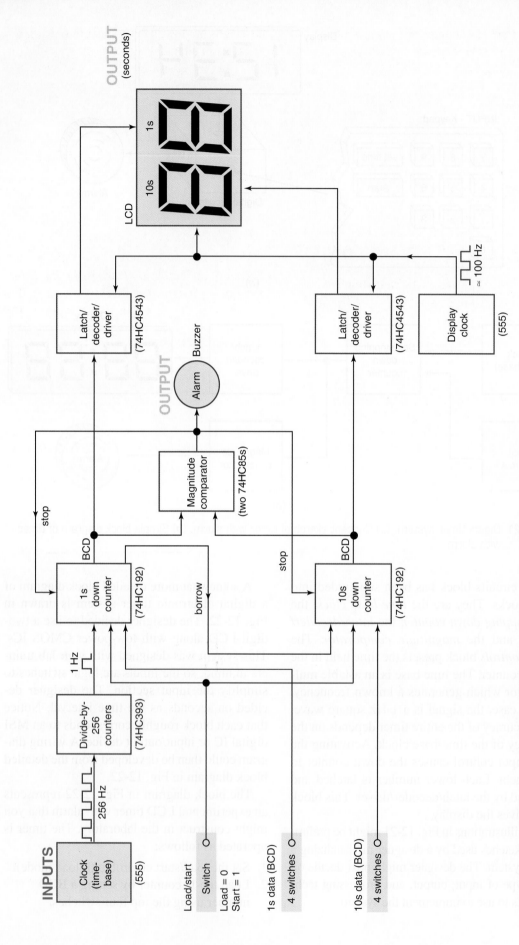

**Fig. 12-22** Detailed block diagram of an experimental LCD timer with alarm.

3. Load the 10s counter by setting a BCD number using the bottom four switches.
4. A two-digit number should now be displayed on the LCD.
5. Move the load-start control to 1 (start down count mode).

The timer will start counting downward in seconds. The LCD shows the time remaining before the alarm sounds. When both counters reach zero, the LCD reads 00 and the alarm will sound. The final step is to disconnect the power to the circuit to turn off the alarm.

A wiring diagram for the *experimental LCD timer circuit* is detailed in Fig. 12-23. Notice that each IC is placed in the same relative position on the wiring diagram as in the block diagram in Fig. 12-22.

Detailed operation of the LCD timer circuit in Figs. 12-22 and 12-23 follows.

## Time Base

The *time-base clock* is a 555 timer IC wired as a free-running MV. It is designed to generate a 256-Hz square wave. The time-base clock in this experimental timer is not very accurate or stable. It can be calibrated by adjusting the value of resistor $R_1$. The nominal value of $R_1$ should be about 20 kΩ.

The second part of the time base is the *divide-by-256-counter block*. The function of this block is to output a 1-Hz signal. The divide-by-256-counter block is actually two 4-bit counters wired together. Figure 12-24 shows the two 4-bit units wired as divide-by-16 counters. Note that the $\overline{CP}$ inputs are clock inputs and only the $Q_D$ outputs are used. The first divide-by-16 counter divides the frequency from 256 to 16 Hz ($256/16 = 16$ Hz). The second counter divides the frequency down to the required 1-Hz output ($16/16 = 1$ Hz).

## Self-Stopping Down Counters

The two 74HC192 decade counters are the 74HCXXX series equivalent to the 74192 TTL IC detailed in Chapter 8. When the load inputs to the 74HC192 counters are activated by a LOW, data at the data inputs ($A$, $B$, $C$, $D$) is immediately transferred into the counter's flip-flops. It then appears at the outputs of the counter ($Q_A$, $Q_B$, $Q_C$, $Q_D$). The data loaded should be in BCD (binary-coded decimal) form. When the load/start control goes HIGH,

the 1-Hz signal activates the count down input of the 1s counter. The count decreases by 1 on each L-to-H transition of the clock pulse. The *borrow out* output of the 1s down counter goes from L-to-H when the 1s counter goes from 0 to 9. This decrements the 10s counter. The down counters are actually wired as a self-stopping down counter because of the *counter stop line* fed back to the CLR input of both 74HC192 counters. When this line goes HIGH, both counters stop at 0000.

## 8-Bit Magnitude Comparator

The 74HC85 4-bit comparators are shown cascaded in Fig. 12-23 to form an *8-bit-magnitude comparator*. Their purpose in this circuit is to detect when the outputs of the counters reach $0000\ 0000_{BCD}$. When both counters reach zero, the output of the 8-bit magnitude comparator ($A = B_{out}$) goes HIGH. This serves two purposes. First, it stops both 74HC192 counters at 0000. Second, the HIGH at the output of the comparator turns transistor $Q_1$ on. This allows current to flow up through the transistor, sounding the buzzer. The diode across the buzzer suppresses transient voltages that may be generated by the buzzer.

## Decoder/Driver

The two 74HC4543 ICs used in the timer circuit serve three purposes. The functions of the 74HC4543 IC are summarized in Fig. 12-25. The latch disable (*LD*) input is permanently tied HIGH in the timer circuit (Fig. 12-23), which disables the latches. The BCD data flows through the latch to the BCD-to-seven-segment decoder. The decoder translates the BCD input to seven-segment code. Finally, the driver circuitry in the 74HC4543 chip energizes the correct segments on the LCD.

The *display clock* shown at the lower right in Fig. 12-23 generates a 100-Hz square wave. This is sent to the common (backplane) connection on the LCD and the Ph inputs of the 74HC4543 ICs. The LCD driver in the 74HC4543 chip sends inverted or 180° out-of-phase signals to the LCD segments that are to be activated. Segments that are not activated receive an in-phase square-wave signal from the LCD driver section of the 74HC4543 IC.

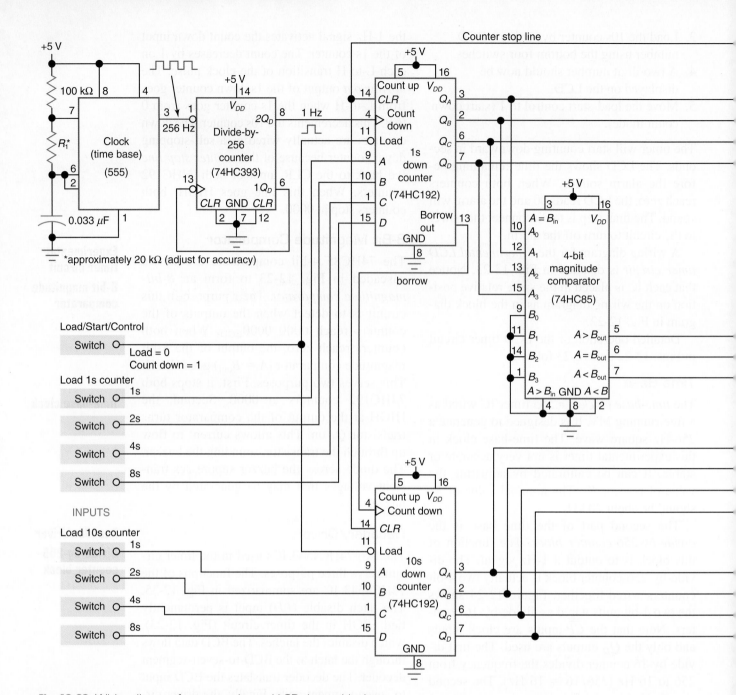

**Fig. 12-23** Wiring diagram for an experimental LCD timer with alarm.

**Experimental
LCD timer with
alarm**

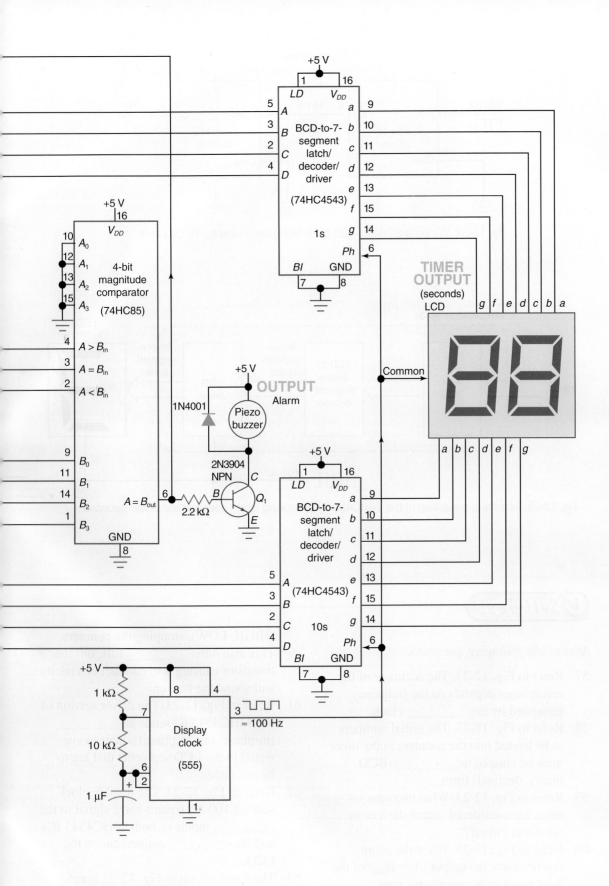

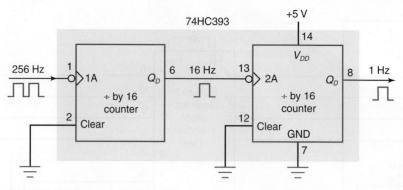

Fig. 12-24 Wiring a divide-by-256 block using two divide-by-16 counters.

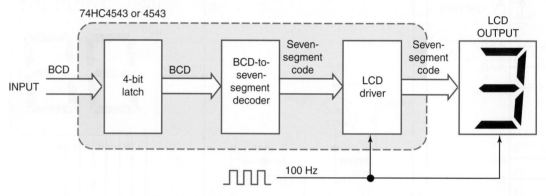

Fig. 12-25 Internal organization of the 74HC4543 IC including latch, decoder, and driver sections.

## Self-Test

*Answer the following questions.*

57. Refer to Fig. 12-23. The accuracy of the entire timer depends on the frequency generated by the _____ clock.

58. Refer to Fig. 12-23. The initial numbers to be loaded into the counters in the timer must be entered in _____ (BCD, binary, decimal) form.

59. Refer to Fig. 12-23. What two components are considered output devices in this timer circuit?

60. Refer to Fig. 12-23. When the count reaches zero, the output ($A = B_{out}$) of the 8-bit magnitude comparator goes _____ (HIGH, LOW). This causes the counter stop line to go _____

(HIGH, LOW), stopping the counters. This also turns _____ (off, on) the transistor causing it to conduct electricity and sound the buzzer.

61. Refer to Fig. 12-23. The driver section of the 74HC4543 IC sends an _____ (in-phase, out-of-phase) square-wave signal to the LCD segments that are to be activated.

62. Refer to Fig. 12-23. The display clock sends a 100-Hz square-wave signal to the _____ inputs of both 74HC4543 ICs and the _____ connection on the LCD.

63. The timer circuit in Fig. 12-23 is calibrated in (minutes, seconds, tenths of a second).

## 12-10 JTAG/Boundary Scan

It is important to be able to test digital systems and subsystems for proper operation, both when manufactured and in the field. Semiconductor manufacturers continue to increase the complexity of digital circuitry built into ICs. Many ICs contain complete digital systems on a single chip. Other technologies such as surface mount technology and multilevel pc boards have both increased the number of components on printed-circuit boards and shrunk their size. This miniaturization has resulted in a loss of access points for testing system and subsystem operation.

In the mid-1980s, the Joint Test Action Group developed a solution to this problem of loss of test point access on pc boards. The solution they developed was a new testing architecture that added testing ability and test access points into integrated circuits. The Institute for Electrical and Electronic Engineers (IEEE) later formalized this solution as Standard 1149.1 Test Access Port and Boundary-Scan Architecture. The boundary-scan architecture is commonly referred to as *boundary scan* or *JTAG*. JTAG is in reference to the group (Joint Test Action Group) that developed the system. Some ICs and printed circuit boards now have this additional testing subsystem built into them.

A simplified JTAG compliant IC is shown in Fig. 12-26. The main JTAG elements are shown in yellow, pink, and red. The black lines in the drawing represent the IC's regular input and output lines. The red lines at the bottom of Fig. 12-26 are the chip's *test access port (TAP)*. The four solid red lines, *TDI (test data input), TDO (test data output), TMS (test mode select),* and *TCK (test clock),* provide a standard 4-wire serial interface for test access to the chip. The dashed line represents the *test reset (TRST)* input, which is an optional fifth wire for resetting the test access port. The other required boundary-scan elements are the instruction register, the bypass register, the TAP controller, and at least one test data register. JTAG compliant ICs may have more than one test data register, however, one of the test data registers must be the boundary-scan register. The boundary-scan register is made up of a series of *boundary-scan cells (BSCs)*. The ten square yellow boxes in Fig. 12-26 represent the boundary-scan cells that make up the boundary-scan register in this example IC. The right side of Fig. 12-26 shows an exploded view of a boundary-scan cell. Notice that a boundary-scan cell may consist simply of two 2-input multiplexers and two flip-flops. The chip's TAP controller controls

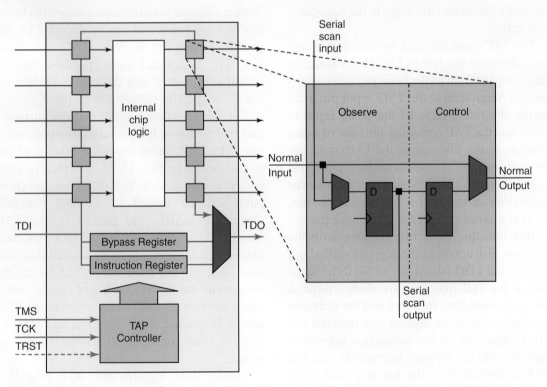

**Fig. 12-26** JTAG compliant IC.

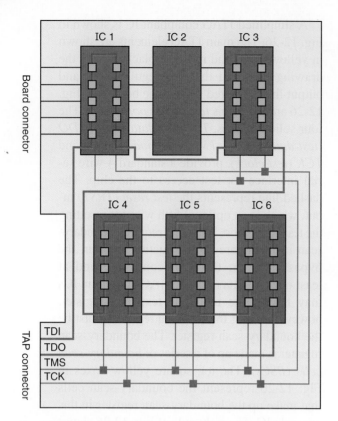

**Fig. 12-27** Simplified pc board with six ICs. Five of the ICs are JTAG compliant using boundary-scan architecture.

the multiplexers and flip-flops in the boundary-scan cells.

The TAP controller and the *instruction register* shown at the left in Fig. 12-26 make up the control section of the boundary-scan architecture. Serial data at the TMS input pin, read during the rising edge of the clock input to TCK, sets the TAP controller into one of many possible *states*. The state of the TAP controller determines if (1) the boundary-scan test system is reset, (2) if the test specified by the instruction in the instruction register is run, (3) if the serial data at the TDI input pin are shifted into the instruction register, with the previous instruction in the register shifted out through the TDO output pin, or (4) if the serial data at the TDI input pin are shifted through one of the test data registers and the previous data in the register shifted out through the TDO output pin. If the instruction register is loaded with the bypass instruction and the TAP controller is in the run test state, then data at the TDI input pin instead pass through

the 1-bit bypass register. The 1-bit bypass register shortens the shifting path of data from other ICs through the chip when the chip is not part of the test being run on a circuit board containing many JTAG ICs. This can significantly speed up the testing of systems and subsystems on pc boards.

JTAG compliant ICs must also implement two other required instructions. The *extest (external test)* instruction allows testing of components and interconnections external to the chip to be tested without the risk of damaging the internal chip logic. When the extest instruction is run, boundary-scan cells at the chip's input pins sample the incoming data, and the BSCs at the output pins output their data to the chip's output pins. During the extest instruction, the TAP controller may prevent the data at the chip's input pins from reaching the internal chip logic in order to prevent damage to the IC. The other required instruction, sample/preload, has two functions. The first function is to take a *snapshot,* or sample, of the data flowing into or out of the internal chip logic without interrupting the normal operation of the chip. The second function of the sample/preload instruction is to preload known test data into the boundary-scan cells of the IC before running other tests. Semiconductor manufacturers may also build in other test types that can be decoded by the instruction register in their JTAG compliant ICs. The additional tests specified by the manufacturer may test the entire system on the IC or specific subsystems of the IC.

Fig. 12-27 shows a simplified circuit board that makes use of the boundary-scan architecture. The TAP connector at the bottom of the circuit board in Fig. 12-27 is typically connected to a computer. Test data from the computer can be loaded in parallel, possibly through a UART, and passed to the TAP serially. Serial data out from TDO (test-data output) are converted back to parallel data and read back into the computer. All of the digital subsystems on this circuit board can be tested using automated tests run by a computer. This testing is quicker, more accurate, and less expensive than conventional testing of circuit boards.

The yellow squares on IC1 and IC3 through IC6 in Fig. 12-27 represent the boundary-scan cells (BSCs) built into the chips, the same as they are represented in the single IC shown in Fig. 12-26. The boundary-scan cells can be used to control or observe the values at each IC's inputs and outputs. Serial test data are shifted in through test data input (TDI) on the rising edge of the clock input at TCK. Serial output is observed at the test data output (TDO) on the falling edge of the clock pulse.

Notice that this circuit board has five JTAG ICs and one non-JTAG IC. The placement of JTAG ICs IC1 and IC3 on either side of the non-JTAG IC allows IC2 (the non-JTAG IC) to be tested for proper operation. To test IC2 for proper system operation, a series of input test data values known as *test vectors* is loaded into the output boundary-scan cells of IC1. IC1 and IC3 are then instructed to run the extest instruction, and the output of IC2 is observed at the input BSCs of IC3. Comparison of the observed inputs at IC3 to the expected outputs of

IC2 will confirm the proper operation of IC2. Some other types of tests that might be performed on this circuit board include (1) verifying the proper operation of each of the JTAG compliant ICs on the board (IC1, IC3–IC6), (2) checking the interconnections (*nets*) between IC4 and IC5 as well as between IC5 and IC6, and (3) observing how the entire system is reacting to the normal system inputs from the board connector.

To summarize, testing is an important part of any complex system. The loss of test access points on printed circuit boards has made boundary scan an important subsystem of many digital systems. The boundary-scan architecture makes testing and correcting problems in digital systems both quicker and easier. Boundary scan is likely to become more widely used as digital systems continue to get more complicated and testing of those systems becomes more difficult because of their miniature size.

## Self-Test

*Answer the following questions.*

64. Boundary-scan testing is commonly known as _____ (four letters) referring to the name of the original group that worked on this method.

65. A series of test values used during a boundary-scan test procedure might be called _____ (arguments, test vectors).

66. The nickname for interconnections in the vocabulary of boundary-scan technology would be _____ (ICs, nets).

67. The test access port input/output connection to a JTAG-compliant pc board uses the acronym _____ (three letters).

68. JTAG is an acronym for _____ in the field of boundary-scan technology.

69. JTAG compliant pc boards are soaked in a special varnish to make them resistant to chemical vapors and moisture. (T or F)

70. JTAG compliant pc boards are becoming more important as digital ICs become _____ (less complex, more complex) and _____ (larger in size, smaller in size).

71. Refer to Fig. 12-27. The small yellow squares on IC1 and IC3–IC6 are called _____ (boundary-scan cells, carbon-zinc cells).

72. On a printed circuit board containing both JTAG and non-JTAG ICs, it may be possible to test non-JTAG ICs for proper operation using boundary scan. (T or F)

## Summary

1. An assembly of digital subsystems connected correctly forms a digital system.

2. Digital systems have six common elements: input, transmission, storage, processing, control, and output.

3. Manufacturers produce ICs that are classified as small-, medium-, large-, very large-, and ultra-large-scale integrations.

4. Electronic games are popular construction projects. Many are simulations of older games like throwing dice.

5. A digital clock and a digital frequency counter are two closely related digital systems. Both make extensive use of counters.

6. Many LSI digital clock chips are available. Most clock ICs need other components to produce a working digital clock.

7. Multiplexing is a commonly used method of driving seven-segment LED displays.

8. All digital systems are basically constructed from AND gates, OR gates, and inverters.

9. A frequency counter is an instrument that accurately counts input pulses in a given time interval and displays it in digital form. It constantly cycles through a reset-count-display sequence.

10. Block diagrams communicate the organization of a digital system. The most detailed block diagrams break the system down to the chip level.

11. The IEEE Standard 1149.1 Test Access Port and Boundary-Scan Architecture (commonly known as either JTAG or boundary scan) specifies standards for imbedding testing and access points in complex, miniature, high-density ICs and pc boards. Testing can be automated for quality-control testing and field troubleshooting.

## Chapter Review Questions

*Answer the following questions:*

12-1. List the six elements found in most digital systems.

12-2. What do the following letters stand for when referring to ICs?
   a. IC    c. MSI    e. VLSI
   b. SSI   d. LSI    f. ULSI

12-3. The term "chip" usually is taken to mean a(n) _____ (IC, sliver of plastic) in digital electronics.

12-4. A _____ (computer, digital wristwatch) is usually based upon a single LSI IC.

12-5. Refer to Fig. 12-3. When the push button _____ is (closed, opened), the display will stop and indicate a random number from 1 to _____ (number) simulating the roll of a single die.

12-6. Refer to Fig. 12-3. This circuit uses _____ (CMOS, TTL) ICs.

12-7. Refer to Fig. 12-3. If a "1" shows on the seven-segment LED display, then outputs _____ (letters) of the 7447 IC are activated with a _____ (HIGH, LOW).

12-8. Refer to Fig. 12-3. When the 74192 IC tries to count upward from 110 to 111, the NAND gate is activated driving the load input _____ (HIGH, LOW). This immediately loads _____ (binary number) into the counter's flip-flops.

12-9. Refer to Fig. 12-5. Two _____ trigger NAND gates and associated resistors and capacitors form the clock section of this digital dice game.

12-10. Refer to Fig. 12-5. Grounding pin 10 of the 4029 IC converts this unit to a(n) _____ (down, up) counter.

12-11. Refer to Fig. 12-5. When the counter's outputs are 110 (HHL), LEDs _____ light. This is caused by the output of the NAND gate going _____ (HIGH, LOW) and transistor $Q_2$ being turned _____ (on, off).

12-12. A bilateral switch is also called a(n) _____ gate.

12-13. A digital clock makes extensive use of _____ (counter, shift register) subsystems.

12-14. A known frequency is the main input to a digital _____ (clock, frequency counter) system.

12-15. Counters are used for counting upward and _____ (shifting data, storing data) in the digital clock system.

12-16. The National Semiconductor MM5314 clock chip _____ (directly drives, multiplexes) the output displays.

12-17. The multiplex oscillator's frequency in Fig. 12-11(a) is set by _____ (connecting an external capacitor and resistor to the correct IC pins; the factory and cannot be changed).

12-18. In practice, the segment driver block shown in Fig. 12-13(b) may consist of _____ (a VLSI chip, seven transistors with associated resistors).

12-19. The multiplexed displays in Fig. 12-15 are _____ (all turned on and off together to save power; turned on and off one at a time in rapid succession).

12-20. The *known frequency* entering the MM5314 clock chip in Fig. 12-15 is _____ Hz. This frequency comes from the _____ (oscillator, transformer).

12-21. Counters are used for counting upward and _____ (counting downward, dividing frequency) in a digital frequency counter.

12-22. The three J-K flip-flops (FF1, FF2, FF3) and the NAND gate in Fig. 12-20 function as a _____ (down counter, frequency divider).

12-23. The 7408 AND gate in Fig. 12-20 serves to _____ (clear, inhibit) the counters.

12-24. The frequency counter in Fig. 12-20 counts from a low of _____ Hz to a high of _____ Hz.

12-25. What IC(s) are being used as wave-shaping circuits in the frequency counter in Fig. 12-20?

12-26. Refer to Fig. 12-20. The unknown frequency is allowed to pass through the control gate for 0.1 s when the count/display waveform goes _____ (HIGH, LOW).

12-27. Refer to Fig. 12-20. The displays are blanked during the _____ portion of the count/display waveform.

12-28. Refer to Fig. 12-23. List two ICs that form the time-base clock section of the LCD timer.

12-29. Refer to Fig. 12-23. List the IC(s) that detect when the count of the timer reaches 00.

12-30. Refer to Fig. 12-23. When the count on the timer reaches 00, the output of the magnitude comparator goes _____ (HIGH, LOW). This turns on transistor $Q_1$, sounding the alarm, and activates the _____ _____ line.

12-31. Refer to Fig. 12-23. When the LCD reads 88, the signals on all of the lines from the 74HC4543 drivers to the displays are _____ (in phase, 180° out of phase) with the signal at the output of the display clock.

12-32. Refer to Fig. 12-23. The accuracy of the entire timer depends on the accuracy of the _____ clock.

12-33. Refer to Fig. 12-21(b). A commercial timer would probably use a(n) _____ controlled oscillator (astable MV) for the time-base clock to assure maximum accuracy.

12-34. Refer to Fig. 12-23. The 74HC4543 ICs have the _____ (decoder, driver, latch) section of the chip disabled in this circuit.

12-35. In common usage, JTAG is also referred to as _____ (boundary scan, joule thermal agent) in the field of digital electronics.

12-36. A testing architecture that includes automated testing and test access points for miniaturized complex pc boards is covered under IEEE _____ (Standard 1149.1, Standard 2000), also sometimes called JTAG.

12-37. Boundary-scan technology is an aftermarket item that can be added to any complex pc board after it has been manufactured. (T or F)

12-38. A JTAG compliant IC would have boundary-scan cells imbedded in the chip. (T or F)

12-39. JTAG is an acronym for _____ in the field of boundary-scan technology.

12-1. List at least five common pieces of equipment that are considered digital systems.

12-2. List at least four devices you used or studied about that are considered digital subsystems.

12-3. How are the segment drivers shown in block form in Fig. 12-13 implemented in a working digital clock (see Fig. 12-15)?

12-4. The oscillator shown at the lower left in Fig. 12-13 is associated with what function within the digital clock?

12-5. Why was the experimental frequency counter shown in Fig. 12-20 included for study when it is not a practical piece of equipment?

12-6. What are some differences between the conceptual version of the digital timer in Fig. 12-21 and the working experimental timer in Fig. 12-23?

12-7. Why would the digital dice game shown in Fig. 12-5 probably be preferred over the simpler version in Fig. 12-3?

12-8. Refer to Fig. 12-5. When the preset pulse line goes _____ (HIGH, LOW), PNP transistor $Q_1$ turns on and the preset enable input to the 4029 counter is activated with a _____ (HIGH, LOW).

12-9. Refer to Fig. 12-5. When the counter's outputs are 100 (HLL), LEDs _____ light. The bilateral switches are closed because of the _____ (HIGH, LOW) on their control inputs. Only transistor _____ ($Q_3$, $Q_4$) is turned on grounding the cathode of LED $D_5$.

## Answers to Self-Tests

1. control
2. input
3. data or digital data
4. 12 to 99
5. 10,000
6. 1960s
7. microcontroller IC
8. microprocessor
9. astable (free-running)
10. LOW, 0001
11. 1, 2, 3, 4, 5, 6
12. 4093
13. 110, 101, 100, 011, 010, 001
14. $Q_5$
15. $D_2$ and $D_3$, closed, $Q_4$
16. bilateral switch or transmission gate
17. frequency divider, count accumulator
18. wave-shaping

19. wave shaping
20. control gate (start/stop control)
21. MOS
22. 60
23. prescale
24. 11 to 19
25. multiplexes
26. $V_{SS}$
27. has internal
28. resistor and capacitor
29. multiplexing
30. digit drivers
31. segment drivers
32. 12
33. $C3$ and $R4$
34. $V_{DD}$
35. 100
36. $C_1$
37. 60 Hz pulsating dc or ac

38. PROM
39. 1.0
40. LOW
41. Schmitt trigger inverters
42. counters
43. 0.1
44. wave-shaping
45. storing or accumulating
46. divide-by-6
47. 0.1, 0.9
48. count
49. reset, positive
50. count, negative
51. Schmitt trigger, wave
52. 10
53. 10, 9990
54. 0.9
55. 60
56. 74192 counter ICs

57. time-base
58. BCD
59. piezo buzzer, LCD (liquid-crystal display)
60. HIGH, HIGH, on
61. out-of-phase
62. Ph (phase), common (back plane)
63. seconds
64. JTAG
65. test vectors
66. nets
67. TAP
68. Joint Test Action Group
69. F
70. more complex, smaller in size
71. boundary-scan cells
72. T

# Chapter 13

# Computer Systems

## Chapter Objectives

*This chapter will help you to:*

1. *Diagram* the general organization of a computer and a microcomputer and *detail* the execution of a program.

2. *Analyze* the operation of a simple microcomputer address decoding system.

3. *Discuss* several aspects of both serial and parallel transmission.

4. *Answer* selected questions about error detection and correction techniques.

5. *Detail* the transmission of data through a computer system.

6. *List* some applications of a microcontroller.

7. *Compare* characteristics of various BASIC Stamp modules and *summarize* the programming of these ICs.

8. *Analyze* a PBASIC program used to drive a BASIC Stamp module.

9. *Draw* a basic block diagram of the sections of a DSP (digital signal processing) system including an A/D converter, memory, DSP section, and D/A converter, and *summarize* the action of each block.

10. *Classify* microprocessor, microcontroller, and digital signal processing systems.

A desktop or laptop personal computer is probably the first device you think of when you think of a computer system. However, computer systems are found in many other devices that we use every day. Automobile engines have embedded computer chips to make them run more efficiently and to help mechanics diagnose problems. Digital cameras, video cameras, and MP3 players have computer processors in them to compress and decompress video and audio signals. Even your cellular phone uses computer chips to send and receive your telephone calls, take pictures, play ringtones, and be a modem for a personal computer. This chapter examines several types of computer systems, some digital subsystems of computer systems, and data transmission.

## 13-1 The Computer

The most complex digital systems include *computers.* Most digital computers can be divided into the five functional sections shown in Fig. 13-1. The input device may be a keyboard, mouse, joystick, graphics tablet, card reader, magnetic tape unit, scanner, network connection, or telephone line. This equipment lets us pass information from *person to machine* (or machine to machine). The input device often must *encode* human language into the binary language of the computer.

The memory section is the storage area for both data and programs. This storage can be supplemented by storage outside the processing unit.

The arithmetic unit is what most people think of as being inside a computer. The arithmetic unit adds, subtracts, multiplies, divides, compares, and does other logic functions.

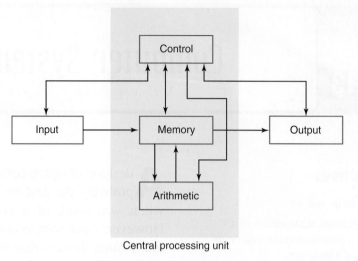

**Central processing unit (CPU)**

**Fig. 13-1** Sections of a digital computer.

Notice that a two-way path exists between the memory and arithmetic sections. In other words, data can be sent to the arithmetic section for action and the results sent back to storage in the memory. The arithmetic unit is sometimes referred to as the ALU (arithmetic-logic unit).

The control section is the nervous system of the computer. It directs all other sections to operate in the proper order and tells the input when and where to place information in the memory. It directs the memory to route information to the arithmetic section and tells the arithmetic section to add. It routes the answer back to the memory and to the output device. It tells the output device when to operate. This is only a sampling of what the control section can do.

The output section is the link between the *machine and a person* (or to a device or network).It can communicate to humans through a printer. It can output information on a CRT or LCD display. Output information can also be placed on bulk storage devices, such as magnetic tape, disk, or optical discs. The output section often must *decode* the language of the computer into human language.

**Computer organization**

The three middle blocks of Fig. 13-1 are often called the CPU. The arithmetic and memory sections and most of the control section are frequently found on a single circuit board. Devices located outside the CPU are often called *peripheral devices.*

**Peripheral devices**

The block diagram of the computer in Fig. 13-1 could well be the diagram for a calculator.

Up to this point the basic systems operate the same. The basic difference between the calculator and computer is *size* and the use of a *stored program* in a computer. Computers are also faster and are multipurpose machines. Figure 13-2 shows that two types of information are put into a computer. One is the program (instructions) telling the control unit how to proceed in solving the problem. This program, which has to be carefully written by a programmer, is stored in the central memory while the problem is being solved. The second type of information fed to the computer is *data,* to be acted on by the computer. Data include the facts and figures needed to solve the problem. Notice that the program information is placed in storage in the memory and used only by the control unit. The data information, however, is directed to various positions within the computer and is processed by the ALU. The data need never go to the control unit. The auxiliary memory is extra memory that may be needed to store partial results in some complex problems. It may not be located in the CPU. Data may be stored in peripheral devices such as a hard drive.

In summary, the computer is organized into five basic functional sections: input, memory, control, ALU, and output. Information fed into the CPU is either program instructions or data to be acted upon. The computer's stored program and size make it different from a calculator.

Computers, one of the most complex of digital systems, are not covered in depth in this section.

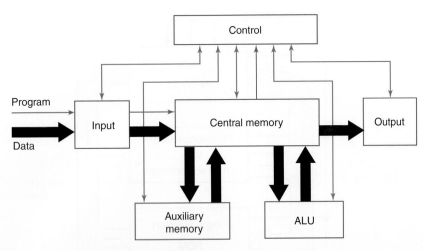

**Fig. 13-2** Flow of program instructions and data in a computer system.

There are entire books about the organization and architecture of computers. Remember, however, that all the circuits in the digital computer are constructed from logic gates, flip-flops, memory cells, and subsystems such as the ones you have studied.

*Answer the following questions.*

1. Devices located outside the computer's CPU are often called _____ devices.

2. List several of the fundamental differences between a computer and a calculator.

3. List the two types of information fed into a digital computer.

## 13-2 The Microcomputer

Computers have been in general use since the 1950s. Formerly, digital computers were large, expensive machines used by governments or large businesses. The size and shape of the digital computer have changed in the past decades as a result of a device called the microprocessor. The *microprocessor* (*MPU,* for "microprocessing unit") is an IC that contains much of the processing capabilities of a larger computer. The MPU is a small but extremely complex VLSI device that is *programmable.* The MPU IC forms the heart of a microcomputer. The *microcomputer* is a *stored-program digital computer.*

The organization of a typical smaller microcomputer system is diagramed in Fig. 13-3. This microcomputer contains all the five basic sections of a computer: the *input* unit, the *control* and *arithmetic* units contained within the MPU, the *memory* units, and the *output* unit.

**Dennis C. Hayes**
In 1978, 28-year-old Dennis C. Hayes formed what became Hayes Microcomputer Products, Inc. When he began, Hayes hand-assembled and soldered products on a borrowed dining room table in his home. Then, in 1981 the Hayes™ Smartmodem™, which was easily integrated into the computer environment, started a communications revolution.

Microprocessor
(MPU)

Stored-program
digital computer

Microcomputer
organization

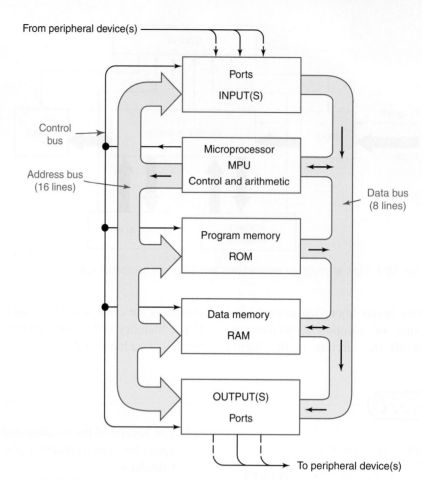

**Fig. 13-3** Block diagram of a microcomputer system.

The MPU controls all the units of the system using the control lines shown at the left in Fig. 13-3. Besides the control lines, the *address bus* (16 parallel conductors) selects a certain memory location, input port, or output port. The *data bus* (eight parallel conductors) on the right in Fig. 13-3 is a *two-way path* for transferring data into and out of the MPU. It is important to note that the MPU can send data to memory or an output port or receive data from memory or an input port.

The microcomputer's ROM commonly contains a program. A *program* is a list of specially coded instructions that tell the MPU *exactly* what to do. The ROM in Fig. 13-3 is the place where the program resides in this example. In actual practice, the ROM contains a start-up or initializing program and perhaps other programs. Separate programs can also be loaded into RAM from auxiliary memory. These are user programs.

The RAM area in Fig. 13-3 is identified in this example as the data memory. Data used in the program reside in this memory.

The CPU and memory sections of the microcomputer are not very useful by themselves.

The CPU must be interfaced with *peripheral devices* for input, output, and storage. Typical peripheral devices used for input, output, and storage on modern microcomputers are diagramed in Fig. 13-4. The keyboard, mouse, and joystick are probably the most common input devices connected to most microcomputers. Several other input devices connected to microcomputers are shown at the left in Fig. 13-4.

The optical disc drive is a popular secondary storage device connected to most microcomputers. Other popular secondary storage devices interfaced with microcomputers are hard and floppy disk drives and to a lesser extent tape drives. The CRT or LCD monitor, printer, and sound systems are the most common output peripheral devices used with typical microcomputers. Other output devices include TVs, plotters, and laser printers.

Computer connections to networks such as the Internet are almost universal. Individual computer users commonly implement these network connections using a *modem (modulator/ demodulator)* which communicates with an

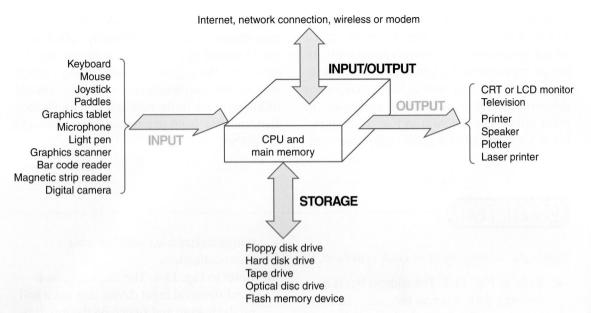

Internet, network connection, wireless or modem

**INPUT/OUTPUT**

OUTPUT

INPUT

CPU and
main memory

Keyboard
Mouse
Joystick
Paddles
Graphics tablet
Microphone
Light pen
Graphics scanner
Bar code reader
Magnetic strip reader
Digital camera

CRT or LCD monitor
Television
Printer
Speaker
Plotter
Laser printer

**STORAGE**

Floppy disk drive
Hard disk drive
Tape drive
Optical disc drive
Flash memory device

**Fig. 13-4** Peripheral devices commonly attached to the CPU of a microcomputer.

**Microcomputer peripheral devices**

Internet service provider over regular home telephone or cable TV lines. The modem is classified as an *input/output* peripheral device in Fig. 13-4. The modem provides two-way communication over the public network called the internet. The modem serves as an output device when transmitting data and an input device when receiving data. The *Internet* is a huge network that links millions of computers world wide. Users of the Internet can find and exchange information, buy and sell products, and even play games.

Another method of securing a connection to the Internet is using a *DSL (digital sub-*

*scriber line)*. A DSL line may be 10 to 100 times faster than a modem and is commonly used by some individuals, telecommuters, and small businesses. Cable TV companies can also provide high-speed internet service in many areas.

Larger organizations commonly use *LANs (local-area networks)* for two-way communication within a building or campus. LANs use private lines and may use one of several protocols for two-way communication between the desk computers and a server (computer with more processing power and memory). A LAN is shown at the left in Fig. 13-5 showing

**LAN**

**DSL**

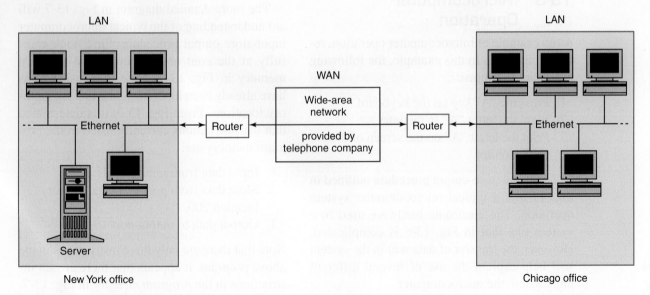

**Fig. 13-5** Use of both LANs and WANs for business communication.

computers connected to the server using the *Ethernet* protocol. A server might be called on for processing, to retrieve/update a file, or for an application. Larger organizations with remote offices may use a *WAN (wide-area network)* to communicate with computers in other cities. The diagram in Fig. 13-5 shows the WAN supplied by a public telephone com-

pany being used for two-way business communication. The *router* (formerly called gateways) shown in Fig. 13-5 is a device that determines the optimal path along which long-range communications traffic should follow to reach its destination. The router may also translate from one transmission protocol to another.

---

**✓ Self-Test**

---

*Supply the missing word in each statement.*

4. Refer to Fig. 13-3. The address bus is a one-way path, whereas the _____ bus is a two-way pathway for information.
5. Refer to Fig. 13-3. The ROM typically holds _____ (data, programs).
6. Refer to Fig. 13-3. The exact memory location, or input/output port, is selected by the MPU's output on the _____ bus and _____ bus.
7. A(n) _____ (modem, scanner) is an input/output peripheral device that enables the microcomputer to send and receive data over telephone lines.
8. Refer to Fig. 13-4. The _____ is probably the most popular peripheral

output device used with low-cost microcomputers.
9. Refer to Fig. 13-4. The _____ is a hand-operated input device that has a ball on the bottom and switch on the top. It is used to control the direct movement of the cursor on the CRT screen.
10. An organization's computer network within a single building that might include a server and the Ethernet protocol is referred to as a _____ (LAN, WAN).
11. A method of accessing the Internet used by small businesses and some individuals that is 10 to 100 times faster than a modem might be a _____ (DSL, SPL).

---

## 13-3 Microcomputer Operation

As an example of microcomputer operation, refer to Fig. 13-6. In this example, the following things are to happen:

1. Press the "A" key on the keyboard.
2. Store the letter "A" in memory.
3. Print the letter "A" on the screen of the CRT monitor.

The input-store-output procedure outlined in Fig. 13-6 is a typical microcomputer system operation. The electronic hardware used in a system like that in Fig. 13-6 is complicated. However, the transfer of data within the system will help explain the use of several different units within the microcomputer.

The more detailed diagram in Fig. 13-7 will aid understanding of the typical microcomputer input-store-output procedure. First, look carefully at the *contents* section of the program memory in Fig. 13-7. Note that instructions have already been loaded into the first six memory locations. From Fig. 13-7, it is determined that the instructions currently listed in the program memory are:

1. Input data from input port 1.
2. Store data from port 1 in data memory location 200.
3. Output data to output port 10.

Note that there are only three instructions in the above program. It appears that there are six instructions in the *program memory* in Fig. 13-7.

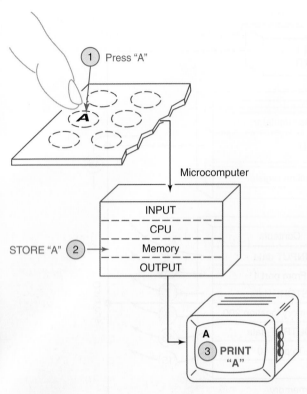

Fig. 13-6 An example of a common input-store-output microcomputer operation.

The reason for this is that instructions are sometimes broken into parts. The first part of instruction 1 above is to input data. The second part tells where the data come from (from port 1). The first, *action* part of the instruction is

called the *operation* and the second part the *operand*. The operation and operand are located in separate memory locations in the program memory in Fig. 13-7. For the first instruction in Fig. 13-7, program memory location 100 holds the input operation while memory location 101 holds the operand (port 1) telling where information will be input from.

Two new sections are identified inside the *MPU* in Fig. 13-7. These two sections are called registers. These special registers are the *accumulator* and the *instruction register.*

The sequence of events happening within the microcomputer in the input-store-output "A" example is outlined in Fig. 13-7. The flow of instructions and data can be followed by keying on the circled numbers in the diagram. Remember that *the MPU is the center of all data transfers and operations.* Refer to Fig. 13-7 for all steps below.

1. The MPU sends out address 100 on the address bus. A control line *enables* the read input on the program memory IC. This step is symbolized in Fig. 13-7 by the encircled number 1.

2. The program memory sends the first instruction (input data) on the data bus, and the MPU receives this coded message. The instruction is transferred to a special memory location within

**Parts of instruction: operation and operand**

**Accumulator**

**Instruction register**

**MPU**

## ABOUT ELECTRONICS

Microcontrollers are small inexpensive "computers on a chip" containing a CPU, RAM, ROM, and input/outputs. The average American interacts with microcontrollers hundreds of times each day in appliances, computers, telephones, security systems, televisions, thermostats, radios, automobiles, "smart cards," and other products.

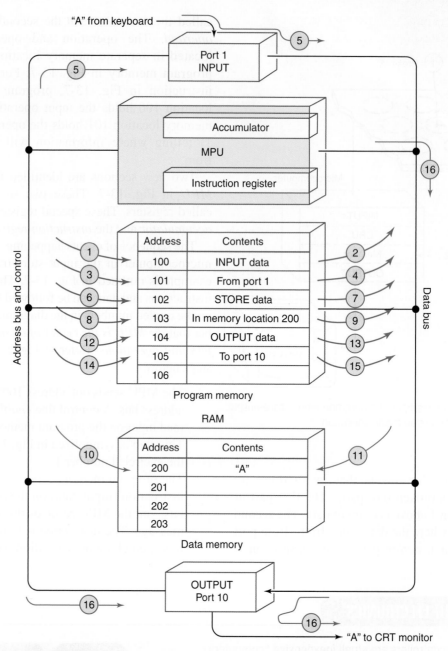

"A" from keyboard

Port 1
INPUT

5

5

Accumulator

MPU

Instruction register

16

Address bus and control

| Address | Contents |
|---------|----------|
| 100 | INPUT data |
| 101 | From port 1 |
| 102 | STORE data |
| 103 | In memory location 200 |
| 104 | OUTPUT data |
| 105 | To port 10 |
| 106 | |

1  3  6  8  12  14

2  4  7  9  13  15

Program memory

Data bus

RAM

| Address | Contents |
|---------|----------|
| 200 | "A" |
| 201 | |
| 202 | |
| 203 | |

10

11

Data memory

OUTPUT
Port 10

16

16

"A" to CRT monitor

**Fig. 13-7** Sequence of microcomputer operations in executing the input-store-output program.

the MPU called the instruction register. The MPU *decodes*, or interprets, the instruction and determines that it needs the operand to the input data instruction.

3. The MPU sends out address 101 on the address bus. The control line enables the read input of the program memory.
4. The program memory places the operand (from port 1) on the data bus. The

operand was located at address 101 in program memory. This coded message (the address for port 1) is received off the data bus and transferred to the instruction register. The MPU now decodes the entire instruction (input data from port 1).

5. The MPU uses the address bus and control lines to the input unit to cause port 1 to open. The coded form for "A"

is transferred to and stored in the accumulator of the MPU.

It is important to note that the MPU always follows a *fetch-decode-execute sequence.* It first fetches the instruction from program memory. Second, the MPU decodes the instruction. Third, the MPU executes the instruction. Try to notice this fetch-decode-execute sequence in the next two instructions. Continue with the program listed in the program memory in Fig. 13-7.

6. The MPU addresses location 102 on the address bus. The MPU uses the control lines to enable the read input on the program memory.
7. The code for the store data instruction is sent on the data bus and received by the MPU, where it is transferred to the instruction register.
8. The MPU decodes the store data instruction and determines that it needs the operand. The MPU addresses the next memory location (103) and enables the program memory read input.
9. The code for "in memory location 200" is placed on the data bus by the program memory. The MPU accepts this operand and stores it in the instruction register. The entire "store data in memory location 200" has been fetched from memory and decoded.
10. The execute process now starts. The MPU sends out address 200 on the address bus and enables the *write* input of the data memory.
11. The MPU sends the information stored in the accumulator on the data bus to data memory. The "A" is received off the data bus and is written into location 200 in data memory. The second instruction has been executed. This store process does not destroy the contents of the accumulator. The accumulator still also contains the coded form of "A."
12. MPU must fetch the next instruction. It addresses location 104 and enables the read input of the program memory.
13. The code for the output data instruction is sent to the MPU on the data bus. The MPU receives the instruction and

transfers it to the instruction register. The MPU decodes the instruction and determines that it needs an operand.
14. The MPU places address 105 on the address bus and enables the read input of the program memory.
15. The program memory sends the code for the operand (to port 10) to the MPU via the data bus. The MPU receives this code in the instruction register.
16. The MPU decodes the entire instruction "output data to port 10." The MPU activates port 10, using the address bus and control lines to the output unit. The MPU sends the code for "A" (still stored in the accumulator) on the data bus. The "A" is transmitted out of port 10 to the CRT monitor.

Most *MPU-based systems* transfer information in a fashion similar to the one detailed in Fig. 13-7. The greatest variations are probably in the input and output sections. Several more steps may be required to get the input and output sections to operate properly.

It is important to notice that the MPU is the center of and controls all operations. The MPU follows the fetch-decode-execute sequence. The actual operations of the MPU system, however, are dictated by the instructions listed in program memory. Instructions are usually performed in sequence (100, 101, 102, and so on).

All three instructions in the example would be fetched, decoded, and executed in a few microseconds or less by most small microcomputers. The advantage of MPU-based systems is their fast operation and flexibility. They are flexible because they can be reprogrammed to perform many tasks.

Microcomputers are complex digital systems containing an MPU IC (or set of ICs), some memory, and inputs and outputs. The MPU chip itself is a complex, highly integrated subsystem that can process instructions at a high rate of speed. It is expected that microcomputers will be a growth industry for many years to come. The last two sections gave only a brief overview of the basic operation and organization of a microcomputer.

Fetch-decode-execute sequence

MPU-based systems

*Supply the missing word or words in each statement.*

12. The action part of a microcomputer instruction is called the _____. The second part of the instruction is called the _____.

13. Refer to Fig. 13-7. Program memory location _____ holds the operation part of the first instruction, whereas location _____ holds the operand part of the instruction.

14. Refer to Fig. 13-7. In this microcomputer, the _____ is the center of all data transfers and operations.

15. The microcomputer's MPU always follows a fetch-_____-_____ sequence when running.

16. Program instructions are usually performed in _____ (random, sequential) order in a microcomputer.

**Microcomputer address decoding**

**High-impedance state**

## 13-4 Microcomputer Address Decoding

Consider the simple 4-bit MPU-based system shown in Fig. 13-8(*a*). This system uses only eight conductors in the address bus and four conductors in the data bus. The RAMs are tiny, 64-bit ($16 \times 4$) units. These RAMs are like the 7489 RAMs you studied earlier.

Two problems become apparent when working with a system like the one shown in Fig. 13-8(*a*). First, how does the MPU select which RAM to read data from when it sends the same 4-bit address to each? Second, how can several devices send data over a common data bus if, generally, outputs of logic devices cannot be tied together? The solutions to both these problems are shown in Fig. 13-8(*b*).

**Address decoder**

The *address decoder* shown in Fig. 13-8(*b*) decodes which RAM is to be used and sends the enabling signal over the chip select line. *Only one chip select line is activated at a time.* The address decoder block consists of familiar combinational logic gates. RAM 0 is selected when the address is 0 through 15. However, RAM 1 is selected when the address is 16 through 31.

**Three-state buffer**

The *three-state buffers* shown in Fig. 13-8(*b*) disconnect the RAM outputs from the data bus when the memory is not sending data. Only one device is allowed to send on the shared data bus at a given time. For this reason, the chip select line is also used to control, or turn on, the three-state buffers. When the three-state buffers are in the turned off mode, it is

**Memory map**

said that the buffer outputs are in their *high-impedance state* and are effectively disconnected from the four data lines at the inputs of the buffers.

The logic circuits used in a simple address decoder are shown in Fig. 13-9. In this example, only when the four address lines ($A_7$ to $A_4$) are all zero is the output of the bottom four-input OR gate LOW. When address lines $A_7$ to $A_4$ are 0000, then RAM 0 is enabled with a LOW at its memory enable ($\overline{ME}$) input.

When the four address lines going into the address decoder in Fig. 13-8 are 0001 ($A_7 = 0$, $A_6 = 0$, $A_5 = 0$, $A_4 = 1$), the top OR gate is activated. The 0001 causes the top OR gate in the address decoder to generate a LOW output, which activates the bottom device-select line. This enables the bottom RAM (RAM 1).

The address decoder in Fig. 13-9 decodes only the four most significant address lines to generate the correct ($\overline{ME}$) logic level. The RAMs internally decode the four least significant address lines ($A_0$ to $A_3$) to locate the exact 4-bit word in RAM.

The MPU-based system in Figs. 13-8 and 13-9 uses eight address lines. This means that the MPU can generate 256 ($2^8$) unique addresses. In the systems in Figs. 13-8 and 13-9, the first 16 addresses are used by RAM 0 while the next 16 addresses are used by RAM 1. It is customary to draw a *memory map* of an MPU-based system. The memory map of our sample system is drawn in Fig. 13-10. This shows that the first 16 (0F in hexadecimal) addresses

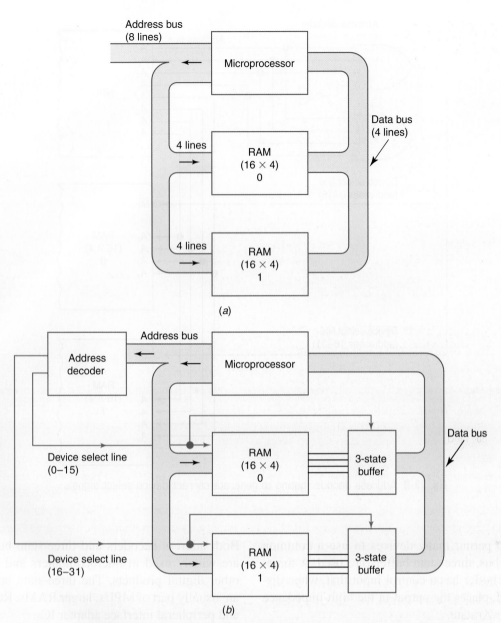

Fig. 13-8 (a) Simplified 4-bit microprocessor interfaced with two 64-bit RAMs. (b) Address
decoder and three-state buffers added to 4-bit microprocessor-based system.

are used by RAM 0. These addresses range
from 0 to 15 (00 to 0F in hexadecimal). The
second 16 addresses are used by RAM 1.
These addresses range from 16 to 31 (10 to 1F
in hexadecimal). The third through sixteenth
groups of addresses are not used in this very
tiny system. It is customary to use hexadecimal notation in specifying addresses in an
MPU-based system.

In Fig. 13-8(b), two blocks are labeled three-
state buffers. The symbol for a buffer is drawn
in Fig. 13-11(a). It has a data input (A) and non-
inverted output (Y). When the control input (C)

is deactivated with a 1, output Y goes to its
high-impedance (high-Z) state and is effectively disconnected from the input.

A commercial version of the three-state
buffer is shown in Fig. 13-11(b). This is the pin
diagram for the *74125 quad three-state buffer
TTL IC*. The truth table for the 74125 IC is
shown in Fig. 13-11(c).

In summary, an address decoder is used to
select *which* device will be connected to the
data bus in an MPU-based system. *Address
decoders* are usually constructed of combinational logic circuits (simple gating circuits).

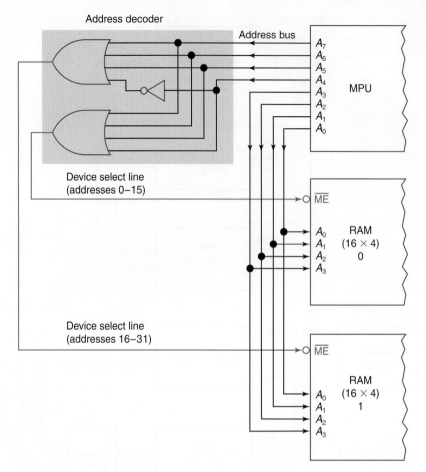

Address decoder

**Fig. 13-9** Address decoder gating to generate correct device select signals.

**Address decoder**

To permit many devices to use a common data bus, three-state buffers are used. A three-state buffer has a control input that, when disabled, places the output in the high-impedance (high-Z) state.

Both address decoders and three-state buffers are widely used in microcomputers and most other digital products. The three-state buffers are usually part of MPUs, larger RAMs, ROMs, and peripheral interface adapter ICs.

 Self-Test

*Supply the missing word or words in each statement.*

17. Refer to Fig. 13-8. The _____ in this system selects which RAM will be used.

18. Refer to Fig. 13-8. When not in use, RAMs are isolated from the data bus by _____.

19. Refer to Fig. 13-9. If the MPU outputs 00001000 on the address bus, RAM

_____ (number) will be activated and storage area _____ (decimal number) located in the RAM will be accessed.

20. Refer to Fig. 13-11. If the control input on the three-state buffer is HIGH, output *Y* is _____ (connected to input *A*; in its high-impedance state).

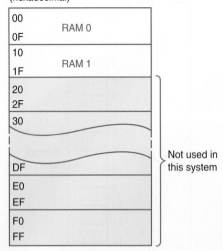

Address
(hexadecimal)

| | |
|---|---|
| 00 | RAM 0 |
| 0F | |
| 10 | RAM 1 |
| 1F | |
| 20 | |
| 2F | |
| 30 | |
| | |
| DF | |
| E0 | |
| EF | |
| F0 | |
| FF | |

Not used in
this system

**Fig. 13-10** Memory map of small microprocessor-based system using two $16 \times 4$ RAMs.

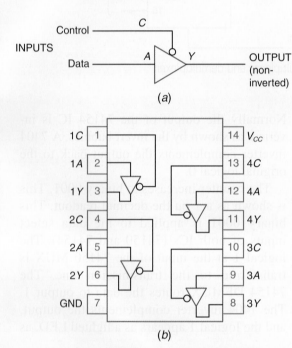

INPUTS

Control —— $C$

Data —— $A$ —— $Y$ —— OUTPUT (non-inverted)

(a)

| | | | |
|---|---|---|---|
| 1C | 1 | 14 | $V_{CC}$ |
| 1A | 2 | 13 | 4C |
| 1Y | 3 | 12 | 4A |
| 2C | 4 | 11 | 4Y |
| 2A | 5 | 10 | 3C |
| 2Y | 6 | 9 | 3A |
| GND | 7 | 8 | 3Y |

(b)

TRUTH TABLE

| INPUTS | | OUTPUT |
|---|---|---|
| C | A | Y |
| L | L | L |
| L | H | H |
| H | X | (Z) |

L = LOW voltage level
H = HIGH voltage level
X = Don't care
(Z) = High impedance (off)

(c)

**Fig. 13-11** (a) Logic symbol for a three-state buffer. (b) Pin diagram for commercial 74125 quad three-state buffer IC. (c) Truth table for 74125 three-state buffer IC.

## 13-5 Data Transmission

Most data in digital systems are transmitted directly through wires and PC boards. Many times bits of data must be transmitted from one place to another. Sometimes the data must be transmitted over telephone lines or cables to points far away. If all the bits in each word were sent at one time over *parallel* wires, the cost and size of these cables would be too expensive and large. Instead, the data are sent over a single wire in *serial* form and reassembled into parallel data at the receiving end. Devices used for sending and receiving serial data are called *multiplexers (MUX)* and *demultiplexers (DEMUX)*.

The basic idea of a MUX and DEMUX is shown in Fig. 13-12. Parallel data from one digital device are changed into *serial* data by the MUX. The serial data are transmitted by a single wire. The serial data are reassembled into parallel data at the output by the DEMUX. Notice the control lines that must also connect the MUX and DEMUX. These control lines keep the MUX and DEMUX synchronized. Notice that the 16 input lines are cut down to only a few transmission lines.

The system in Fig. 13-12 works in the following manner. The MUX first connects input 0 to the serial data transmission line. The bit is then transmitted to the DEMUX, which places this bit of data at output 0. The MUX and DEMUX proceed to transfer the data at input 1 to output 1, and so on. The bits are transmitted one bit at a time.

A MUX works much like a single-pole, many-position rotary switch, as shown in Fig. 13-13. Rotary switch 1 shows the action of a MUX. The DEMUX operates like rotary switch 2 in Fig. 13-13. The mechanical control in this diagram makes sure input 5 on SW 1 is delivered to output 5 on SW 2. Notice that the mechanical switches in Fig. 13-13 permit data to travel in either direction. Being made from logic gates, MUXs and DEMUXs permit data to travel only from input to output, as in Fig. 13-12.

You may have used a MUX before. The other name for MUX is *data selector*. DEMUXs are sometimes called *distributors* or *decoders*. The term "distributor" describes the action of SW 2 in Fig. 13-13, as it distributes

Data
transmission

Serial data

Multiplexers
[MUX]
Demultiplexers
[DEMUX]

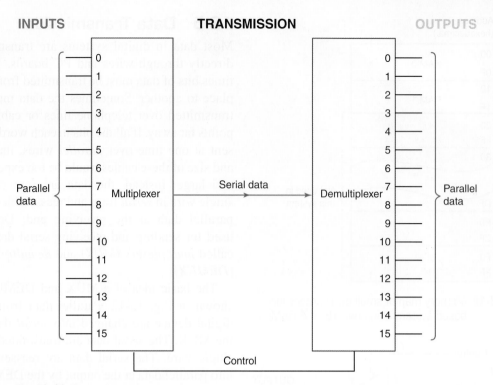

Fig. 13-12 Serial data transmission using a multiplexer and demultiplexer.

the serial data first to output 1, then to output 2, then to output 3, and so forth.

Figure 13-14 is a detailed wiring diagram of an experimental transmission system using the MUX/DEMUX arrangement. A word (16 bits long) is entered at the inputs (0 to 15) of the *74150 MUX IC*. The 7493 counter starts at binary 0000. This would be shown as 0 on the seven-segment display. With the data select inputs (*D, C, B, A*) of the 74150 MUX at 0000, the data are taken from input 0, which is shown as a logical 0. The logical 0 is transferred to the *74154 DEMUX IC*, where it is routed to output 0.

Normally the output of the 74154 IC is inverted, as shown by the invert bubbles. A 7404 inverter complements the output back to the original logical 0.

The counter increases to binary 0001. This is shown as a 1 on the decimal readout. This binary 0001 is applied to the data select inputs of both ICs (74150 and 74154). The logical 1 at the input of the 74150 MUX is transferred to the transmission line. The 74154 DEMUX routes the data to output 1. The 7404 inverter complements the output, and the logical 1 appears as a lighted LED, as

**74150 MUX IC**

**74154 DEMUX IC**

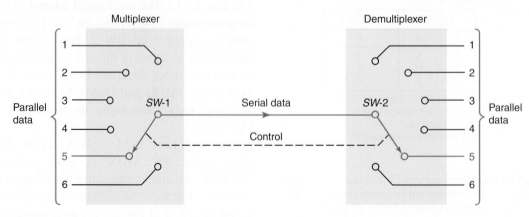

Fig. 13-13 Rotary switches act like multiplexers and demultiplexers.

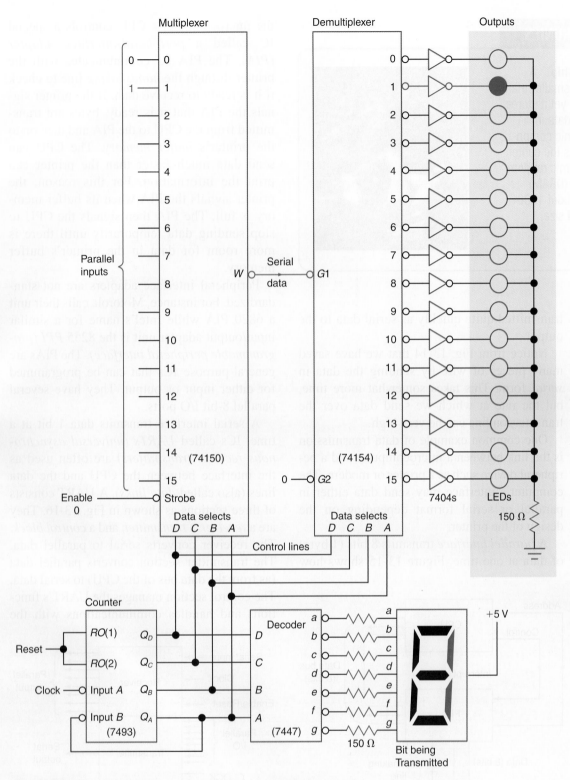

**Fig. 13-14** Wiring diagram for an experimental transmission system.

shown in the diagram. The counter continues to scan each input of the 74150 IC and transfer the contents to the output of the 74154. Notice that the counter must count from binary 0000 to 1111 (16 counts) to transfer just

one parallel word from the input to the output of this system. The seven-segment LED readout provides a convenient way of keeping track of which input is being transmitted. If the clock is pulsed very fast, the parallel data can be

Peripheral-
interface
adapter [PIA]

Handshaking

Buffer memory

PPI
[programmable
peripheral
interface]

UART [universal
asynchronous
receiver-
transmitters]

Data links

Parallel
interface

the microcomputer's CPU controls a special IC called a *peripheral-interface adapter (PIA)*. The PIA IC communicates with the printer through the *handshaking* line to check if it is ready to receive data. If the printer signals the PIA that it is ready, bytes are transmitted from the CPU to the PIA and then on to the printer's *buffer memory*. The CPU can send data much faster than the printer can print the information. For this reason, the printer signals the PIA when its buffer memory is full. The PIA then signals the CPU to stop sending data temporarily until there is more room for data in the printer's buffer memory.

Peripheral interface adapters are not standardized. For instance, Motorola calls their unit a 6820 PIA while Intel's name for a similar input/output adapter unit is the 8255 *PPI (programmable peripheral interface)*. The PIAs are general-purpose ICs that can be programmed for either input or output. They have several parallel 8-bit I/O ports.

A serial interface transmits data 1 bit at a time. ICs called *UARTs (universal asynchronous receiver-transmitters)* are often used as the interface between the CPU and the data lines (also called *data links*). A UART consists of three sections as shown in Fig. 13-16. They are a *receiver*, a *transmitter*, and a *control block*. The receiver converts serial to parallel data. The transmitter section converts parallel data (as from the data bus of the CPU) to serial data. The control section manages the UART's functions and handles communications with the

transmitted quite quickly as serial data to the output.

Notice from Fig. 13-14 that we have saved many pieces of wire by sending the data in *serial* form. This takes somewhat more time, but the rate at which we send data over the transmission line can be very high.

One common example of data transmission is the link between a microcomputer and a peripheral device such as a printer or modem. The computer's interface may send data either in parallel or serial format depending on the design of the printer.

A *parallel interface* transmits 8 bits (1 byte) of data at one time. Figure 13-15 shows how

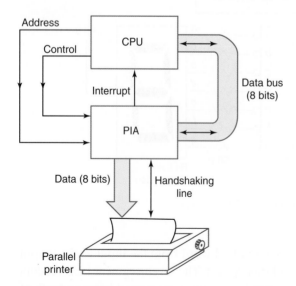

**Fig. 13-15** Parallel data transmission from the CPU to printer using a peripheral-interface adapter [PIA] IC.

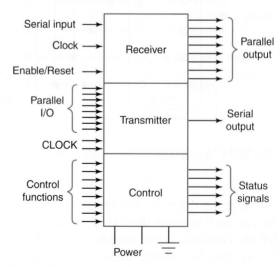

**Fig. 13-16** Block diagram of a typical UART.

CPU and the peripheral device. The UART also encodes and decodes the serial signal including start, stop, and parity bits.

The speed at which serial data is transmitted is called the baud rate. The *baud rate* is the number of bits per second being transmitted through a data link. The baud rate is *not* the same as the number of characters or words transmitted per second. Common baud rates might be 110, 300, 1200, 2400, 9600, 19,200, and 38,400.

The signal levels found in data lines are many times defined by standards. Two *serial interface standards* are the EIA RS-232C standard and the older 20-mA current loop teletype standard.

Two common *parallel interfaces* are the Centronics standard and the IEEE-488 standard. The Centronics standard is used between many microcomputers and printers. The IEEE-488 interface is used between computers and scientific instrumentation.

**Serial interface standards**

**Baud rate**

**Parallel interface**

# ✔ Self-Test

*Answer the following questions.*

21. Refer to Fig. 13-12. A(n) _____ changes parallel data to serial data, whereas a(n) _____ changes serial data to parallel data for transmission.
22. Refer to Fig. 13-14. The 7493 IC is used to sequence the data selects from 0000 through _____ (binary number).
23. Refer to Fig. 13-15. A complex chip called a(n) _____ is used to output

parallel data to the printer in some microcomputer systems.
24. An LSI IC used for asynchronous data transmission is called a(n) _____.
25. A measure of the speed of serial data transmission is called the _____ rate.
26. The EIA RS-232C standard might be used for _____ (parallel, serial) interfacing between a microcomputer and a peripheral device.

## 13-6 Detecting Errors in Data Transmissions

Digital equipment, such as a computer, is valuable to people because it is fast and *accurate*. To help make digital devices accurate, special *error detection* methods are used. You can well imagine an error creeping into a system when data is transferred from place to place.

To detect errors, we must keep a constant check on the data being transmitted. To check accuracy we can generate and transmit an extra *parity bit*. Figure 13-17 shows such a system. In this system three parallel bits ($A$, $B$, and $C$) are being transmitted over a long distance. Near the input they are fed into a *parity bit generator* circuit. This circuit generates what

**Error detection**

**Parity bit**

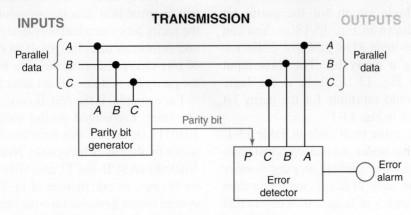

**Fig. 13-17** Error-detection system using a parity bit.

## Table 13-1 Truth Table for Parity Bit Generator

| Inputs | | | Output |
|---|---|---|---|
| Parallel data | | | Parity bit |
| C | B | A | P |
| 0 | 0 | 0 | 0 |
| 0 | 0 | 1 | 1 |
| 0 | 1 | 0 | 1 |
| 0 | 1 | 1 | 0 |
| 1 | 0 | 0 | 1 |
| 1 | 0 | 1 | 0 |
| 1 | 1 | 0 | 0 |
| 1 | 1 | 1 | 1 |

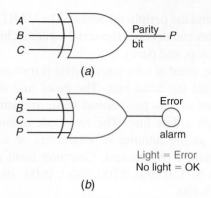

Light = Error
No light = OK

**Fig. 13-18** (a) Parity bit generator circuit. (b) Error detector circuit.

is called a parity bit. The parity bit is transmitted with the data, and near the output the results are checked. If an error occurs during transmission, the *error detector* circuit sounds an alarm. If all the parallel data is the same at the output as it was at the input, no alarm sounds.

Table 13-1 will help you understand how the error-detection system works. This table is really a truth table for the parity bit generator in Fig. 13-17. Notice that the inputs are labeled *A, B,* and *C* for the three data transmission lines. The output is determined by looking across a horizontal row. We want an *even number of 1s* in each row (zero 1s, or two 1s, or four 1s). Notice that row 1 has no 1s. Row 2 has a single 1 plus the parity bit 1. Row 2 has two 1s. As you look down Table 13-1 you will notice that each horizontal row contains an even number of 1s. Next, the truth table is converted to a logic circuit. The logic circuit for the parity bit generator is drawn in Fig. 13-18(a). You can see that a three-input XOR gate will do the job for generating a parity bit. The three-input XOR gate in Fig. 13-18, then, is the logic circuit you would substitute for the parity bit generator block in Fig. 13-17.

Look at the entire truth table in Table 13-1. We can see that under normal circumstances each horizontal row contains an *even number* of 1s. Were an error to occur, we might then have an *odd number* of 1s appear. A circuit that gives a logical 1 output any time an odd number

**XOR gate used for parity bit generation and error detection**

of 1s appear is shown in Fig. 13-18(b). A four-input XOR gate would detect an odd number of 1s at the inputs and turn on the alarm light. Figure 13-18(b) diagrams the logic circuit that can substitute for the error-detector block in Fig. 13-17.

The parity bit can be generated for longer words such as a 7-bit ASCII character. For instance, the ASCII code for T is 1010100 (from Table 6-3). If transmitted using an even parity bit, an extra 1 would have to be added (to get an even number of 1s or four 1s). Another example, the ASCII code for S is 1010011 (from Table 6-3). If transmitted using an even parity bit, an extra 0 would have to be added (four 1s already). A seven-input XOR gate would generate the correct even parity bit for 7-bit ASCII characters. An 8-bit XOR gate at the receiver end would serve as an error detector circuit (H = error, L = no error). Either an *even* or *odd parity bit* may be transmitted or received. An XNOR gate is used to generate an odd parity bit.

The parity bit system is a simple way to detect an error in a data transmission. However, the parity bit system can only detect errors if an odd number of bits changes. If an even number of bits changes during the data transmission, the parity bit system will not detect the error.

For example, if the ASCII code 1010100 for the letter T changed during transmission to 1010111 (letter W), this error would *not be detected* by the parity bit system. Notice that both 1010100 (ASCII for T) and 1010111 (ASCII for W) have an odd number of 1s. The parity bit system would generate no error message in this example.

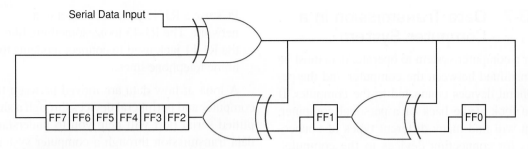

Serial Data Input

| FF7 | FF6 | FF5 | FF4 | FF3 | FF2 | | FF1 | | FF0 |

**Fig. 13-19** A CRC-8 checksum generator circuit.

One common system used to check for multiple bit errors during transmission is the *Cyclic Redundancy Check,* or *CRC.* The CRC system adds several extra bits to the end of the transmitted data. The extra bits enable the system to detect almost all transmission errors. Systems that use the CRC to detect errors may add 8, 16, or 32 bits to the data. These are commonly referred to as CRC-8, CRC-16, or CRC-32 systems.

The cyclic redundancy check creates a unique code, or *checksum,* for the data by shifting the data through a special shift register circuit made up of flip-flops with XOR gates inserted at specific locations. The example in Fig. 13-19 shows a circuit for generating one possible CRC-8 checksum. After all the data have been shifted into the serial data input of the circuit, the shift register (FF0–FF7) holds the 8-bit checksum for the data.

In a CRC error detection system, both the transmitter and the receiver implement the same circuit. The transmitter uses the circuit to generate the checksum. At the receiver, the received data are passed through the checksum circuit. After all the data have been received, the checksum at the receiver is compared to the checksum sent by the transmitter. If the checksums match, the data were successfully transmitted. If the checksums do not match, the receiver requests the transmitter to resend the data.

**CRC**

The use of parity bits or CRCs in a system only warns if there was an error during transmission. These systems do not automatically correct errors. Some systems such as the *Hamming code* both detect and correct errors in transmission. Codes such as the Hamming code are known as *error-correcting codes.* Other methods of ensuring accuracy in data transmissions have also been developed.

**Checksum**

---

**✓ Self-Test**

*Supply the missing word or words in each statement.*

27. Refer to Table 13-1. This is the truth table for an _____ (even, odd) parity bit generator.
28. Refer to Fig. 13-17. The parity bit generator block could be replaced with a three-input _____ gate, whereas the error detector block could be replaced with a four-input _____ gate.

29. Using even parity, what bit would be transmitted with the 7-bit ASCII code 1011000 as a parity bit?
30. Using odd parity, what bit would be transmitted with the 7-bit ASCII code 1011000 as a parity bit?
31. A seven-input _____ (AND, XOR) gate will generate an even parity bit for a 7-bit ASCII code.
32. When dealing with error detection in data transmissions, the acronym CRC stands for _____.

## 13-7 Data Transmission in a Computer System

For a computer system to operate, data must be transmitted between the computer and the peripheral devices connected to the computer. If you look at the back of a personal computer, you will find many different types of connectors for connecting devices to the computer. These connectors are often called *ports*. Some of the most common ports are:

**Keyboard port:** A dedicated port specifically for connecting a keyboard. Most current PCs use the mini-DIN PS/2 style connector (6 pins).

**Mouse port:** A dedicated port specifically for connecting a mouse. Many PCs use the mini-DIN style connector. Some use the USB style connector.

**Video port:** The port used for connecting the computer to a display monitor, often a CRT or LCD display. This is usually a 15-pin VGA (video graphics adapter) port or a new DVI (digital video interface) port.

**Serial port:** This port has a 9-pin D-shaped (DB) connector. It is one of the oldest ports on a computer and is used for connecting many types of devices. The serial port is not available on some newer PCs.

**Parallel port:** This port has a 25-pin DB (data bus) connector. It is also one of the oldest ports on a computer. It is often used to connect printers to the computer. It sends data 8 bits at a time.

**Audio ports:** A computer may have two or more of these ports. These are commonly 3.5 mm audio connectors. At least one port is typically used for audio output to headphones or speakers. Another port is often used to connect a microphone for audio input.

**USB port:** The USB (universal serial bus) port is found on most modern computers and is replacing the older serial port as the preferred port for connecting peripheral devices to a computer. The USB ports on computers are typically small, rectangular ports (about 4.5 mm × 12 mm).

**Ethernet port:** This port is used to connect computers to networks at high speed. Many PCs use a RJ-45 jack to connect to a network. The RJ-45 looks something like the RJ-11 jack used to connect modems to home telephone lines.

A look at how data are moved between the computer and peripheral devices through a simplified serial port will help you to understand data transmission through a computer system. Section 13-5 introduced the UART as an IC commonly used as a serial interface to the CPU. In personal computers, the UART is the IC used to control the serial port. UARTs are *full duplex* devices because they can send and receive data at the same time.

Fig. 13-20 is a block diagram of a UART used in a simple computer system for both the input and output ports. Data coming into the UART are the input data to the computer system. Data sent out of the UART are the output data. Notice that the UART in this sketch shows four registers within it. They are the *transmit-data register,* the *receive-data register,* the *control register,* and the *status register.* By reading and writing values to these registers, the MPU is able to control data flowing in and out of the UART. The UART makes it easier for the MPU to send and receive data to peripheral devices because it allows the MPU to treat reading and writing data to peripheral devices almost the same as reading and writing data to RAM memory. In microcomputer systems, registers in chips used to control data flowing in and out of ports are mapped into a memory space, the same as the RAM memory is mapped in Fig. 13-10. Although the UART in Fig. 13-20 shows four registers, it is common for the transmit-data register and receive-data register to share the same memory location. Thus, the memory map for the UART in this figure requires only three memory locations. The MPU signals the UART which data register it wants to use via the R/W control line. If the MPU signals a WRITE operation on the data register memory location of the UART, then data are sent to the transmit-data register. If a READ operation is signaled to the same memory location, then data are read from the receive-data register.

Table 13-2 details where the registers of the UART of Fig. 13-20 exist relative to a base memory location. As an example, let's say the designer of the computer system of Fig. 13-20

More on PC ports at www. howstuffworks. com

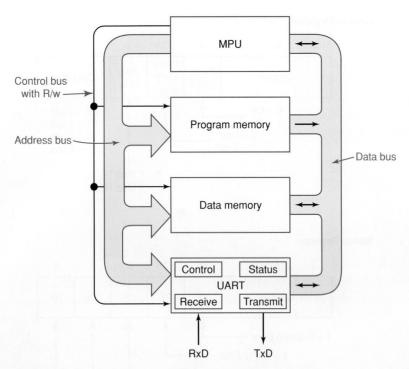

**Fig. 13-20** Block diagram of a UART in a computer system.

| Table 13-2 | |
| --- | --- |
| Register | Offset |
| Receive data | 00h |
| Transmit data | 00h |
| Status | 01h |
| Control | 02h |

placed the UART at a base memory location of 300. In this configuration, the MPU accesses the status register by reading from memory location 301 and accesses the control register by reading or writing to memory location 302. Fig. 13-21 provides more details about the control and status registers. Each of these two registers is subdivided into a number of bits, and each bit has special meaning. The special meanings assigned to each bit of these registers are also shown in Fig. 13-21.

Now let's walk through an example of the computer system of Fig. 13-20 setting up the UART and sending the ASCII code for the letter A (1100001 binary) to a peripheral attached to the serial port. The data are to be

sent using even parity and one stop bit. Following the same fetch-decode-execute sequence described in Section 13-3, the MPU runs a program that instructs the MPU to do the following:

Load the value 00011010 binary (1A hex) into the MPU's accumulator from program memory

Place the memory address of 302 on the address bus

Place the value in the accumulator on the data bus

Place the WRITE signal on the R/W control line

A look at the bits of the value (00011010) just sent to the control register of the UART by the instructions above shows that the UART is now configured to transmit and receive data using seven data bits with even parity and one stop bit. Bit 1 of the control register is 1 and bit 0 is 0. From Fig. 13-21($a$), we see that this combination of bits instructs the UART to transmit and receive seven bits of data at a time. Bit 2 of the control registers is set to 0 for one stop bit. Bit 4 is 1, which instructs the UART to use parity encoding, and bit 3 is set to one for even

**Control Register:**

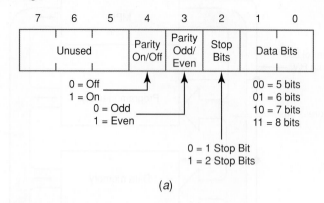

(a)

**Status Register:**

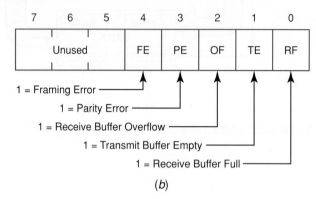

(b)

**Fig. 13-21** (a) UART control-register details. (b) UART status-register details.

parity. Continuing on, the program being run instructs the MPU to:

Load 01100001 binary (the ASCII code for the letter A) into the accumulator from program memory

Place the memory address of 300 on the address bus

Place the value of the accumulator on the data bus

Place the WRITE signal on the R/W control line

The UART now has enough information to begin transmitting the character A without any further action from the MPU. However, the instructions from the program that the MPU is running still instructs the MPU to read the status register (memory location 301) occasionally to check to see that the data were sent. The program performs this check by looking at bit 1 of the status register. If bit 1 of the status register is 1, the transmit-data

buffer is empty indicating that the data were sent.

Fig. 13-22 shows the serial data sent out on the TxD (transmit data) line of the UART. Notice that 10 bits were transmitted. The UART did the work of adding the start bit, parity bit, stop bit, and transmitting the data without any further instruction from the MPU. Also notice that the seven bits representing the letter A are reversed. Data sent out from the UART are sent least-significant bit first. The UART also receives data with the least-significant bit first.

When the UART receives data, it automatically strips off the start, stop, and parity bits. The UART also checks the parity of the data received. If the parity does not match the expected parity, the error is indicated, or *flagged,* in the parity error bit of the status register. In fact, four of the five *flags* in the status register are for use when receiving data. In addition to the parity error (PE) flag, there are the

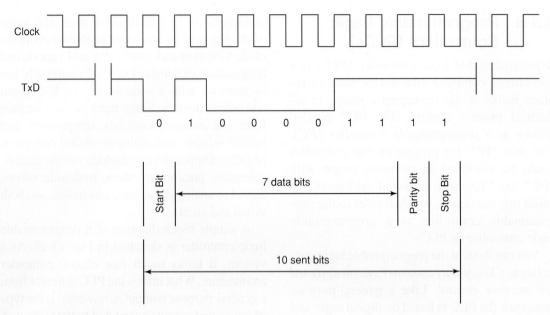

Clock

TxD

0 1 0 0 0 0 1 1 1 1

Start Bit

7 data bits

Parity bit

Stop Bit

10 sent bits

**Fig. 13-22** Serial data sent out on TxD of the UART.

receive-buffer-full (RF) flag, the overflow (OF) flag, and the framing-error (FE) flag. When the UART has received a byte of data, it sets the receive-buffer-full flag to 1, indicating that it has data ready to be read by the MPU. If the UART receives another byte of data before the MPU has read the previously received byte, then it sets the overflow flag to 1, indicating that data were lost. A framing error occurs when the UART doesn't receive the expected number of bits for its current configuration. In our example, this might happen if the peripheral device were configured to send two stop bits instead of one. In this case, the UART would receive 11 bits instead of the expected 10 bits and would notify the MPU of the error by setting the framing-error flag.

The UART in this example was intentionally kept simple to show you how data flow through the computer system to peripherals. UARTs used in personal computers typically have additional registers that allow for greater control over things such as the baud rate. However, the steps used to set up and control those UARTs are the same as what was presented here.

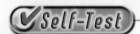

*Answer the following questions.*

33. Connectors on computers that are used to connect to external devices are often called _____ .

34. _____ are ICs that are used to control data flowing into and out of the serial port of a personal computer.

35. UARTs are _____ (half, full) duplex devices.

36. Registers in ICs that are used for interfacing external devices to a computer are special, so they are not mapped into the computer's memory space. (T or F)

37. The MPU must do the work of adding the start, stop, and parity bits to the data before sending them to the UART for transmission. (T or F)

38. A _____ (port, flag) is another name of a bit used to indicate status or an error condition.

## 13-8 Programmable Logic Controllers (PLCs)

**Programmable logic controller (PLC)**

A *programmable logic controller (PLC)* is a specialized computer like device used to replace banks of electromagnetic relays in industrial process control. The PLC is also known as a *programmable controller (PC)*. The title "PC" for programmable controller could be confused in common usage with "PC" used to mean personal computer. To avoid this confusion, we shall refer to the programmable controller as a programmable logic controller or PLC.

You can think of the programmable logic controller as a *heavy-duty computer system designed for machine control.* Like a general-purpose computer, the PLC is based on digital logic and can be field-programmed. The programming language is a bit different because the purpose of the PLC is to control machines. The PLC is used to time and sequence functions that might be required in assembly lines, robots, and chemical processing. It is designed to deal with the harsh conditions of the industrial environment; some of the physical environment problems could include vibration and shock, dirt and vapors, and temperature extremes. The PLC commonly has to interface with a wide variety of both input and output devices. Some input devices include limit and pressure switches, temperature and optical sensors, and analog-to-digital converters (ADCs). Output devices include relays, motors, solenoids, pneumatic valves, hydraulic valves, digital to analog converters, and indicators (both visual and aural).

A simple block diagram of a programmable logic controller is sketched in Fig. 13-23. As a system, it looks much like classic computer architecture. What makes the PLC different from a general-purpose computer, however, is the type of inputs and outputs connected to the system. A PC system commonly has a keyboard or mouse for primary input while the PLC must interface with sensors, which detect the machine's action. The primary output from a PC is a monitor or printer, whereas the PLC must drive motors and

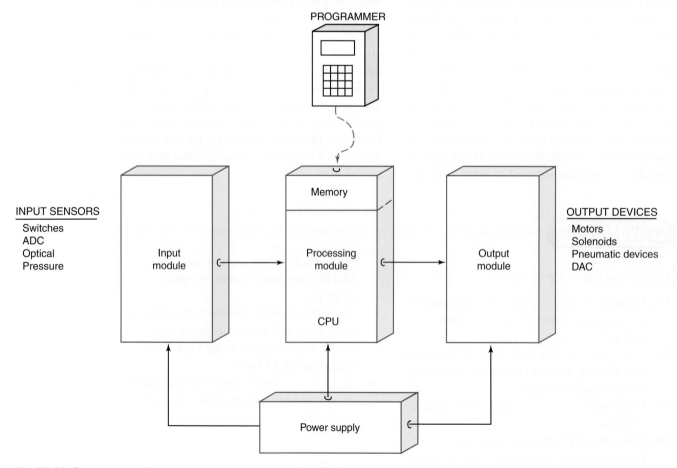

**Fig. 13-23** Organization of a programmable logic controller (PLC).

solenoids. Notice from Fig. 13-23 that the programmer is shown as a separate module, which may or may not be connected to the processing unit. The programming device in Fig. 13-23 can be connected when an update is needed in the PLC and disconnected when the task is finished. Semiconductor memory devices within the processing unit of the PLC hold the machine or process control program. In small PLCs, the input and output modules can be part of the device. In larger systems, the processing module, input module, output module, and power supplies are housed in separate heavy-duty, industrial-style enclosures. The programmer can be a dedicated terminal, general-purpose computer, or hand-held programming device.

The *processing unit* of the PLC typically contains a CPU (a microprocessor) and semiconductor memory devices such as RAM and EEP-ROM or EPROM. The CPU communicates with memory and input/output (I/O) modules via the typical address, control and data buses. The input sensors and output devices are hardwired to the input and output modules. The architecture of the PLC and a PC look very much alike. Many PLCs have a simple machine-control language built permanently into their memory. The PLC programming language is simpler than the languages used to program general-purpose computers. Programmable logic controllers can be reprogrammed by the electricians and technicians that maintain the other industrial electrical-electronic devices in a factory or plant. The *instruction set* for a specific PLC may contain as few as 15 to as many as 100 instructions. Besides the normal arithmetic and logic functions associated with computer CPUs, specialized instructions are needed to sense and control output devices and to do the following tasks:

Examine an input bit for ON condition

Examine an input bit for OFF condition

Turn ON and latch an output

Turn OFF and latch an output

Turn ON for a certain time, then turn off

Programmable logic controllers are closely associated with relay logic or hard-wired logic used prior to their introduction in the 1970s. A *relay ladder diagram* is a graphic method of describing how a circuit works. A *ladder logic diagram* is a graphic programming language developed from the relay ladder diagram and is

## EXAMPLE 13-1

Two limit switches (LS) connected in series are used to control a solenoid (SOL). ( *from Frank Petruzella, Programmable Logic Controllers, 2nd ed., New York: Glencoe/ McGraw-Hill, 1998*)

Relay schematic

Ladder logic program

Gate logic

Boolean equation: $AB = Y$

useful in programming a PLC. Some examples of equivalent relay ladder diagrams, relay logic diagrams, and logic gate diagrams are illustrated in examples below. These examples are from the excellent textbook, *Programmable Logic Controllers,* 2nd edition by Frank Petruzella, Glencoe/McGraw-Hill. Notice that each of the three types of diagrams have their own symbols and conventions. Each of the types of diagrams were developed by various manufacturers and users to suit their needs. For instance, the relay schematics were developed before digital logic as we know it became popular. The ladder logic diagrams were developed directly from the relay schematics used earlier.

Example 13-1 shows two input switches in series and the output device is a solenoid valve. The relay schematic is shown at the top, the ladder logic diagram in the center, and the familiar logic gate diagram near the bottom. We recognize that two switches in series is an AND situation as shown at the bottom with a Boolean expression of $AB = Y$. Note the symbols used in the relay schematic, ladder logic diagram, and gate logic are different but each represents the same task—the AND function.

Example 13-2 shows two input switches in parallel, and the output device is a solenoid

**Relay ladder diagram**

**Ladder logic diagram**

## EXAMPLE 13-2

Two limit switches (LS) connected in parallel are used to control a solenoid (SOL). (*from Frank Petruzella, Programmable Logic Controllers, 2nd ed., New York: Glencoe/McGraw-Hill, 1998*)

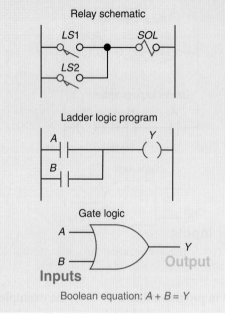

Relay schematic

Ladder logic program

Gate logic

Boolean equation: $A + B = Y$

## EXAMPLE 13-3

Two limit switches (LS) connected in parallel are placed in series with a relay contact (CR) are used to control pilot lamp (PL). (*from Frank Petruzella, Programmable Logic Controllers, 2nd ed., New York: Glencoe/McGraw-Hill, 1998*)

Relay schematic

Ladder logic program

Gate logic

Boolean equation: $(A + B)C = Y$

## EXAMPLE 13-4

Two limit switches (LS) connected in series with each other and in parallel with a third limit switch are used to control a warning horn (H). (*from Frank Petruzella, Programmable Logic Controllers, 2nd ed., New York: Glencoe/McGraw-Hill, 1998*)

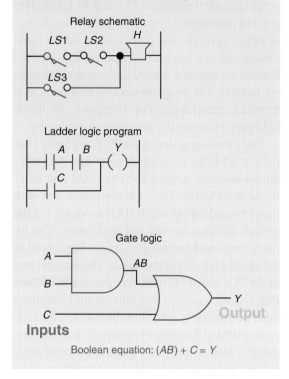

Relay schematic

Ladder logic program

Gate logic

Boolean equation: $(AB) + C = Y$

valve. The relay schematic is shown at the top, the ladder logic diagram in the center, and the logic gate diagram near the bottom. We recognize that two switches in parallel is an OR situation as shown near the bottom with a Boolean expression of $A + B = Y$.

Example 13-3 shows two input switches in parallel with a normally open relay contact in series with both, and the output device is a green pilot light. The relay schematic is shown at the top, the ladder logic diagram in the center, and the logic gate diagram near the bottom. We recognize that two switches in parallel ($A$ and $B$) is an OR situation which feeds a series relay contact (AND situation). Both the Boolean expression $(A + B)C = Y$ and logic gate diagram are shown near the bottom.

Example 13-4 shows two input switches ($A$ and $B$) in series with each other, and both are in parallel with a single switch ($C$). The output device in this example is a warning horn. The relay schematic is shown at the top, the ladder logic diagram in the center, and the familiar logic gate diagram near the bottom. We recognize that two switches (*A* and *B*) are in series, which is an AND situation while switch *C* is parallel with the two switches. Again, remember that the relay schematic, ladder logic diagram, and gate logic diagrams all represent the same logic function as described by the Boolean expression $(AB) + C = Y$.

In summary, a programmable logic controller (PLC) is a heavy-duty computer system used to replace older relay logic. PLCs are used in factories and plants to control machines, material handling, and chemical processing. PLCs are built to withstand the more harsh environ-ment of a factory, warehouse, or processing plant. The language used to program a PLC has specialized instructions for evaluating inputs and generating outputs. Some PLC languages are based directly on relay ladder diagrams. Because the processing unit (CPU) of the PLC is a microprocessor, it can also perform arithmetic and logic functions as well as data handling and branch and subroutine calls typical of general-purpose computer languages. Some manufacturers of PLCs are Allen-Bradley Company, Cincinnati Milcron Company, Eaton Corporation (Cutler-Hammer products), Gould Inc., Honeywell, Inc., Square D Company, Texas Instruments, and Westinghouse Electric Company.

## Self-Test

*Answer the following questions.*

39. The programmable controller (PC) is also commonly known as the _____ _____ _____ or PLC.
40. A programmable logic controller (PLC) is a heavy-duty computer system designed for _____ (general-purpose office use, machine control in factories).
41. Once programmed, inputs to a PLC would probably come from devices such as _____ (keyboard and mouse; limit switches, pressure switches, temperature and optical sensors).
42. Typically a PLC has a programming module connected to it _____ (always, occasionally during reprogramming).

43. Refer to Fig. 13-23. The power supply, processing, input, and output sections are referred to as modules because they are sometimes physically housed in separate enclosures in larger systems. (T or F)
44. The programming language used with PLCs is commonly _____ (less, more) complex than general-purpose computer languages.
45. Given the relay schematic in Fig. 13-24, draw the ladder logic program that might be used with a PLC for this circuit.
46. Given the relay schematic in Fig. 13-24, draw a logic gate equivalent of this circuit using AND and OR symbols.
47. Given the relay schematic in Fig. 13-24, write the Boolean expression that describes the logic function of this circuit.

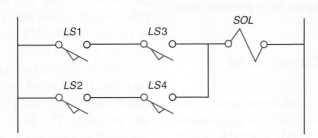

**Fig. 13-24** Relay schematic diagram.

## 13-9　Microcontrollers

**Microcontroller**

A *microcontroller* is considered to be a "computer on a chip." A single-package microcontroller contains a central processing unit (CPU), semiconductor memory (RAM for data memory and read-only memory for program memory), a clock generator, and input/output capabilities. The read-only memory used to store programs in a microcontroller can take the form of ROM, EPROM, EEPROM, or even flash EEPROM. Microcontrollers are low-cost, programmable, electronic devices that can be embedded in inexpensive appliances. Microcontrollers are popular in consumer products because of their extremely low cost: a simple microcontroller IC may cost only a few dollars. The features of microcontrollers vary widely; some are faster, some contain more memory, some have more input/output ports, and some have other characteristics that may be necessary for a specific application. The term "microcomputer" or "small computer" might be used to describe the microcontroller but is *not* common usage. The term "microcontroller" fits the jobs these small "computers on a chip" perform, which are *control functions.* Microcontrollers are not used as general-purpose computers.

Microcontrollers were developed shortly after their larger relatives, microprocessors. The same companies that developed microprocessor chips (for instance Intel and Motorola) also developed a line of microcontrollers. The first 8-bit microcontrollers appeared in the late 1970s, and some of these are still in use today. Microcontrollers sell in large volumes. Motorola, for example, recently announced that it had shipped its two billionth 68HC05 microcontroller. Microcontrollers are embedded in many everyday consumer products, such as cars, toys, TVs, VCRs, microwave ovens, and PC keyboards. For instance, a modern automobile may contain 10 microcontrollers whereas a hightech home may contain many more than that.

### Microcontrollers Compared to Microprocessors

**SOIC package**
**SSOP package**

When compared to a microprocessor-based system, a microcontroller has less semiconductor memory (RAM, ROM, EPROM, and/or EEPROM), is lower in cost, and uses less printed circuit board space. Microcontrollers commonly address only a limited size memory. Microcontrollers usually have fewer commands in their instruction set than microprocessors. Microcontrollers typically are programmed to do several limited tasks efficiently and are usually not reprogrammed. Microcontroller programs are commonly held in read-only memory. Microcontroller-based systems rarely have complex input/output devices attached, such as keyboards, disk drives, printers, and monitors. Manufacturers support both their microcontroller and microprocessor product with software development tools and application notes (examples of typical applications).

Manufacturers of microcontrollers produce a wide variety of low-cost programmable devices. For instance, a recent *IC Master* (a manual that lists many of the world's ICs) used more than 60 pages to list the various microcontrollers available from manufacturers.

### A Family of Microcontrollers

The chart in Fig. 13-25 illustrates a "family of microcontrollers" from Microchip Technology, Inc. This family of devices is described by the manufacturer as the EPROM/ROM-based 8-bit CMOS Microcontroller Series. The PIC16C5X device is listed on the left side of the chart with columns showing some of the important characteristics of these low-cost microcontrollers. The operating frequencies of these units allow them to execute instructions very quickly. The *program memory size* is given in words (word size equals 12 bits for the 16C5X series) and is stored in either ROM or EPROM. The *data memory* or *RAM size* is very small ranging from 24 to 73 bytes. Because microcontrollers are *control devices,* they typically have many IC pins dedicated to either input or output (I/O pins). The number of I/O pins for the PIC16C5X microcontrollers range from 12 to 20. These I/O pins can be programmed to be either *inputs* or *outputs.* The PIC16C5X series of microcontroller ICs are CMOS devices and operate on low voltages. All of the ICs are available in a variety of packages including the traditional DIP (dual in-line package), *SOIC* (small-outline IC), and *SSOP* (shrink small-outline package). The SOIC and SSOP packages are small surface-mount packages. Remember

| | Clock | Memory | | | Peripherals | | | Features | |
|---|---|---|---|---|---|---|---|---|---|
| | Maximum Frequency of Operation (MHz) | EPROM | ROM | RAM Data Memory (bytes) | Timer Module(s) | I/O Pins | Voltage Range (Volts) | Number of Instructions | Packages |
| | Program Memory (12 bit words) | | | | | | | | |
| PIC16C52 | 4 | 384 | — | 25 | TMR0 | 12 | 2.5–6.25 | 33 | 18-pin DIP, SOIC |
| PIC16C54 | 20 | 512 | — | 25 | TMR0 | 12 | 2.5–6.25 | 33 | 18-pin DIP, SOIC; 20-pin SSOP |
| PIC16C54A | 20 | 512 | — | 25 | TMR0 | 12 | 2.0–6.25 | 33 | 18-pin DIP, SOIC; 20-pin SSOP |
| PIC16CR54A | 20 | — | 512 | 25 | TMR0 | 12 | 2.0–6.25 | 33 | 18-pin DIP, SOIC; 20-pin SSOP |
| PIC16C55 | 20 | 512 | — | 24 | TMR0 | 20 | 2.5–6.25 | 33 | 28-pin DIP, SOIC; SSOP |
| PIC16C56 | 20 | 1K | — | 25 | TMR0 | 12 | 2.5–6.25 | 33 | 18-pin DIP, SOIC; 20-pin SSOP |
| PIC16C57 | 20 | 2K | — | 72 | TMR0 | 20 | 2.5–6.25 | 33 | 28-pin DIP, SOIC; SSOP |
| PIC16CR57B | 20 | — | 2K | 72 | TMR0 | 20 | 2.5–6.25 | 33 | 28-pin DIP, SOIC; SSOP |
| PIC16C58A | 20 | 2K | — | 73 | TMR0 | 12 | 2.0–6.25 | 33 | 18-pin DIP, SOIC; 20-pin SSOP |
| PIC16CR58A | 20 | — | 2K | 73 | TMR0 | 12 | 2.5–6.25 | 33 | 18-pin DIP, SOIC; 20-pin SSOP |

All PIC 16/17 Family devices have Power-On Reset, selectable Watchdog Timer, selectable code protect and high I/O current capability.

**Fig. 13-25** General specifications for the PIC16C5X family of microcontrollers. (*Courtesy of Microchip Technology, Inc.*)

that microcontrollers are embedded CPUs in everyday devices and the small package ICs are ideal for "hiding" them inside of products.

The PIC16C5X series of microcontrollers features RISC architecture using only 33 instructions in their instruction set. *RISC* means *reduced instruction set computing* as opposed to *CISC (complex instruction set computing)*. RISC CPUs have fewer instructions but execute them faster. CISC CPUs have more instructions in their instruction set with some of these instructions executing complex tasks. The RISC architecture was developed to speed up the processors, but for complex operations many instructions are needed.

### The PIC16C55 Microcontroller

As an example, the 28-pin DIP diagram for the PIC16C55 microcontroller is reproduced in Fig. 13-26(*a*). A description of the IC's pins is detailed in the chart in Fig. 13-26(*b*). Note especially from the pin diagram and pin-out descriptions the great number of I/O pins. They are organized into three ports (*A*, *B*, and *C*). Port A (4-bit port) consists of I/O pins RA0-RA3, while ports *B* and *C* are each 8-pin ports. Individual I/O pins can be programmed to be either an input or output.

### Using a Microcontroller

A simple application of the PIC16C55 microcontroller is shown in the schematic diagram reproduced in Fig. 13-27(*a*) on page 457. The PIC16C55 has been programmed by Chaney Electronics, Inc. to display various light patterns on the nine-row by 10-column LED display board. A schematic of the 9 × 10 LED display is detailed in Fig. 13-27(*b*). To light the upper-left red LED on the display, the top row must go LOW (Y1 = 0) while the left column must go HIGH (X1 = 1). To light the entire horizontal row of yellow LEDs in the middle of the display in Fig. 13-27(*b*), all column inputs must be HIGH whereas only row 5 input must go LOW (Y5 = 0). You can see that the driver for this display must have 18 output pins, which are available using the 16C55 microcontroller. Two I/O pins of the 16C55 IC are programmed to be inputs ($RA_0$ and $RA_1$) in this design. These pins (6 and 7) can be held LOW or HIGH based on the position of switches *S2* and *S3*. The input conditions caused by switches *S2* and *S3* cause the microcontroller to execute one of four possible programs which produce four unique light displays. Switches *S4* and *S5* change the *RC* timing circuit, which is connected to the *CLKIN* input of the IC. Various positions of *S4*

**RISC**

**CISC**

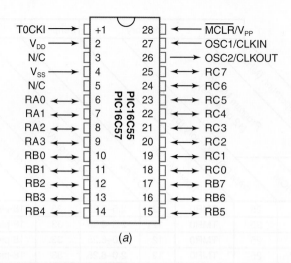

(a)

**PIC16C55/C57 PINOUT DESCRIPTION**

| Name | DIP, SOIC No. | SSOP No. | I/O/P Type | Input Levels | Description |
|---|---|---|---|---|---|
| RA0 | 6 | 5 | I/O | TTL | Bidirectional I/O port |
| RA1 | 7 | 6 | I/O | TTL | |
| RA2 | 8 | 7 | I/O | TTL | |
| RA3 | 9 | 8 | I/O | TTL | |
| RB0 | 10 | 9 | I/O | TTL | Bidirectional I/O port |
| RB1 | 11 | 10 | I/O | TTL | |
| RB2 | 12 | 11 | I/O | TTL | |
| RB3 | 13 | 12 | I/O | TTL | |
| RB4 | 14 | 13 | I/O | TTL | |
| RB5 | 15 | 15 | I/O | TTL | |
| RB6 | 16 | 16 | I/O | TTL | |
| RB7 | 17 | 17 | I/O | TTL | |
| RC0 | 18 | 18 | I/O | TTL | Bidirectional I/O port |
| RC1 | 19 | 19 | I/O | TTL | |
| RC2 | 20 | 20 | I/O | TTL | |
| RC3 | 21 | 21 | I/O | TTL | |
| RC4 | 22 | 22 | I/O | TTL | |
| RC5 | 23 | 23 | I/O | TTL | |
| RC6 | 24 | 24 | I/O | TTL | |
| RC7 | 25 | 25 | I/O | TTL | |
| T0CKI | 1 | 2 | I | ST | Clock input to Timer0. Must be tied to $V_{SS}$ or $V_{DD}$, if not in use, to reduce current consumption. |
| $\overline{MCLR}/V_{PP}$ | 28 | 28 | I | ST | Master clear (reset) input/programming voltage input. This pin is an active low reset to the device. Voltage on $\overline{MCLR}/V_{PP}$ must not exceed $V_{DD}$ to avoid unintended entering of programming mode. |
| OSC1/CLKIN | 27 | 27 | I | ST | Oscillator crystal input/external clock source input. |
| OSC2/CLKOUT | 26 | 26 | O | — | Oscillator crystal output. Connects to crystal or resonator in crystal oscillator mode. In *RC* mode, *OSC2* pin outputs *CLKOUT* which has 1/4 the frequency of *OSC1*, and denotes the instruction cycle rate. |
| $V_{DD}$ | 2 | 3,4 | P | — | Positive supply for logic and I/O pins. |
| $V_{SS}$ | 4 | 1,14 | P | — | Ground reference for logic and I/O pins. |
| N/C | 3,5 | — | — | — | Unused, do not connect |

Legend: I = input, O = output, I/O = input/output,
P = power, — = Not Used, TTL = TTL input,
ST = Schmitt trigger input

(b)

**Fig. 13-26** PIC16C55 microcontroller IC. (*a*) Pin diagram (DIP or SOIC packages only). (*b*) Pinout description. (*Courtesy of Microchip Technology, Inc.*)

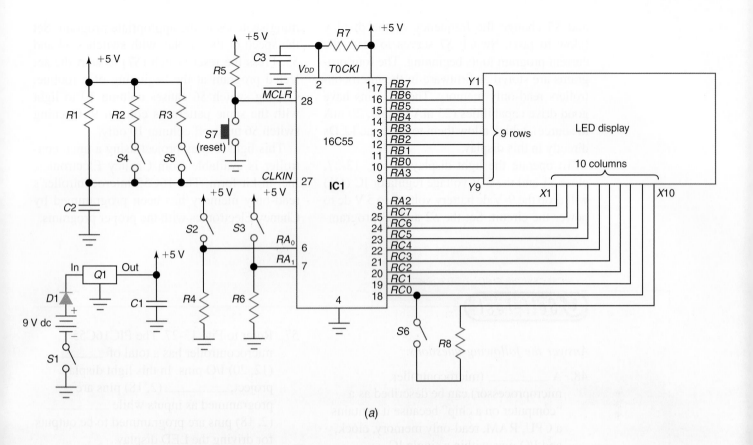

*(a)*

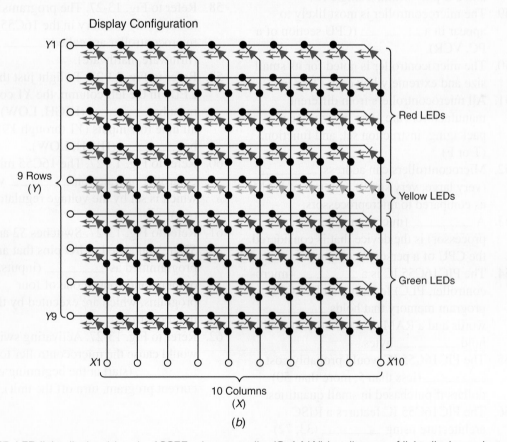

Display Configuration

Y1

9 Rows
(Y)

Red LEDs

Yellow LEDs

Green LEDs

Y9

X1

X10

10 Columns
(X)

*(b)*

**Fig. 13-27** LED light display driven by 16C55 microcontroller IC. (*a*) Wiring diagram of light display project. (*b*) Wiring diagram for nine-row by 10-column LED display. *(Courtesy of Chaney Electronics, Inc.)*

and *S5* change the frequency of the display (slow to fast). Switch *S7* serves to reset the current program to its beginning. The four programs are stored as firmware in the microcontrollers read-only memory. The I/O pins have good drive capabilities (25 mA to sink, 20 mA to source), which allow them to drive the LEDs directly in this display.

To operate the light display in Fig. 13-27, close *S1* and the 5-V voltage regulator IC (*Q1*) will drop the 9-V dc battery voltage to 5 V dc to power the circuit. Set the *S2* and *S3* program-ming switches to the appropriate program. Set the speed of the display with switches *S4* and *S5*. Press the reset switch (*S7*) to start the selected program at the beginning of a routine. Closing switch *S6* causes column 10 to light with the same pattern as column 9. Opening switch *S6* turns off column 10 only.

This light display project using a microcontroller is available from Chaney Electronics, Inc. in kit form. The 16C55 microcontroller's read-only memory has been programmed by Chaney Electronics with the proper programs.

## Self-Test

*Answer the following questions.*

48. A _____ (microcontroller, microprocessor) can be described as a "computer on a chip" because it contains a CPU, RAM, read-only memory, clock, and I/O pins within a single IC.
49. The microcontroller is most likely to appear in a _____ (CPU section of a PC, VCR).
50. The microcontroller is noted for its small size and extremely low cost. (T or F)
51. All microcontrollers from different manufacturers are alike in size, speed, packaging, instruction set, and function. (T or F)
52. Microcontrollers can address _____ (very large, very small) amounts of RAM as compared to microprocessors.
53. A _____ (microcontroller, microprocessor) is the device that is considered the CPU of a personal computer.
54. The PIC16C55 IC is a _____ (microcontroller, PLC) featuring an EPROM program memory that holds _____ words and a RAM data memory that will hold _____ bytes.
55. The PIC16C55 IC would probably cost _____ (less than 5, more than 50) dollars if purchased in small quantities.
56. The PIC16C55 IC features a RISC architecture using _____ (33, 72) instructions in its instruction set.

57. Refer to Fig. 13-27. The PIC16C55 microcontroller has a total of _____ (12, 20) I/O pins. In this light display project, _____ (2, 18) pins are programmed as inputs while _____ (2, 18) pins are programmed to be outputs for driving the LED display.
58. Refer to Fig. 13-27. The programs held in read-only memory in the 16C55 microcontroller are called _____ (firmware, hardware).
59. Refer to Fig. 13-27. To light just the nine LEDs in the left column, the *X*1 column must be _____ (HIGH, LOW) while all nine row inputs (*Y*1 through *Y*9) must be _____ (HIGH, LOW).
60. Refer to Fig. 13-27. The 16C55 microcontroller operates on _____ volts, which is set by the voltage regulator IC (*Q1*).
61. Refer to Fig. 13-27. Switches *S2* and *S3* are connected to I/O pins that are programmed as _____ (inputs, outputs) and control one of four programs, which are executed by the microcontroller.
62. Refer to Fig. 13-27. Activating switch *S7* would cause the microcontroller to _____ (start at the beginning of the current program, turn off the unit).

## 13-10 The Basic Stamp Microcontroller Modules

One of the most popular microcontrollers used in technical training is the BASIC Stamp by Parallax, Inc. The popularity of the BASIC Stamp modules is due to their ease of programming especially for beginners. BASIC Stamp modules are small (about the size of a postage stamp) and fairly inexpensive. Parallax also encourages the educational use of the BASIC Stamp modules with free downloads of both PBASIC editor software and educational materials from their educational website (www.stampsinclass.com).

Two versions of BASIC Stamp modules are sketched in Fig. 13-28. The modules are the BASIC Stamp 1 and BASIC Stamp 2. Programming of either of the BASIC Stamp modules is accomplished employing a modern PC using the correct *PBASIC* (Parallax BASIC) editor program. When the student has finished typing the PBASIC program on the PC it is then downloaded via the proper output port of the PC to the BASIC Stamp module. The *PBASIC interpreter* software translates from the downloaded code to machine code to operate the microcontroller. The cable from the PC can then be disconnected and the program remains in memory on the BASIC Stamp module. The downloaded program then resides in EEPROM where it is executed starting from the be-

ginning of the program each time the power is turned on to the BASIC Stamp module. The downloaded program is held in EEPROM even if the power to the BASIC Stamp module is turned off. An old program in EEPROM will be written over if a new program is downloaded from the PC. After programming the BASIC Stamp module, the microcontroller unit would operate independently. Notice that the simple BASIC Stamp 1 uses the parallel port (printer port) of the PC. The larger BASIC Stamp 2 receives downloading from the serial port of the PC.

The BASIC Stamp 1 module (BS1) shown in Fig. 13-28 is a small printed-circuit board packaged as a 14-pin SIP (single-in-line-package) IC measuring about 0.4 in wide by 1.4 in long. The BS1 module is powered by a 9V battery. An onboard dc voltage regulator drops the voltage to 5 V dc for use by the microcontroller and memory ICs. The main IC on the BS1 module is a custom PIC16C56 microcontroller chip with the PBASIC 1 interpreter in firmware. Because the PIC16C56's memory is used by the PBASIC interpreter, a separate 256 byte *program memory* is provided. The program memory, implemented using an EEPROM, can hold about 75 instructions. The BASIC Stamp 1 module has eight input/output (I/O) pins. The I/O pins used to control your device are digital in nature. Several special inputs/outputs include those for pulses, sound, PWM (pulse-width modulation)

**BASIC Stamp**

**PBASIC interpreter**

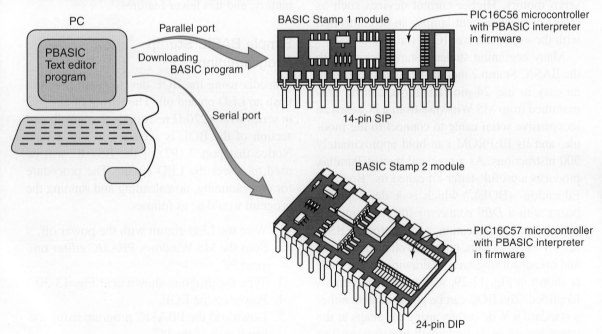

Fig. 13-28 Downloading PBASIC programs to BS1 or BS2 modules.

PC

PBASIC Text editor program

Parallel port

Downloading BASIC program

Serial port

BASIC Stamp 1 module

PIC16C56 microcontroller with PBASIC interpreter in firmware

14-pin SIP

BASIC Stamp 2 module

PIC16C57 microcontroller with PBASIC interpreter in firmware

24-pin DIP

BASIC Stamp information and downloads are available at www. parallaxinc.com and/or www. stampsinclass. com.

output, and potentiometer input. Besides the microcontroller, 256-byte EEPROM, and voltage-regulator ICs, the BS1 module also houses a ceramic resonator and a special reset IC.

The BASIC Stamp 2 module (BS2) shown in Fig. 13-28 is a small printed-curcuit board packaged as a 24-pin DIP IC. The BS2 module is powered by a 9V battery. An onboard dc voltage regulator drops the voltage to 5V dc for use by the microcontroller and memory ICs. The main IC on the BS2 module is a custom PIC16C57 microcontroller chip with the PBASIC 2 interpreter in firmware. Because the PIC16C57's memory is used by the PBASIC interpreter, a separate 2048 byte program memory is provided. The program memory, implemented using an EEPROM, can hold about 500 instructions. The BASIC Stamp 2 module has 16 input/output pins. The I/O pins used to control your device are digital in nature. Special inputs/outputs include those for pulses, PWM (pulse-width modulation) output, potentiometer input, X-10 appliance control, touchtone output, sound, and frequency measurement. Besides the microcontroller, 2048-byte EEPROM, and voltage-regulator ICs; the BS2 module also houses a ceramic resonator and a special reset IC along with some transistor buffers.

The current-drive capabilities of the BASIC Stamp modules are good ranging from 20 mA to 30 mA. This is enough to drive digital logic or even devices such as LEDs, piezo buzzers, or servo motors. Higher current devices such as relays or incandescent lamps can be controlled with the use of a driver IC or transistor.

Many beginning students start working with the BASIC Stamp 2 module because it comes in an easy to use 24-pin DIP form, can be programmed from MS Windows from a PC, uses an inexpensive serial cable to connect to the module, and its EEPROM can hold approximately 500 instructions. As a practical matter, Parallax produces a useful starter kit called its "Board of Education- (BOE)," which is a development board with a *DB9 connector* for programming and serial communication, a socket for the BS2-IC, 9 V battery snaps, onboard voltage regulator, and breadboarding area. A drawing of the BOE is shown in Fig. 13-29, with some key sections identified. The BOE can be powered with either a standard 9 V dc battery using the snaps at the upper left or an ac-to-dc wall transformer. The solderless-breadboarding area is used by students to wire their projects. I/O port connections from the BS2-IC are available just left of the solderless breadboard for easy connection using 22 gauge-solid wire. Power connections ($V_{dd}$ = +5V dc, and $V_{ss}$ = ground) are available above the solderless breadboard in Fig. 13-29. Serial downloading of PBASIC programs from your PC gain access to the BOE through the DB9 connector at the lower left.

The voltage regulator near to top of the BOE in Fig. 13-29 shifts the input voltage level to +5V (labeled $V_{dd}$) and negative ground (labeled $V_{ss}$). The connectors 12, 13, 14, and 15 are hobby servo connectors. The three-position slide switch near the bottom turns power off (position 0), power ON with servo ports OFF (position 1), or power ON with servo ports ON (position 2). The reset button will restart the program that is held in the program memory of the BS2-IC module. A top view of the BS2-IC is shown at the lower left on the BOE development board. The DB9 connector at the lower left in Fig. 13-29 is used from downloading from a PC via the serial port. The header X2 is for connecting the I/O ports of the BS2-IC to circuits on the solderless breadboard using 22-gauge wire. The XI header is a connector for add-on modules that are available from Parallax, Inc.

A less-expensive version of the BOE is called the BASIC Stamp HomeWork Board by Parallax, Inc. The HomeWork board features the BS2 module, accepts power from only a 9V battery, and has fewer features.

## Simple BASIC Stamp Programming

Consider using the BOE development board to flash an LED on and off. The wiring of an LED in series with a 220 Ω resistor on the breadboard section of the BOE is sketched in Fig. 13-30. Notice that port 7 (P7) of the BS2-IC will be used to power the LED circuit. The procedure for programming, downloading and running the program would be as follows:

1. Wire the LED circuit with the power off.
2. Start the MS Windows *PBASIC editor* on your PC.
3. Type the program shown near Fig. 13-30.
4. Power on the BOE.
5. Download the PBASIC program using the serial port of the PC.
6. Disconnect the serial download cable.
7. Turn BOE power off and then on.

**DB9 connector**

**PBASIC editor**

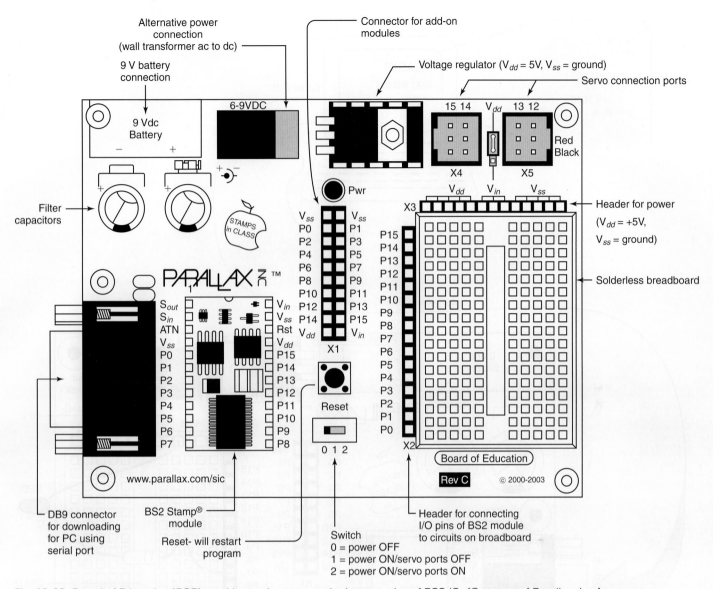

Fig. 13-29 Board of Education [BOE] used by students to study the operation of BS2-IC. *[Courtesy of Parallax, Inc.]*

8. The program will start at the beginning which will (1) make port 7 of the BS2 an output, (2) turn the LED on, (3) pause 1 second (1000 msec), (4) turn the LED off, (5) pause 1 second (1000 msec), and go to the beginning of the loop titled **blink:**.

The LED will blink continuously until the power is turned off.

The simple PBASIC blink program in Fig. 13-30 is detailed below. Notice the use of *remark* statements. In PBASIC, remark statements start

```
'Blinking LED 1          'Title of PBASIC program (Fig. 13-30)

output 7                 'Configure I/O port 7 as an output

blink:                   'Label for loop
   out 7 = 0             'Output 7 to logical 0 which turns on LED
   pause 1000            'Pause (do nothing) for 1000 milliseconds
   out 7 = 1             'Output 7 to logical 1 which turns off LED
   pause 1000            'Pause for 1000 msec
goto blink               'Go back to beginning of loop called blink
```

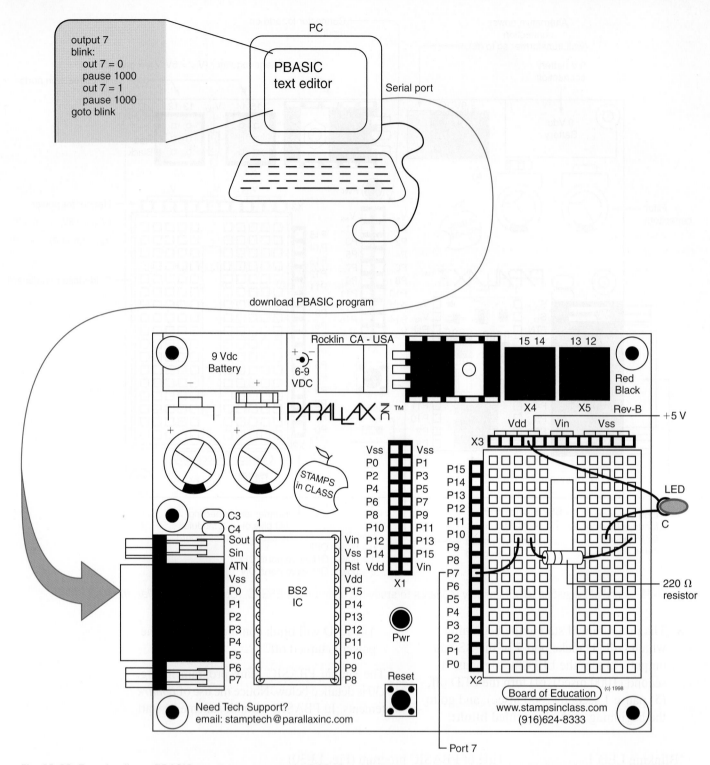

```
output 7
blink:
    out 7 = 0
    pause 1000
    out 7 = 1
    pause 1000
goto blink
```

PC

PBASIC
text editor

Serial port

download PBASIC program

Rocklin CA - USA

9 Vdc
Battery

6-9
VDC

15 14   13 12

Red
Black

X4       X5       Rev-B

+5 V

PARALLAX ™

STAMPS
in CLASS

Vdd   Vin   Vss

X3

LED

C

220 Ω
resistor

C3
C4
Sout
Sin
ATN
Vss
P0
P1
P2
P3
P4
P5
P6
P7

1

BS2
IC

Vin
Vss
Rst
Vdd
P15
P14
P13
P12
P11
P10
P9
P8

Vss    Vss
P0     P1
P2     P3
P4     P5
P6     P7
P8     P9
P10    P11
P12    P13
P14    P15
Vdd    Vin

X1

Pwr

Reset

P15
P14
P13
P12
P11
P10
P9
P8
P7
P6
P5
P4
P3
P2
P1
P0

X2

Board of Education    (c) 1998

www.stampsinclass.com
(916)624-8333

Need Tech Support?
email: stamptech@parallaxinc.com

Port 7

**Fig. 13-30** Downloading a PBASIC program to cause the BS2-IC to blink the LED circuit. Note the use of an older version of BOE.

with an apostrophe ('). These remark statements are not executed by the BS2 module but appear in the PBASIC editor listing to help humans understand the program.

The first line of the program starts with an apostrophe marking it is a remark statement. In this case **'Blinking LED 1** is the name of the program. Line 2 of the program is **output 7** which causes I/O port 7 to be configured as an output. This is clarified in the remark section of line 2. Line 3 (**blink:**) of the program is the name of the upcoming loop. The semicolon (:) after a word (such a **blink:** in this program) is interpreted by the microcontroller to be a *label*. The label (**blink:** in this example) is a reference point in the programming that can be referred to by other PBASIC commands. Line 4 (**out7 = 0**) causes the BS2 module to drive output P7 LOW which turns on the student wired LED shown in Fig. 13-30. Line 5 (**pause 1000**) causes the BS2 module to do nothing for about 1 second (1000 ms). This means the LED is ON for about 1 second. Line 6 (**out7 = 1**) causes the BS2 module to drive output P7 HIGH which turns off the LED. Line 7 (**pause 1000**) causes another 1s delay which means the LED is off during this time. Line 8 (**goto blink**) causes the program jump back to beginning of the loop labeled **blink:** and repeat the sequence. Remark statements are not required in PBASIC programs but are customary for titles and explaining program operation. Remark statements are very useful for beginners.

An LED and 220 Ω resistor were wired between the P7 I/O pin of the BS2 module and the positive rail ($V_{dd}$) of the power supply in Fig. 13-30. A simplified schematic diagram of this arrangement is shown in Fig. 13-31(*a*). The LED shown in Fig. 13-31(*a*) will be used as an output device in the next program.

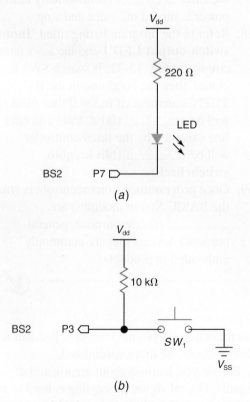

(a)

(b)

**Fig. 13-31** (*a*) Output LED wired to P7 port of BS2 module. (*b*) Input switch wired to P3 of BS2 module.

A pushbutton switch is being used as an input device (sensor) to I/O pin P3 of the BS2 module in Fig. 13-31(*b*). You will notice that the pushbutton is wired as an active-low switch. When the switch is open, I/O port P3 is HIGH and when the switch is closed input P3 goes LOW.

A program using both the LED output and pushbutton switch input connected to a BS2 module (hardware is diagrammed in Fig. 13-31) is listed below. When downloaded into a BS2 module, the PBASIC program constantly checks if the switch is closed. If the switch is open, the **switchcheck:** loop repeats. However, if the switch is closed the microcontroller will jump to the **blink:** routine and flash the LED. After flashing the LED on and off once the program returns to the **switchcheck:** routine. The LED will flash on and off the entire time the input switch SW1 is closed.

The IF-THEN statement in the **switchcheck:** routine might need some explanation. In this example the condition **if in3 = 0** (if input 3 is LOW) is evaluated by the microcontroller as either true or false. If the condition (**if in3 = 0** or in English if switch SW1 is closed) is true then the microcontroller causes a jump to the routine labeled **blink.** However, if the condition (**if in3 = 0** or in English if switch SW1 is closed) is false then the microcontroller continues to the next line of the program which is the **goto switchcheck** command. IF-THEN commands are very important in the operation of microcontroller programs because they are the decision-making statements.

Microcontrollers, such as the BASIC Stamp modules, can respond to a variety of inputs such as switches, or variations in light, temperature, position, voltage, or resistance. The microcontroller's outputs can drive a variety of devices such as LEDs, piezo buzzers, speakers, displays (LED or LCD), relays, servos, or stepper motors. BASIC Stamp modules, especially when premounted on a work surface such as Parallax's Board of Education (BOE), allow for an easy introduction to programming and using microcontrollers.

Remember that microcontrollers are tiny computer-like devices that are imbedded into products. Microcontrollers can respond to a limited number of inputs and control several output devices. After initial programming, most microcontrollers are single-use controllers as opposed to general-purpose personal computers.

```
'Input switch-output LED 1        'Title of PBASIC program (Fig. 13-31)

output 7                          'Configure I/O P7 as output
out 7 = 1                         'Output 7 to 1 turning off LED
input 3                           'Configure I/O P3 as input

switchcheck:                      'Label for switch checking routine
   if in 3 = 0 then blink         'If input 3 = 0 (switch closed) then go to blink routine

goto switchcheck                  'Check switch again (if input 3 = 1)

blink:                            'Label for LED blink routine
   out 7 = 0                      'Output 7 to 0 which turns on LED
   pause 500                      'Pause for 500 ms
   out 7 = 1                      'Output 7 to 1 which turns off LED
   pause 500                      'Pause for 500ms
goto switchcheck                  'Go back and check switch again
```

## Self-Test

*Answer the following questions.*

63. A BASIC Stamp module contains at least a _____ (microcontroller, microprocessor), voltage regulator, ceramic resonator, reset IC, and EEPROM chip for holding downloaded programs.

64. A BASIC Stamp 2 IC is programmed using a PBASIC editor program on your PC which is downloaded to the module via the _____ (parallel, serial) port of your computer.

65. BASIC Stamp modules accept programs download from a PC while a(n) _____ (interpreter, sequencer) program in firmware translates from the PBASIC high-level language to machine language used by the microcontroller.

66. Refer to Fig. 13-29. The $V_{dd}$ pin of the BASIC Stamp 2 module is connected to _____ (+5 V, ground), while a pin labeled P7 would be a(n) _____ (I/O port, power connection).

67. Refer to the program listing titled **'Blinking LED 1.** This program along with the circuit in Fig. 13-30 will blink the LED _____ (continuously until power is turned off, once and stop).

68. Refer to the program listing titled **'Input switch-output LED 1** and the associated circuits in Fig. 13-31. If switch SW1 is closed, then the condition in the IF-THEN statement (if in3 = 0 then blink) will be _____ (false, true) and next line executed by the microcontroller will be _____ (**blink:, goto switchcheck**).

69. Once programmed, microcontrollers (like the BASIC Stamp modules) are _____ (single-purpose, general-purpose) devices that are commonly embedded in products.

## 13-11 Digital Signal Processing

DSP

*Digital signal processing (DSP)* has become a very popular field in digital electronics. DSP is used in many pieces of equipment such modems, DVD players, MP3 players, and cellular phones. While these are similar to the microcomputer systems you have just studied, DSP systems are more specialized.

Earlier you learned about analog and digital signals. Digital signal processing is used to analyze and modify digital signals. Many digital signals used in DSP systems are acquired from

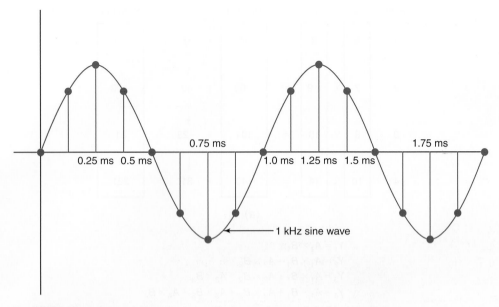

**Fig. 13-32** Sampling a 1 kHz sine wave during analog-to-digital conversion.

analog signals using a process known as sampling. *Sampling* can be thought of as taking snapshots of an analog signal's voltage at fixed or *discrete* points in time. For this reason, digital signals are also called *discrete-time signals*. An example of a discrete-time signal acquired by sampling a 1 kHz analog sine wave is shown in Fig. 13-32. The blue line represents the analog signal while the green dots and vertical lines describe the digital signal. The green dots/vertical lines are a "snapshot" of the analog signal at a given time and are digitized (converted into a number).

A simplified block diagram of a DSP system is shown in Fig. 13-33. The *A/D converter* (analog-to-digital converter) performs the sampling and converts the analog signal into a digital signal. The binary numbers from the A/D converter are stored in memory and used by the *digital signal processor*. The DSP performs many calculations which modifies the signal. The output of the DSP section is routed to the *D/A converter* (digital-to-analog converter). The D/A converter alters the signal from its digital to an analog form.

The most common type of calculation performed by the digital signal processor is known as the *sum-of-products*. An example of a sum-of-products calculation is detailed in Fig. 13-34(*a*). In each row, the number from column A is multiplied by the number in column B to yield the result (product) in the product column. The right column in Fig. 13-34(*a*) is the sum of the products. Notice that for each row, the product of the multiplication for that row is added together, or *summed,* with the sum of the previous products. In DSP, the word *accumulate* is used for the process of adding the new product to the total of the previous products. This process of *multiplying and accumulating* is known as *MAC.* DSP systems need to perform thousands if not millions of these calculations per second. The mathematical equations that represent the calculations for each row of Fig. 13-34(*a*) are written in Fig. 13-34(*b*). Engineers who work with DSP use a shortened form of these equations shown in Fig. 13-34(*c*). You may encounter equations like one in Fig. 13-34(*c*) if you work in the DSP field.

Digital signal processors are specialized microprocessors that are optimized to quickly perform the repetitive calculations required by the MAC process. DSPs can perform many

**Sum-of-products calculation**

**MAC**

**A/D converter**

**D/A converter**

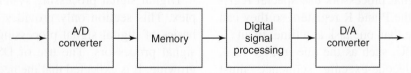

**Fig. 13-33** Simplified block diagram of a DSP system.

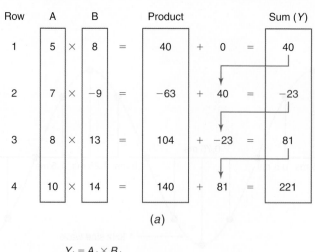

| Row | A | | B | | Product | | Sum (Y) |
|-----|---|---|---|---|---------|---|---------|
| 1 | 5 | × | 8 | = | 40 | + 0 = | 40 |
| 2 | 7 | × | −9 | = | −63 | + 40 = | −23 |
| 3 | 8 | × | 13 | = | 104 | + −23 = | 81 |
| 4 | 10 | × | 14 | = | 140 | + 81 = | 221 |

(a)

$$Y_1 = A_1 \times B_1$$
$$Y_2 = A_1 \times B_1 + A_2 \times B_2$$
$$Y_3 = A_1 \times B_1 + A_2 \times B_2 + A_3 \times B_3$$
$$Y_4 = A_1 \times B_1 + A_2 \times B_2 + A_3 \times B_3 + A_4 \times B_4$$

(b)

$$Y_i = \sum_{i=1}^{4} A_i \cdot B_i$$

(c)

**Fig. 13-34** Sum-of-products DSP calculations.
(a) An example (b) Formulas (c) Another formula.

**MIPS**

*millions instructions per second (MIPS).* The basic architecture of a DSP is detailed in Fig. 13-35. Data are read into the separate *program memory* and *data memory* sections of the DSP. Samples from the A/D converter (see Fig. 13-33), which were stored in memory, are read into the data memory of the DSP. Fixed numbers called

**Coefficients**

*coefficients,* which are designed to change the digital signal in a specific way, are read into the program memory. The numbers from program- and data-memory are multiplied together in the *multiplier* and stored in the *P register.* The *accumulator* then adds the results of the products stored in the P register with the previous sum of the products and stores the new result in the *R register.* Each output of the accumulator, stored in the R register, is a new output sample of the processed digital signal. The output is sent to the D/A converter to be changed back to an analog signal.

Digital signal processors use special registers such as the P and R registers so they can work on more than one task at a time. In an elementary MPU, such as the one in Fig. 13-4, each fetch-decode-execute sequence must finish before the next one can begin. DSPs,

**Pipelining**

however, can fetch a new instruction while the previous instruction is being decoded and another instruction is being executed. This process of starting a new task before the current one is finished is known as *pipelining.* Due to pipelining, the DSP in Fig. 13-35 will be performing all of the following tasks at the same time:

1. Writing the contents of the R register to the D/A converter,
2. Accumulating a new resulting sum of previous products,
3. Multiplying two numbers from the program- and data-memories, and
4. Fetching a new sample into the data memory.

Pipelining is also used in most modern microprocessors. Pipelining may take various forms in microprocessors or DSPs but its purpose is to speed up the execution of instructions.

Digital signal processing systems are complex. This section only provides some of the basics of digital signal processing and digital signal processors. The use of DSP has been growing. It is expected that the need for people familiar with DSP will continue to grow.

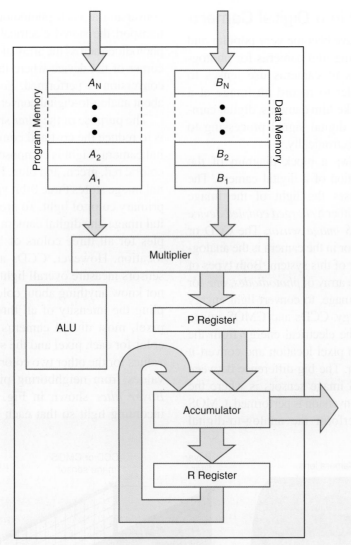

Fig. 13-35 Architecture of a digital signal processor.

---

*Answer the following questions.*

70. In the DSP field, another name that might be used for a digital signal is a _____ signal.

71. During digital signal processing, what type of calculation is common?

72. In the DSP field, MAC stands for the process of _____ _____ _____.

73. In a digital signal processor or other microprocessor, starting a task before the previous one is finished is called _____.

74. A digital signal processor (DSP) is an IC designed for high-speed data manipulation used in audio, communications, image manipulation and other data acquisition and data control applications. (T or F)

## 13-12 DSP in a Digital Camera

Digital cameras have become very popular and are rapidly replacing film cameras for photography. Both types of cameras use lenses to focus light in order to record an image at a point in time. Unlike film cameras, digital cameras make use of digital signal processing to store the image electronically.

Fig. 13-36 shows a block diagram of the image capture section of a digital camera. The camera lens focuses the light of the image through a filter to either a *charged coupled device (CCD)* or a *CMOS image sensor.* The CCD or CMOS image sensor in the camera is the analog-to-digital converter of this system. Both types of light sensors use an array of *photodiodes,* one for each *pixel* in the image, to convert light energy into electrical energy. CCDs and CMOS image sensors measure the electrical energy from the photodiode at each pixel location and convert it to a digital number. The big difference between CCDs and CMOS image sensors is where the analog to digital conversion is performed. CMOS image sensors perform the analog-to-digital

conversion at each photodiode location. CCDs transport the stored electrical charge from each photodiode across the array of photodiodes to one corner of the device where the analog-to-digital conversion is performed. You will learn more about analog-to-digital conversion in Chapter 14.

The purpose of the *filter* shown in Fig. 13-36 is to reduce the cost and complexity of the digital camera. Light is composed of three primary colors: red, green, and blue. Each pixel in a digital image stores three 8-bit values, one for each primary color of light. To properly create a digital image, the digital camera has to create samples for all three colors of light at each pixel location. However, CCDs and CMOS image sensors measure overall light intensity; they do not know anything about color. Instead of sampling the intensity of all three colors for each pixel, most digital cameras sample only one color for each pixel and the system later inserts values for the other two colors by looking at the values from neighboring pixels. The special *Bayer filter* shown in Fig. 13-36 filters the incoming light so that each photodiode in the

**Charged coupled device (CCD)**

**CMOS image sensor**

**Bayer filter**

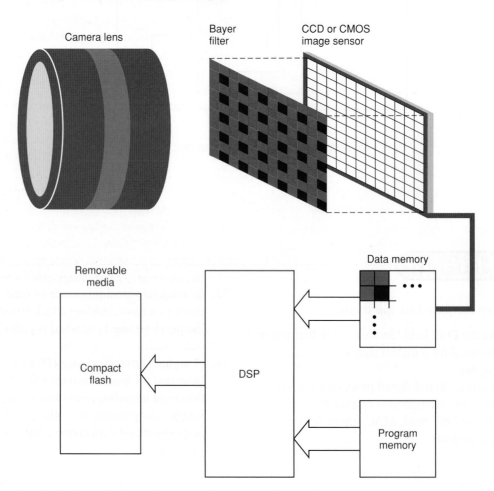

**Fig. 13-36** Block diagram of image capture in a digital camera.

image sensor sees only one color of light. The color of each square in the Bayer filter is the color of light that is allowed to pass through that filter to the photodiode of the image sensor behind it. You can see from Fig. 13-36 that the pattern of the Bayer filter is rows of red and green filters alternating with rows of blue and green filters. The reason that there are more green filters than red and blue filters is that the human eye is more sensitive to the green component of light than it is to the other two colors.

The output of the image sensor block of Fig. 13-36 is a series of bytes containing the sampled digital values from each photodiode in the image sensor. This series of sampled bytes is the digital signal input to the rest of the system. The samples are stored in temporary memory and read into the data input of the digital signal processor. Recall that a stored digital image needs three values, one for each primary color, for each pixel. At this point in the system, the digital signal has only one sampled value for each pixel. The digital signal processor in Fig. 13-36 runs an *algorithm,* or sequence of steps, to calculate the two missing color values for each pixel based on the sampled values from neighboring pixels. The process of calculating and inserting new data values from known existing values is called *interpolation.* Coefficients for the interpolation algorithm are the input to the program memory of the digital signal processor. The output from the digital signal processing block of Fig. 13-36 is a series of three 8-bit values for the red, green, and blue components of each pixel in the image. When the image is viewed, these three values are recombined to recreate the image.

Some cameras may store this RGB (for red, green, blue) output of the DSP block directly to a removable memory card in the camera as shown in Fig. 13-36. However, digital images can be very large. For example, a five megapixel digital camera has five million image sensors, each measuring only one color of light for each pixel of the image. So each picture taken with a five megapixel camera would require 15 megabytes of memory. For this reason, most digital cameras also do some further processing of the digital signal. Digital cameras often compress the digital image so that it takes up less space when stored on the compact flash or other removable media card in the camera. The most common compression algorithm used by digital cameras is called JPEG after the group that created it, the Joint Photographic Experts Group. When compression is done, the original RGB output of the DSP block may once again be stored in temporary memory and read back into the DSP's data memory path. The program memory for the DSP is changed so that it contains coefficients used by the JPEG compression algorithm. The output of the DSP block is the JPEG compressed image that is then stored on the removable memory card.

**JPEG**

The digital-to-analog conversion block of a digital signal processing system is not shown in Fig. 13-36. The digital-to-analog conversion for this system may happen in many places once the digital camera stores the image. The D/A conversion happens when the image is viewed on the LCD screen of the camera. If the image is moved to a personal computer, the D/A conversion will happen when the image is viewed on a monitor attached to the computer. The D/A conversion will also occur when the image is printed on a printer or taken to a photo developer that can print digital images.

This section examined some digital signal processing that occurs within a digital camera. Digital cameras use several digital signal processing algorithms to provide many features. Many other devices, such as cell phones, digital video cameras, and MP3 players, also use digital signal processing. Digital signal processing has become a part of many systems that we use in our everyday lives.

## ✓ Self-Test

*Answer the following questions.*

75. Image sensors in digital cameras use _____ (solar cells, photodiodes) to convert light energy into electrical energy.

76. Green light is _____ (blocked from passing through, allowed to pass through) a green light filter.

77. _____ (Inspiration, Interpolation) is the process of calculation and inserting new values from known existing values.

78. Digital images store values for the three primary colors of light, which are _____, _____ and _____.

# Chapter 13 Summary and Review

1. The computer is one of the most complex digital systems. It is unique because of its adaptability, vast size, high speed, and stored program.

2. The microcomputer is slower and less expensive than its larger counterparts. The microcomputer is a microprocessor-based digital system.

3. The microcomputer makes extensive use of both ROM and RAM for internal storage. Floppy, optical, and hard disks are widely used for secondary bulk storage. Microcomputers support many peripheral input and output devices including networks.

4. Microprocessing unit instructions are composed of the operation and operand parts. The MPU follows the fetch-decode-execute sequence when running a program.

5. Combinational logic gates can be used for microcomputer address decoding.

6. Three-state devices, such as buffers, must be used when several memories and microprocessors transfer information over a common data bus.

7. Multiplexers and demultiplexers can be used for data transmission. More complex UARTs may also be used for serial data transmission.

8. Data transmission can be either serial or parallel. Various interface ICs are available for sending and receiving parallel and serial data.

9. Errors occurring during data transmission can be detected using parity bits or cyclic redundancy checks.

10. A programmable logic controller (PLC) is a rugged computer system used in factories, warehouses, and chemical plants to control machines. PLCs are replacing hard-wired relay logic for machine control.

11. Ladder relay schematics, ladder relay logic diagrams, logic gate diagrams, and Boolean expressions can all be used to describe a control logic problem.

12. A microcontroller is a "computer on a chip" embedded in many everyday devices. Microcontrollers contain a CPU, a small RAM (data memory), read-only memory (program memory containing firmware), a clock, and I/O pins.

13. Microcontrollers are produced in huge quantities at very low prices.

14. BASIC Stamp modules allow students and others to easily program and download code for directing the action of a microcontroller.

15. The PBASIC high-level language is used to program BASIC Stamp modules.

16. A digital signal processor (DSP) is a specialized microprocessor designed for high-speed data manipulation used in audio, communications, image manipulation, and other data acquisition and data control applications.

17. DSP devices are commonly used as part of a system including A/D converters, memory, DSP, and D/A converters.

*Answer the following questions.*

13-1. The CPU of a computer contains what three sections?

13-2. The _____ (ALU, MUX) section of a computer performs calculations and logic functions.

13-3. The more complex digital system is a _____ (computer, digital multimeter).

13-4. An IC called a(n) _____ is the heart of the CPU of a microcomputer.

13-5. Refer to Fig. 13-3. The parts of a microcomputer system are connected by control lines, a(n) _____ bus, and a two-way _____.

13-6. The input-store-print operation shown in Fig. 13-6 required three instructions, which use _____ bytes of program memory.

13-7. Classify these microcomputer peripheral devices as input, output, storage, or input/output units:
   a. CRT monitor       f. laser printer
   b. floppy disk drive  g. hard disk drive
   c. keyboard          h. plotter
   d. mouse             i. ethernet
   e. modem

13-8. Microcomputer memory addresses are commonly listed in _____ (Gray code, hexadecimal).

13-9. What do the following letters stand for when referring to a microcomputer system?
   a. CPU       c. PPI
   b. PIA       d. UART

13-10. The baud rate is the number of _____ per second being transmitted serially through a data link.

13-11. The IEEE-488 standard is a common _____ (parallel, serial) interface standard for the data link between a computer and scientific instrumentation.

13-12. Draw the logic symbol and truth table for a three-state buffer.

13-13. A MUX/DEMUX system converts parallel input data to _____ (asynchronous, serial) data for transmission.

13-14. A MUX/DEMUX system operates somewhat like two _____ (rotary, three-way) switches.

13-15. Errors in transmission can be detected by using a _____ (parity bit, rotary switch).

13-16. An _____ (AND, XOR) gate can detect an odd number of 1s at its input.

13-17. The programmable controller is also commonly known as the _____ _____ _____, or PLC.

13-18. A programmable logic controller (PLC) is a heavy-duty computer system designed for _____ (general-purpose office use; machine control in factories, warehouses, and chemical plants).

13-19. Once programmed, inputs to a _____ (microcomputer, PLC) would probably come from devices such as limit switches, pressure switches, and temperature and optical sensors.

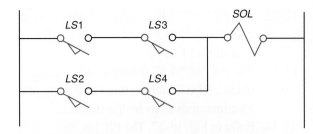

Fig. 13-37 Relay schematic diagram.

13-20. A programming module is always connected to a programmable logic controller, because program changes need to be made each moment. (T or F)

13-21. Refer to Fig. 13-23. The power supply, processing, input, and output sections are referred to as modules because they are sometimes physically housed in separate enclosures in larger systems. (T or F)

13-22. Given the relay schematic in Fig. 13-37, draw the ladder logic program that might be used with a PLC for this circuit.

13-23. Given the relay schematic in Fig. 13-37, write the Boolean expression that describes the logic function of this circuit.

13-24. A _____ (microcontroller, programmable logic controller) can be described as a "computer on a chip" because it contains a CPU, RAM, read-only memory, clock, and I/O pins within a single IC.

13-25. Microcontrollers are most likely to appear _____ (in the CPU section of a PC, embedded in electronic devices in your automobile).

13-26. The microcontroller is noted for its small size and _____ (high, very low) cost.

13-27. Program memory in a microcontroller is held in a read-only memory device and is _____ (constantly, rarely) reprogrammed.

13-28. Microcontrollers are manufactured in _____ (large, small) volumes.

13-29. Compared to microprocessors, microcontrollers commonly address _____ (very large, very small) amounts of RAM (data memory).

13-30. A microcontroller is the device that is considered the CPU of a personal computer. (T or F)

13-31. The PIC16C55 IC is a _____ (microcontroller, PLC) featuring an EPROM program memory that holds 512 words and a RAM data memory that will hold 24 bytes.

13-32. The PIC16C55 IC would probably cost less than five dollars if purchased in small quantities. (T or F)

13-33. The PIC16C55 IC features a(n) _____ (CISC, RISC) architecture employing only 33 commands in its instruction set.

13-34. Refer to Fig. 13-27. The PIC16C55 microcontroller has a total of _____ (12, 20) I/O pins. In this light-display project, _____ (2, 18) pins are programmed as inputs and _____ (2, 18) pins are programmed to be outputs for driving the LED display.

13-35. Refer to Fig. 13-27. The programs held in read-only memory in the 16C55 microcontroller may be called firmware. (T or F)

13-36. Refer to Fig. 13-27. To light just the 10 red LEDs across the top of the display, the Y1 row input must be _____ (HIGH, LOW), and all 10 column inputs (X1 through X10) must be _____ (HIGH, LOW).

13-37. Refer to Fig. 13-27. Switches S2 and S3 are connected to I/O pins that are programmed as _____ (inputs, outputs) and control which one of four programs is executed by the microcontroller.

13-38. Refer to Fig. 13-27. Activating switch S7 would cause the microcontroller to _____ (reset at the beginning of the current program, turn off the unit).

13-39. Refer to Fig. 13-27. The PIC16C55 microcontroller's I/O pins have enough drive capability to drive the LED display directly. (T or F)

13-40. The BASIC Stamp 2 is packaged as a 24-pin DIP and contains several components including a custom PIC16C57 _____

(microcontroller, PLC) with PBASIC interpreter in firmware.

13-41. The _____ (EEPROM, ROM) IC on the BASIC Stamp module is used for program memory.

13-42. Programming in PBASIC is done on a _____ (PC, small 12-key keypad) and then downloaded to the BASIC Stamp module.

13-43. The _____ (parallel, serial) port of a PC is used for downloading to the BASIC Stamp 2 module.

13-44. Refer to Fig. 13-29. Parallax's Board of Education development board is used by students when studying and experimenting with the _____ (BS1-IC, BS2-IC).

13-45. Refer to the PBASIC program titled 'Blinking LED 1'. The purpose of the first pause 1000 line of code would be to wait 1 second (1000 ms) for an input from an attached switch. (T or F)

13-46. The acronym DSP stands for _____.

13-47. In DSP, a digital signal might be referred to as a _____ (discrete-time, random-time) signal.

13-48. Refer to Fig. 13-33. Sampling should be performed by the _____ (A/D converter, DSP) section of the system.

13-49. The most common type of calculation performed by a digital signal processor is known as _____ (multiply-and-shift left, sum-of-products).

13-50. Digital signal processors can perform _____ (a few thousand, millions of) instructions per second.

13-51. Refer to Fig. 13-33. The output of the DSP block is digital in nature and the D/A converter changes this to a(n) _____ (analog, multiplexed) signal.

## Critical Thinking Questions

13-1. Draw a block diagram of the organization of the five main sections of a computer. Show the flow of *program* information and *data* through the system.

13-2. Why do PLCs simulate relay logic so closely?

13-3. Given the relay schematic in Fig. 13-38, draw the ladder logic program that might be used with a PLC for this circuit.

13-4. Given the relay schematic in Fig. 13-38, draw the logic gate diagram for this circuit (use AND

and OR symbols) and write the Boolean expression that describes the logic of the circuit.

13-5. Describe how you might light all of the red LEDs (top four rows) on the display shown in Fig. 13-27(b).

13-6. At the option of your instructor, use circuit simulation software to (1) draw a four-row by four-column LED matrix [something like that shown in Fig. 13-27(b)], (2) use a word generator

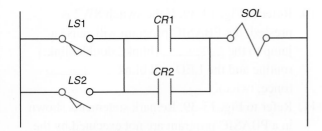

Fig. 13-38 Relay schematic diagram.

to program light patterns on the 4 × 4 LED display, (3) operate your 4 × 4 LED display, and (4) show your instructor your light pattern generator.

13-7. Refer to Fig. 13-39. The inputs in this problem are both _____ (active-HIGH, active-LOW) switches and the output LED is activated (turned on) with a _____ (HIGH, LOW).

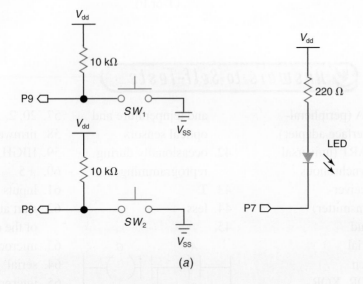

(a)

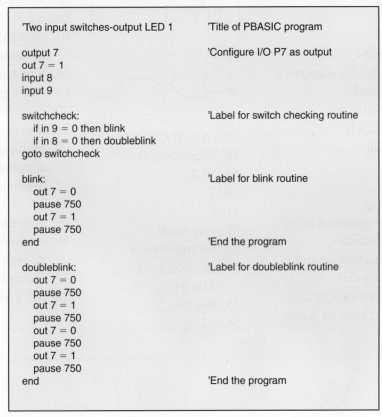

```
'Two input switches-output LED 1        'Title of PBASIC program

output 7                                'Configure I/O P7 as output
out 7 = 1
input 8
input 9

switchcheck:                            'Label for switch checking routine
    if in 9 = 0 then blink
    if in 8 = 0 then doubleblink
goto switchcheck

blink:                                  'Label for blink routine
    out 7 = 0
    pause 750
    out 7 = 1
    pause 750
end                                     'End the program

doubleblink:                            'Label for doubleblink routine
    out 7 = 0
    pause 750
    out 7 = 1
    pause 750
    out 7 = 0
    pause 750
    out 7 = 1
    pause 750
end                                     'End the program
```

(b)

Fig. 13-39 Problem using BASIC Stamp 2. (a) External inputs and output. (b) PBASIC program listing.

13-8. Refer to Fig. 13-39. Line 4 in the PBASIC program (input 8) causes I/O port 8 of the BASIC Stamp 2 to be configured as a(n) _____ (input, time delay port).

13-9. Refer to Fig. 13-39. If the switch SW1 is pressed, the PBASIC program will cause a jump to the _____ (blink:, doubleblink:) routine and the LED will blink _____ (once, twice).

13-10. Refer to Fig. 13-39. If the switch SW2 is pressed, the PBASIC program will cause a jump to the _____ (blink:, doubleblink:) routine and the LED will blink _____ (once, twice).

13-11. Refer to Fig. 13-39. Remark statements shown in a PBASIC program are not executed by the microcontroller in the BASIC Stamp module. (T or F)

## ✔ Answers to Self-Tests

1. peripheral
2. size, stored program, speed, multipurpose
3. program information and data
4. data
5. programs
6. address, control
7. modem
8. CRT or LCD monitor
9. mouse
10. LAN
11. DSL
12. operation, operand
13. 100, 101
14. MPU (microprocessor)
15. decode-execute
16. sequential
17. address decoder
18. three-state buffers or tristate buffers
19. 0, 8
20. in its high-impedance state
21. multiplexer, demultiplexer
22. 1111

23. PIA (peripheral-interface adapter)
24. UART (universal asynchronous receiver-transmitter)
25. baud
26. serial
27. even
28. XOR, XOR
29. 1
30. 0
31. XOR
32. Cyclic Redundancy Check
33. ports
34. UARTs
35. full
36. F
37. F
38. flag
39. programmable logic controller
40. machine control in factories
41. limit switches, pressure switches,

and temperature and optical sensors
42. occasionally during reprogramming
43. T
44. less
45.

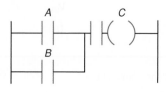

46.

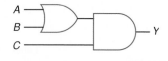

47. (A + B)C = Y
48. microcontroller
49. VCR
50. T
51. F
52. very small
53. microprocessor
54. microcontroller, 512, 24
55. less than 5
56. 33

57. 20, 2, 18
58. firmware
59. HIGH, LOW
60. +5
61. inputs
62. start at the beginning of the current program
63. microcontroller
64. serial
65. interpreter
66. +5 V, I/O port
67. continuously until power is turned off
68. true, blink:
69. single-purpose
70. discrete-time
71. sum-of-products
72. Multiplying and accumulating
73. pipelining
74. T
75. photodiodes
76. is allowed to pass through
77. Interpolation
78. red, green, blue

# Connecting with Analog Devices

## Chapter Objectives

*This chapter will help you to:*

1. Discuss analog-to-digital and digital-to-analog conversion.

2. *Design* an op-amp circuit with a given gain.

3. *Analyze* the operation of several elementary D/A converter circuits.

4. *Answer* selected questions on a counter-ramp A/D converter with voltage comparator.

5. *Discuss* the operation of an elementary digital voltmeter (A/D converter).

6. *Identify* several other A/D converters including the ramp and successive-approximation types.

7. *List* several specifications associated with commercial A/D converters.

8. *Answer* selected questions about the commercial ADC0804 A/D converter IC.

9. *Analyze* several digital light meter systems that are based on the ADC0804 A/D converter IC.

10. *Analyze* a digitizer circuit with a thermistor as the temperature transducer and a Schmitt trigger device used as an elementary A/D converter.

To this point in our study, most data entering or leaving a digital system have been digital information. Many digital systems, however, have *analog* inputs that vary *continuously* between two voltage levels. In this chapter, we discuss the *interfacing* of analog devices to digital systems.

**Interfacing**

Most real-world information is analog. For instance, time, speed, weight, pressure, light intensity, and position measurements are all *analog in nature.*

The digital system in Fig. 14-1 has an analog input. The voltage varies continuously from 0 to 3 V. The *encoder* is an electronic device that converts the analog signal to digital information. The encoder is called an *analog-to-digital converter* or *A/D converter.* The A/D converter, then, converts analog information to digital data.

**Analog-to-digital converter**

**A/D converter**

The digital system diagramed in Fig. 14-1 also has a *decoder.* This decoder is a special type: it converts the digital information from the digital processing unit to an analog output. For instance, the analog output may be a continuous voltage change from 0 to 3 V. We call this decoder a *digital-to-analog converter* or *D/A converter.* The D/A converter, then, decodes digital information to analog form.

**D/A converter**

The entire system in Fig. 14-1 might be called a *hybrid system* because it contains both digital and analog devices. The encoders and decoders that convert from analog to digital and digital to analog are called *interface devices* by engineers and technicians. The word "interface" is generally used when referring to a device or circuit that converts from one mode of operation to another. In this case we are converting between analog and digital data.

**Hybrid system**

Note that the input block in Fig. 14-1 refers to an analog voltage ranging from 0 to 3 V. This

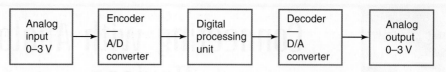

Fig. 14-1 A digital system with analog input and analog output.

DIGITAL
INPUT

ANALOG
OUTPUT

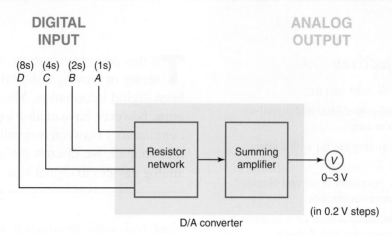

Fig. 14-2 Block diagram of a D/A converter.

**Transducer**

**Resistor network**

**Summing amplifier (scaling amplifier)**

voltage could be produced by a transducer. A *transducer* is defined as a device that converts one form of energy to another. For instance, a photocell could be used as an input transducer to give a voltage proportional to light intensity. In this example, light energy is being converted into electrical energy by the photocell. Other transducers might include microphones, speakers, strain gauges, photoresistive cells, temperature sensors, potentiometers, and hall-effect sensors.

## 14-1 D/A Conversion

Refer to the D/A converter in Fig. 14-1. Let us suppose we want to convert the binary from the processing unit to a 0- to 3-V output. As with any decoder, we must first set up a truth table of all the possible situations. Table 14-1 shows four inputs (D, C, B, A) into the D/A converter. The inputs are in binary form so the exact value of the inputs is not important. Each 1 is about +3 to +5 V. Each 0 is about 0 V. The outputs are shown as voltages in the rightmost column in Table 14-1. According to the table, if binary 0000 appears at the input of the D/A converter, the output is 0 V. If binary 0001 is the input, the output is 0.2 V. If binary 0010 appears at the input, then the output is 0.4 V. Notice that for each row you progress downward in Table 14-1, the analog output increases by 0.2 V.

A block diagram of a D/A converter is shown in Fig. 14-2. The digital inputs (D, C, B, A) are at the left. The decoder consists of two sections: the *resistor network* and the *summing*

| Table 14-1 | Truth Table for D/A Converter | | | | |
|---|---|---|---|---|---|
| | Digital inputs | | | | Analog output |
| | D | C | B | A | Volts |
| Row 1 | 0 | 0 | 0 | 0 | 0 |
| Row 2 | 0 | 0 | 0 | 1 | 0.2 |
| Row 3 | 0 | 0 | 1 | 0 | 0.4 |
| Row 4 | 0 | 0 | 1 | 1 | 0.6 |
| Row 5 | 0 | 1 | 0 | 0 | 0.8 |
| Row 6 | 0 | 1 | 0 | 1 | 1.0 |
| Row 7 | 0 | 1 | 1 | 0 | 1.2 |
| Row 8 | 0 | 1 | 1 | 1 | 1.4 |
| Row 9 | 1 | 0 | 0 | 0 | 1.6 |
| Row 10 | 1 | 0 | 0 | 1 | 1.8 |
| Row 11 | 1 | 0 | 1 | 0 | 2.0 |
| Row 12 | 1 | 0 | 1 | 1 | 2.2 |
| Row 13 | 1 | 1 | 0 | 0 | 2.4 |
| Row 14 | 1 | 1 | 0 | 1 | 2.6 |
| Row 15 | 1 | 1 | 1 | 0 | 2.8 |
| Row 16 | 1 | 1 | 1 | 1 | 3.0 |

*amplifier.* The output is shown as a voltage reading on the voltmeter at the right.

The resistor network in Fig. 14-2 must take into account that a 1 at input *B* is worth twice as much as a 1 at input *A*. Also, a 1 at input *C* is worth four times as much as 1 at input *A*. Several arrangements of resistors are used to do this job. These circuits are called *resistive ladder networks*.

The summing amplifier in Fig. 14-2 takes the output voltage from the resistor network and amplifies it the proper amount to get the voltages shown in the rightmost column of Table 14-1. The summing amplifier typically uses an IC unit called an *operational amplifier*. An operational amplifier is often simply called an *op amp*. The summing amplifier is also called a *scaling amplifier*.

The special decoder called a D/A converter consists of two parts: a group of resistors forming a resistive ladder network and an op amp used as the summing amplifier.

---

### ✔ Self-Test

*Supply the missing word in each statement.*

1. A special encoder that converts from analog to digital information is called a(n) _____.
2. A special decoder that converts from digital to analog information is called a(n) _____.
3. A D/A converter consists of a(n) _____ network and a(n) _____ amplifier.
4. The name "op amp" stands for _____.

5. Refer to Fig. 14-2 and Table 14-1. If the binary input to the D/A converter is $0111_2$, the analog output will be _____ volts.
6. Refer to Fig. 14-2 and Table 14-1. If the binary input to the D/A converter is $1111_2$, the analog output will be _____ volts.
7. Refer to Fig. 14-2 and Table 14-1. If the binary input increases from 0001 to 0010, the analog output will increase by _____ volts.

---

## 14-2 Operational Amplifiers

The special amplifiers called *op amps* are characterized by high input impedance, low output impedance, and a variable voltage gain that can be set with external resistors. The symbol for an op amp is shown in Fig. 14-3(*a*). The op amp shown has two inputs. The top input is labeled an *inverting input*. The inverting input is shown by the minus sign ($-$) on the symbol. The other input is labeled a *noninverting input*. The noninverting input is shown by the plus sign ($+$) on the symbol. The output of the amplifier is shown on the right of the symbol.

The operational amplifier is almost never used alone. Typically, the two resistors shown in Fig. 14-3(*b*) are added to the op amp to set the voltage gain of the amplifier. Resistor $R_{in}$ is called the input resistor. Resistor $R_f$ is called the feedback resistor. The *voltage gain* of this amplifier is found by using the simple formula

$$A_v \text{ (voltage gain)} = \frac{R_f}{R_{in}}$$

Suppose the values of the resistors connected to the op amp are $R_f = 10 \text{ k}\Omega$ and $R_{in} = 10 \text{ k}\Omega$. Using our voltage gain formula, we find that

$$A_v = \frac{R_f}{R_{in}} = \frac{10,000}{10,000} = 1$$

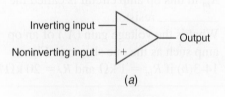

(*a*)

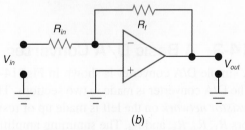

(*b*)

**Fig. 14-3** Operational amplifier. (*a*) Symbol. (*b*) With input and feedback resistors for setting gain.

The gain of the amplifier is 1. In our example, if the input voltage at $V_{in}$ in Fig. 14-3(b) is 5 V, the output voltage at $V_o$ is 5 V. The inverting input is being used, and so if the input voltage is +5 V, then the output voltage is −5 V. The voltage gain of the op amp can also be calculated using the formula

$$A_v = \frac{V_{out}}{V_{in}}$$

The voltage gain for the circuit above is then

$$A_v = \frac{V_{out}}{V_{in}} = \frac{5}{5} = 1$$

The voltage gain is again found to be 1.

Suppose the *input and feedback resistors* are 1 kΩ and 10 kΩ, as shown in Fig. 14-4. What is the voltage gain for this circuit? The voltage gain is calculated as

$$A_v = \frac{R_f}{R_{in}} = \frac{10{,}000}{1000} = 10$$

The voltage gain is 10. If the input voltage is +0.5 V, then the voltage at the output is how many volts? If the gain is 10, then the input voltage of 0.5 V times 10 equals 5 V. The output

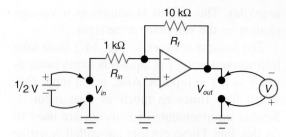

**Fig. 14-4** Amplifier circuit using an op amp.

voltage is −5 V, as measured on the voltmeter in Fig. 14-4.

You have seen how the voltage gain of an op amp can be changed by changing the ratio between the input and feedback resistors. You should know how to set the gain of an operational amplifier by using different values for $R_{in}$ and $R_f$.

In summary, the op amp is part of a D/A converter; it is used as a *summing amplifier* in the converter. The gain of the op amp is easily set by the ratio of the input and feedback resistors.

**Input and feedback resistors**

**Summing amplifier**

---

## Self-Test

*Answer the following questions.*

8. Refer to Fig. 14-3(b). The resistor labeled $R_f$ in this op amp circuit is called the _____ resistor.
9. Refer to Fig. 14-3(b). The resistor labeled $R_{in}$ in this op amp circuit is called the _____ resistor.
10. What is the voltage gain ($A_v$) of an op amp such as the one shown in Fig. 14-3(b) if $R_{in} = 1$ kΩ and $R_f = 20$ kΩ?

11. What is the output voltage ($V_o$) from the op amp in question 10 if the input voltage is +0.2 V?
12. What is the voltage gain ($A_v$) of an op amp such as the one shown in Fig. 14-3(b) if $R_{in} = 5$ kΩ and $R_f = 20$ kΩ?
13. What is the output voltage ($V_o$) from the op amp in question 12 if the input voltage is +1.0 V?

---

## 14-3 A Basic D/A Converter

A simple D/A converter is shown in Fig. 14-5. The D/A converter is made in two sections. The *resistor network* on the left is made up of resistors $R_1$, $R_2$, $R_3$, and $R_4$. The summing amplifier on the right consists of an op amp and a feedback resistor. The input ($V_{in}$) is 3 V applied

**Resistor network**

to switches *D*, *C*, *B*, and *A*. The output voltage ($V_o$) is measured on a voltmeter. Notice that the op amp requires a dual power supply: a +10-V supply and a −10-V supply.

With all switches at GND (0 V), as shown in Fig. 14-5, the input voltage at point *A* is 0 V and the output voltage is 0 V. This corresponds to

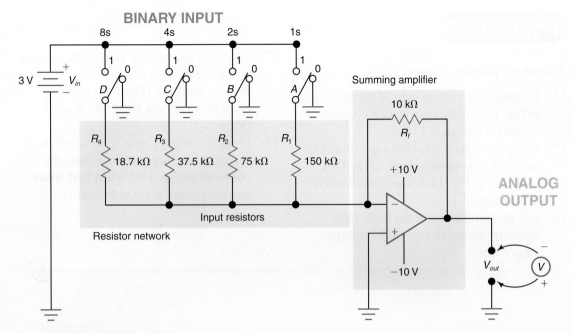

**BINARY INPUT**

Fig. 14-5 A D/A converter circuit.

row 1, Table 14-1. Suppose we move switch $A$ to the logical 1 position in Fig. 14-5. The input voltage of 3 V is applied to the op amp. We next calculate the gain of the amplifier. The gain is dependent upon the feedback resistor ($R_f$), (which is 10 kΩ), and the input resistor ($R_{in}$), which is the value of $R_1$, or 150 kΩ. Using the gain formula, we have

$$A_v = \frac{R_f}{R_{in}} = \frac{10,000}{150,000} = 0.066$$

To calculate the output voltage, we multiply the gain by the input voltage as shown here:

$$V_{out} = A_v \times V_{in} = 0.066 \times 3 = 0.2 \text{ V}$$

The output voltage is 0.2 V when the input is binary 0001. This satisfies the requirements of row 2, Table 14-1.

Let us now apply binary 0010 to the D/A converter in Fig. 14-5. Only switch $B$ is moved to the logical 1 position, applying 3 V to the op amp. The gain is

$$A_v = \frac{R_f}{R_{in}} = \frac{10,000}{75,000} = 0.133$$

Multiplying the gain by the input voltage gives us 0.4 V. The 0.4 V is the output voltage. This satisfies row 3, Table 14-1.

Notice that for each binary count in Table 14-1 the output voltage of the D/A converter

increases by 0.2 V. This increase occurs because of the increased voltage gain of the op amp as we switch in different resistors ($R_1$, $R_2$, $R_3$, $R_4$). If only resistor $R_4$ from Fig. 14-5 were connected by activating switch $D$, the gain would be

$$A_v = \frac{R_f}{R_{in}} = \frac{10,000}{18,700} = 0.535$$

The gain multiplied by the input voltage of 3 V gives 1.6 V at the output of the op amp. This is what is required by row 9, Table 14-1.

When all switches are activated (at logical 1) in Fig. 14-5, the op amp puts out the full 3 V because the gain of the amplifier has increased to 1.

Any input voltage up to the limits of the operational amplifier power supply ($\pm 10$ V) may be used. More binary places may be added by adding switches. If a 16s place value switch is added in Fig. 14-5, it needs a resistor with half the value of resistor $R_4$. Its value would then have to be 9350 Ω. The value of the feedback resistor would also be changed to 5 kΩ. The input would then be a 5-bit binary number; the output would still be an analog output varying from 0 to $-3.1$ V (in 0.1 V steps).

Trying to expand D/A converter in Fig. 14-5 to many bits results in an impractical range of resistor values and poor accuracy.

*Answer the following questions.*

14. Calculate the voltage gain of the op amp in Fig. 14-5 when only switch *C* (the 4s switch) is at logical 1.

15. Using the voltage gain from question 14, calculate the output voltage of the D/A converter in Fig. 14-5 when only switch *C* is at logical 1.

16. List two limitations of the basic D/A converter shown in Fig. 14-5 for large binary words.

17. Calculate the voltage gain of the op amp in Fig. 14-5 when both switches *A* and *B* are at logical 1 [Hint: use parallel resistance formula $R_T = (R_1 \times R_2)/(R_1 + R_2)$].

18. Using the voltage gain from question 17, calculate the output voltage from the D/A converter in Fig. 14-5 when both input switches *A* and *B* are at logical 1.

## 14-4 Ladder-Type D/A Converters

Digital-to-analog converters consist of a resistor network and a summing amplifier. Figure 14-6 diagrams a type of resistor network that provides the proper weighting for the binary inputs. This resistor network is sometimes called the *R-2R ladder network.* The advantage of this arrangement of resistors is that only two values of resistors are used. Resistors $R_1$, $R_2$, $R_3$, $R_4$, and $R_5$ are 20 k$\Omega$ each. Resistors $R_6$, $R_7$, $R_8$, and $R_f$ are each 10 k$\Omega$. Notice that all the horizontal resistors on the "ladder" are exactly twice the value of the vertical resistors, hence the title R-2R ladder network.

The summing amplifier in Fig. 14-6 is the same one used in the last section. Again notice the use of the dual power supply on the op amp.

The operation of this D/A converter is similar to the basic one in the last section. Table 14-2 details the operation of this D/A converter. Notice that we are using an input voltage of 3.75 V on this converter. Each binary count increases the analog output by 0.25 V, as shown in the rightmost column of Table 14-2. Remember that each 0 on the input side of the table means 0 V applied to that input. Each 1 on the input side of the table means 3.75 V applied to that input. The input voltage of 3.75 V is used because this is very close to the output of TTL

counters and other ICs you may have used. The inputs (*D*, *C*, *B*, *A*) in Fig. 14-6, then, could be connected directly to the outputs of a TTL IC and operate according to Table 14-2. In actual practice, however, the outputs of a TTL IC are not accurate enough; they have to be put

### Table 14-2 Truth Table for D/A Converter

| Binary inputs | | | | Analog output |
|---|---|---|---|---|
| 8s | 4s | 2s | 1s | |
| *D* | *C* | *B* | *A* | Volts |
| 0 | 0 | 0 | 0 | 0 |
| 0 | 0 | 0 | 1 | 0.25 |
| 0 | 0 | 1 | 0 | 0.50 |
| 0 | 0 | 1 | 1 | 0.75 |
| 0 | 1 | 0 | 0 | 1.00 |
| 0 | 1 | 0 | 1 | 1.25 |
| 0 | 1 | 1 | 0 | 1.50 |
| 0 | 1 | 1 | 1 | 1.75 |
| 1 | 0 | 0 | 0 | 2.00 |
| 1 | 0 | 0 | 1 | 2.25 |
| 1 | 0 | 1 | 0 | 2.50 |
| 1 | 0 | 1 | 1 | 2.75 |
| 1 | 1 | 0 | 0 | 3.00 |
| 1 | 1 | 0 | 1 | 3.25 |
| 1 | 1 | 1 | 0 | 3.50 |
| 1 | 1 | 1 | 1 | 3.75 |

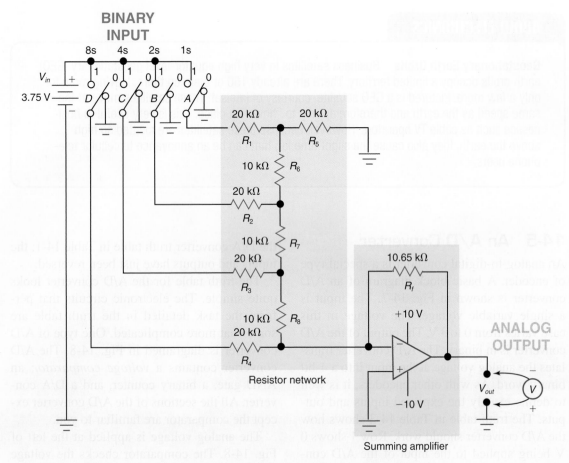

BINARY
INPUT

8s    4s    2s    1s

$V_{in}$ $+$ 3.75 V

D    C    B    A

20 kΩ    20 kΩ
$R_1$    $R_5$

10 kΩ  $R_6$

20 kΩ
$R_2$

10 kΩ  $R_7$

20 kΩ
$R_3$

10 kΩ  $R_8$

20 kΩ
$R_4$

Resistor network

10.65 kΩ
$R_f$

+10 V

−10 V

$V_{out}$

ANALOG
OUTPUT

V

Summing amplifier

**Fig. 14-6** A D/A converter circuit using an R-2R ladder resistor network.

through a level translator to get a very precise voltage output.

More binary places (16s, 32s, 64s, and so on) can be added to the D/A converter in Fig. 14-6. Follow the pattern of resistor values shown in this diagram when adding place values.

Two types of special decoders called digital-to-analog converters have been covered. The R-2R ladder-type D/A converter has some advantages over the more basic unit. The heart of the D/A converter consists of the resistor network and the summing amplifier.

✓ Self-Test

*Answer the following questions.*

19. The digital-to-analog converter in Fig. 14-6 is a(n) _____ -type D/A converter.
20. Refer to Fig. 14-6. The gain of the op amp is greatest when all input switches are at logical _____ (0, 1).
21. Refer to Fig. 14-6 and Table 14-2. The gain of the op amp is *least* when switch _____ (A, B, C, D) is the only switch at logical 1.

22. Refer to Fig. 14-6 and Table 14-2. If the binary input to the D/A converter is 1011, the analog output voltage is _____ volts.
23. Refer to Fig. 14-6 and Table 14-2. If the binary input to the D/A converter increases from 0111 to 1000, the analog output voltage increases by _____ volts.

A/D converter

Voltage
comparator

## 14-5   An A/D Converter

An analog-to-digital converter is a special type of encoder. A basic block diagram of an A/D converter is shown in Fig. 14-7. The input is a single variable voltage. The voltage in this case varies from 0 to 3 V. The output of the A/D converter is in binary. The A/D converter translates the analog voltage at the input into a 4-bit binary word. As with other encoders, it is well to define exactly the expected inputs and outputs. The truth table in Table 14-3 shows how the A/D converter should work. Row 1 shows 0 V being applied to the input of the A/D converter. The output is binary 0000. Row 2 shows a 0.2-V input. The output is binary 0001. Notice that each increase of 0.2 V increases the binary count by 1. Finally, row 16 shows that when the maximum 3 V is applied to the input, the output reads binary 1111. Notice that the truth table in Table 14-3 is just the reverse of

the D/A converter truth table in Table 14-1; the inputs and outputs have just been reversed.

The truth table for the A/D converter looks quite simple. The electronic circuits that perform the task detailed in the truth table are somewhat more complicated. One type of A/D converter is diagramed in Fig. 14-8. The A/D converter contains a *voltage comparator,* an AND gate, a binary counter, and a D/A converter. All the sections of the A/D converter except the comparator are familiar to you.

The analog voltage is applied at the left of Fig. 14-8. The comparator checks the voltage coming from the D/A converter. If the analog input voltage at *A* is *greater than* the voltage at input *B* of the comparator, the clock is allowed to *increase* the count of the 4-bit counter. The count on the counter increases until the feedback voltage from the D/A converter becomes greater than the analog input voltage. At this

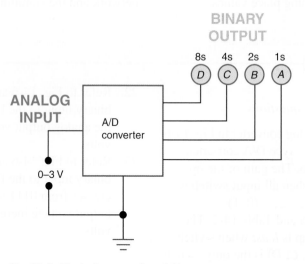

**Fig. 14-7** Block diagram of an A/D converter.

## Table 14-3 Truth Table for A/D Converter

| Analog input | Binary output | | | |
|---|---|---|---|---|
| | 8s | 4s | 2s | 1s |
| Volts | D | C | B | A |
| Row 1    0 | 0 | 0 | 0 | 0 |
| Row 2    0.2 | 0 | 0 | 0 | 1 |
| Row 3    0.4 | 0 | 0 | 1 | 0 |
| Row 4    0.6 | 0 | 0 | 1 | 1 |
| Row 5    0.8 | 0 | 1 | 0 | 0 |
| Row 6    1.0 | 0 | 1 | 0 | 1 |
| Row 7    1.2 | 0 | 1 | 1 | 0 |
| Row 8    1.4 | 0 | 1 | 1 | 1 |
| Row 9    1.6 | 1 | 0 | 0 | 0 |
| Row 10   1.8 | 1 | 0 | 0 | 1 |
| Row 11   2.0 | 1 | 0 | 1 | 0 |
| Row 12   2.2 | 1 | 0 | 1 | 1 |
| Row 13   2.4 | 1 | 1 | 0 | 0 |
| Row 14   2.6 | 1 | 1 | 0 | 1 |
| Row 15   2.8 | 1 | 1 | 1 | 0 |
| Row 16   3.0 | 1 | 1 | 1 | 1 |

point the comparator stops the counter from advancing to a higher count. Suppose the input analog voltage is 2 V. According to Table 14-3, the binary counter increases to 1010 before it is stopped. The counter is reset to binary 0000, and the counter starts counting again.

Now for more detail on the A/D converter in Fig. 14-8. Let us assume that there is a logical 1 at point $X$ at the output of the comparator. Also assume that the counter is at binary 0000. Assume, too, that 0.55 V is applied to the analog input. The 1 at point $X$ enables the AND gate, and the first pulse from the clock appears at the CLK input of the counter. The counter advances its count to 0001. The 0001 is displayed on the lights in the upper right of Fig. 14-8. The 0001 is also applied to the D/A converter.

According to Table 14-1, a binary 0001 produces 0.2 V at the output of the D/A converter. The 0.2 V is fed back to the $B$ input of the comparator. The comparator checks its inputs. The $A$ input is higher (0.55 V as opposed to 0.2 V), and so the comparator puts out a logical 1. The 1 enables the AND gate, which lets the next clock pulse through to the counter. The counter advances its count by 1. The count is now 0010. The 0010 is applied to the D/A converter.

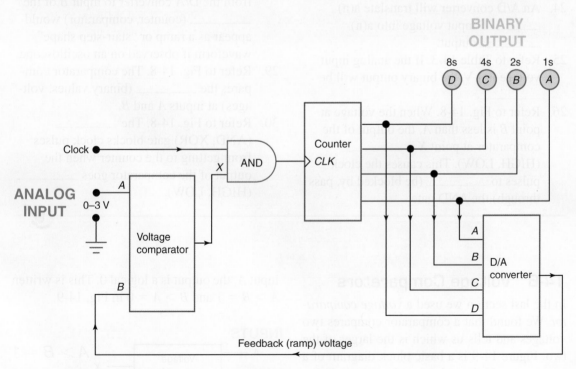

Fig. 14-8 Block diagram of a counter-ramp-type A/D converter.

According to Table 14-1, a 0010 input produces a 0.4-V output. The 0.4 V is fed back to the B input of the comparator. The comparator again checks the B input against the A input; the A input is still larger (0.55 V as opposed to 0.4 V). The comparator outputs a logical 1. The AND gate is enabled, letting the next clock pulse reach the counter. The counter increases its count to binary 0011. The 0011 is applied back to the D/A converter.

According to Table 14-1, a 0011 input produces a 0.6 V output. The 0.6 V is fed back to the B input of the comparator. The comparator checks input A against input B; for the first time the B input is larger than the A input. The comparator puts out a logical 0. The logical 0 disables the AND gate. No more clock pulses can reach the counter. The counter stops at binary 0011. A look at row 4, Table 14-3, shows that 0.6 V gives the readout of binary 0011. Our A/D converter has worked according to the truth table.

**Counter-ramp A/D converter**

**Ramp**

If the input analog voltage were 1.2 V, the binary output would be 0110, according to Table 14-3. The counter would have to count from binary 0000 to 0110 before being stopped by the comparator. If the input analog voltage were 2.8 V, the binary output would be 1110. The counter would have to count from binary 0000 to 1110 before being stopped by the comparator. Notice that it does take some time for the conversion of the analog voltage to a binary readout. However, in most cases the clock runs fast enough so that this time lag is not a problem.

You now should appreciate why we studied the D/A converter before the A/D converter. This *counter-ramp A/D converter* is fairly complex and needs a D/A converter to operate. The term "ramp" in the name for this converter refers to the gradually increasing voltage from the D/A converter that is fed back to the comparator. A 4-bit converter produces a staircase waveform. When enough bits are used, the waveform approaches a smooth ramp.

### Self-Test

*Supply the missing word or words in each statement.*

24. An A/D converter will translate a(n) _____ input voltage into a(n) _____ output.
25. Refer to Table 14-3. If the analog input voltage is 1 V, the binary output will be _____.
26. Refer to Fig. 14-8. When the voltage at point B is less than A, the output of the comparator at point X is _____ (HIGH, LOW). This causes the clock pulses to _____ (be blocked by, pass through) the AND gate.

27. The unit diagramed in Fig. 14-8 is a(n) _____ -type A/D converter.
28. Refer to Fig. 14-8. The feedback voltage from the D/A converter to input B of the _____ (counter, comparator) would appear as a ramp or "stair-step shape" waveform if observed on an oscilloscope.
29. Refer to Fig. 14-8. The comparator compares the _____ (binary values, voltages) at inputs A and B.
30. Refer to Fig. 14-8. The _____ (AND, XOR) gate blocks clock pulses from getting to the counter when the output of the comparator goes _____ (HIGH, LOW).

## 14-6 Voltage Comparators

**Voltage comparator**

In the last section we used a *voltage comparator*. We found that a comparator compares two voltages and tells us which is the larger of the two. Figure 14-9 is a basic block diagram of a comparator. If the voltage at input A is larger than at input B, the comparator gives a logical 1 output. If the voltage at input B is larger than at input A, the output is a logical 0. This is written $A > B = 1$ and $B > A = 0$ in Fig. 14-9.

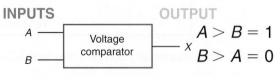

**Fig. 14-9** Block diagram of a voltage comparator.

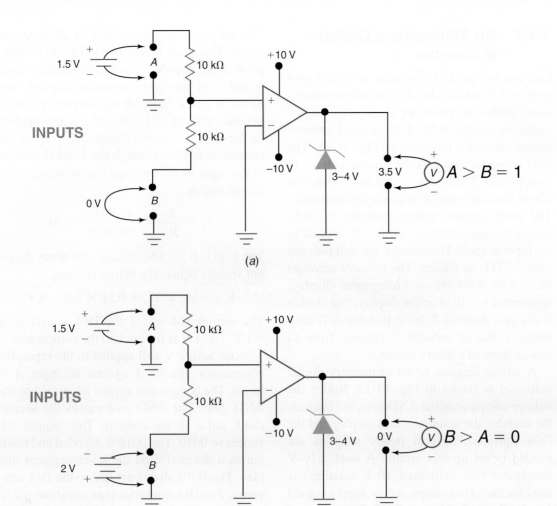

**Fig. 14-10** Voltage comparator circuit. (a) With greater voltage at input A. (b) With greater voltage at input B.

**Voltage comparator circuit**

**Op amp**

**Zener diode**

The heart of a voltage comparator is an *op amp*. Figure 14-10(a) shows a comparator circuit. Notice that input A has 1.5 V applied and input B has 0 V applied. The output voltmeter reads about 3.5 V, or a logical 1.

Figure 14-10(b) shows that the input B voltage has been increased to 2 V. Input A is still at 1.5 V. Input B is larger than input A. The output of the comparator circuit is about

0 V (actually the voltage is about −0.6 V), or a logical 0.

The comparator in the A/D converter in Fig. 14-8 works exactly like this unit. The *zener diode* in the comparator in Fig. 14-10 is there to clamp the output voltage at about +3.5 or −0.6 V. Without the zener diode the output voltages would be about +9 and −9 V. The +3.5 and −0.6 V are more compatible with TTL ICs.

## Self-Test

*Answer the following questions.*

31. The comparator block shown in Fig. 14-8 compares two _____ (binary numbers, decimal numbers, dc voltages).

32. A voltage comparator circuit can be constructed using a(n) _____ IC, several resistors, and a zener diode.

33. Refer to Fig. 14-10. When input B increases and becomes higher than input A, the output of the op amp will change from _____ (HIGH, LOW) to _____ (HIGH, LOW).

## 14-7 An Elementary Digital Voltmeter

One use for an A/D converter is in a *digital voltmeter*. You have already used all the subsystems needed to make an elementary digital voltmeter system. A block diagram of a simple digital voltmeter is shown in Fig. 14-11. The A/D converter converts the analog voltage to binary form. The binary is sent to the decoder, where it is converted to a seven-segment code. The seven-segment readout indicates the voltage in decimal numbers. With 7 V applied to the input of the A/D converter, the unit puts out binary 0111, as shown. The decoder activates lines *a* to *c* of the seven-segment display; segments *a* to *c* light on the display. The display reads as a decimal 7. Note that the A/D converter is also an encoder; it encodes from an analog input to a binary output.

A wiring diagram of an elementary digital voltmeter is shown in Fig. 14-12. Notice the voltage comparator, the AND gate, the counter, the decoder, the seven-segment display and the D/A converter. Several power supplies are needed to set up this circuit. A dual ±10-V supply (or two individual 10-V supplies) is used for the 741 op amps. A 5-V supply is used for the 7408, 7493, and 7447 TTL ICs and the seven-segment LED display. A 0- to 10-V variable dc power supply is also needed for the analog input voltage.

Let us assume a 2-V input to the analog input of the digital voltmeter in Fig. 14-12. Reset the counter to 0000. The comparator checks inputs *A* and *B*; *A* is larger ($A = 2$ V, $B = 0$ V). The comparator output is a logical 1. This 1 enables the AND gate. The pulse from the clock passes through the AND gate.

The pulse causes the counter to advance one count. The count is now 0001. The 0001 is applied to the decoder. The decoder enables lines *b* and *c* of the seven-segment display; segments *b* and *c* light on the display, giving a decimal readout of 1. The 0001 is also applied to the D/A converter. About 3.2 V from the counter is applied through the 150-kΩ resistor to the input of the op amp. The voltage gain of the op amp is

$$A_v = \frac{R_f}{R_{in}} = \frac{47{,}000}{150{,}000} = 0.31$$

The gain is 0.31. The voltage gain times the input voltage equals the output voltage:

$$V_{out} = A_v \times V_{in} = 0.31 \times 3.2 = 1 \text{ V}$$

The output voltage of the D/A converter is −1 V. The 1 V is fed back to the comparator.

Now, with 2 V still applied to the input, the comparator checks *A* against *B*; input *A* is larger. The comparator applies a logical 1 to the AND gate. The AND gate passes the second clock pulse to the counter. The counter advances to 0010. The 0010 is decoded and reads out as a decimal 2 on the seven-segment display. The 0010 also is applied to the D/A converter. The D/A converter puts out about −2 V, which is fed back to the *B* input of the comparator.

The display now reads 2. The 2 V is still applied to input *A* of the comparator. The comparator checks *A* against *B*; *B* is just slightly larger. Output *X* of the comparator goes to logical 0. The AND gate is disabled. No clock pulses reach the counter. The count has stopped at 2 on the display. This is the voltage applied at the analog input.

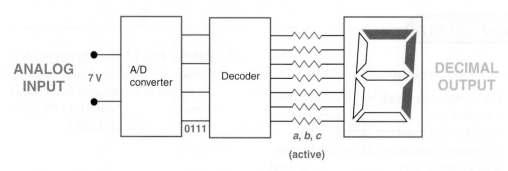

**Fig. 14-11** Block diagram of an elementary digital voltmeter.

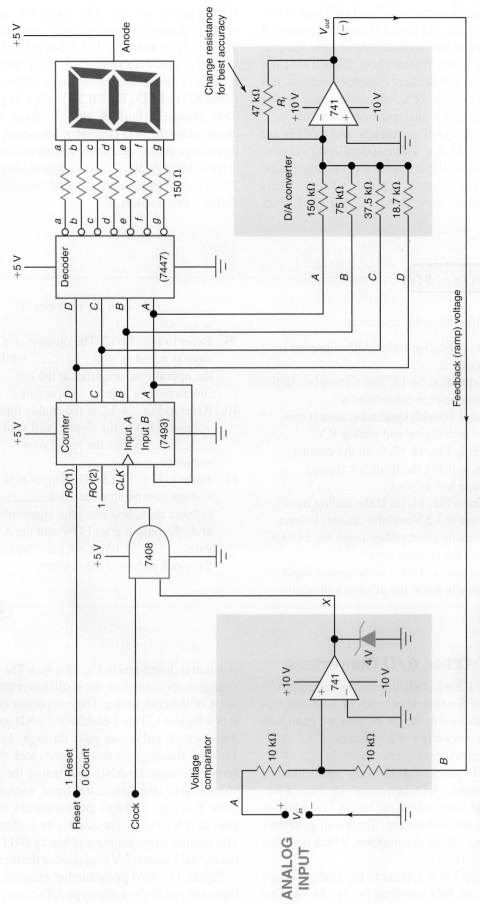

Fig. 14-12 Wiring diagram of an elementary digital voltmeter.

**Elementary
digital voltmeter**

The digital voltmeter in Fig. 14-12 is an experimental circuit. The circuit is included because it demonstrates the fundamentals of how a digital voltmeter works. It shows how SSI and MSI ICs can be used to build more complex functions. It is a simple example of a *hybrid electronic system* containing both digital and analog devices.

Modern digital voltmeters and DMMs are based on LSI ICs. These specialized A/D converters are available from many manufacturers. *Large-scale-integrated digital voltmeter chips* include all the active devices on a single CMOS

IC. Included are the A/D converter, seven-segment decoders, display drivers, and a clock. The ICL7106 and ICL7107 3½-digit A/D converters are two examples of these complex devices. They will directly drive either LCD (7106 IC) or LED (7107 IC) 3½-digit displays. They feature a built-in clock, voltage references, accurate A/D converter, auto-zero, high input impedance, decoders, and direct display drivers for 3½-digit seven-segment displays. These chips can be used in digital voltmeters or digital thermometers.

**Hybrid electronic system**

**LSI digital voltmeter IC**

## ✓ Self-Test

*Answer the following questions.*

34. One application for an A/D converter is in a(n) _____.
35. Refer to Fig. 14-12. The elementary digital voltmeter is considered a _____ (digital, hybrid) system because it contains both digital and analog ICs.
36. Refer to Fig. 14-12. With the counter *reset* to 0000, the feedback (ramp) voltage will be about _____ V.
37. Refer to Fig. 14-12. If the analog input voltage is 3.5 V and the counter is reset, how many clock pulses reach the 7493 IC before the counter stops?
38. Refer to Fig. 14-12. If the analog input voltage is 4.6 V, the display will read

_____ V after the reset/count sequence.
39. Refer to Fig. 14-12. The op amp on the right is wired as a(n) _____ while the operational amplifier at the left functions as a voltage comparator.
40. Refer to Fig. 14-12. If the analog input voltage is 8.5 V, the display will read _____ V after the reset/count sequence.
41. Refer to Fig. 14-12. When input *B* of the voltage comparator becomes _____ (greater than, less than) the input voltage at *A*, the output goes LOW and the AND gate _____ (does not pass, passes) the clock pulses to the counter.

## 14-8 Other A/D Converters

In Sec. 14-5 we studied the counter-ramp A/D converter. Several other types of A/D converters are also used; in this section we shall discuss two other types of converters.

A *ramp A/D converter* is shown in Fig. 14-13. This A/D converter works very much like the counter-ramp A/D converter in Fig. 14-8. The *ramp-generator* at the left in Fig. 14-13 is the only new subsystem. The ramp generator produces a *sawtooth waveform,* which is shown in Fig. 14-14(*a*).

Suppose 3 V is applied to the analog voltage input of the A/D converter in Fig. 14-13. This

**Ramp A/D converter**

**Ramp-generator**

**Sawtooth waveform**

situation is diagramed in Fig. 14-14(*a*). The ramp voltage starts to increase but is still lower than input *A* of the comparator. The comparator output is at a logical 1. This 1 enables the AND gate so that a clock pulse can pass through. In Fig. 14-14(*a*) the diagram shows three clock pulses getting through the AND gate before the ramp voltage gets larger than the input voltage. At point *Y* in Fig. 14-14(*a*) the comparator output goes to a logical 0. The AND gate is disabled. The counter stops counting at binary 0011. The binary 0011 means 3 V is applied at the input.

Figure 14-14(*b*) gives another example. The input voltage to the ramp-type A/D converter is

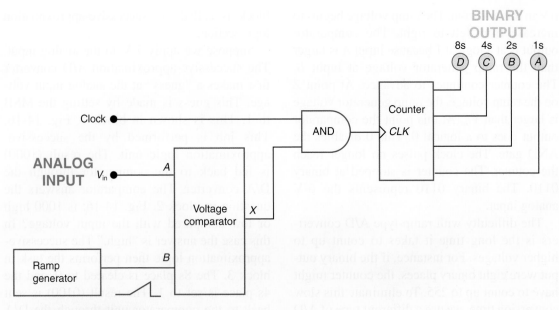

**Fig. 14-13** Block diagram of a ramp-type A/D converter.

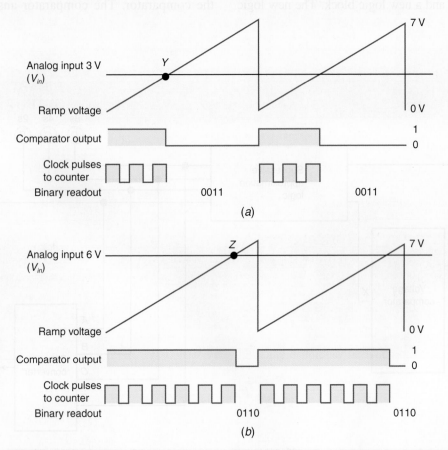

**Fig. 14-14** Ramp-type A/D converter waveforms (*a*) With 3 V applied.
(*b*) With 6 V applied.

**Ramp-type A/D converter**

6 V in this situation. The ramp voltage begins to increase from left to right. The comparator output is at a logical 1 because input $A$ is larger than the ramp generator voltage at input $B$. The counter continues to advance. At point $Z$ on the ramp voltage, the ramp generator voltage is larger than $V_{in}$. At this point the comparator output goes to a logical 0. This 0 disables the AND gate. The clock pulses no longer reach the counter. The counter is stopped at binary 0110. The binary 0110 represents the 6-V analog input.

The difficulty with ramp-type A/D converters is the long time it takes to count up to higher voltages. For instance, if the binary output were eight binary places, the counter might have to count up to 255. To eliminate this slow conversion time, we use a different type of A/D converter. A converter that cuts down on conversion time is a *successive-approximation A/D converter*.

A block diagram of a successive-approximation A/D converter is shown in Fig. 14-15. The converter consists of a voltage comparator, a D/A converter, and a new logic block. The new logic

block is called the successive-approximation logic section.

Suppose we apply 7 V to the analog input. The successive-approximation A/D converter first makes a "guess" at the analog input voltage. This guess is made by setting the MSB to 1. This is shown in block 1, Fig. 14-16. This job is performed by the successive-approximation logic unit. The result (1000) is fed back to the comparator through the D/A converter. The comparator answers the question in block 2, Fig. 14-16: is 1000 high or low compared with the input voltage? In this case the answer is "high." The successive-approximation logic then performs the task in block 3. The 8s place is cleared to 0, and the 4s place is set to 1. The result (0100) is sent back to the comparator unit through the D/A converter. The comparator next answers the question in block 4: is 0100 high or low compared with the input voltage? The answer is "low." The successive-approximation logic then performs the task in block 5. The 2s place is set to 1. The result (0110) is sent back to the comparator. The comparator answers the

**Successive-
approximation
A/D converter**

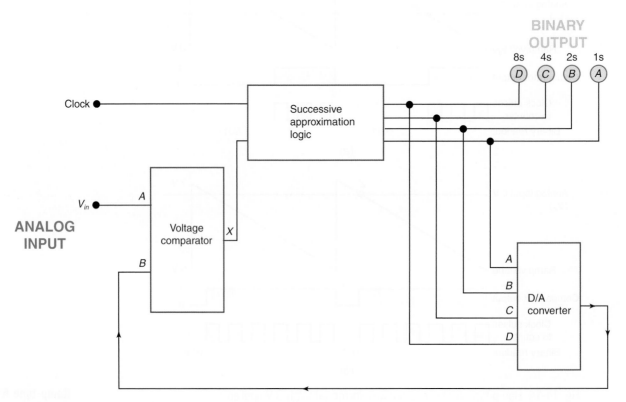

**Fig. 14-15** Block diagram of a successive-approximation-type A/D converter.

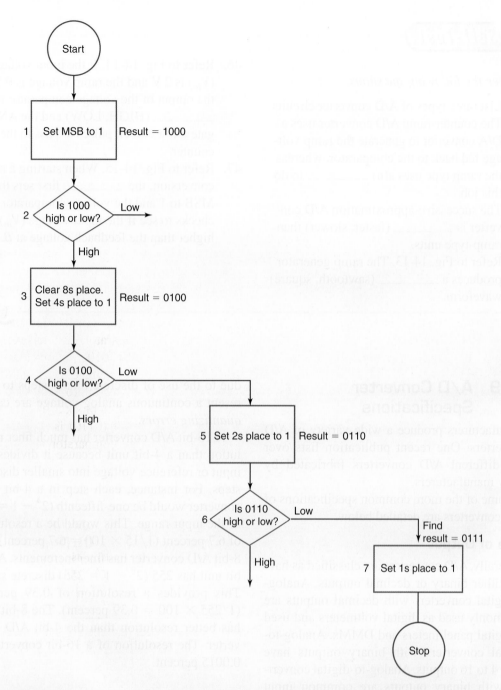

**Fig. 14-16** Flowcharting the operation of the successive-approximation-type A/D converter.

question in block 6: is 0110 high or low compared with the input voltage? The answer is "low." The successive-approximation logic then performs the task in block 7. The 1s place is set to 1. The final result is binary 0111. This stands for the 7 V applied at the input of the A/D converter.

Notice in Fig. 14-16 that the items in the blocks are performed by the successive-approximation logic unit. The questions are answered by the comparator. Also notice that the task performed by the successive-approximation logic depends upon whether the answer to the previous question is "low" or "high" (see blocks 3 and 5).

The advantage of the successive-approximation A/D converter is that it takes fewer guesses to get the answer. The *digitizing* process is thus faster. The successive-approximation A/D converter is very widely used.

*Answer the following questions.*

42. List three types of A/D converter circuits.
43. The counter-ramp A/D converter uses a D/A converter to generate the ramp voltage fed back to the comparator, whereas the ramp type uses a(n) _____ to do this job.
44. The successive-approximation A/D converter is _____ (faster, slower) than ramp-type units.
45. Refer to Fig. 14-13. The ramp generator produces a _____ (sawtooth, square) waveform.

46. Refer to Fig. 14-14. If the input voltage ($V_{in}$) is 2 V and the ramp voltage is 0 V, the output of the voltage comparator is _____ (HIGH, LOW) and the AND gate allows clock pulses to pass to the counter.
47. Refer to Fig. 14-15. When starting a new conversion, the _____ first sets the MSB to 1 and the voltage comparator checks to see if the input voltage ($V_{in}$) is higher than the feedback voltage at *B*.

**A/D converter specifications**

## 14-9  A/D Converter Specifications

Manufacturers produce a wide variety of A/D converters. One recent publication lists over 300 different A/D converters fabricated by many manufacturers.

Some of the more common specifications of A/D converters are detailed below.

### Type of Output

Generally, A/D converters are classified as having either binary or decimal outputs. Analog-to-digital converters with decimal outputs are commonly used as digital voltmeters and used in digital panel meters and DMMs. Analog-to-digital converters with binary outputs have from 4 to 16 outputs. Analog-to-digital converters with binary outputs are common input devices to microprocessor-based systems. The latter are sometimes referred to as μP-type A/D converters.

**μP-type A/D converters**

### Resolution

**Resolution**

The *resolution* of an A/D converter is given as the *number of bits* at the output for a binary-type unit. For decimal-output A/D converters (used in DMMs), the resolution is given as the *number of digits* in the readout (such as $3\frac{1}{2}$ or $4\frac{1}{2}$). Typical A/D converters with binary outputs have resolutions of 4, 6, 8, 10, 12, 14, and 16 bits. The errors that occur

**Accuracy**

due to the use of discrete binary steps to represent a continuous analog voltage are called *quantizing errors*.

A 16-bit A/D converter has much finer resolution than a 4-bit unit because it divides the input or reference voltage into smaller discrete steps. For instance, each step in a 4-bit A/D converter would be one-fifteenth ($2^4 - 1 = 15$) of the input range. This would be a resolution of 6.7 percent ($1/15 \times 100 = 6.7$ percent). An 8-bit A/D converter has finer increments. An 8-bit unit has 255 ($2^8 - 1 = 255$) discrete steps. This provides a resolution of 0.39 percent ($1/255 \times 100 = 0.39$ percent). The 8-bit unit has better resolution than the 4-bit A/D converter. The resolution of a 16-bit converter is 0.0015 percent.

### Accuracy

The resolution of an A/D converter can be thought of as the inherent "digital" error due to the discrete steps available at the output of the IC. Another source of error in an A/D converter might be an analog component, such as the comparator. Other errors might be introduced by the resistor network. The overall precision of an A/D converter is called the *accuracy* of the A/D converter IC.

The accuracies of typical A/D converter ICs with binary outputs range from $\pm\frac{1}{2}$ LSB to $\pm2$ LSB. Those with decimal outputs might range from 0.01 accuracy to 0.05 percent accuracy.

## Conversion Time

The *conversion time* is another important specification of an A/D converter. It is the time it takes for the IC to convert the analog input voltage to binary (or decimal) data at the outputs. Typical conversion times range from 0.05 to 100,000 μs for A/D converter ICs with binary outputs. Conversion times for A/D converters with decimal outputs are somewhat longer and might typically be 200 to 400 ms.

## Other Specifications

Four other common characteristics given for A/D converters are the power supply voltage, output logic levels, input voltage, and maximum power dissipation. Power supply voltages are commonly +5 V. However, some A/D converter ICs operate on voltages from +5 to +15 V. The output logic levels are either TTL, CMOS, or tristate. The input voltage range is commonly 5 V. Maximum power dissipation for an A/D converter IC might be in a range from about 15 to 3000 mW.

---

### Self-Test

*Supply the missing word or number in each statement.*

48. An A/D converter with binary outputs is sometimes referred to as a _____ (meter, μP)-type unit.
49. The _____ of an A/D converter is given as the number of bits at the output of a binary-type unit.
50. An 8-bit A/D converter has greater resolution than a _____ (4, 12)-bit chip.

51. A typical conversion time for an A/D converter might be about _____ (110 μs, 1s).
52. A typical A/D converter might have a maximum power dissipation of about _____ (850 mW, 10 μW).
53. The conversion time for meter-type A/D converters is _____ (longer, shorter) than for μP-type units.

---

## 14-10 An A/D Converter IC

A commercial A/D converter IC will be featured in this section. Figure 14-17(*a*) shows the pin diagram for an *ADC0804 8-bit A/D converter IC*. The table in Fig. 14-17(*b*) lists the name and function of each pin on the ADC0804 IC.

The ADC0804 A/D converter was designed to interface directly with older 8080, 8085, or Z80 microprocessors. Some pin labels on the ADC0804 IC correspond to pins on popular microprocessors. For instance, the ADC0804 uses $\overline{RD}$, $\overline{WR}$, and $\overline{INTR}$ as pin labels which correspond to the *RD*, *WR*, and *INTR* pins on the older 8085 microprocessor. The ADC0804 can also be interfaced with other older 8-bit microprocessors such as the 6800 and 6502. The $\overline{CS}$ control input to the ADC0804 A/D converter receives its signal (chip select) from the microprocessor address-decoding circuitry.

The ADC0804 is a CMOS 8-bit *successive-approximation A/D converter*. It has three-state outputs so that it can interface directly with a microprocessor-based system data bus. The ADC0804 has binary outputs and features a short conversion time of only 100 μs. Its inputs and outputs are both MOS- and TTL-compatible. It has an on-chip clock generator. The on-chip generator does need two external components (resistor and capacitor) to operate. The ADC0804 IC operates on a standard +5-V dc power supply and can encode input analog voltages ranging from 0 to 5 V.

The ADC0804 A/D converter IC can be tested using the circuit shown in Fig. 14-18. The function of the circuit is to encode the difference in voltage between $V_{in}(+)$ and $V_{in}(-)$ compared to the reference voltage (5.12 V in this example) to a corresponding binary value. The resolution of the ADC0804 IC is 8 bits or

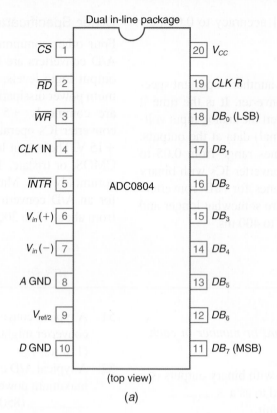

Dual in-line package

| | | | |
|---|---|---|---|
| $\overline{CS}$ | 1 | 20 | $V_{CC}$ |
| $\overline{RD}$ | 2 | 19 | CLK R |
| $\overline{WR}$ | 3 | 18 | $DB_0$ (LSB) |
| CLK IN | 4 | 17 | $DB_1$ |
| $\overline{INTR}$ | 5 | 16 | $DB_2$ |
| $V_{in}(+)$ | 6 | 15 | $DB_3$ |
| $V_{in}(-)$ | 7 | 14 | $DB_4$ |
| A GND | 8 | 13 | $DB_5$ |
| $V_{ref/2}$ | 9 | 12 | $DB_6$ |
| D GND | 10 | 11 | $DB_7$ (MSB) |

ADC0804

(top view)

(a)

ADC0804 A/D converter IC

| Pin No. | Symbol | Input/Output or Power | Description |
|---|---|---|---|
| 1 | $\overline{CS}$ | Input | Chip select line from $\mu$P-control |
| 2 | $\overline{RD}$ | Input | Read line from $\mu$P-control |
| 3 | $\overline{WR}$ | Input | Write line from $\mu$P-control |
| 4 | CLK IN | Input | Clock |
| 5 | $\overline{INTR}$ | Output | Interrupt line goes to $\mu$P interrupt input |
| 6 | $V_{in}(+)$ | Input | Analog voltage (positive input) |
| 7 | $V_{in}(-)$ | Input | Analog voltage (negative input) |
| 8 | A GND | Power | Analog ground |
| 9 | $V_{ref/2}$ | Input | Alternate voltage reference (+) |
| 10 | D GND | Power | Digital ground |
| 11 | $DB_7$ | Output | MSB data output |
| 12 | $DB_6$ | Output | Data output |
| 13 | $DB_5$ | Output | Data output |
| 14 | $DB_4$ | Output | Data output |
| 15 | $DB_3$ | Output | Data output |
| 16 | $DB_2$ | Output | Data output |
| 17 | $DB_1$ | Output | Data output |
| 18 | $DB_0$ | Output | LSB data output |
| 19 | CLK R | Input | Connect external resistor for clock |
| 20 | $V_{CC}$ (or ref) | Power | Positive of 5-V power supply and primary reference voltage |

(b)

**ADC0804 A/D converter IC**

Fig. 14-17 ADC0804 A/D converter IC. (a) Pin diagram. (b) Pin labels and functions.

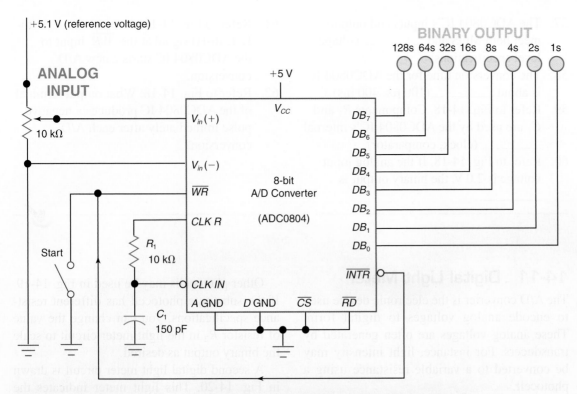

+5.1 V (reference voltage)

ANALOG INPUT

10 kΩ

Start

$R_1$ 10 kΩ

$C_1$ 150 pF

+5 V

$V_{CC}$

$V_{in}(+)$

$V_{in}(-)$

$\overline{WR}$

CLK R

CLK IN

A GND    D GND    $\overline{CS}$    $\overline{RD}$

8-bit A/D Converter

(ADC0804)

$DB_7$
$DB_6$
$DB_5$
$DB_4$
$DB_3$
$DB_2$
$DB_1$
$DB_0$

$\overline{INTR}$

BINARY OUTPUT
128s 64s 32s 16s 8s 4s 2s 1s

**Fig. 14-18** Wiring diagram for a test circuit using the ADC0804 CMOS A/D converter IC.

0.39 percent. This means that for each 0.02 V (5.1 V × 0.39 percent = 0.02 V) increase in voltage at the analog inputs, the binary count increases by 1.

The "start switch" in Fig. 14-18 is first closed and then opened to start this free-running A/D converter. It is "free-running" because it continuously converts the analog input to digital outputs. The start switch should be left open once the A/D converter is operating. The $\overline{WR}$ input can be thought of as a clock input with the interrupt output $(\overline{INTR})$ pulsing the $\overline{WR}$ input at the end of each analog-to-digital conversion. A L-to-H transition of the signal at the $\overline{WR}$ input starts the A/D converter process. When the conversion is finished, the binary display is updated and the $\overline{INTR}$ output emits a negative pulse. The negative interrupt pulse is fed back to clock the $\overline{WR}$ input and it initiates another

A/D conversion. The circuit in Fig. 14-18 will perform about 5,000 to 10,000 conversions per second. The conversion rate of the ADC0804 is high because it uses the successive-approximation technique in the conversion process.

The resistor $(R_1)$ and capacitor $(C_1)$ connected to the CLK R and CLK IN inputs to the ADC0804 IC in Fig. 14-18 cause the internal clock to operate. The data outputs (DB7-DB0) drive the LED binary displays. The data outputs are active HIGH three-state outputs.

What is the binary output in Fig. 14-18 if the analog input voltage is 1.0 V? Recall that each 0.02 V equals a single binary count. Dividing 1.0 V by 0.02 V equals 50 in decimal. Converting decimal 50 to binary equals $00110010_2$. The output indicators will show binary 00110010 (LLHHLLHL).

**✔ Self-Test**

*Supply the missing word in each statement.*

54. The ADC0804 A/D converter is manufactured using _____ (CMOS, TTL) technology.

55. The ADC0804 IC is a _____ (meter, microprocessor)-type A/D converter.

56. The ADC0804 is an A/D converter with a resolution of _____.

57. The ADC0804 IC's inputs and outputs meet both MOS and _____ voltage-level specifications.
58. The conversion time for the ADC0804 IC is about _____ (100 μs, 400 ms).
59. Refer to Fig. 14-18. Components $R_1$ and $C_1$ are used by the ADC0804 IC's internal _____ (clock, comparator).
60. Refer to Fig. 14-18. If the analog input voltage is 2.0 V, the binary output is _____ .

61. Refer to Fig. 14-18. An _____ (H-to-L, L-to-H) signal at the $\overline{WR}$ input to the ADC0804 IC starts a new A/D conversion.
62. Refer to Fig. 14-18. What output terminal of the ADC0804 IC produces a negative pulse immediately after each A/D conversion?

## 14-11 Digital Light Meter

**Digital light meter**

The A/D converter is the electronic device used to encode analog voltages to digital form. These analog voltages are often generated by transducers. For instance, light intensity may be converted to a variable resistance using a photocell.

A schematic diagram for a basic *digital light meter* is drawn in Fig. 14-19. The ADC0804 IC is wired as a free-running A/D converter as in the last section. The push-button switch is pressed only once to start the A/D converter. The analog input voltage is being measured across resistor $R_2$. The photo-

**Transducer**

cell ($R_3$) is the light sensor or *transducer* in this circuit. As the light intensity increases, the resistance of the photocell ($R_3$) decreases. Decreasing the resistance of $R_3$ causes an increase in current through series resistances $R_2$ and $R_3$. The increased current through $R_2$ causes a proportional increase in the voltage drop across the resistor. The voltage drop across $R_2$ is the analog input voltage to the A/D converter. An increase in the analog input voltage causes an increase in the reading at the binary outputs.

**Cadmium sulfide photocell**

The *cadmium sulfide photocell* used in Fig. 14-19 is a variable resistor. As the intensity of the light striking the photocell increases, its resistance decreases. The photocell shown in Fig. 14-19 might have a maximum resistance of about 500 kΩ and a minimum of about 100 Ω. The cadmium sulfide photocell is most sensitive in the green-to-yellow portion of the

**Photoresistor**
**Photoresistive cell**

light spectrum. The photocell is also referred to as a *photoresistor,* cds photocell, or a *photoresistive cell.*

Other photocells may be used in Fig. 14-19. If the substitute photocell has different resistance specifications, you can change the value of resistor $R_2$ in the light meter circuit to scale the binary output as desired.

A second digital light meter circuit is drawn in Fig. 14-20. This light meter indicates the relative brightness of the light striking the photocell in decimal (0 to 9). The new light meter is similar to the circuit in Fig. 14-19. The new light meter has a clock added to the circuit. The clock consists of a 555 timer IC, two resistors, and a capacitor wired as an astable MV. The clock generates a TTL output with a frequency of about 1 Hz. This means the analog input voltage is only converted into digital form one time per second. The very low conversion rate keeps the output from "jittering" between two readings on the seven-segment LED display.

The 7447A IC decodes the four MSBs (DB7, DB6, DB5, DB4) from the output of the ADC0804 A/D converter. The 7447A IC also drives the segments on the seven-segment LED display. The seven 150-Ω resistors between the 7447A IC and seven-segment LED display limit the current through an "on" segment to a safe level.

As in the previous circuit (Fig. 14-19), the output of the new light meter may have to be scaled so that low light reads 0 and high light intensity reads 9 on the seven-segment LED display. The value of resistor $R_2$ can be changed to scale the output. If $R_2$ is substituted with a lower-value resistor, the decimal output will read lower for the same light intensity. However, if the resistance value of $R_2$ is increased, the output will read higher.

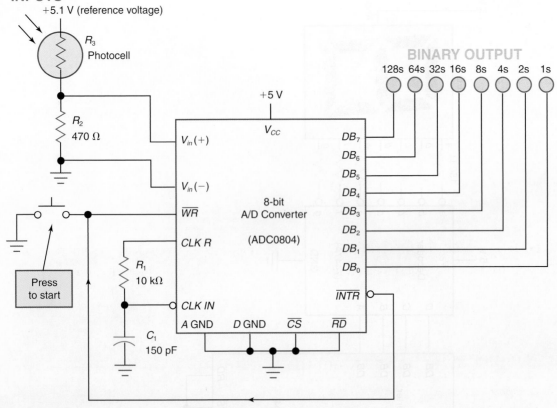

**INPUTS**

+5.1 V (reference voltage)

$R_3$ Photocell

$R_2$ 470 Ω

$V_{in}(+)$

+5 V

$V_{cc}$

$V_{in}(-)$

$\overline{WR}$

8-bit
A/D Converter

(ADC0804)

CLK R

$R_1$ 10 kΩ

Press to start

CLK IN

$A$ GND   $D$ GND   $\overline{CS}$   $\overline{RD}$

$C_1$ 150 pF

**BINARY OUTPUT**

128s 64s 32s 16s 8s 4s 2s 1s

$DB_7$
$DB_6$
$DB_5$
$DB_4$
$DB_3$
$DB_2$
$DB_1$
$DB_0$

$\overline{INTR}$

**Fig. 14-19** Wiring diagram for a digital light meter using binary outputs.

## ✓ Self-Test

*Supply the missing word or number in each statement.*

63. Refer to Fig. 14-19. As the light intensity striking the surface of the photocell increases, the binary value at the output of the light meter circuit _____ (decreases, increases).

64. Refer to Fig. 14-19. As the light intensity striking the surface of the photocell increases, the resistance of the photocell _____ (decreases, increases).

65. Refer to Fig. 14-20. If current through series resistances $R_2$ and $R_3$ increases, the analog input voltage to the A/D converter _____ (decreases, increases).

66. Refer to Fig. 14-20. The conversion rate of the ADC0804 IC in this digital light meter circuit is about _____ (1, 400) A/D conversion(s) per second.

67. Refer to Fig. 14-20. Substituting $R_2$ with a resistor of a lower ohmic value would cause the output display to read _____ (higher, lower) for the same light intensity.

68. Refer to Fig. 14-20. The part labeled $R_3$ in the light meter circuit is a _____ (transducer, transformer) that converts light intensity into a variable resistance.

69. Refer to Fig. 14-20. The component labeled $R_3$ is a cadmium _____.

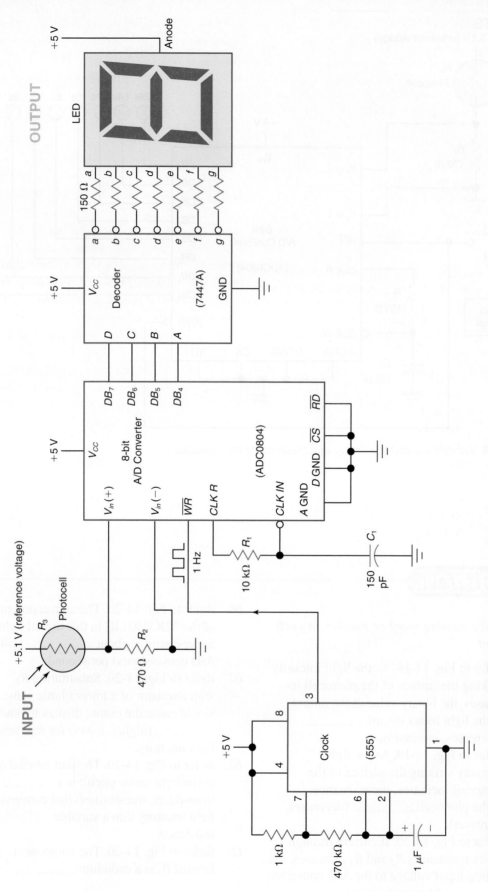

**Fig. 14-20** Wiring diagram for a digital light meter circuit using a decimal display.

**Digital light meter circuit**

## 14-12 Digitizing Temperature

The A/D converter could be used to convert an analog temperature to a digital quantity. A digital thermometer is one example of the use of an A/D converter in digitizing temperature. Devices other than an A/D converter might also be used to convert an analog temperature to a digital form.

As a general definition, to *digitize* means to convert an analog measurement into digital units or digital pulses. The A/D converter is one example of a *digitizer*. In this section the digitizer will be an elementary Schmitt trigger inverter.

A simple circuit for digitizing temperature is shown in Fig. 14-21. The digitizing device is a simple Schmitt trigger inverter (74LS14 IC). A *thermistor* is the temperature transducer. A thermistor is a *temperature-sensitive resistor*. As the temperature of the thermistor increases its resistance will decrease. Thermistors are said to have a *negative temperature coefficent* while most metals (like copper) have a positive temperature coefficent.

Recall that the switching threshold of the 74LS14 Schmitt trigger inverter is about 1.7 V when the input voltage is increasing. Because of hysteresis, the switching threshold of the Schmitt trigger inverter is lower or about 1 V when the input voltage is decreasing.

As the temperature of the thermistor in Fig. 14-21 increases its resistance will decrease. This will cause the voltage at the input of the Schmitt trigger inverter to increase (see voltmeter). When the temperature increases the voltage at the input of the inverter will finally exceed about +1.7 V and the output of the inverter will snap from HIGH to LOW as indicated on the logic probe.

Furthermore, as the temperature of the thermistor in Fig. 14-21 decreases its resistance will increase. This causes the voltage at the input of the inverter to decrease (see voltmeter). When the temperature decreases below the threshold voltage of about +1 V, the output of the Schmitt trigger inverter will snap from LOW to HIGH as indicated on the logic probe. The potentiometer shown in Fig. 14-21 allows the user some adjustment as to what temperatures the digitizer circuit triggers to HIGH or to LOW. In other words, potentiometer $R_1$ is used for calibration.

In the example in Fig. 14-21, it is said that we have digitized the temperature. In this example digitizing takes the form of generating either a HIGH or LOW. This example is like the sensing function of a thermostat. The I/O pins of a microcontroller (like the BASIC Stamp) can digitize analog data much like the circuit in Fig. 14-21 when I/O pins are used as inputs. After

**Digitize**

**Thermistor**

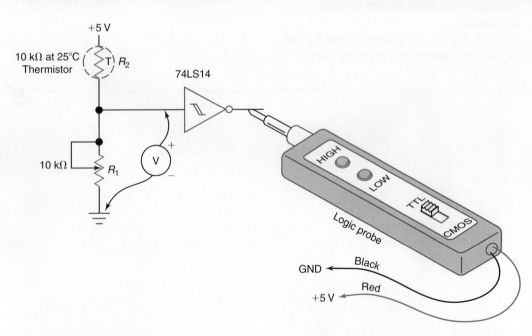

**Fig. 14-21** Using a thermistor to sense temperature and a Schmitt trigger inverter to digitize the analog input.

sensing the HIGH or LOW input, a microcontroller can then be programmed to respond to the higher or lower temperature.

A thermistor, such as the one used in Fig. 14-21, is constructed by sintering combinations of metallic oxides into different shapes. A common shape for a thermistor is a small teardrop to which leads are attached. The metallic oxides commonly used in manufacturing thermistors include those of titanium, iron, copper, cobalt and nickel. A common thermistor you may use in lab will have a resistance value of 10 kΩ at a temperature of 25°C. This same thermistor might have a resistance of 28 kΩ at 0°C and 1 kΩ at 100°C.

The advantage of the thermistor is that it is simple, inexpensive and easy to interface. One disadvantage of the thermistor is that it has a *nonlinear temperature-vs-resistance charac-*

*teristic.* This nonlinear characteristic makes the thermistor difficult to use as the thermal sensor in a thermometer application.

Many more expensive *linear thermal sensors* are available in IC form that can be used as sensors in thermometers. These include the three-terminal LM34 and LM35 Temperature Sensors from National Semiconductor and the two-terminal AD592 Precision IC Temperature Transducer from Analog Devices. More complex ICs like the 8-pin DIP DS1620 Digital Thermometer and Thermostat IC include more functions including sensing temperature, converting the temperature into a 9-bit word, a 3-wire serial interface, and programmable thermostatic controls. The DS1620 is especially useful when used in conjuction with a microcontroller (such as the BASIC Stamp). The DS1620 is manufactured by Dallas Semiconductor.

![Self-Test]

*Answer the following questions.*

70. Refer to Fig. 14-21. The device that can be classified as a *temperature transducer* is the _____ (Schmitt trigger inverter, thermistor).

71. Refer to Fig. 14-21. The device that can be classified as the *digitizer* is the _____ (Schmitt trigger inverter, thermistor).

72. Refer to Fig. 14-21. Potentiometer $R_1$ can be used to calibrate the digitizer circuit. (T or F)

73. Refer to Fig. 14-21. If the temperature of the thermistor is greatly increased, its resistance _____ (decreases, increases, stays the same) and the voltage at the input to the inverter _____ (decreases, increases). This increasing thermistor temperature causes the output of the Schmitt trigger inverter to snap from _____ (HIGH to LOW, LOW to HIGH).

# Chapter 14 Summary and Review

## Summary

1. Special interface encoders and decoders are used between analog and digital devices. These are called D/A converters and A/D converters.
2. A D/A converter consists of a resistor network and a summing amplifier.
3. Operational amplifiers are used in D/A converters and comparators. Gain can be easily set with external resistors on the op amp.
4. Several different resistor networks are used for weighting the binary input to a D/A converter.
5. Common A/D converters are the counter-ramp, ramp-generator, and successive-approximation types.
6. A voltage comparator compares two voltages and determines which is larger. An operational amplifier is the heart of the comparator.
7. Common specifications used for A/D converters include such characteristics as type of output, resolution, accuracy, conversion time, power supply voltage, output logic levels, input voltage, and power dissipation.
8. The ADC0804 IC is a CMOS 8-bit A/D converter. It features fast conversion times, microprocessor compatibility, three-state outputs, TTL level inputs and outputs, and an on-chip clock.
9. A photocell can be used as a transducer to drive an A/D converter in a digital light meter circuit.
10. A thermistor (temperature-sensitive resistor) can be used as temperature transducer. The thermistor has a nonlinear temperature-vs-resistance characteristic.
11. A Schmitt trigger device can be used as a very elementary digitizer.
12. An A/D converter is at the heart of a digital voltmeter. Most commercial digital voltmeters and DMMs use complex meter-type A/D converter LSI ICs.

## Chapter Review Questions

*Answer the following questions.*

14-1. An A/D converter is a special type of _____ (decoder, encoder).

14-2. A D/A converter is a(n) _____ (decoder, encoder).

14-3. The _____ (A/D, D/A) converter digitizes analog information.

14-4. The _____ (A/D, D/A) converter translates from binary to an analog voltage.

14-5. A D/A converter consists of a(n) _____ network and a summing _____.

14-6. The term "operational amplifier" is frequently shortened to _____.

14-7. The voltage gain of the operational amplifier in Fig. 14-3(*b*) is determined by dividing the value of _____ ($R_f$, $R_{in}$) by the value of _____ ($R_f$, $R_{in}$).

14-8. Draw a symbol for an operational amplifier. Label the inverting input with a minus sign and the noninverting input with a plus sign. Label the output. Label the +10-V and −10-V power supply connections.

14-9. Refer to Fig. 14-4. What is the gain ($A_v$) of the op amp in this diagram if $R_{in} = 1$ k$\Omega$ and $R_f = 20$ k$\Omega$?

14-10. Refer to Fig. 14-4. With the input voltage at +½ V, the output voltage is _____ (+, −)5 V. This is because we are using the _____ (inverting, noninverting) input of the op amp.

14-11. Refer to Fig. 14-5. What is the voltage gain of the op amp in this circuit with only switch $A$ at logical 1?

14-12. Refer to Fig. 14-5. What is the combined resistance of parallel resistors $R_1$ and $R_2$ if both switches $A$ and $B$ are at logical 1?

14-13. Refer to Fig. 14-5. What is the gain ($A_v$) of the op amp with switches $A$ and $B$ at logical 1? (Use the resistance value from question 14-12.)

14-14. Refer to Fig. 14-5. What is the output voltage when binary 0011 is applied to the D/A converter? (Use the $A_v$ from question 14-14.)

14-15. The arrangement of resistors in Fig. 14-6 is called the _____ ladder network.

14-16. A HIGH, or logical 1, from a TTL device is about _____ (0, 3.75, 8.5) V.

14-17. The _____ (A/D, D/A) converter is the more complicated electronic system.

14-18. Refer to Fig. 14-8. If point $X$ is at a logical _____ (0, 1), the counter advances one count as a pulse comes from the clock.

14-19. Refer to Fig. 14-8. If input $B$ of the comparator has a higher voltage than input $A$, the AND gate is _____ (disabled, enabled).

14-20. The primary component in a voltage comparator is a(n) _____ (counter, op amp).

14-21. Refer to Fig. 14-12. This digital voltmeter uses a _____ (counter-ramp, successive-approximation) A/D converter.

14-22. The _____ (ramp, successive-approximation) A/D converter is faster at digitizing information.

14-23. Devices such as microphones, speakers, strain gauges, photocells, temperature sensors, and potentiometers convert one form of energy to another and are generally called _____.

14-24. An A/D converter with binary outputs might be classified as a _____ (meter, microprocessor)-type unit.

14-25. Refer to Fig. 14-18. What is the resolution of the ADC0804 A/D converter?

14-26. A(n) _____ (8, 16)-bit A/D converter has a lower quantization error and is considered more "accurate."

14-27. Conversion times are somewhat longer for _____ (meter, microprocessor)-type A/D converters.

14-28. The ADC0804 (Fig. 14-17) has _____ (binary, decimal) outputs.

14-29. The A/D converter wired in Fig. 14-18 performs about _____ (3, 5,000 to 10,000) A/D conversions per second.

14-30. Refer to Fig. 14-18. If the analog input voltage is 3.0 V, the binary output is _____.

14-31. Refer to Fig. 14-20. Decreasing the light intensity striking $R_3$ causes the resistance of the photocell to _____ (decrease, increase).

14-32. Refer to Fig. 14-20. Decreasing the light intensity striking the photocell causes the decimal output to _____ (decrease, increase).

14-33. Refer to Fig. 14-20. If current through series resistances $R_2$ and $R_3$ decreases, the analog input voltage to the A/D converter _____ (decreases, increases).

14-34. Thermistors can be used as temperature transducers but have a nonlinear temperature-vs-resistance characteristic which makes them difficult to use as a thermal sensor in a thermometer. (T or F)

14-35. Refer to Fig. 14-21. As the temperature of the thermistor decreases the voltage at the input to the 74LS14 inverter _____ (decreases, increases) due to the _____ (decreased, increased) resistance of the thermal sensor $R_2$.

14-36. Refer to Fig. 14-21. As the temperature of the thermistor decreases greatly the output of the Schmitt trigger inverter will snap from _____ (HIGH-to-LOW, LOW-to-HIGH).

14-37. Refer to Fig. 14-21. Potentiometer $R_1$ is used for _____ (digitizing, calibration) in this simple A/D converter circuit.

14-1. Calculate the gain of the op amp circuit in Fig. 14-4 if $R_{in} = 1$ k$\Omega$ and $R_f = 5$ k$\Omega$. Using the calculated gain, what is the output voltage ($V_{out}$) if $V_{in} = 0.5$ V?

14-2. Refer to Fig. 14-5.

    a. What is the combined resistance of parallel resistors $R_2$ and $R_3$ if both switches $B$ and $C$ are at logical 1?

    b. Using the calculated resistance, what is the gain ($A_v$) of the op amp with switches $B$ and $C$ at a logical 1?

    c. What is the output voltage when binary 0110 is applied to the inputs of the D/A converter (use calculated $A_v$)?

14-3. Compare Tables 14-1 and 14-2. Explain the difference between the data in the two tables.

14-4. List the four sections of a counter-ramp A/D converter circuit.

14-5. List the four sections of a ramp-type A/D converter circuit.

14-6. Compare the D/A converter resistor networks in Figs. 14-5 and 14-6. Why would the R-2R ladder resistor network in Fig. 14-6 be easier to expand from four to eight binary inputs?

14-7. Refer to Fig. 14-8. What would be the *resolution* of this A/D converter?

14-8. A digital voltmeter is one application of a(n) _____ (A/D, D/A) converter.

14-9. At the option of your instructor, use circuit simulation software to (1) draw the 4-bit D/A converter using the R-2R ladder resistor network and op amp detailed in Fig. 14-22, (2) operate the 4-bit D/A converter circuit, and (3) show instructor your working D/A converter.

14-10. At the option of your instructor, use circuit simulation software to (1) draw a 5-bit D/A converter using the R-2R ladder resistor network and op amp something like the unit in Fig. 14-22, (2) operate the 5-bit D/A converter circuit, and (3) show the instructor your working 5 bit D/A converter.

14-11. At the option of your instructor, use circuit simulation software to (1) draw the generic 8-bit A/D converter circuit (with binary output) detailed in Fig. 14-23, (2) operate the 8-bit A/D converter circuit, and (3) show the instructor your working A/D converter.

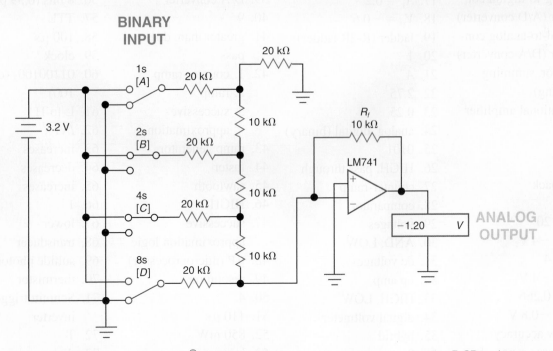

**Fig. 14-22** Electronics Workbench® simulation circuit of a D/A converter circuit using R-2R resistor network and op amp (scaling amplifier).

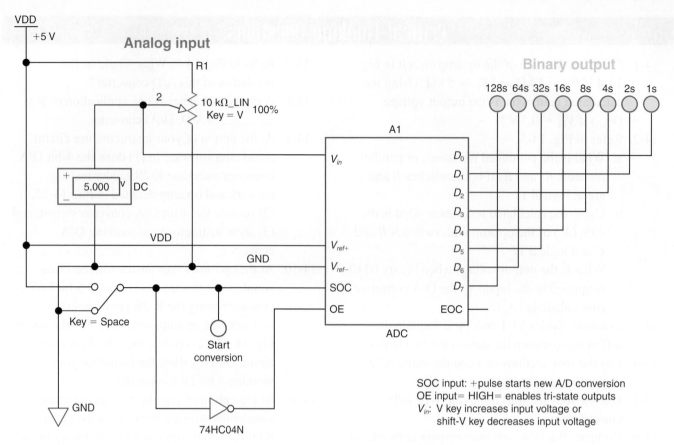

**Fig. 14-23** A/D converter circuit with 8-bit binary readout. (Prepared using MultiSIM 8).

## Answers to Self-Tests

1. analog-to-digital converter (A/D converter)
2. digital-to-analog converter (D/A converter)
3. resistor, summing (scaling)
4. operational amplifier
5. 1.4
6. 3.0
7. 0.2
8. feedback
9. input
10. $A_v = 20$
11. $V_o = -4\,V$
12. $A_v = 4$
13. $V_o = -4\,V$
14. $A_v = 0.266$
15. $V_o = -0.8\,V$
16. 1. low accuracy
    2. a large range of resistor values needed

17. $A_v = 0.2$
18. $V_o = -0.6$
19. ladder (R-2R ladder)
20. 1
21. $A$
22. 2.75
23. 0.25
24. analog, digital (binary)
25. 0101
26. HIGH, pass through
27. counter-ramp
28. comparator
29. voltages
30. AND, LOW
31. dc voltages
32. op amp
33. HIGH, LOW
34. digital voltmeter
35. hybrid
36. 0
37. four
38. 5

39. D/A converter
40. 9
41. greater than, does not pass
42. 1. counter-ramp
    2. ramp
    3. successive-approximation
43. ramp generator
44. faster
45. sawtooth
46. HIGH
47. successive-approximation logic
48. μP (microprocessor)
49. resolution
50. 4
51. 110 μs
52. 850 mW
53. longer
54. CMOS
55. microprocessor

56. 8 bits (0.39 percent)
57. TTL
58. 100 μs
59. clock
60. $01100100_2$ (decimal 100)
61. L-to-H
62. $\overline{INTR}$
63. increases
64. decreases
65. increases
66. 1
67. lower
68. transducer
69. sulfide photocell
70. thermistor
71. Schmitt trigger inverter
72. T
73. decreases, increases, HIGH-to-LOW

# Appendix A
## Solder and the Soldering Process

## From Simple Task to Fine Art

Soldering is the process of joining two metals together by the use of a low-temperature melting alloy. Soldering is one of the oldest known joining techniques, first developed by the Egyptians in making weapons such as spears and swords. Since then, it has evolved into what is now used in the manufacturing of electronic assemblies. Soldering is far from the simple task it once was; it is now a fine art, one that requires care, experience, and a thorough knowledge of the fundamentals.

The importance of having high standards of workmanship cannot be overemphasized. Faulty solder joints remain a cause of equipment failure, and because of that, soldering has become a *critical skill*.

The material contained in this appendix is designed to provide the student with both the fundamental knowledge and the practical skills needed to perform many of the high-reliability soldering operations encountered in today's electronics.

Covered are the fundamentals of the soldering process, the proper selection, and the use of the soldering station.

The key concept in this appendix is *high-reliability soldering*. Much of our present technology is vitally dependent on the reliability of countless, individual soldered connections. High-reliability soldering was developed in response to early failures with space equipment. Since then the concept and practice have spread into military and medical equipment. We have now come to expect it in everyday electronics as well.

## The Advantage of Soldering

Soldering is the process of connecting two pieces of metal together to form a reliable electrical path. Why solder them in the first place? The two pieces of metal could be put together with nuts and bolts, or some other kind of mechanical fastening. The disadvantages of these methods are twofold. First, the reliability of the connection cannot be assured because of vibration and shock. Second, because oxidation and corrosion are continually occurring on the metal surfaces, electrical conductivity between the two surfaces would progressively decrease.

A soldered connection does away with both of these problems. There is no movement in the joint and no interfacing surfaces to oxidize. A continuous conductive path is formed, made possible by the characteristics of the solder itself.

## The Nature of Solder

Solder used in electronics is a low-temperature melting alloy made by combining various metals in different proportions. The most common types of solder are made from tin and lead. When the proportions are equal, it is known as 50/50 solder—50 percent tin and 50 percent lead. Similarly, 60/40 solder consists of 60 percent tin and 40 percent lead. The percentages are usually marked on the various types of solder available; sometimes only the tin percentage is shown. The chemical symbol for tin is Sn; thus Sn 63 indicates a solder which contains 63 percent tin.

Pure lead (Pb) has a melting point of 327°C (621°F); pure tin, a melting point of 232°C (450°F). But when they are combined into a 60/40 solder, the melting point drops to 190°C (374°F)—lower than either of the two metals alone.

Melting generally does not take place all at once. As illustrated in Fig. A-1, 60/40 solder begins to melt at 183°C (361°F), but it has not fully melted until the temperature reaches 190°C (374°F). Between these two temperatures, the solder exists in a plastic (semiliquid) state—some, but not all, of the solder has melted.

The plastic range of solder will vary, depending on the ratio of tin to lead, as shown in Fig. A-2. Various ratios of tin to lead are shown across the top of this figure. With most ratios, melting begins at 183°C (361°F), but the full melting temperatures vary dramatically. There is one ratio of tin to lead that has no plastic state and is known as *eutectic solder*. This ratio is 63/37 (Sn 63) and it fully melts and solidifies at 183°C (361°F).

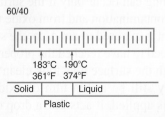

**Fig. A-1** Plastic range of 60/40 solder. Melt begins at 183°C (361°F) and is complete at 190°C (374°F).

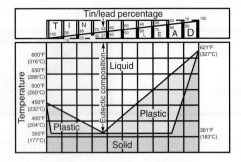

**Fig. A-2** Fusion characteristics of tin/lead solders.

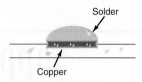

**Fig. A-3** The wetting action. Molten solder dissolves and penetrates a clean copper surface, forming an intermetallic bond.

The solder most commonly used for hand soldering in electronics is the 60/40 type, but because of its plastic range, care must be taken not to move any elements of the joint during the cool-down period. Movement may cause a disturbed joint. Characteristically, this type of joint has a rough, irregular appearance and looks dull instead of bright and shiny. It is unreliable and therefore one of the types of joints that is unacceptable in high-reliability soldering.

In some situations, it is difficult to maintain a stable joint during cooling, for example, when wave soldering is used with a moving conveyor line of circuit boards during the manufacturing process. In other cases it may be necessary to use minimal heat to avoid damage to heat-sensitive components. In both of these situations, eutectic solder is the preferred choice, since it changes from a liquid to a solid during cooling with no plastic range.

## The Wetting Action

To someone watching the soldering process for the first time, it looks as though the solder simply sticks the metals together like a hot-melt glue, but what actually happens is far different.

A chemical reaction takes place when the hot solder comes into contact with the copper surface. The solder dissolves and penetrates the surface. The molecules of solder and copper blend together to form a new metal alloy, one that is part copper and part solder and that has characteristics all its own. This reaction is called *wetting* and forms the intermetallic bond between the solder and copper (Fig. A-3).

Proper wetting can occur only if the surface of the copper is free of contamination and from oxide films that form when the metal is exposed to air. Also, the solder and copper surfaces need to have reached the proper temperature.

Even though the surface may look clean before soldering, there may still be a thin film of oxide covering it. When solder is applied, it acts like a drop of water on an oily surface because the oxide coating prevents the solder from coming into contact with the copper. No reaction takes place, and the solder can be easily scraped off. For a good solder bond, surface oxides must be removed during the soldering process.

## The Role of Flux

Reliable solder connections can be accomplished only on clean surfaces. Some sort of cleaning process is essential in achieving successful soldered connections, but in most cases it is insufficient. This is due to the extremely rapid rate at which oxides form on the surfaces of heated metals, thus creating oxide films which prevent proper soldering. To overcome these oxide films, it is necessary to utilize materials, called *fluxes,* which consist of natural or synthetic rosins and sometimes additives called activators.

It is the function of flux to remove surface oxides and keep them removed during the soldering operation. This is accomplished because the flux action is very corrosive at or near solder melt temperatures and accounts for the flux's ability to rapidly remove metal oxides. It is the fluxing action of removing oxides and carrying them away, as well as preventing the formation of new oxides, that allows the solder to form the desired intermetallic bond.

Flux must activate at a temperature lower than solder so that it can do its job prior to the solder flowing. It volatilizes very rapidly; thus it is mandatory that the flux be activated to flow onto the work surface and not simply be volatilized by the hot iron tip if it is to provide the full benefit of the fluxing action.

There are varieties of fluxes available for many applications. For example, in soldering sheet metal, acid fluxes are used; silver brazing (which requires a much higher temperature for melting than that required by tin/lead alloys) uses a borax paste. Each of these fluxes removes oxides and, in many cases, serves additional purposes. The fluxes used in electronic hand soldering are the pure rosins, rosins combined with mild activators to accelerate the rosin's fluxing capability, low-residue/no-clean fluxes, or water-soluble fluxes. Acid fluxes or highly activated fluxes should never be used in electronic work. Various types of flux-cored solder are now in common use. They

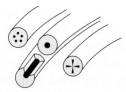

**Fig. A-4** Types of cored solder, with varying solder-flux percentages.

provide a convenient way to apply and control the amount of flux used at the joint (Fig. A-4).

## Soldering Irons

In any kind of soldering, the primary requirement, beyond the solder itself, is heat. Heat can be applied in a number of ways—conductive (e.g., soldering iron, wave, vapor phase), convective (hot air), or radiant (IR). We are mainly concerned with the conductive method, which uses a soldering iron.

Soldering stations come in a variety of sizes and shapes, but consist basically of three main elements: a resistance heating unit; a heater block, which acts as a heat reservoir; and the tip, or bit, for transferring heat to the work. The standard production station is a variable-temperature, closed-loop system with interchangeable tips and is made with ESD-safe plastics.

## Controlling Heat at the Joint

Controlling tip temperature is not the real challenge in soldering; the real challenge is to control the *heat cycle* of the work—how fast the work gets hot, how hot it gets, and how long it stays that way. This is affected by so many factors that, in reality, tip temperature is not that critical.

The first factor that needs to be considered is the *relative thermal mass* of the area to be soldered. This mass may vary over a wide range.

Consider a single land on a single-sided circuit board. There is relatively little mass, so the land heats up quickly. But on a double-sided board with plated-through holes, the mass is more than doubled. Multilayered boards may have an even greater mass, and that's before the mass of the component lead is taken into consideration. Lead mass may vary greatly, since some leads are much larger than others.

Further, there may be terminals (e.g., turret or bifurcated) mounted on the board. Again, the thermal mass is increased, and will further increase as connecting wires are added.

Each connection, then, has its particular thermal mass. How this combined mass compares with the mass of the iron tip, the "relative" thermal mass, determines the time and temperature rise of the work.

With a large work mass and a small iron tip, the temperature rise will be slow. With the situation reversed, using a large iron tip on a small work mass, the temperature rise of the work will be much more rapid—even though the *temperature of the tip is the same.*

Now consider the capacity of the iron itself and its ability to sustain a given flow of heat. Essentially, irons are instruments for generating and storing heat, and the reservoir is made up of both the heater block and the tip. The tip comes in various sizes and shapes; it's the *pipeline* for heat flowing into the work. For small work, a conical (pointed) tip is used, so that only a small flow of heat occurs. For large work, a large chisel tip is used, providing greater flow.

The reservoir is replenished by the heating element, but when an iron with a large tip is used to heat massive work, the reservoir may lose heat faster than it can be replenished. Thus the *size* of the reservoir becomes important: a large heating block can sustain a larger outflow longer than a small one.

An iron's capacity can be increased by using a larger heating element, thereby increasing the wattage of the iron. These two factors, block size and wattage, are what determine the iron's recovery rate.

If a great deal of heat is needed at a particular connection, the correct temperature with the right size tip is required, as is an iron with a large enough capacity and an ability to recover fast enough. *Relative thermal mass,* then, is a major consideration for controlling the heat cycle of the work.

A second factor of importance is the *surface condition* of the area to be soldered. If there are any oxides or other contaminants covering the lands or leads, there will be a barrier to the flow of heat. Then, even though the iron tip is the right size and has the correct temperature, it may not supply enough heat to the connection to melt the solder. In soldering, a cardinal rule is that a good solder connection cannot be created on a dirty surface. Before attempting to solder, the work should always be cleaned with an approved solvent to remove any grease or oil film from the surface. In some cases pretinning may be required to enhance solderability and remove heavy oxidation of the surfaces prior to soldering.

A third factor to consider is *thermal linkage*—the area of contact between the iron tip and the work.

Figure A-5 shows a cross-sectional view of an iron tip touching a round lead. The contact occurs only at the point indicated by the "X," so the linkage area is very small, not much more than a straight line along the lead.

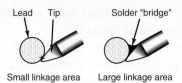

Lead   Tip         Solder "bridge"

Small linkage area      Large linkage area

**Fig. A-5** Cross-sectional view (left) of iron tip on a round lead. The "X" shows point of contact. Use of a solder bridge (right) increases the linkage area and speeds the transfer of heat.

The contact area can be greatly increased by applying a small amount of solder to the point of contact between the tip and workpiece. This solder heat bridge provides the thermal linkage and assures rapid heat transfer into the work.

From the aforementioned, it should now be apparent that there are many more factors than just the temperature of the iron tip that affect how quickly any particular connection is going to heat up. In reality, soldering is a very complex control problem, with a number of variables to it, each influencing the other. And what makes it so critical is *time*. The general rule for high-reliability soldering on printed circuit boards is to apply heat for no more than 2 s from the time solder starts to melt (wetting). Applying heat for longer than 2 s after wetting may cause damage to the component or board.

With all these factors to consider, the soldering process would appear to be too complex to accurately control in so short a time, but there is a simple solution—the *workpiece indicator* (WPI). This is defined as the reaction of the workpiece to the work being performed on it—a reaction that is discernible to the human senses of sight, touch, smell, sound, and taste.

Put simply, workpiece indicators are the way the work talks back to you—the way it tells you what effect you are having and how to control it so that you accomplish what you want.

In any kind of work, you become part of a closed-loop system. It begins when you take some action on the workpiece; then the workpiece reacts to what you did; you sense the change, and then modify your action to accomplish the result. It is in the sensing of the change, by sight, sound, smell, taste, or touch, that the workpiece indicators come in (Fig. A-6).

For soldering and desoldering, a primary workpiece indicator is *heat rate recognition*—observing how fast heat flows into the connection. In practice, this means observing the rate at which the solder melts, which should be within 1 to 2 s.

This indicator encompasses all the variables involved in making a satisfactory solder connection with minimum heating effects, including the capacity of the iron and its tip temperature, the surface conditions, the thermal linkage between tip and workpiece, and the relative thermal masses involved.

If the iron tip is too large for the work, the heating rate may be too fast to be controlled. If the tip is too small, it may produce a "mush" kind of melt; the heating rate will be too slow, even though the temperature at the tip is the same.

A general rule for preventing overheating is "Get in and get out as fast as you can." That means using a heated iron you can react to—one giving a 1- to 2-s dwell time on the particular connection being soldered.

## Selecting the Soldering Iron and Tip

A good all-around soldering station for electronic soldering is a variable-temperature, ESD-safe station with a pencil-type iron and tips that are easily interchangeable, even when hot (Fig. A-7).

The soldering iron tip should always be fully inserted into the heating element and tightened. This will allow for maximum heat transfer from the heater to the tip.

The tip should be removed daily to prevent an oxidation scale from accumulating between the heating

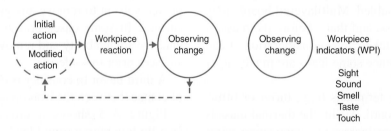

**Fig. A-6** Work can be viewed as a closed-loop system (left). Feedback comes from the reaction of the workpiece and is used to modify the action. Workpiece indicators (right)—changes discernible to the human senses—are the way the "work talks back to you."

element and the tip. A bright, thin tinned surface must be maintained on the tip's working surface to ensure proper heat transfer and to avoid contaminating the solder connection.

The plated tip is initially prepared by holding a piece of flux-cored solder to the face so that it will tin the surface when it reaches the lowest temperature at which solder will melt. Once the tip is up to operating temperature, it will usually be too hot for good tinning, because of the rapidity of oxidation at elevated temperatures. The hot tinned tip is maintained by wiping it lightly on a damp sponge to shock off the oxides. When the iron is not being used, the tip should be coated with a layer of solder.

## Making the Solder Connection

The soldering iron tip should be applied to the area of maximum thermal mass of the connection being made. This will permit the rapid thermal elevation of the parts being soldered. Molten solder always flows toward the heat of a properly prepared connection.

When the solder connection is heated, a small amount of solder is applied to the tip to increase the thermal linkage to the area being heated. The solder is then applied to the opposite side of the connection so that the work surfaces, not the iron, melt the solder. Never melt the solder against the iron tip and allow it to flow onto a surface cooler than the solder melting temperature.

Solder, with flux, applied to a cleaned and properly heated surface will melt and flow without direct contact with the heat source and provide a smooth, even surface, feathering out to a thin edge (Fig. A-8). Improper soldering will exhibit a built-up, irregular appearance and poor filleting. The parts being soldered must be held rigidly in place until the temperature decreases to solidify the solder. This will prevent a disturbed or fractured solder joint.

Selecting cored solder of the proper diameter will aid in controlling the amount of solder being applied to the connection (e.g., a small-gauge solder for a small connection; a large-gauge solder for a large connection).

## Removal of Flux

Cleaning may be required to remove certain types of fluxes after soldering. If cleaning is required, the flux residue should be removed as soon as possible, preferably within 1 hour after soldering.

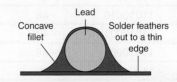

**Fig. A-7** Cross-sectional view of a round lead on a flat surface.

See www. paceworldwide. com for information on lead free solders and soldering

# Appendix B
## Formulas and Conversions

| Two's Complement Number Conversion Chart | | | | | | | |
|---|---|---|---|---|---|---|---|
| **2's Comp** | **Decimal** | **2's Comp** | **Decimal** | **2's Comp** | **Decimal** | **2's Comp** | **Decimal** |
| 11111111 | −1 | 11011111 | −33 | 10111111 | −65 | 10011111 | −97 |
| 11111110 | −2 | 11011110 | −34 | 10111110 | −66 | 10011110 | −98 |
| 11111101 | −3 | 11011101 | −35 | 10111101 | −67 | 10011101 | −99 |
| 11111100 | −4 | 11011100 | −36 | 10111100 | −68 | 10011100 | −100 |
| 11111011 | −5 | 11011011 | −37 | 10111011 | −69 | 10011011 | −101 |
| 11111010 | −6 | 11011010 | −38 | 10111010 | −70 | 10011010 | −102 |
| 11111001 | −7 | 11011001 | −39 | 10111001 | −71 | 10011001 | −103 |
| 11111000 | −8 | 11011000 | −40 | 10111000 | −72 | 10011000 | −104 |
| 11110111 | −9 | 11010111 | −41 | 10110111 | −73 | 10010111 | −105 |
| 11110110 | −10 | 11010110 | −42 | 10110110 | −74 | 10010110 | −106 |
| 11110101 | −11 | 11010101 | −43 | 10110101 | −75 | 10010101 | −107 |
| 11110100 | −12 | 11010100 | −44 | 10110100 | −76 | 10010100 | −108 |
| 11110011 | −13 | 11010011 | −45 | 10110011 | −77 | 10010011 | −109 |
| 11110010 | −14 | 11010010 | −46 | 10110010 | −78 | 10010010 | −110 |
| 11110001 | −15 | 11010001 | −47 | 10110001 | −79 | 10010001 | −111 |
| 11110000 | −16 | 11010000 | −48 | 10110000 | −80 | 10010000 | −112 |
| 11101111 | −17 | 11001111 | −49 | 10101111 | −81 | 10001111 | −113 |
| 11101110 | −18 | 11001110 | −50 | 10101110 | −82 | 10001110 | −114 |
| 11101101 | −19 | 11001101 | −51 | 10101101 | −83 | 10001101 | −115 |
| 11101100 | −20 | 11001100 | −52 | 10101100 | −84 | 10001100 | −116 |
| 11101011 | −21 | 11001011 | −53 | 10101011 | −85 | 10001011 | −117 |
| 11101010 | −22 | 11001010 | −54 | 10101010 | −86 | 10001010 | −118 |
| 11101001 | −23 | 11001001 | −55 | 10101001 | −87 | 10001001 | −119 |
| 11101000 | −24 | 11001000 | −56 | 10101000 | −88 | 10001000 | −120 |
| 11100111 | −25 | 11000111 | −57 | 10100111 | −89 | 10000111 | −121 |
| 11100110 | −26 | 11000110 | −58 | 10100110 | −90 | 10000110 | −122 |
| 11100101 | −27 | 11000101 | −59 | 10100101 | −91 | 10000101 | −123 |
| 11100100 | −28 | 11000100 | −60 | 10100100 | −92 | 10000100 | −124 |
| 11100011 | −29 | 11000011 | −61 | 10100011 | −93 | 10000011 | −125 |
| 11100010 | −30 | 11000010 | −62 | 10100010 | −94 | 10000010 | −126 |
| 11100001 | −31 | 11000001 | −63 | 10100001 | −95 | 10000001 | −127 |
| 11100000 | −32 | 11000000 | −64 | 10100000 | −96 | 10000000 | −128 |

# Glossary of Terms and Symbols

| Term | Definition | Symbol or Abbreviation |
|------|------------|------------------------|
| Access time | In memories, the time it takes to retrieve a piece of data from storage. | |
| Active **HIGH** input | Digital input that executes its function when a HIGH is present. | |
| Active **LOW** input | Digital input that executes its function when a LOW is present. | |
| Active-matrix display | A high-quality expensive color LCD using active-matrix technology, which involves the use of thin-film transistors. For contrast see *passive-matrix display*. | |
| A/D converter | Device for converting an analog voltage into a proportional digital quantity. Types include a microprocessor-compatible device with binary outputs or meter-type with decimal outputs. | ADC |
| Adder | Combinational logic circuit which generates sum and carry outputs from any set of binary inputs. Half-adders and full-adders are two fundamental adder circuits. | |
| Address | In computer systems, a number that represents a unique storage location. | |
| Alphanumeric | Consisting of numbers, letters and other characters. ASCII is a common alphanumeric code. | |
| Ampere | Base unit of current. | A |
| Analog | A branch of electronics dealing with infinitely varying quantities. Also referred to as *linear electronics*. | |
| Analog to digital | Conversion of an analog signal to a digital quantity such as binary. | A/D |
| AND gate | Basic combinational logic device where all inputs must be HIGH for the output to be HIGH. | |
| Angular velocity | Another method of describing the speed of rotation of a shaft or other object. | |
| Anode | Positive section of a device such as a diode or LED. | |
| Arithmetic logic unit | Part of central processing unit of computer that processes data using arithmetic and logic operations. | ALU |
| American Standard Code for Information Interchange | One of the most widely used alphanumeric codes. | ASCII |
| Astable multivibrator | Device that oscillates between two stable states. Commonly called a free-running clock or multivibrator. | |
| Asynchronous | In digital circuits, meaning that operations are not executed in step with the clock. | |
| Base | Center section of a bipolar transistor used to control current flow from the emitter to collector. | |
| BASIC | An easy-to-learn high-level programming language commonly used to teach beginning programming. An acronym for *beginners all-purpose symbolic instruction code*. | BASIC |
| Baud | Unit of signal transmission speed in telecommunications equal to the number of discrete events per second. | Bd |

| Term | Definition | Symbol or Abbreviation |
|------|-----------|------------------------|
| BCD counter | A 4-bit counter that commonly counts from binary 0000 to 1001 and resets to 0000. | |
| Binary | Base 2 number system using numbers 0 or 1. | |
| Binary-coded decimal | A common code in which each decimal digit (0–9) is represented by a 4-bit group. | BCD |
| Bistable multivibrator | A device having two stable states but it must be triggered to jump from one to the other. Also called a *flip-flop*. | |
| Bit | A single binary digit (0 or 1). Useful in representing on-off switching in digital circuits. An acronym for *binary digit*. | |
| Block diagram | A drawing using labeled blocks for functional sections of an electronic system. | |
| Boolean algebra | Mathematical system for representing logical statements. Very useful in digital electronics. | |
| Boolean expression | Mathematical representation of a logic function. Function could also be described using a truth table or logic circuit diagram. | $AB + C'D = Y$ |
| Boundary-scan technology | A system of imbedding test points in silicon during the design process for ease of testing for quality control and field testing. See *JTAG*. | JTAG |
| Broadside loading | Parallel loading. | |
| Bubble | On logic symbol, it means an active LOW input or output. | |
| Buffer | Special solid state device used to increase the drive current at the output. Noninverting buffer has no logical function. | |
| Bus | In a computer system, parallel conductors used for communication between CPU, memories, and perpherial devices. Most systems have an address bus, data bus, and control bus. | |
| Byte | An 8-bit group that is commonly used to represent a number or code in computers and digital electronics. | |
| Cache memory | In computers, an extremely fast, very expensive SRAM unit used to store frequently needed or recently used data. The cache memory is the bridge between the ultrafast processor and the much slower main/hard drive/CD-ROM memory. Cache memory is commonly referred as L1 (primary) or L2 (secondary). | |
| Cascading | Generally, the series connection of electronic devices with the output of the first feeding the input of the second. Term is used in both linear and digital electronics. | |
| Cathode | Negative section of a device such as a diode or LED. | |
| Cathode-ray tube | Vacuum tube used in televisions, video monitors, and most oscilloscopes to display images. | CRT |
| CD-R | A compact disc that you can record data once using a CD burner on a standard PC. CD-R is an acronym for *compact disc-recordable*. | CD-R |
| CD-ROM | A read-only mass storage device based on the compact disc. | |
| CD-RW | A compact disc that you can rewrite data many times using your computer system. CD-RW is an acronym for *compact disc-rewritable*. | CD-RW |
| Cell | In memories, a single storage element. | |
| Central processing unit | In computer system, the logic unit that performs logic arithmetic, control functions, and is the center of most data transfers. | CPU |

| Term | Definition | Symbol or Abbreviation |
|------|-----------|------------------------|
| Charge-coupled device | Image sensor using a light-sensitive array of photocells based on capacitorlike semiconductor devices. Used in digital cameras, scanners, camcorders, and other scientific imaging equipment. See *CMOS image sensor* for an alternative technology. | CCD |
| Chip | An integrated circuit. | |
| Clock | Signal generated by an oscillator used to provide timing for a digital system such as a computer. | |
| CMOS image sensor | An image sensor using a light-sensitive array of photocells much like the CCD but less expensive to fabricate. Used in less-expensive digital cameras and cell phones. See also *charge-coupled device*. | |
| Collector | The region of a bipolar transistor that receives the flow of current carriers. | |
| Combinational logic | Use of logic gates to produce desired output immediately. No memory or latching characteristics. | |
| Complementary metal-oxide semiconductor | A popular technology for manufacturing ICs, which features extremely low-power consumption. Uses opposite polarity field-effect transistors in its design. | CMOS |
| CPLD | A specific programmable logic device much like the GAL only for much larger scale logic problems. An acronym for *complex programmable logic device*. | CPLD |
| Current | Movement of charge in a specified direction. Base unit is ampere. | |
| Current sinking | Conventional current flow into LOW output of digital device. Current is "sinking" to ground. | |
| Current sourcing | Conventional current flow from HIGH output into load. Output is "sourcing" current. | |
| Cylinder | On a hard disk drive, a series of identical tracks on various platters. | |
| D/A converter | Device for converting a digital quantity into a proportional analog voltage. | DAC |
| Data selector | Combination logic block that selects one-of-*X* data inputs and connects that information to the output. Also called a multiplexer. | |
| Decoder | A logic device that translates from binary code to decimal. Generally, it translates processed data into a digital system to another format such as alphanumeric. | |
| Decrement | To decrease the count by 1. | |
| Demultiplexer | Combination logic block that distributes data from single input to one-of-*X* outputs. Also called a distributor. Can change serial to parallel data. | DEMUX |
| D flip-flop | Flip-flop with at least set and reset modes of operation. Also called a data or delay flip-flop. | |
| Digital | Branch of electronics dealing with discrete signal levels. Signals are commonly HIGH or LOW and may be represented by binary numbers. | |
| Digital potentiometer | An electronic device comparable to a traditional potentiometer with resistance outputs variable in discrete steps. The wiper position can be stored in EEPROM when the power is turned off. Digital input pulses control the movement of the wiper. Also referred to as a | |

| Term | Definition | Symbol or Abbreviation |
|------|-----------|------------------------|
| | solid-state potentiometer or non-volatile (NV) digital potentiometer. | |
| Digital signal processor | A specialized microprocessor-like device that can be programmed to condition and enhance signals (eliminate noise, increase frequency response, etc.). Commonly used in conjunction with A/D- and D/A-converters. | DSP |
| Digital to analog | Conversion of a digital signal to its analog equivalent, such as a voltage. | D/A |
| Digital Versatile Disc | A popular very high-capacity optical disc which looks like a traditional CD. It can store from about 4.7 GB to about 17 GB of video, audio, or computer data. Also referred to as a digital video disc. | DVD |
| Digitize | To convert an analog signal into digital units or pulses. See A/D converter. | |
| DIMM | In computer technology, a modern RAM memory board holding many SDRAM memory chips used on the latest PCs. An acronym for dual-in-line memory module. See also slightly older SIMM. | DIMM |
| DIN connector | Connectors used on computers following the standards of the German association DIN (Deutsche Industrie Norm). | DIN |
| Diode | Two-terminal semiconductor device. They usually allow current to flow in only one direction. | |
| Discrete time signal | Another name for a digital signal especially used in DSP applications where digital inputs are commonly a sampling of an analog input. | |
| Display multiplexing | To light multiple alphanumeric displays by sequentially activating them one at a time in rapid succession so they appear to be turned on continuously. Display multiplexing saves components and cost. | |
| Double data rate SDRAM | A synchronous dynamic RAM that is faster than regular SDRAM. | DDR SDRAM |
| Drive | Generally in computers, it refers to a mass storage device such as a floppy disk drive, hard disk drive, optical drive, or even a solid-state drive. Usually an electromagnetic or optical device which moves mass storage media under read/write heads. | |
| Dynamic RAM | Extremely common random-access (read/write) memory device whose memory cells need refreshing many times per second. A volatile memory. Compare with *SDRAM* and *RDRAM*. | DRAM |
| Dual-in-line package | Popular packaging method for ICs. | DIP |
| Edge triggering | In synchronous devices such as flip-flops, the exact time the device is activated such as on the rising (positive-edge) or falling edge (negative-edge) of the clock pulse. | |
| 8421 BCD code | Four-bit BCD code with weighting of 8, 4, 2, and 1. See binary-coded decimal. | |
| Electrically erasable programmable read-only memory | A nonvolatile memory that can be programmed, electrically erased, and reprogrammed. Flash memories are a type of EEPROM. | EEPROM |
| Enable | To activate a function or input to a digital circuit. The opposite of disable. | |
| Encoder | A logic device that translates from decimal to another code such as binary. Generally, it translates input information to a code useful to digital circuits. | |

| Term | Definition | Symbol or Abbreviation |
|---|---|---|
| Emitter | The region of a bipolar transistor that sends the current carriers to the collector. | |
| Even parity | In data transmission, sending a parity bit that will make the number of 1s in a group even. | |
| Extended Binary-Coded Decimal Interchange Code | An 8-bit alphanumeric code used mainly on mainframe computers. | EBCDIC |
| Fan-out | Output drive characteristic of logic device. The number of inputs of the logic family that can be driven by a single output. | |
| Ferroelectric RAM | A semiconductor nonvolatile RAM with good access speed that allows in-circuit programming. FeRAM memory cells are based on ferroelectric capacitors and MOS transistors. | FeRAM or FRAM |
| Field-effect transistor | Type of transistor where gate terminal controls the resistance of a semiconducting channel. | |
| Firmware | Computer programs and data held permanently in nonvolatile memory devices such as ROMs. See also *hardware* and *software*. | |
| Flash memory | A newer nonvolatile memory similar to the EEPROM. Its outstanding characteristics include very high density (small memory cell), low power, and nonvolatile but rewritable. | |
| Flip-flop | Basic sequential logic device having two stable states. Can serve as a memory device. Also called a bistable multivibrator. | |
| Floating input | An input not held HIGH or LOW which may "float" either HIGH, LOW or in between. Can cause problems. | |
| FPLD | A specific programmable logic device something like the CPLD, but containing simpler cells allowing more flexibility during the design process. An acronym for field programmable logic device. | FPLD |
| Frequency divider | A logic block that divides the input waveform's frequency by a certain number (such as divide-by-10). Counters commonly perform this function. | |
| Full-adder | Digital circuit with three inputs for carry in and two bits with sum and carry out outputs. | |
| Gain | A ratio of output to input. May be measured in terms of voltage, current, or power. Also known as amplification. | |
| Gate | Basic combinational logic device which performs a specific logic function (AND, OR, NOT, NAND, NOR). | |
| Generic array logic | A specific programmable logic device (PLD) with an array of ANDs that can be reprogrammed, a fixed array of ORs gates. | GAL |
| Glitch | An unwanted current or voltage spike that usually commonly reoccurs but not regularly. | |
| GND | Label for negative of power supply in TTL ICs and some CMOS ICs. Common ground. | |
| Half-adder | Digital circuit that will add two bits and output a sum and carry. Cannot handle carry inputs. | |
| Hall-effect sensor | A transducer that converts an increasing or decreasing magnetic field into a proportional varying voltage These sensors are commonly packaged as *Hall-effect switches* featuring a digital output (HIGH or LOW). | |

| Term | Definition | Symbol or Abbreviation |
|------|-----------|------------------------|
| Hardware | In computer technology, the physical components of a computer system. See also *software* and *firmware*. | |
| Hertz | The base unit of frequency. One cycle per second. | Hz |
| Hexadecimal | Base 16 number system using characters 0 thru 9, A, B, C, D, E, and F. Used to represent binary numbers 0000 through 1111. | Hex |
| Hysteresis | Unequal switching thresholds exhibited by some logic circuits making their outputs "snap action." Schmitt trigger logic devices exhibit this feature. | |
| IEEE | Institute for Electrical and Electronic Engineers. | |
| Increment | To increase the count by 1. | |
| Input/Output | A connection to a digital device that can be programmed to serve as either an input or output. Very common on many complex devices including microcontrollers. | I/O |
| Instruction set | The complete set of commands responded to by a microprocessor, microcontroller, or PLC. | |
| Interfacing | The design of interconnections between circuits that shift the levels of voltage and current to make them compatible. | |
| Integrated circuit | Combination of many electronic components in a compact package that functions as an analog, digital, or hybrid circuit. Classified as to levels of circuit complexity (SSI, MSI, LSI, VLSI, or ULSI). | IC |
| Inverter | Basic logic function where the output is always opposite the input. Also called the NOT function. | |
| JEDEC | Joint Electron Device Engineering Council. | |
| J-K flip-flop | Flip-flop with at least set, reset, toggle and hold modes of operation. Very adaptable. | |
| JTAG | In common use, JTAG refers to the boundary-scan method of imbedding test points in silicon during the design process for automated testing. An acronym for Joint Test Action Group responsible for developing the IEEE STD 1149.1 Test Access Port and Boundary Scan Architecture. See also boundary-scan technology. | JTAG |
| Karnaugh map | A graphic method of reducing Boolean expressions to simpler forms. | K maps |
| Latch | Fundamental binary storage device. Also called a flip-flop. | |
| Large-scale integration | Used by some manufacturers to indicate the complexity of an integrated circuit. LSI usually means having a complexity of from 100 to 9999 gates. | LSI |
| Least significant bit | The bit position in a binary number with the least weight. | LSB |
| Light-emitting diode | Special PN junction that gives off light when current flows through it. Has lens to focus the light. | LED |
| Liquid-crystal display | Very low power display technology used in most battery operated devices. Nematic fluid in display changes reflectivity when energized changing display from silver to black characters. Color LCDs are also available. | LCD |
| Logic analyzer | An expensive test instrument that can sample and store many channels of digital information. | |
| Logic diagram | A schematic showing interconnection between logic devices like gates, flip-flops, etc. | |
| Logic family | A group of totally compatible digital ICs that can be interconnected with no interfacing problems. Common | |

| Term | Definition | Symbol or Abbreviation |
|------|-----------|------------------------|
| | examples are the 7400 series TTL, 74HC00 series CMOS, and 4000 series CMOS. | |
| Logic function | The logical task needed to be performed. It might be represented by the name (such as AND), a logic symbol, a Boolean expression (such as $AB = Y$), and/or a truth table. | |
| Logic levels | In digital electronics, voltage ranges at which inputs to digital devices interpret signal as HIGH, LOW, or undefined. Voltage thresholds may be different for various logic families. | |
| Logic probe | Simple service tool which indicates logical 0s, logical 1s, or pulses in digital circuits. | |
| Logic subfamilies | Groups of related digital ICs that have similar characteristics but may vary in speed, power dissipation, and current drive capabilities. Examples might be 7400-, 74LS00-, 74F00-, 74ALS00-, and 74AS00-series TTL ICs. In some applications you may be able to substitute between subfamilies. | |
| Logic symbols | Two systems are used in the U.S. Traditional representations using the unique shaped logic gate symbols. The newer IEEE symbols using rectangle boxes. | |
| Magnetic core memory | Older nonvolatile read/write memory system using ferrite cores as memory cells. | |
| Magnetoresistive RAM | A semiconductor nonvolatile RAM with excellent access speed, allows in-circuit programming and low power, and features high density. MRAM memory cells are based on a transistor and a magnetic tunnel junction (MTJ). | MRAM |
| Magnitude comparator | A combinational logic block that compares two binary inputs $A$ and $B$ and activates one of three outputs ($A > B$, $A = B$, or $A < B$). | |
| Maxterm Boolean expression | See product-of-sums. | |
| Medium-scale integration | Used by some manufacturers to indicate the complexity of an integrated circuit. MSI usually means having a complexity of from 12 to 99 gates. | MSI |
| Memory card | Packaging method for arrays of memory devices (such as flash memories). The cards are commonly about the size of a thick credit card with edge connectors. See PCMCIA. | |
| Metal oxide semiconductor | Technology used in the fabrication of integrated circuits using metal and an oxide (silicon dioxide) as an important part of the devices' structure. | |
| Microcontroller | An inexpensive IC which contains a tiny processor, limited RAM, ROM, and I/O. A small computer on a chip. They are usually embedded in a product. | |
| Microprocessor | An IC which forms the CPU of most microcomputers. | MPU |
| Minterm Boolean expression | See sum-of-products. | |
| Minuend | The number the subtrahend is being subtracted from. | |
| Monostable multivibrator | Emits a single pulse when triggered. Also called a one-shot multivibrator. | |
| Most significant bit | The bit position in a binary number with the most weight. | MSB |
| Multiplex | In driving displays, to turn on/off one of several displays each for a short time in turn at a high enough frequency | |

| Term | Definition | Symbol or Abbreviation |
|------|-----------|------------------------|
| | so they appear to be lit continuously. In general, transmitting several signals over common lines. | |
| Multiplexer | Combinational logic block selects one-of-X inputs and directs the information to a single output. Also called a data selector. Can change parallel to serial data. | MUX |
| Multivibrator circuits | Classified as bistable (flip-flops), monostable (one-shots), and astable (free-running clocks). | MV |
| NAND gate | Basic combinational logic device where all inputs must be HIGH for the output to be LOW. A not AND circuit. | |
| Nibble | One half a byte. A 4-bit binary word. | |
| Noise | In digital electronics, it is unwanted voltages induced in connecting wires and PC board traces that might affect input logic levels and therefore outputs in circuits. | |
| Noise immunity | A digital circuit's insensitivity to undesired voltages or noise. Also called noise margin in digital circuits. | |
| Nonvolatile memory | Memory which retains data even if the power is turned off. | |
| Nonvolatile RAM | Read/write memory that will hold its data even when the power is turned off. | |
| NOR gate | Basic combinational logic device where all inputs must be LOW for the output to be HIGH. A not OR circuit. | |
| NOT | Basic combinational logic device where the output is always the opposite from the input. Also called an inverter. | |
| Octal | Base 8 number system using characters 0 thru 7. | |
| Odd parity | In data transmission, sending a parity bit that will make the number of 1s in a group odd. | |
| Ohm | The base unit of resistance. | $\Omega$ |
| 1s complement | To convert from 1s complement to binary number invert each bit. | |
| Open collector | Digital circuit output which has no internal path to the positive of the power supply. Commonly used with an external pull-up resistor. | |
| Operational amplifier | An adaptable amplifier with inverting and non-inverting inputs featuring high input- and low output-impedance, and very high gain. Gain can be set by external components. | op amp |
| Optical disc drive | Very high capacity mass storage device which commonly stores data as surface pits. Reading is done by directing laser bean at pits/no pits and detecting the light bouncing from the reflective disc. Other optical recording methods are also used. | |
| Optoisolator | An interface device used to electrically isolate input from output by using a light beam to transfer data. | |
| OR gate | Basic combinational logic device where the output goes HIGH when any or all inputs are HIGH. | |
| Oscillator | Electronic circuit that generates AC waveforms from a DC source. | |
| Oscilloscope | Test instrument that plots time against voltage drawing a graph or waveform on the screen. Oscilloscopes are available in either analog or digital models. Also called a scope. | |
| Parallel data | Transmission of data in groups at the same time over multiple lines. | |
| Parity | A system used to detect errors in binary data transmission. | |
| Parity bit | An extra bit sent with data bits to check for errors in transmission. | |

| Term | Definition | Symbol or Abbreviation |
|------|------------|------------------------|
| Passive-matrix display | An low-resolution LCD which is satisfactory for low-cost monochrome displays but not good for high-quality color LCDs. For contrast see *active-matrix display*. | |
| PC | Commonly means personal computer but it may also be used to refer to a programmable controller or programmable logic controller. | |
| PCMCIA | Personal Computer Memory Card International Association sets standards for memory cards. | |
| Phase-change technology | Used in DVD − RW and DVD + RW optical discs. A phase-change alloy is employed for reading, writing, and erasing information. The tiny "pit" and "no pit" areas of optical disc are either dark/non-reflective if the alloy is in its amorphous state or reflective if the phase-change alloy is in its crystalline state. These discs are rewritable. | |
| Photo resistive cell | A photo sensitive resistor whose resistance decreases as the light striking the unit increases. A cadmium sulfide photo cell or photo resistor. | Cds |
| Pipelining | In computer terminology, a method of speeding up processing by fetching and decoding instructions ahead of time so the next instruction is waiting to be executed immediately. Also called prefetching. | |
| Plastic leaded chip carrier | A type of surface mount IC package with leads bent under the case. | PLCC |
| Platter | In a hard disk drive, a single hard disk. The drive may contain a stack of platters to increase storage capacity. | |
| Port | In computers and microcontrollers, the circuits used to transfer data in and out of the system. | I/O |
| Product-of-sums | The form of a Boolean expression that looks like this: $(A + B)(C + D) = Y$. Implemented using an OR/AND logic diagram. Also called a maxterm Boolean expression. | |
| Program | List of instructions which tells computer what to do. May be written in a variety of computer languages. | |
| Programmable array logic | A specific PLD containing an array of ANDs which are programmable with a fixed OR array. | PAL |
| Programmable logic controller | A specialized heavy-duty computer system used for process control in factories, chemical plants, and warehouses. Closely associated with traditional relay logic. Also called a programmable controller (PC). | PLC |
| Programmable logic device | A generic name for a group of specific programmable logic devices including PALs, GALs, CPLDs, and FPLDs. | PLD |
| Programmable read-only memory | Nonvolatile memory which is programmed once by the user or distributor. | PROM |
| Propagation delay | The time it takes the output of a digital device to change state after the input is activated. Usually measured in nanoseconds. | |
| Pull-up resistor | A resistor connected to the positive of the power supply to hold a point in the circuit HIGH when it is inactive. | |
| Pulse-width modulation | Information is placed on a digital signal by increasing and decreasing the width (duration) of pulses. Used to drive hobby servo motors. Also referred to as pulse-duration modulation. | PWM |

| Term | Definition | Symbol or Abbreviation |
|---|---|---|
| Radix | The base of a number. | |
| Random-access memory | Memory organization allowing for easy access to each bit, byte or word. RAM is commonly used to mean semiconductor read/write memory. | RAM |
| RDRAM | In computer technology, an extremely fast dynamic RAM. Acronym for Rambus dynamic RAM. Compare with DRAM, SDRAM. | RDRAM |
| Register | A group of temporary memory cells (such as flip-flops) for temporary storage that have a common purpose. For instance, a register might have a name (such as DIRS in a popular microcontroller) and have a specific width (such as 8- or 16-bits). | |
| Relay | Electrical device which uses the force of an electromagnet to open/close contacts. Used for heavy-duty switching and isolation of circuits. | |
| Resistance | Opposition to current flow. Measured in ohms. | R |
| Read | The process of sensing and retrieving data from a memory cell or cells. | |
| Read-only memory | Non-volatile memory which is not usually changed once it is programed. ROM commonly used to refer to mask-programmable read-only memory. | ROM |
| Reset condition | In a flip-flop, the normal output ($Q$) has been reset or cleared to 0. | |
| Rewritable optical disc | A very large capacity optical disc that can be rewritten to many times. Some versions are called PD rewritable optical disc or CD-E (compact disc erasable). | CD-E |
| RIMM | Rambus DRAM memory packaging for use in computers comparable to the DIMM. RIMM cannot be interchanged with DIMM. | |
| Ripple counter | Simple binary counter where the changing state of the LSB flip-flop triggers the clock input of the next, etc. A time delay results from the rippling of the count from LSB to MSB. | |
| Ring counter | A recirculating shift register which is loaded with a pattern of 1s (such as a single 1) which continue to circulate around in the circle or repeated clock pulses. | |
| R-S flip-flop | Flip-flop with at least set, reset, and hold modes of operation. Fundamental latching (memory) circuit. | |
| Sampling | To measure a signal level at discrete times. Widely used in DSP while digitizing an analog input at discrete times. | |
| Schmitt trigger | A circuit that exhibits hysteresis and is useful in signal conditioning in digital electronics. May be used to digitize an analog input. | |
| SDRAM | In computer technology, a very fast dynamic RAM. Acronym for synchronous dynamic RAM. Compare with DRAM and RDRAM. | SDRAM |
| Semiconductor | Elements having four valence electrons and electrical properties between those of conductors and insulators. | |
| Sequential logic | A logic circuit whose logic states depend on asynchronous and synchronous inputs. Exhibit memory characteristics. | |
| Serial data | The transmission of data one bit at a time. | |
| Servo | General term for a motor whose either angular position or speed can be precisely controlled by a servo loop which uses feedback from the output back to the input for control. | |

| Term | Definition | Symbol or Abbreviation |
|---|---|---|
| Set condition | In a flip-flop, the normal output ($Q$) has been set to 1. | |
| Seven-segment display | Numeric display with seven segments. May be implemented with LED, LCD or VF technologies. A few letters can also be displayed for indicating hexadecimal numbers. | (seven-segment display symbol with segments labeled *a*, *b*, *c*, *d*, *e*, *f*, *g*) |
| Shift register | A sequential logic block made up of flip-flops that allows parallel or serial loading and serial or parallel outputs as well as shifting bit by bit. | |
| Signal | The information transmitted within, to, and from electronics circuits. | |
| Silicon | A semiconductor element used in the manufacture of most solid-state devices such as diodes, transistors, and integrated circuits. | |
| SIMM | In computer technology, a RAM memory board holding many memory chips used on fairly recent PCs. An acronym for single-in-line memory module. See also more modern DIMM. | SIMM |
| Small-outline DIMM | Compact memory module packaging for use in laptop-sized computers. One example is the 200-pin DDR SDRAM SO DIMM. | SO DIMM |
| Small-scale integration | Used by some manufacturers to indicate the complexity of an integrated circuit. SSI usually means having a complexity of less than 12 gates. | SSI |
| Software | Computer programs that instruct the hardware. Two main classifications of software are applications (like a word processor or game) and operating systems. Other categories might include network software and programming software. See also hardware and firmware. | |
| SOIC | Smaller package for IC than DIP packaging. Used in SMT. An acronym for small-outline integrated circuit. | (SOIC package symbol) |
| Solenoid | An actuator which converts electrical energy into linear motion. It is constructed as a hollow coil with a sliding iron core. In operation the spring-loaded iron core is "sucked into" the coil when current flows in the coil. | (solenoid symbol) |
| Solid-state drive | A nonvolatile read/write memory device which would function like the hard disk drive in a computer system but consist of semiconductor memory (perhaps flash memory cards). Might be used to save power and weight in tiny portable systems. | |
| Source | Terminal of a field-effect transistor that sends current carriers to the drain. | (FET symbol) S |
| Static RAM | Common random-access (read/write) memory device which stores data in a flip-flop like cell. Volatile memory. | SRAM |
| Stepper motor | A DC motor that jogs in short uniform angular movements in either direction given the proper digital signals. Common step angles might be 1.8°, 3.6°, 7.5°, and 15°. Two types are permanent-magnet and variable-reluctance stepper motors. | (Stepper motor symbol) Stepper motor |
| Subtrahend | The number being subtracted from the minuend. | |
| Successive approximation | In D/A and A/D converters, a technique used to decrease conversion time. | |
| Sum-of-products | The form of a Boolean expression that looks like this: $AB + CD = Y$. Implemented using an AND/OR logic diagram. Also called a minterm Boolean expression. | |

| Term | Definition | Symbol or Abbreviation |
|---|---|---|
| Surface-mount technology | SMT covers all aspects of the manufacturing techniques, equipment, and parts (surface-mount devices or SMDs) used in soldering electronic components to the surface of a printed circuit board. | Plastic-leaded chip carrier (PLCC) Small-outline package (SOT) Chip component — Solder — Solder — Circuit board |
| Synchronous | In digital circuits, meaning that operations are executed in step with the clock. | |
| T flip-flop | Short for toggle flip-flop. The output toggles to the opposite logic state on repeated clock pulses. Very useful in digital counter circuits. | T |
| Thermistor | A thermally sensitive resistor used as a heat sensor. | |
| Three-state output | Condition of outputs on certain digital ICs which includes three possible states including HIGH, LOW, or high impedance. Also commonly referred to as Tristate® (trademark of National Semiconductor). | |
| Toggle | To change the opposite logic state. A pulse that changes logic circuits state to opposite condition. A mode of operation in a flip-flop where the output goes to the opposite state on each successive clock pulse. | |
| Transducer | In electronics, a device that converts from one form of energy to another, such as a photocell converting from light to electricity or a speaker converting from electrical to mechanical/acoustic (sound) energy. | |
| Transistor | A solid-state amplifying or controlling device which commonly has three leads. | |
| Transistor-transistor logic | A type of digital IC fabricated using bipolar junction transistors. | TTL |
| 2s complement | Notation commonly used to indicate sign and magnitude of a number using only 0s and 1s. To form 2s complement, take 1s complement of binary and add 1. Helpful when using binary adders for binary subtraction. | |
| Trigger | A pulse that causes a logic device to be activated or change states. | |
| Truth table | Tabular listing of all inputs and resultant output conditions for a logic function or circuit. | A B \| Y<br>0 0 \| 0<br>0 1 \| 0<br>1 0 \| 0<br>1 1 \| 1 |
| 2s complement subtraction | Method of subtraction using a 2s complement subtrahend added to the minuend. Used so adders can be used to perform subtraction. | |
| Ultra large-scale integration | Used by some manufacturers to indicate the complexity of an integrated circuit. ULSI usually means having a complexity of 100,000 or more gates. | ULSI |
| Universal shift register | Register with many features including serial-in/out, parallel-in/out, hold, and shift right or left. | |
| USB port | USB is an acronym for Universal Serial Bus. A modern general-purpose serial port for transmitting data from a microcomputer to peripherals such as external printers, modems, mouses, keyboards, portable drives (optical, magnetic), or flash memory modules. The USB port provides power to the device and can be plugged in or disconnected when the computer is turned on. | 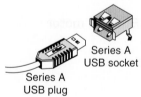<br>Series A USB socket<br>Series A USB plug |
| Vacuum fluorescent display | Low voltage triode vacuum tube display which commonly glows green (without filters). | VF |
| $V_{CC}$ | Label for positive of power supply in TTL ICs and some CMOS ICs (commonly +5V). | |

| Term | Definition | Symbol or Abbreviation |
|---|---|---|
| $V_{DD}$ | Label for positive of power supply in many but not all CMOS ICs (+3 to +18V). | |
| $V_{SS}$ | Label for negative of power supply on many but not all CMOS ICs. | |
| Very large-scale integration | Used by some manufacturers to indicate the complexity of an integrated circuit. VLSI usually means having a complexity of from 10,000 to 99,999 gates. | VLSI |
| Volatile memory | Memory that can store data only as long as power is applied. | |
| Volt | Base unit voltage. | V |
| Voltage | Electrical pressure. | V |
| Voltage comparator | An op amp circuit that compares a positive voltage input ($A$) with a negative voltage input ($B$) and indicates with a logic output which input is higher. | |
| Waveforms | A graphic representation of voltage versus time as might be viewed on an oscilloscope. | |
| Winchester drive | Historical name for a hard disk drive. | |
| Word | In computer terminology, a group of bits that are processed as a single unit. The exact definition of a word depends on the system. Word sizes of 16- or 32-bits are common. | |
| Write | The process of recording data in a memory cell or cells. | |
| Write-once read-many | An optical CD recordable disc can be recorded on once using your PC and it then is permanent like a CD-ROM. | WORM |
| XNOR gate | Basic combinational logic device where an even number of HIGH inputs generates a HIGH output. A not XOR gate. | |
| XOR gate | Basic combinational logic device where an odd number of HIGH inputs generates a HIGH output. | |

# Credits

**Pg. x** (left) Cindy Lewis; (right) Lou Jones/Getty Images; **Pg. 3** (left) Courtesy Simpson Electric Co; (right) Courtesy Fluke Corporation. Reproduced with permission; **Pg. 4** (top left) file photo; (top right) © Fred Wilson/Getty Images; (bottom left) Courtesy Apple Computers; (bottom right) Courtesy Apple Computers; **Pg. 5** © Laurent Gillieron/ Keystone/epa/Corbis; **Pg. 11** Courtesy of Dynalogic 1.800.246.4907; **Pg. 54** © Mary Evans Picture Library; **Page 64** International Telecommunication Union & Inmarsat; **Pg. 87** Courtesy Braun; **Pg. 105** Courtesy Intel Corp & Sandia National Laboratories. Photo by Randy Montoya; **Pg. 119** © Corbis; **Pg. 120** © Corbis; **Pg. 204** © AP/Wide World Photos; **Pg. 207** Courtesy Fluke Corporation; **Pg. 232** Photomicrograph by Leo Deriak/Lucent Technologies; **Pg. 240** Courtesy Alpine Electronics; **Pg. 260** Brian Quintard/Lawrence Livermore National Laboratory; **Pg. 302** Photomicrograph by John Carnivale/ Lucent Technologies; **Pg. 304** Courtesy Microvision; **Pg. 325** Courtesy Thales Navigation; **Pg. 352** Courtesy Yamaha Consumer Electronics; **Pg. 373** Courtesy Handspring, Inc.; **Pg. 374** Courtesy Seagate Technology; **Pg. 429** Hayes Microcomputer Products; **Pg. 433** Courtesy Motorola; **Pg. 442** PhotoLink/PhotoDisc.

# Index